ELEMENTS OF DISCRETE MATHEMATICS

ELEMENTS OF DISCRETE MATHEMATICS

Ronald E. Prather
Syracuse University

HOUGHTON MIFFLIN COMPANY BOSTON
Dallas Geneva, Illinois Lawrenceville, New Jersey Palo Alto

To Jacqueline

Illustrations by Jean-Paul Marchand

Printed in the U.S.A.
Library of Congress Catalog Card Number: 85-81288
ISBN: 0-395-35165-0
ABCDEFGHIJ-MP-89876

CONTENTS

Preface *xi*

Chapter One ***INTUITIVE SET THEORY*** **1**

1.1 The Notion of a Set *2*
Sets and Membership, *2* Russell's Paradox, *5* Constructible Sets, *6* Set-Builder Notation, *7*
1.2 The Calculus of Sets *10*
Subsets of a Set, *11* Set Operations, *12* Characterization of Union and Intersection, *14* Equality of Sets, *15* Properties of the Set Operations, *15* Set Complementation, *16* Generalized Unions and Intersections, *19*
1.3 Functions and Relations *23*
Definition of a Function, *23* Sequences, *26* Properties of Functions, *28* Composition of Functions, *30* Set Products, *33* The Notion of a Relation, *34*
1.4 Equivalence Relations and Partitions *38*
Relations on a Set, *38* Equivalence Relations, *40* Equivalence Classes, *43* Characterization of Equivalence Relations, *45*
1.5 Elementary Counting Techniques *47*
Rules of Product and Sum, *48* Permutations and Combinations, *51* Binomial Coefficients, *56* Sets of the Same Cardinality, *58* The Powerset Construction, *58* An Important Combinatorial Principle, *60*

Chapter Two ***DEDUCTIVE MATHEMATICAL LOGIC*** **62**

2.1 Axiomatic Mathematical Systems *64*
Primitive Terms, *64* Axioms of Order, *65* Logical Implication, *67* Law of Contraposition, *69* Extension by Definition, *70* Euclidean Geometry, *71*
2.2 Proof Techniques *78*
Implication and Equivalence, *78* Forward and Backward, *80*

Law of Detachment, *84* Logical Connectives, *85*
Proof by Cases, *86*

2.3 Propositional Rules *89*
Rules of Implication, *90* Conjunction and Disjunction Rules, *91*
Rules of Negation, *93* Definition of Equivalence, *94*
The Logical Calculus, *95*

2.4 Deductive Logic *98*
Summary of Rules, *99* Formalized Proof Technique, *99*
Conventions on Indentation, *101* An Abbreviated Format, *102*
Development of Proofs, *104*

2.5 Derived Rules *108*
Rules Involving Implication, *109* Rules Involving Conjunction, *110* Rules Involving Disjunction, *111* Rules Involving Negation, *112* Rules Involving Equivalence, *114*

2.6 Constructive vs. Classical Rules *116*
Rules of Double Negation, *117* Law of the Excluded Middle, *118*
Constructive vs. Nonconstructive Proofs, *119*
Extended Classical Laws, *120* Brief Historical Comment, *124*

2.7 The Statement Calculus *127*
Well-formed Formulas, *128* Characteristic Tables, *130*
Tautology and Truth Table, *133* Semantic Implication, *135*
Consistency and Adequacy, *136*

Chapter Three ***DISCRETE NUMBER SYSTEMS*** **142**

3.1 The Natural Numbers *144*
Axioms for the Natural Numbers, *144* Definition of Arithmetic Operations, *146* Laws of Integer Arithmetic, *148*
Extension of the Number System, *151* Ordering of the Integers, *152*

3.2 Induction and Recursion *154*
The Principle of Mathematical Induction, *155* Recursive Definition of Functions, *158* The Summation Symbol, *160*
Recursive Solution of Problems, *162* Recursive Definition of Sets, *164*

3.3 Divisibility of Integers *167*
Integer Division, *168* Greatest Common Divisors, *170*
Further Divisibility Questions, *172*

3.4 Base Conversion *175*
Positional Notation, *176* Conversion Between Binary and Decimal, *177* Conversion of Fractional Parts, *179*

3.5 The Rational Numbers *183*
Construction of the Rationals, *184* Arithmetic Operations, *186*

Arithmetic Laws, *187* Questions of Cardinality, *188* Ordering of the Rationals, *190*

3.6 Computer Arithmetic *192*

Normalized Exponential Systems, *193* Internal Representation, *194* Arithmetic Operations, *196* Errors of Computation, *197* The Boolean Number System, *199*

Chapter Four *THE NOTION OF AN ALGORITHM* 202

4.1 Problem-solving Principles *204*

Three Sample Problems, *204* Guiding Principles, *206* Definition of an Algorithm, *210* Elementary Processes, *211* State of a Computation, *212*

4.2 Toward an Algorithmic Language *215*

Compound Processes, *216* Top-Down Methodology, *220* Derived Constructs, *222* Pseudolanguage Guidelines, *225*

4.3 Extensions of the Language *231*

Subscripted Variables, *232* A Sorting Algorithm, *237* Algorithms as Modules, *238*

4.4 Proofs of Correctness *245*

Testing vs. Verification, *246* Analysis of Elementary Processes, *248* Deductive Rules, *249* Correctness of the Euclidean Algorithm, *255*

4.5 Recursive Algorithms *259*

Recursive Functions, *259* Divide and Conquer, *262*

4.6 Introduction to Computational Complexity *264*

Running Time and Growth Rate, *265* Computation of Time Complexity, *269*

Chapter Five *POLYNOMIAL ALGEBRA* 276

5.1 Polynomial Calculus *278*

Basic Terminology, *278* Addition and Multiplication, *280* Polynomial Evaluation, *285*

5.2 Graphs of Polynomials *289*

Graph of a Function, *290* Limiting Values and Intercepts, *292* Synthetic Division, *294* Roots of Polynomials, *296* Approximate Roots, *302* Algebraic and Computable Numbers, *304*

5.3 Interpolation Theory *309*

Prescribed Roots, *310* Straight Line Equation, *311* Polynomial Interpolation, *312*

5.4 Factorization of Polynomials *318*
Quotients and Remainders, *319* Auxiliary Algorithms, *320*
Polynomial Factorization, *324* An Irreducibility Criterion, *328*
5.5 Rational Functions *331*
Graphs of Rational Functions, *332* Multiplicity of Roots and Poles, *334* Powers as Denominators, *335* Partial Fraction Expansion, *337*
5.6 Difference and Summation Calculus *342*
Difference Operator, *342* Factorial Polynomials, *345*
Newton Expansion, *347* Polynomial Conversion, *351*
The Fundamental Theorem, *352*
5.7 Independence and Rank *357*
Linear Combinations of Polynomials, *358* Independent Polynomials, *359* Coordinate Vector Spaces, *362* Rank of a Collection, *364*

Chapter Six ***GRAPHS AND COMBINATORICS*** **372**

6.1 Graphs and Subgraphs *374*
Representation of Graphs, *374* Complete Graphs, *377*
Vertex Degrees, *380* Paths and Circuits, *381*
Connectivity, *382*
6.2 Circuit Rank *386*
Network Analysis, *386* Independent Circuits, *389* Circuit Rank Formula, *393*
6.3 Trees *397*
Characterization of Trees, *398* Spanning Trees, *399*
Optimal Trees, *402* Labeled Binary Trees, *404*
6.4 Planar Graphs *415*
Planarity, *416* The Euler Formula, *418* Tests for Planarity, *420*
6.5 Isomorphism and Invariants *426*
The Notion of Isomorphism, *426* Graphical Invariants, *429*
The Coloring Problem, *435*
6.6 Covering Problems *438*
Definition of Covering Problems, *439* The Covering Algebra, *443*
Covering and Domination Numbers, *446*
Independence and Clique Numbers, *447* The Chromatic Number, *451*
6.7 Directed Graphs *455*
Digraphs as Relations, *455* Matrix Computations, *458*
Reachability, *463*

6.8 Path Problems *470*
Eulerian and Hamiltonian Paths, *471* Shortest Paths in Graphs, *474* Shortest Paths in Digraphs, *476*

Solutions to Selected Exercises *482*

Index *520*

PREFACE

Elements of Discrete Mathematics presents an elementary survey of the methods of reasoning and the objects of study encountered in modern *discrete* (as opposed to the classical *continuous*) mathematics. To some extent, this text should be viewed as a replacement for the author's earlier book, *Discrete Mathematical Structures for Computer Science* (Houghton Mifflin Company, 1976). In keeping with recent trends, however, the treatment here is considerably less formal, so as to be suitable for beginning college students.

A group of leading mathematics and computer science educators have called for the revamping of the traditional freshman-sophomore mathematics program to permit the inclusion of discrete mathematical topics at an early stage. This material has long been recognized as essential in developing the requisite background for beginning computer science students. More recently, there has been a growing awareness that this same material is equally beneficial to mathematics majors and others as well. Against this backdrop, the Alfred P. Sloan Foundation has sponsored a number of innovative curriculum development programs, designed to ensure that discrete mathematics will occupy a role of equal importance to that of the calculus in the first two years of undergraduate mathematics.

Traditionally, the first two years of college mathematics have consisted of a year of differential and integral calculus, followed by a linear algebra–differential equations–multivariable calculus sequence at the sophomore level, with virtually no discrete mathematics included. Under Sloan Foundation support, a new curriculum was developed at the University of Denver that combined the discrete and the continuous mathematics from the very beginning of the freshman year. This was accomplished by (1) collapsing the traditional first-year calculus by a factor of two-thirds, making room for an elementary discrete mathematics course, (2) introducing algorithmic material into the standard calculus segments, and (3) distributing the traditional sophomore linear algebra over the differential equations course and a second course in discrete mathematics.

This new two-year curriculum is outlined below, with three course titles per year (for academic quarters) and with topic headings of the new discrete mathematics courses listed under the first-quarter titles. In the freshman sequence, one must understand that the term *calculus* is being used in the broader sense as "methods of reckoning or calculating."

The Freshman Sequence

Discrete Calculus

Chapter 1 Intuitive Set Theory
Chapter 2 Deductive Mathematical Logic
Chapter 3 Discrete Number Systems
Chapter 4 The Notion of an Algorithm

Differential Calculus

Integral Calculus

The Sophomore Sequence

Linear Algebra and Combinatorial Mathematics

Chapter 5 Polynomial Algebra
Chapter 6 Graphs and Combinatorics

Linear Algebra and Differential Equations

Multivariable Calculus

Traditional texts on calculus, linear algebra, and differential equations were used in the "classical" segments of the pilot program. The topic headings of the new discrete mathematics courses have become the chapter titles for our book.

It is not necessary that *Elements of Discrete Mathematics* be used in precisely the context outlined above. We see this text as meeting somewhat different objectives from one institution to another. For instance, in a semester-long companion to a classical calculus sequence, one might study Chapters 1 through 4, together with one or the other of our concluding "applications" chapters. The brief discussion that follows outlines the intent of the various chapters, commenting on the anticipated connections with a student's high school mathematics background and with his or her collateral studies in continuous mathematics, so that an instructor will be able to relate the text to individual needs.

We have taken the view that an introduction to mathematical logic must act as a cornerstone to the new treatment of discrete mathematics. On the other hand, the elements of set theory are generally agreed to be easier than those of mathematical logic for a student beginning the study of (abstract) mathematics, and this accounts for the ordering of our two opening chapters. Throughout these discussions, we have provided a more "constructive" presentation than is normally seen. Thus, a set doesn't just happen to exist; it is constructed by some mechanical means. In this way, we make connection with the student's real world computing experience, from high school and from everyday life. Similarly, in Chapter 2, we view the "deductive" logical systems as far superior pedagogically to the more usual axiomatic or truth-tabular treatments. The deductive systems lead to a natural method of constructive proof technique. Here, it is anticipated that significant connections can be made to the student's previous exposure to formal proofs in high school geometry.

In Chapter 3, we develop a familiarity and facility with the more common discrete number systems, so that they can be exploited in the algorithmic examples of Chapter 4 and elsewhere. Some of this material (for example, integers and the rational number system) will be familiar to students from high school mathematics, whereas some of it (computer arithmetic, for instance) will probably be new. This blending of the new with the old acts as a useful learning device, here and elsewhere throughout the text.

If the ideas of mathematical logic provide a foundation for our studies, the notion of an algorithm must be central to the development, acting as the glue that bonds all the topics together. Various concrete instances of algorithms are introduced informally throughout the text. But in Chapter 4, we provide a more formal treatment. A generic pseudocode language is developed for the phrasing of the algorithms themselves. At the same time, sufficient preparations have been made (in the study of sets, logic, number systems, and so on) so that reasonably sophisticated treatments can be given of such important topics as the analysis of algorithmic complexity and proofs of correctness.

Chapter 5 on polynomial algebra provides an excellent opportunity to build directly on the student's high school algebra course. On the other hand, we treat much of this material from an algorithmic standpoint that is new to the student. But Chapter 5 has an added advantage in providing a convenient transition from discrete to continuous mathematics (through the appearance of irrational numbers), a feature that the instructor may wish to exploit in outlining a course sequence.

Chapter 6 offers a fairly comprehensive survey of the methods of combinatorial mathematics, particularly those relating to graph theory. This is the one chapter that has been carried over from the author's earlier text, with but few modifications. As far as can be determined, this treatment had much to do with the success of the earlier book.

In the broadest sense, a calculus is a way of reckoning, a method of computing, whether through the use of a pile of pebbles or with the abstract machinery of a theory of integration. It matters not whether we are dealing with a discrete domain or an uncountable continuum. And so, both historically and conceptually, "the Calculus" tells only a part of the story. Our text looks at the other side of the coin.

The writing of any text is mainly a solitary endeavor. But in the present case, there are a number of individuals who have made significant contributions at various stages of the development. Particularly, the author would like to acknowledge the financial support of the Sloan Foundation; the encouragement of his colleague, Professor Herbert J. Greenberg; and the many helpful suggestions of his teaching assistant, J. Paul Myers, Jr. In addition, the author wishes to express his appreciation to the reviewers of this book:

Donald Bushaw, Washington State University; Kenneth Kalmanson, Montclair State College; Wulf Rehder, San Jose State University; C. L. Lui, University of Illinois at Urbana-Champaign; Anthony Ralston, State

University of New York at Buffalo; Thomas Moore, Bridgewater State College; John J. Neuhauser, Boston College; E. G. Whitehead, Jr., University of Pittsburgh; Robert Crawford, Western Kentucky University; Peter Grogono, Concordia University; Subrata Dasgupta, S.W. Louisiana State University; and Robert Mueller, Colorado State University.

Finally, thanks should be conveyed to all of the University of Denver students participating in the development of the new curriculum. Theirs will no doubt prove to have been the biggest contribution of all.

R. E. P.

ELEMENTS OF DISCRETE MATHEMATICS

Chapter One

INTUITIVE SET THEORY

Contents

1.1 The Notion of a Set
1.2 The Calculus of Sets
1.3 Functions and Relations
1.4 Equivalence Relations and Partitions
1.5 Elementary Counting Techniques

Hermann Weyl (1885–1955)

Hermann Weyl was born in Elmshorn, Germany, and was educated at the Universities of Munich and Göttingen. From 1913 to 1930 he was professor of mathematics at the Technische Hochschule in Zurich, except for a year at Princeton. He later returned to Princeton to join the Institute for Advanced Study, where he retired in 1952. Weyl has been called an "ornament to our time." His powerful intellect served to enrich mathematics as well as philosophy and the natural sciences with new concepts and insights seldom paralleled in modern times. He is remembered particularly for his writings on the philosophies of science and mathematics, and for his distinguished contributions to group theory, function theory, quantum mechanics, topology, and relativity, in which areas he excelled as few have done in any one field alone.

Writings

Mathematics has a creative definition *at its disposal, through which new ideal objects can be generated. A special case is the process of* definition by abstraction.

In looking at a flower I can mentally isolate the abstract feature of color as such. The act of abstraction would be primary while the statement that two flowers have the same color 'red' would be based on it; whereas in mathematical abstraction it is the equality which is primary, while the feature with regard to which there is equality comes second and is derived from the equality relation.

With every property p(x) which is meaningful for the objects x of a given category we correlate a set, namely 'the set of objects x having the property p.' Thus we may speak of the set of all even numbers, or the set of all points on a given line. The conception that such a set can be obtained by assembling its individual elements should by all means be rejected. To say that we know a set means only that we are given a property characteristic of its elements. Only in the case of a finite set do we have, in addition to such general *description, the possibility of* individual *description which would exhibit each of its elements. [Formally, by the way, the latter mode of description is a special case of the former.]*

If the concept of a set is understood in this way, then the creative description is seen to be nothing but the transition from a property to a set, so that the mathematical construction of new classes of ideal objects can quite generally be characterized as set formation. *Now there is no longer anything objectionable in describing the circle about O through A as the set of all points P whose distance from O equals OA, or the color of an object as the set of objects having the same color, or the cardinal number 5 as the set of all those aggregates which are numerically equivalent to the exhibited aggregate of the fingers of my right hand. But it is an illusion—in which Dedekind, Frege, and Russell indulged for a time, because they apparently conceived of a 'set' after all as a collection—to think that thereby a concrete representation of the ideal objects has been achieved. On the contrary, it is through the principle of creative definition that the meaning of the general set concept is elucidated as well as safeguarded against false interpretations.*

Philosophy of Mathematics and Natural Science **(Weyl)**

Sec. 1.1 THE NOTION OF A SET

A set is a collection of (mathematical) objects, constructed with a definite membership requirement in mind.

This opening section should build directly on the student's elementary and high school introduction to sets and set notation.

Goals

After studying this section, you should be able to:

- Describe a set as a collection of objects enclosed in braces.
- Use the *membership symbol* properly.
- Understand and discuss *Russell's Paradox.*
- Describe sets with the *set-builder* notation.
- Characterize sets by giving the *defining property* of their members.

Sets and Membership

We can define a **set** intuitively as a collection of *objects*, particularly mathematical objects of one kind or another. Thus we may conceive of the set as consisting of the even integers between 0 and 100, or in another context, the set of all circles about a certain center in the plane. We may consider the set of primary colors or a set of three abstract objects, denoted simply by the symbols a, b, and c. At the outset, we would hope that the reader could view this notion of a set as being entirely general, as our list of illustrations should indicate.

Ordinarily we use capital letters to name the sets we introduce. If a set S consists of only a finite number of objects, then it is possible, though not always practical, to enumerate or list all the objects, enclosing them in braces as in Example 1.1.

Example 1.1 We may write

$$S = \{a, b, c\}$$

to express the set of abstract objects a, b, c, described previously. This notation indicates that a, b, c, and only these three elements, are members of the set S. The order of appearance of the objects within the braces does not matter. Thus $\{a, c, b\}$ and $\{c, b, a\}$ are simply two other representations of this same set. ■

Note that our naming convention is completely arbitrary. We could just as well have called this set T rather than S. By the same token, it would be difficult to continue to use different names for all the various sets we plan to introduce. Thus if we use S as a name for a different set on another occasion, it should not be confused with $S = \{a, b, c\}$ as above; the context will make the meaning clear. As a matter of fact, S is often used as kind of a generic name for an arbitrary set in general discussions in which we are thinking of any set whatsoever.

By way of illustration, and in order to introduce a further matter of notation, we let $|S|$ denote the size or **cardinality** of an arbitrary but finite set S—that is, the number of objects in S. For instance, $|S| = 3$ if S is the set given in Example 1.1. Evidently, it is just a matter of counting; that is,

$$a \longleftrightarrow 1$$
$$b \longleftrightarrow 2$$
$$c \longleftrightarrow 3$$

to establish a correspondence with the first three counting numbers of our everyday experience. If such a correspondence can be achieved using only the counting numbers from 1 to n, then S is said to be a **finite set** (of cardinality or size n). Otherwise, S is an **infinite set**.

The words *collection* or *class* will occasionally be used as synonyms for *set*. An object of a set will sometimes be called a **member** or an **element** of the set in question. Moreover, we introduce a special symbol $\in$, the *membership symbol* to refer to this special relationship of objects as members of a given set. Thus we could write

$$a \in S$$

read "*a is a member of S*," to describe the situation in Example 1.1, where the object a indeed belongs to the set S. On the other hand, the statement "$d \in S$" would not be correct in this context. In fact, we write $d \notin S$ (negating the membership symbol) in such situations.

With a judicious use of ellipses (generally three dots), we can extend the bracing notation used above to represent large finite, even infinite, sets.

Example 1.2 We can write

$$T = \{0, 2, 4, \ldots, 100\}$$

and the meaning would more than likely be clear. Note that this is the set that was described in words at the outset, namely, "the set consisting of the even integers between (and including) 0 and 100." ■

Example 1.3 The set of all integers can be denoted

$$Z = \{0, -1, 1, -2, 2, \ldots\}$$

The ellipses indicate that the listing just goes on and on, without end. As a matter of fact, it might be more illustrative to write

$$Z = \{\ldots, -2, -1, 0, 1, 2, \ldots\}$$

to indicate the ordering of the integers. The positive integers or **natural numbers** as they are called by mathematicians, form a closely related set. (We also refer to this collection as the set of *counting numbers*.) We can express the natural numbers as the set

$$N = \{1, 2, 3, \ldots\}$$

making use of the ellipses once more to indicate that this set of *all* the positive integers again goes on and on, unendingly. Note that every member of N is also a member of Z, but not conversely; that is, there are members of Z (zero and all the negative integers) that are not found in N. ■

Some of the more important sets have letter names that are somewhat standard, for example, N and Z in Example 1.3. By standard we mean that the symbols N and Z ordinarily will not be used to denote any other sets that we might introduce; they are, in effect, *reserved* to always have the meaning given in Example 1.3. The same can be said for the finite sets in Example 1.4.

Example 1.4 For each positive integer n, we define a corresponding finite set

$$\mathbf{n} = \{1, 2, \ldots, n\}$$

consisting precisely of the integers from 1 to n. Thus $\mathbf{3} = \{1, 2, 3\}$, $\mathbf{7} = \{1, 2, 3, 4, 5, 6, 7\}$, and so on. Of course, the reader should be cautioned that these boldface identifiers $\mathbf{n}$ are not integers themselves, but only names of sets closely related to the integers n. Another finite set with a reserved name is the following:

$$B = \{0, 1\}$$

used primarily in logical discussions, where 0 and 1 are identified with the terms *false* and *true*, respectively. ■

In certain contexts the members of a finite set are considered as *symbols* of an *alphabet*; that is, a character set of some kind or another. In these cases we will often use A as a generic name for such a set (A stands for alphabet).

Example 1.5 We might have A denote the lowercase letters of our ordinary alphabet; that is,

$$A = \{a, b, c, \ldots, z\}$$

Or, we may use a special name on occasion—for example, **P** for the alphabet for the Pascal programming language, with symbols divided into groupings as shown in Table 1.1. Note our inclusion of three multicharacter symbols (**div**, **mod**, **in**), not characters in the usual sense. Nevertheless they are so treated. ■

TABLE 1.1

Groupings	*Symbols*
Lowercase letters	a b c ... z
Uppercase letters	A B C ... Z
Digits	0 1 2 3 4 5 6 7 8 9
Arithmetic operators	$+$ $-$ $*$ $/$ **div** **mod**
Logical operators	$\vee$ $\wedge$ $\neg$
Relational operators	$<$ $\leq$ $=$ $\neq$ $\geq$ $>$ **in**
Parentheses	( ) [ ] { }
Punctuation symbols	. , : ; ' " ! ?
Special symbols	¢ $ # @ % ↑ &
Blank	□

Sets may be members of other sets, as Example 1.6 indicates.

Example 1.6 If S is the set of Example 1.1 (and N and Z have their standard meanings), we may consider the following set:

$$T = \{S, N, Z, d\}$$

Even though N and Z are infinite sets, T is finite as a set, and in fact its cardinality is $|T| = 4$. Note as well that $d \in T$ but $3 \notin T$. What is true is that $3 \in N$ (or Z) and $N \in T$. That is quite different from saying "3 is a member of T" (which is false). ■

This game can go on and on. The set T (in Example 1.6) in turn can be a member of another set. In this way, vast hierarchies of sets can be introduced at will. So, as we begin to see, the theory of sets is not nearly as simple as it might have seemed to be at first. This is brought home most forcefully in the discussion to follow.

Russell's Paradox

We have seen that it is possible for a given set to have another set as one of its members. But can a set have itself as a member? Rather absurd, you might say, unless perhaps we are willing to consider "the set of all abstract ideas," or some equally bizarre concoction. If such examples are bothersome to the reader's sensibility, he or she need not fear. We are only going to consider here those sets that are, in fact, not members of themselves.

In so doing, however, we will arrive at one of the most interesting paradoxes in the history of mathematics, due to the English philosopher and mathematician, Lord Bertrand Russell (1872–1970). We are asked to imagine a set R that has as its members all those sets that are not members of themselves. Seemingly there is no problem here. We know that sets can have other sets as members. And we only consider the usual sets, that is, those that are *not* members of themselves, as members of R. But we now ask (with Russell) the question: *Is R a member of itself*?

Suppose it is. Then, according to the way that R was introduced, R cannot be a member of itself after all! (Do you see why?) On the other hand, suppose that R is (a set that is) *not* a member of itself. Then again, because of the sense of the definition of R, R is indeed a member of R, that is, of itself!

How are we to settle this matter? Either R is a member of itself or it is not. We don't expect there to be any other possibilities. Yet, both of these possibilities ($R \in R$ or $R \notin R$) have been shown to be self-contradictory. The popular consensus among current-day mathematicians is that sets such as R should not be permitted to exist. They should be rooted out of any sensible set theory. There are a number of ways to accomplish this. Perhaps the simplest approach is the "constructive" one, in which we are assured that sets such as R could not possibly arise.

Constructible Sets

We began this whole introductory section with the simple intuitive idea of a set as a collection of (mathematical) objects. After studying Russell's Paradox, however, we see that some care must be taken. We don't want to become too precise, for that would defeat our overall purpose of considering sets in an entirely informal and intuitive way. As we have indicated, the simplest solution is to agree that a *set* is the totality of all (mathematical) objects *constructed* in accordance with a certain prescribed membership requirement. That is, we are to give a "prescription" for constructing (a typical member of) the set. Certainly a prescription could not be given for Russell's set R. Yet, this understanding will be sufficient for our purposes and should not lead to any further difficulties.

Example 1.7 For any finite set, the prescription is clear. For example, if we want someone to consider the set

$$S = \{a, b, c\}$$

of Example 1.1, we have only to provide them with the list, a, b, and c. In general, we simply construct a list, then construct a member of the set by selecting an object from the list. Then someone can decide whether or not a given object belongs to the set in question by a simple reference to the list. Clearly then, the constructibility of any finite set is hardly in question. ■

Example 1.8 Let's consider the infinite set

$$N = \{1, 2, 3, \ldots\}$$

of Example 1.3, and ask, "How do we construct the typical member?" By what "prescription" is this done? Obviously, it's just a matter of counting. We count up from 1 to any integer n whatsoever; when we stop counting, we have constructed a typical member of the set N. ■

With any set that we wish to introduce in an argument, the procedure of Example 1.8 should be followed (at least, to ourselves). We should be sure that we are capable of providing a means to construct (the typical element of) the set in question. In keeping with Weyl's analysis of the situation in the chapter-opening reading, we will not necessarily have assembled all the members in a *closed* collection of any sort; but we will have elucidated the essential character of its individual members. With this much understood, we will rarely refer to constructions hereafter. Yet, when we use the term *set* from now on, it is the meaning introduced here that we have in the back of our mind.

Set-Builder Notation

It happens, however, that there is a simple mechanism for accomplishing our desired purpose; namely, to be able to freely introduce new sets at will, without fear that we are creating a contradictory situation. Echoing Weyl's words once again, we merely refer to a property p that objects of a given category may or may not satisfy. And we thereby introduce the set of objects satisfying p, a set that we denote in the following form:

$$P = \{x: p(x)\}$$

We read this expression as "P is the set of all objects x for which $p(x)$ holds; that is, p holds for x." This is known as the **set-builder notation**. It is understood that

$$x \in P \quad \text{just in case} \quad p(x) \quad \text{holds}$$

that is, the two statements "$x \in P$" and "$p(x)$ holds" (p holds for x) are then viewed as entirely equivalent. Note the underlying assumption in all of this; namely, that the objects x are drawn from (remembering Weyl's "category") some constructible "universe of discourse" (a set) U. We can amend the notation just used, writing instead

$$P = \{x: x \in U \quad \text{and} \quad p(x)\}$$

for added emphasis. Moreover, it is further assumed that with each member $x \in U$, it is possible, in principle, to answer *yes* or *no* to the question: Does x satisfy the property p; that is, does $p(x)$ hold? In this way, the constructibility of the new set P is assured.

Example 1.9 If we are willing to suspend our usual assumption that only mathematical objects are to be considered, we can introduce the set

$$P = \{x: x \text{ was a U.S. president}\}$$

and the members are quite clear. Note that this is much easier than writing down all the members in a list, as we would have had to do in the old bracing notation. And yet, it is equally precise. Presumably, the "universe of discourse" here is the set of all persons, living or dead. And we may still decide that Nixon $\in P$ and Ghandi $\notin P$, because Nixon was indeed a U.S. president (and Ghandi was not). ∎

Example 1.10 Consider the set T of Example 1.2:

$$T = \{0, 2, 4, \ldots, 100\}$$

Using the set builder notation, we can describe this same set as

$$T = \{n \colon n \text{ even and } 0 \leq n \leq 100\}$$

Note the unstated assumption here that the "universe" is probably $U = Z$. Note further the unimportance of the particular "dummy argument" used; that is, we have written n where the reader probably expected to see an x. By the same token, we do not have to refer to a set constructed with the set builder as if it were necessarily named P. Here we have used T (to be in agreement with the name used previously). ∎

Finally, note that the *defining property* p may be expressed in any unambiguous language, combining precise English with mathematical notation if that is convenient. The point is, we are able to use the defining property to decide which objects are in our set and which are not.

Example 1.11 For the defining property of Example 1.10, that is,

$$p(n) \colon n \text{ is even and } 0 \leq n \leq 100$$

we must remember that we are already assuming that candidates are integers, as if the definition of T had read:

$$T = \{n \colon n \in Z \text{ and } n \text{ is even and } 0 \leq n \leq 100\}$$

We then can easily check that $5 \notin T$ (5 is odd) and $104 \notin T$ $(104 > 100)$, whereas $4 \in T$ and $94 \in T$. ∎

Example 1.12 The set of Example 1.1 can be written as follows:

$$S = \{x \colon x = a \quad \text{or} \quad x = b \quad \text{or} \quad x = c\}$$

and from this observation, it becomes clear that the set-builder notation is a generalization of the "bracing" notation that we used previously for describing finite sets. Again, this was already anticipated by Weyl's preliminary remarks. Here, the intended "universe" is not quite clear, but in this instance, that could hardly be a problem in deciding the membership question. ∎

EXERCISES 1.1

1 Describe each of the following sets (named $A, B, \ldots, H$, respectively) as a collection of objects enclosed in braces or with the use of the set-builder notation, whichever is more convenient.

(a) All postwar (World War II) United States presidents

(b) All English words beginning with the letter *a*

(c) All integers between (and including) 0 and 1000

(d) All even integers

(e) All New England states

(f) All mountain peaks over 10,000 feet in elevation

(g) All primary colors

(h) All musical keys

2 Describe each of the following sets in words; that is, do the reverse of Exercise 1.

(a) $A = \{m: m = 4n, n \in Z\}$

(b) $B = \{\text{Matthew, Mark, Luke, John}\}$

(c) $C = \{x: ax^2 + bx + c = 0\}$

(d) $D = \{x: x \in Z, x^2 = 1\}$

(e) $E = \{a, b, c, \ldots z\}$

(f) $F = \{\vee, \wedge, \neg\}$

(g) $G = \{n: n \in N, n^2 = 1\}$

(h) $H = \{n: n \in N, n > 1000\}$

3 For the sets of Exercise 1, which of the following represent(s) a proper use of the membership symbol? Why or why not?

(a) Nixon $\in A$

(b) aqua $\in B$

(c) $-1 \in C$

(d) $-6 \notin D$

(e) Colorado $\in E$

(f) Pikes Peak $\in F$

(g) aqua $\in G$

(h) $C \in H$

4 For the sets of Exercise 2, which of the following represent(s) a proper use of the membership symbol? Why or why not?

(a) $0 \in A$

(b) Jean Paul $\notin B$

(c) $-b \in C$

(d) $-1 \in D$

(e) $y \in E$

(f) $\neg \in F$

(g) $-1 \in G$

(h) $1000 \in H$

5 For each given "universe of discourse," provide an example of a set (named $A, B, \ldots, H$, respectively) whose members belong to this universe.

(a) Living persons

(b) Birds

(c) Coins

(d) Nations

(e) Colors

(f) Television programs

(g) Integers

(h) French words

6 A situation is said to be *paradoxical* if it seems that one of the two statements:

"p" or "not p"

must hold in relation to the given situation, and yet

"p" implies "not p" and "not p" implies "p"

Identify such a statement p in reference to:

(a) Russell's Paradox

(b) The Barber's Paradox (See Exercise 7)

7 (Barber's Paradox) A barber in a certain village shaves all those people who do not shave themselves. Does the barber shave himself? Explain why this situation is paradoxical (See Exercise 6).

8 Determine the cardinality of each of the following sets.

(a) A in Exercise 1 **(b)** C in Exercise 1

(c) E in Exercise 1 **(d)** H in Exercise 1

(e) B in Exercise 2 **(f)** C in Exercise 2

(g) E in Exercise 2 **(h)** G in Exercise 2

9 Which of the sets in Exercise 1 are infinite?

10 Which of the sets in Exercise 2 are infinite?

Sec. 1.2 *THE CALCULUS OF SETS*

According to Webster, a "calculus" is a method of calculating or reckoning with symbols.

In this section, we introduce operations on sets that are somewhat analogous to the addition, multiplication, and so on, that we have grown accustomed to using in reference to numerical calculations. At the same time, it should be noted that these set operations form the fundamental structure of the "set data type" found in the modern computer programming languages, for example, in Pascal. For this reason alone, the present study is quite important to the student's development. Many of the ideas presented here will reappear in different guises throughout the text, particularly in the treatment of mathematical logic (Chapter 2). For this reason as well, they should be studied quite carefully.

Goals

After studying this section, you should be able to:

- Understand the notion of *set containment.*
- State the definitions of the various set operations.
- State the principles that characterize *unions* and *intersections.*
- Compute unions, intersections, and complements of given sets.

- Understand the definition of *equality for sets.*
- Discuss the usefulness of the idea of an *empty set.*
- State and verify the fundamental properties of the set operations.
- Discuss the notion of a *partition of a set.*

Subsets of a Set

Throughout this whole section we will assume that the objects in our sets are all to be found in some universe U. Ordinarily, this kind of assumption means that we have in mind some particular domain of discourse, some area of study in which the objects in our universal set are the only objects under consideration. We may be interested in discussing properties of the integers (and then, $U = Z$), or perhaps our universe is that of all living persons. Whatever the field of investigation, an appropriate "universe" of objects will easily come to mind. Then, as we have said, the objects in each of the sets we introduce will belong to this particular universe.

Suppose A and S are sets (of objects from our universe U). Then we will say that A is a **subset** of S if every member of A is also a member of S (a situation that we denote symbolically in the form $A \subseteq S$ and pictorially in Figure 1.1). Equivalently, we say that the set A is **contained** in the set S or that S **contains** A. Occasionally, we will describe this latter situation by saying that S is a **superset** of A (and write $S \supseteq A$), just to turn the relationship the other way around. Note that in spite of the phrasing, "every member of A," we rarely have an exhaustive checking to be performed in verifying that $A \subseteq S$, since the members of A will ordinarily have been characterized—for example, as those objects satisfying a certain property. If we are then able to see that any and, hence, every element of A also belongs to S, then our task is complete.

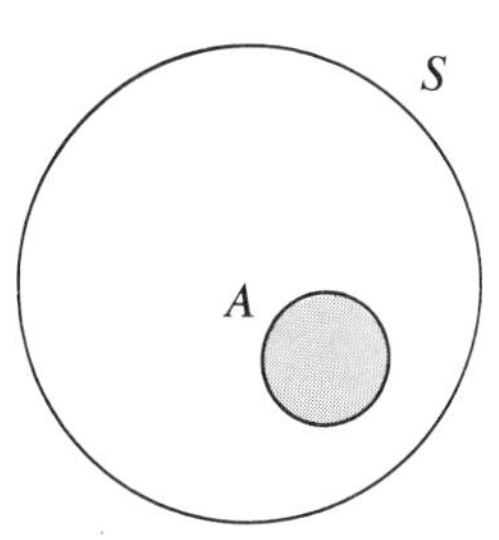

FIGURE 1.1

Example 1.13 If N and Z are the sets of Example 1.3, then we may indeed write $N \subseteq Z$, because every element of N is a positive integer, and hence, an integer (that is, a member of Z). ■

Example 1.14 With finite sets, only a visual inspection may be necessary to determine whether one set is a subset of another. Suppose our universe is

$$U = \{1, 2, 3, 4, 5, 6, 7, 8, 9, 10\}$$

and we are considering the three sets

$$S = \{1, 3, 5, 7, 9\} \qquad A = \{1, 5, 7\} \qquad B = \{1, 3, 5, 6, 7\}$$

Then clearly $A \subseteq S$, but $B \nsubseteq S$ because $6 \in B$, whereas $6 \notin S$ (so we cannot say that *every* element of B is also a member of S). Note that we have $A \subseteq B$ as well as $A \subseteq S$. ■

We find it helpful (although beginning students find it to be puzzling at first) to introduce in any universe of discourse, a set having no members at all. Naturally enough, we call this set the **empty set**, and regardless of the universe under consideration, we always use the symbol $\varnothing$ to represent this set. If we recall the notion of "constructibility," it follows that the empty set is constructed by doing nothing. In some computer programming languages, for example, Pascal, the empty set is denoted by $\{\ \}$, that is, as a set of braces with nothing inside. This $\{\ \}$ notation can be a helpful visualization of the concept of the empty set. There are technical reasons for introducing such a set, reasons that become clear as we proceed. As you would expect, however, the empty set is quite special in its properties in relation to other sets.

Example 1.15 For any set S (regardless of the universe) we have

$$\varnothing \subseteq S \quad \text{and} \quad S \subseteq S$$

The second property comes as no surprise. Surely every member of S is a member of S! But the first of these relationships can only be understood if the reader is willing to be somewhat pedantic in the use of logical reasoning. We are asked to see that "every member of $\varnothing$ is also a member of S." Now remember, the empty set has no members. So we can say anything we like about them, for example, "every member of the empty set is green-eyed and fuzzy," and we can never be challenged. It is this line of reasoning that makes the statement $\varnothing \subseteq S$ (for any S) a valid one. ■

Set Operations

We have been accustomed, since elementary school, to using the operations of arithmetic—addition, multiplication, and so on—with numbers in order to solve a variety of problems. We can apply a somewhat analogous algebraic system to sets. Of course, the type of problems we then consider are quite different, but still, we have a "calculus of sets" at our disposal, that is, a method of reckoning that is most useful in reasoning with sets.

The definitions that we are about to give make clearer the advantages of the set-builder notation introduced in Section 1.1. In fact, exactly what we are doing is building new sets from old ones. Thus if A and B are sets, the derived collection

$$A \cup B = \{x: x \in A \quad \text{or} \quad x \in B\}$$

is called the **union** of A and B, whereas

$$A \cap B = \{x: x \in A \quad \text{and} \quad x \in B\}$$

is called the **intersection** of A and B. In words, $A \cup B$ consists of all objects that are found either in A or in B (or in both), whereas $A \cap B$ is the collection of those objects that appear in both A and B. See Figure 1.2.

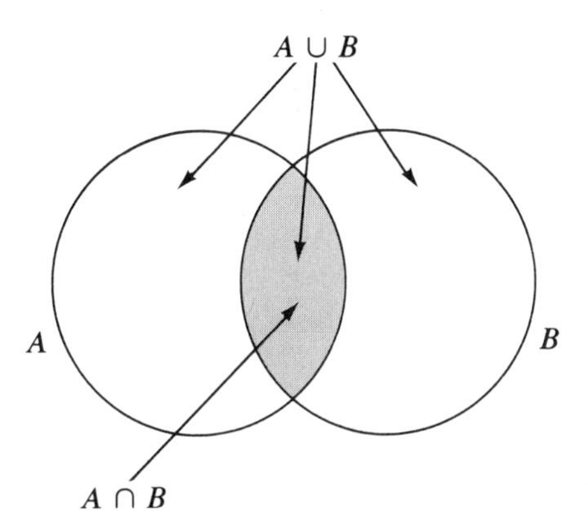

FIGURE 1.2

Example 1.16 As in Example 1.14, suppose that our universe is

$$U = \{1, 2, 3, 4, 5, 6, 7, 8, 9, 10\}$$

and let A and B be given as follows:

$$A = \{1, 2, 3, 7, 10\} \qquad B = \{2, 4, 7, 9\}$$

Then the union and intersection would be, respectively,

$$A \cup B = \{1, 2, 3, 4, 7, 9, 10\}$$
$$A \cap B = \{2, 7\}$$

Note that if A had been the set $A = \{1, 3, 5, 8\}$, then there would have been no members common to both A and B. In that case, it would make sense to write $A \cap B = \varnothing$, and in such instances the reader can begin to appreciate the usefulness of the notion of an empty set. ■

Whenever we have two sets A, B with $A \cap B = \varnothing$, we say that the sets are **disjoint**. More generally, suppose we have a whole collection

$$\mathscr{A} = \{A_1, A_2, \ldots\}$$

of sets A_i. As is customary when treating a collection (set) of sets, we use a more elaborate script letter as a name for the collection. If the collection is finite, it is more common to write $\mathscr{A} = \{A_1, A_2, \ldots, A_n\}$ for clarity. But in any case, the sets of $\mathscr{A}$ are said to be **mutually disjoint** if all pairs of distinct sets chosen from $\mathscr{A}$ are disjoint, that is, if $A_i \cap A_j = \varnothing$ for all $i \neq j$. This notion plays an important role in the definition of a partition given later in this section.

Characterization of Union and Intersection

We can make use of our "set containment" relationship to precisely characterize the set unions and intersections (see Exercises 14 and 15). The important principles (a) and (b) that follow state formally and precisely that the union of two sets is a "smallest" set, containing each of them, whereas their intersection is a "largest" set, contained in each.

Characterization of Set Unions and Intersections For any sets A and B,

(a) $A \subseteq A \cup B$ and $B \subseteq A \cup B$. Any set C satisfying $A \subseteq C$ and $B \subseteq C$ will contain $A \cup B$

(b) $A \cap B \subseteq A$ and $A \cap B \subseteq B$. Any set C satisfying $C \subseteq A$ and $C \subseteq B$ will be contained in $A \cap B$

In spite of the rather heavy use of symbolism, a closer inspection reveals that these properties are in fact quite easily verified.

Example 1.17 Consider the last phrase in principle (b). If indeed we have $C \subseteq A$ and $C \subseteq B$, then we can show that such a set C is contained in $A \cap B$ as follows. Letting $c \in C$ be any element of C, we would have $c \in A$ and $c \in B$ (because $C \subseteq A$ and $C \subseteq B$, respectively). By the definition of intersections, we must have $c \in A \cap B$. Since c is any member of C (and we conclude that $c \in A \cap B$), we have indeed shown that $C \subseteq A \cap B$, as required. Note that principle (b) indeed has the effect of saying that $A \cap B$ is, in some sense, the "largest" set that is contained in both A and B. ■

Example 1.18 If A and B are as originally given in Example 1.16, then $C = \{2\}$ is a set that is contained in both A and B. However, set $\{2, 7\}$ is "larger," and that is what is so special about the intersection. In fact, we can say that $C \subseteq A$ and

$C \subseteq B$ for each of the following sets C (and no others), recalling Example 1.15 in the first instance.

$$C = \varnothing \qquad C = \{2\} \qquad C = \{7\} \qquad C = \{2, 7\}$$

Among these, $C = \{2, 7\} = A \cap B$ is the largest. ■

Equality of Sets

Intuitively, we would expect two sets to be considered as equal if they each contain exactly the same members. It happens that we can use our set-containment relation once more to say the same thing more precisely. Thus if A and B are sets, we will agree that the sets are *equal* (and we naturally write $A = B$) if both $A \subseteq B$ and $B \subseteq A$. Notice what this definition entails. It requires that every member of A be also a member of B, and conversely, that each member of B be a member of A as well. Certainly that agrees with our intuition.

Properties of the Set Operations

We require a precise understanding of set equality if we are to examine the properties of our set operations more carefully. In so doing, we will find that the union and intersection operations behave, in part anyway, as do the operations of addition and multiplication to which we have all grown accustomed since our earliest school days. In this sense, the reader should feel quite familiar with some of the properties of set operations in the following list.

S1a.	$A \cup A = A$	**S1b.**	$A \cap A = A$
S2a.	$A \cup B = B \cup A$	**S2b.**	$A \cap B = B \cap A$
S3a.	$(A \cup B) \cup C = A \cup (B \cup C)$	**S3b.**	$(A \cap B) \cap C = A \cap (B \cap C)$
S4a.	$A \cup (A \cap B) = A$	**S4b.**	$A \cap (A \cup B) = A$
S5a.	$A \cup (B \cap C) = (A \cup B) \cap (A \cup C)$	**S5b.**	$A \cap (B \cup C) = (A \cap B) \cup (A \cap C)$

These are known collectively, in pairs, as the **idempotent**, **commutative**, **associative**, **absorption**, and **distributive** laws, respectively. As we have indicated, some of these will look familiar if the reader mentally replaces $\cup$ by $+$ and $\cap$ by $\times$. In particular, the commutative (S2) and associative (S3) laws should be familiar, as should the second of the distributive laws (S5b). Note however, that (S5a) is not true of arithmetic; that is, we do not write

$$a + (b \times c) = (a + b) \times (a + c)$$

in dealing with numerical quantities. In fact, it is precisely this law, together with those of (S1) and (S4) that should be given the most attention, inasmuch as they are probably new to most of you.

Example 1.19 Suppose that we want to verify the identity (S4a). It is important to remember that the equal sign denotes a set equality, so we must appeal to the definition of set equality (each side of the equation must be shown to be contained in the other). The containment

$$A \subseteq A \cup (A \cap B)$$

is easy enough to establish, for it follows from principle (a) given earlier—with B replaced by $A \cap B$. To establish the reverse inclusion, namely

$$A \cup (A \cap B) \subseteq A$$

we let x be any member of $A \cup (A \cap B)$. Then we must have $x \in A$ or $x \in A \cap B$. If the first is true, then we are finished already; if the second is the case, then we have $x \in A$ again, according to principle (b) listed earlier. ■

Example 1.20 In (S5a) we must show that for arbitrary sets A, B, C we have

$$A \cup (B \cap C) \subseteq (A \cup B) \cap (A \cup C)$$

and vice versa. If $x \in A \cup (B \cap C)$, then $x \in A$ or $x \in B \cap C$. In the first instance, we have both $x \in A \cup B$ and $x \in A \cup C$ by virtue of principle (a) given previously. Finally, $x \in (A \cup B) \cap (A \cup C)$ by the definition of intersection. In the second instance, we have $x \in B$ and $x \in C$, so that again by principle (a), it follows that $x \in A \cup B$ and $x \in A \cup C$. (Note that in this use of principle (a), we are appealing to (S2a) as well, but we may assume that this property has already been verified.) In any case, we then have x in the intersection as before. We leave the verification of the reverse inclusion as an exercise for the student. ■

Set Complementation

To complete our description of the calculus of sets, we introduce a kind of set subtraction. Suppose A and B are sets, and we want to speak of the elements of B that are not in A (see Figure 1.3) as if we had subtracted them out. For this purpose, we can define an appropriate **set difference** by writing

$$B \sim A = \{x\colon x \in B \quad \text{and} \quad x \notin A\}$$

For the special case in which B is the whole universe U (see Figure 1.4), we refer to the **complement** of A, writing (rather than $U \sim A$)

$$\sim A = \{x\colon x \notin A\}$$

with the additional requirement $x \in U$ understood.

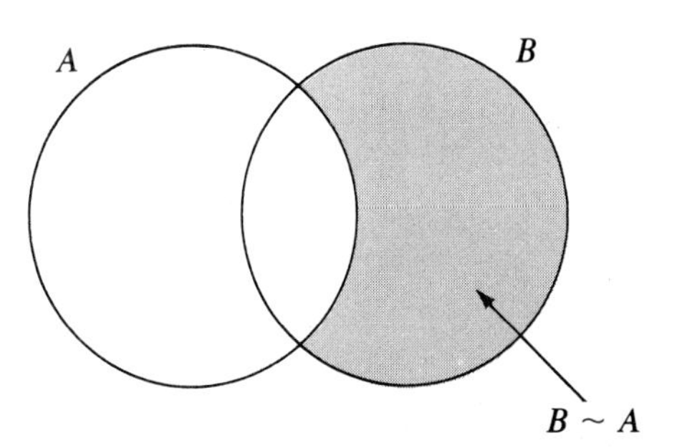

FIGURE 1.3

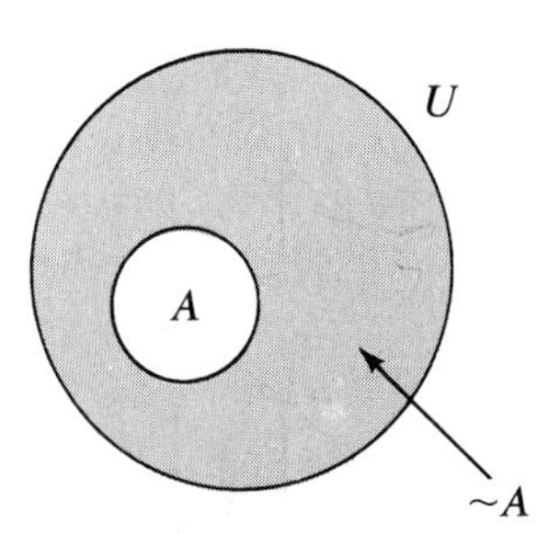

FIGURE 1.4

Example 1.21 Assuming U, A, B are the sets originally given in Example 1.16, we may compute

$$B \sim A = \{4, 9\} \quad \text{whereas} \quad \sim A = \{4, 5, 6, 8, 9\}$$

noting that $\sim A$ consists precisely of those elements of U that are not in A. It is in this sense that $\sim A$ is complementary to A, as is described most succinctly in properties (S8) that follow. ■

The introduction of the complementation operator (together with our previous discussions of the universe and the empty set) forces an enlargement of our list of properties, so as to include the following identities, for any set A:

S6a. $A \cup \varnothing = A$ **S6b.** $A \cap \varnothing = \varnothing$

S7a. $A \cup U = U$ **S7b.** $A \cap U = A$

S8a. $A \cup (\sim A) = U$ **S8b.** $A \cap (\sim A) = \varnothing$

S9a. $\sim(A \cup B) = \sim A \cap \sim B$ **S9b.** $\sim(A \cap B) = \sim A \cup \sim B$

S10. $\sim(\sim A) = A$

The dual identities (S9) are called **DeMorgan's Laws** (after the nineteenth century English mathematician and logician, Augustus deMorgan). Once again, the reader must be reminded that such general identities as these cannot be verified by merely examining a picture such as Figure 1.4. We must make use of the precise definition of set equality as given earlier.

Example 1.22 In verifying property (S8a), we have $A \subseteq U$ and $\sim A \subseteq U$ (because U is, after all, our universe). It follows that

$$A \cup (\sim A) \subseteq U$$

because of the last phrase of principle (a) given previously. Conversely, to show that the reverse inclusion

$$U \subseteq A \cup (\sim A)$$

also holds, we consider any element $x \in U$. Most likely, the reader is willing to assume that there can only be two possibilities—either $x \in A$ or $x \notin A$. However, $x \in \sim A$ in the case of the latter, so $x \in A \cup (\sim A)$ by the definition of set union. ■

Example 1.23 Suppose we are considering 10 runners in a race, identified by numbers drawn from the universe

$$U = \{1, 2, 3, 4, 5, 6, 7, 8, 9, 10\}$$

Let the sets

$$A = \{1, 3, 4, 8\} \qquad B = \{3, 5, 9\} \qquad C = \{1, 2, 4, 6, 9, 10\}$$

represent those who developed cramps, those who became short of breath, and those who managed to finish the race, regardless of their condition, respectively. Suppose we try to write a kind of set expression and then try to solve for the set of runners who developed cramps *or* shortness of breath *and* didn't finish. Remembering that *or* signifies union and that *and* signifies intersection, we are led to write:

$$\begin{aligned}(A \cup B) \cap \sim C &= (\{1, 3, 4, 8\} \cup \{3, 5, 9\}) \cap \sim \{1, 2, 4, 6, 9, 10\} \\ &= \{1, 3, 4, 5, 8, 9\} \cap \{3, 5, 7, 8\} \\ &= \{3, 5, 8\}\end{aligned}$$

noting that the negation *didn't* signifies complementation. We make the observation that the whole calculation could have been performed quite differently using property (S5b) together with (S2b). Thus we could have written:

$$\begin{aligned}(A \cup B) \cap \sim C &= (A \cap \sim C) \cup (B \cap \sim C) \\ &= \{3, 8\} \cup \{3, 5\} \\ &= \{3, 5, 8\}\end{aligned}$$

with the same result as before. ■

Generalized Unions and Intersections

In many situations, it is useful to generalize the operations of set union and intersection so that they might apply to arbitrarily many operands, rather than just two. For this purpose it is well to analyze the elementary operations once again to see just what is involved. In a union (of two sets) we collect all objects that are found in *some* one (or the other, or both) of the sets. On the other hand, in the intersection we include only those objects that are found in *all* (that is, both) of the sets. Now suppose that we have more than two sets, say a whole collection

$$\mathscr{A} = \{A_1, A_2, \ldots\}$$

of sets A_i that can be either finite or infinite in number. Then, in view of the new phrasing of unions and intersections we have just provided, the following definitions would seem to be appropriate. We can express the *generalized union and intersection*, respectively, of the sets in $\mathscr{A}$ as follows:

$$\bigcup A_i = \{x : x \in A_i \text{ for some } i\}$$
$$\bigcap A_i = \{x : x \in A_i \text{ for all } i\}$$

If the collection is finite, say $\mathscr{A} = \{A_1, A_2, \ldots, A_n\}$, then it is more common to write

$$A_1 \cup A_2 \cup \cdots \cup A_n = \{x : x \in A_i \text{ for some } i, 1 \le i \le n\}$$
$$A_1 \cap A_2 \cap \cdots \cap A_n = \{x : x \in A_i \text{ for all } i, 1 \le i \le n\}$$

for added clarity. The reader will note that if there are only two sets as before, then these definitions agree with the original ones.

But the generalized definitions are most useful on a number of occasions. By way of illustration, consider the intuitive idea of a partition—that is, the way we cut a pie or the way we split up a country after a major war. In all such cases we are considering a collection of subsets of a given set, call it S (see Figure 1.5).

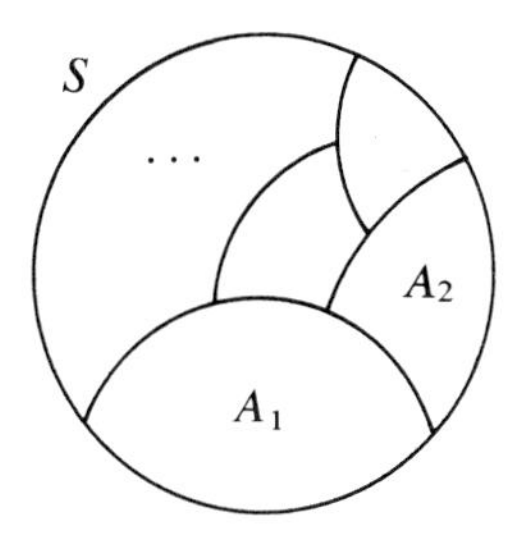

FIGURE 1.5

Formally, we can agree to say that a **partition** of S is a collection

$$\mathscr{A} = \{A_1, A_2, \ldots\}$$

of nonempty subsets $A_i \subseteq S$ with

1. $\bigcup A_i = S$ (the subsets *cover* S)
2. $A_i \cap A_j = \varnothing$ $(i \neq j)$ (the subsets are mutually disjoint)

Note that in verifying that the subsets cover S, only the inclusion $S \subseteq \bigcup A_i$ needs to be checked to see that every member of the set S is found in one or the other of the subsets A_i. The reverse inclusion (necessary for equality) follows automatically from the fact that the A_i are subsets of S.

Example 1.24 Assume that $S = \{1, 2, 3, 4, 5, 6, 7, 8, 9\}$. Then the subsets

$$A_1 = \{1, 2, 5, 6\} \qquad A_2 = \{3, 7, 9\} \qquad A_3 = \{4, 8, 9\}$$

do not form a partition of S, because $A_2 \cap A_3 = \{9\} \neq \varnothing$, noting the disjointness requirement (2). If instead, $A_2 = \{3, 7\}$, then the subsets do form a partition of S. The partition is denoted by

$$\begin{aligned} \mathscr{A} &= \{A_1, A_2, A_3\} \\ &= \{\{1, 2, 5, 6\}, \{3, 7\}, \{4, 8, 9\}\} \end{aligned}$$

where it is clearly seen that we are dealing with a set of sets. However, if $A_2 = \{3\}$, then again we do not have a partition, because there is then a member (7) of S that is found in none of the subsets of the collection. ■

Example 1.25 To show that the notion of a partition is not restricted to a finite number of subsets, consider the sets of positive integers:

$$A_i = \{n\colon 10(i - 1) < n \leq 10i\} \qquad (i \geq 1)$$

More specifically, we can write:

$$\begin{aligned} A_1 &= \{1, 2, 3, 4, 5, 6, 7, 8, 9, 10\} \\ A_2 &= \{11, 12, 13, 14, 15, 16, 17, 18, 19, 20\} \\ A_3 &= \{21, 22, 23, 24, 25, 26, 27, 28, 29, 30\} \end{aligned}$$

and so on, and the reader will begin to see that indeed, the collection

$$\mathscr{A} = \{A_1, A_2, A_3, \ldots\}$$

forms a partition of N (see Exercise 13). ■

EXERCISES 1.2

1 Given the sets described in Exercise 1 of Section 1.1, show that:

(a) $C \nsubseteq D$ **(b)** A, B are disjoint

(c) $D \subseteq Z$ **(d)** $A \cup (B \cap C) = A$

2 Given the sets described in Exercise 2 of Section 1.1, show that:

(a) $G \subseteq D$ **(b)** G, H are disjoint

(c) $A \cup H \nsubseteq N$ **(d)** A, H are not disjoint

3 Given the universe $U = \{a, b, c, d, e, f, g\}$ with subsets

$$A = \{a, b, d, f\} \qquad B = \{a, c, f, g\} \qquad C = \{b, e, g\}$$

determine the sets:

(a) $A \cap B$ **(b)** $A \cup C$

(c) $(\sim A) \cup B$ **(d)** $(A \cup B) \cap C$

(e) $(A \cup (\sim B)) \cap C$ **(f)** $A \cap B \cap C$

(g) $A \cap C$ **(h)** $B \cup (\sim C)$

4 For arbitrary sets A, B verify the opening phrase of

(a) Principle (a), the characterization of set unions:

$$A \subseteq A \cup B \quad \text{and} \quad B \subseteq A \cup B$$

(b) Principle (b), the characterization of set intersections:

$$A \cap B \subseteq A \quad \text{and} \quad A \cap B \subseteq B$$

5 Verify the last phrase of principle (a), the characterization of set unions:

$$\text{If } A \subseteq C \text{ and } B \subseteq C, \text{ then } A \cup B \subseteq C.$$

6 For arbitrary sets A,B,C verify the identities:

(a) Property (S5b) **(b)** Property (S4b)

(c) Property (S3a) **(d)** Property (S2b)

(e) Property (S9b) **(f)** Property (S1a)

(g) Property (S8b) **(h)** Property (S7a)

(i) Property (S10) **(j)** Property (S6b)

7 Show for arbitrary sets A, B, C, D that if $A \subseteq C$ and $B \subseteq D$ then

$$A \cap B \subseteq C \cap D$$

8 Refer to Example 1.23 and write a set expression and then solve for the set of runners who:

(a) Developed cramps *and* shortness of breath but nevertheless (*and*) finished the race.

(b) Developed cramps *and* shortness of breath *or* didn't finish the race.

9 A boy needs a subject for a biographical report on a North American. He would like to write about a sports personality. He would prefer to write about a female personality, unless the subject is wealthy or famous. He does not want to write about a U.S. citizen unless that person is wealthy and famous. Consider the following sets of persons:

$$A = \{\text{sports personalities}\} \qquad D = \{\text{wealthy persons}\}$$
$$B = \{\text{North Americans}\} \qquad E = \{\text{famous persons}\}$$
$$C = \{\text{females}\} \qquad F = \{\text{U.S. citizens}\}$$

Write a set expression representing the suitable subjects for the report.

10 Given the universe $U = \{1, 2, 3, 4, 5, 6, 7, 8, 9, 10\}$ and the sets

$$A = \{3, 6, 9, 10\} \qquad D = \varnothing$$
$$B = \{2, 4, 6, 8\} \qquad E = \{2, 3, 4, 5\}$$
$$C = \{3, 5, 7\} \qquad F = \{1, 2, 3, 7, 8, 9\}$$

determine the sets:

(a) $A \cup B$ **(b)** $(A \cup B) \cap D$

(c) $(C \cap E) \cup F$ **(d)** $A \cap B$

(e) $(D \cup E) \sim F$ **(f)** $(A \sim B) \cup (B \sim A)$

11 Write a set expression representing the shaded portion of Figure 1.6 for two sets A, B.

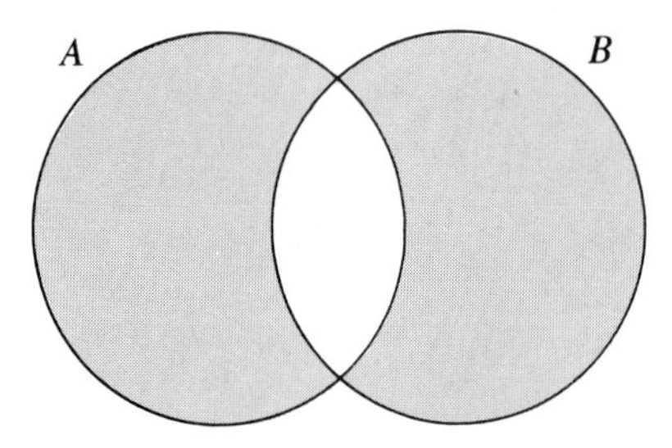

FIGURE 1.6

12 List all of the various partitions of each set:

(a) $\{a, b, c\}$ **(b)** $\{1, 2, 3, 4\}$

13 Verify that the sets A_i of Example 1.25 indeed form a partition of N.

14 Show that if D is a set that satisfies the following properties, then $D = A \cup B$.

$$A \subseteq D, B \subseteq D. \textit{ Any set } C \textit{ satisfying}$$
$$A \subseteq C \text{ and } B \subseteq C \textit{ contains } D$$

15 State and prove an analogous statement (to that of Exercise 14) for set intersection.

Sec. 1.3 FUNCTIONS AND RELATIONS

A function is a rule associating a unique object with each given (
a relation may associate more than one object with another.

In arithmetic and algebra, numbers (and symbols that stand for numbers) are the main objects of study. In discrete mathematics and in calculus as well, we often find that the objects we want to study are the relationships between members of one set and another. These sets may be sets of numbers, but not necessarily. Certain of these relationships are of quite a special nature; we call them *functions*, and because they are special, we give them our first attention. From this preliminary discussion alone, it is clear that the notion of a set will loom quite large in the background of our development, and thus we are building directly on the material discussed in the two opening sections.

Goals

After studying this section, you should be able to:

- Define the notion of a *function.*
- Provide rules (equations) to define *numerical functions* and *sequences.*
- Give rules (listings) to define *finite functions* and *sequences.*
- Discuss the idea of a *word* on an alphabet.
- Give precise definitions for the terms *one-to-one* and *onto.*
- Define and use the idea of a *composite function.*
- Decide whether a given function is one-to-one, onto, *invertible.*
- Determine the *inverse of a function* (if it is invertible).
- Describe the concept of a *set product.*
- Define the notion of a *relation* and contrast it with that of a function, abstractly and by example.

Definition of a Function

We would like to continue to pursue the "constructive" aspect of set theory as we now move from sets to functions. Whereas a function is indeed viewed as a relationship of one set to another, we would like to insist that such relationships be constructed by a definite *rule* given in advance. This point of view is closely allied to the questions of computability and feasibility from which a modern practical treatment of discrete mathematics must spring. It is not as though the set relationships already exist in some ideal mathematical world, just waiting to be discovered; we (or our computing agents) construct them to serve a practical purpose. Thus we wish to avoid the more abstract treatment of the subject that has confounded all too many beginning students.

Let's assume that S and T are sets. We then agree that a **function f from S to T** is a rule that associates with each object of S a unique object in T. This association is symbolized in one or the other of the forms

$$f: S \to T \quad \text{or} \quad S \xrightarrow{f} T$$

whichever is more convenient at the time. In order to emphasize that it is the function that is of primary importance, S is called the **domain** of the function f, and T is called its **range**.

If the function $f: S \to T$ associates the unique object $t \in T$ with the object $s \in S$, then we either write

$$f(s) = t \quad \text{or} \quad s \xrightarrow{f} t$$

(often omitting the symbol f in the latter), somewhat in correspondence with the two styles of notation given previously. In the symbolism $f(s) = t$, it should be emphasized that there is no multiplication intended, and in fact, the left-hand member is read "f of s," so as to make this clear. In this connection any object s in the domain S is called an **argument** of the function $f: S \to T$, and the associated object t in the range T is called the **value** of the function (at s), or sometimes, the **image** of s.

It is perhaps unfortunate that the idea of a function requires so much accompanying notation and terminology. The concept is sufficiently abstract on its own. But the student will find, after seeing a few examples, that our previous experience and our common sense are enough to "carry the day." In particular, if S and T are sets of numbers, then it is often possible to specify a rule $f: S \to T$ in the form of an equation, and in such cases, all of the terminology and notation will seem quite familiar.

Example 1.26 Suppose $S = T = Z$ (the integers), and we consider the function

$$f: Z \to Z$$

defined by the equation

$$f(x) = x^2 + 2x + 1$$

In writing $f: Z \to Z$ we only signify that we intend the domain of the function f to be the set Z of all the integers (and its range also to be Z). It is important to remember that this notation alone does not specify a function. A rule must be given, and in the present case, this is done with an equation. Certainly it will make sense to consider any integer as an argument in the right-hand side of the previous equation, and thus Z is acceptable as the domain of f. Moreover, if we substitute an integer x into the equation, we certainly obtain another integer $f(x)$, so it is reasonable to think of Z as the range of f as well. As we have said, the equation becomes the rule for computing the integer value $f(x)$ that f associates with a given argument, the integer x. Thus we may write:

$$\begin{aligned}
f(-3) &= (-3)^2 + 2(-3) + 1 = 4 \\
f(-2) &= (-2)^2 + 2(-2) + 1 = 1 \\
f(-1) &= (-1)^2 + 2(-1) + 1 = 0 \\
f(0) &= (0)^2 + 2(0) + 1 = 1 \\
f(1) &= (1)^2 + 2(1) + 1 = 4
\end{aligned}$$

etc., or alternatively, we may write:

$$\begin{aligned}
-3 &\to 4 \\
-2 &\to 1 \\
-1 &\to 0 \\
0 &\to 1 \\
1 &\to 4
\end{aligned}$$

etc. In any case, the reader will see that the equation provides a mechanical means for computing any value $f(x)$ that we wish. ■

Example 1.27 Suppose S and T are the sets

$$S = \{1, 2, 3, 4\} \qquad T = \{a, b, c, d, e\}$$

and we wish to construct a function $f: S \to T$. Since everything is finite here, we can give a rule by simply listing each x and $f(x)$ in pairs, however we choose, for example,

$$\begin{aligned}
1 &\to e \\
2 &\to c \\
3 &\to b \\
4 &\to c
\end{aligned}$$

Then the rule is simply this: read down the list to find x on the left, then follow the arrow to the right and read off $f(x)$; that is, we have $f(2) = c$, $f(3) = b$, etc. As far as we know, there is no special significance to this particular function, and as a matter of fact, if nothing else, this example shows the complete generality of the whole concept. ■

Our first two examples should make one thing quite clear. The name, f, in both instances of a function is only a name; another identifier, say g, would do just as well. As in the case of the naming of sets, it is important to become accustomed to the use of the same name in different settings. The context will eliminate any possibility for confusion. Moreover, we use the generic name f for functions in much the same way that we use S in speaking of general sets. Again, by analogy with our conventions for the naming of sets, a special function will sometimes have a name reserved for that meaning only, as in Example 1.28.

Example 1.28 Suppose, as in Example 1.26, that our domain and range is the set Z of integers. This does not necessarily mean that we will be able to state the rule of a function in the form of an equation. Consider the definition

$$f(x) = \begin{cases} -x & \text{if } x \text{ is negative} \\ x & \text{otherwise} \end{cases}$$

This particular rule will be recognized immediately as the *absolute value* function, commonly denoted mathematically as $|x|$, or as abs(x) in some computer programming languages. Thus, we would normally have introduced this function by writing:

$$|x| = \begin{cases} -x & \text{if } x \text{ is negative} \\ x & \text{otherwise} \end{cases}$$

instead of calling it f. ■

Sequences

If a function has the set N of natural numbers as its domain, then it is said to be a **sequence** (or, to be precise, an **infinite sequence**). More specifically, a function $x:N \to X$ is said to be a **sequence (or infinite sequence) with values in the set X.** At the same time, the **terms** of the sequence, that is, the function values $x(1), x(2)$, etc. are often denoted instead with subscripts as x_1, x_2, and so on.

Example 1.29 The function

$$x(n) = (-1)^n$$

establishes a sequence $x:N \to Z$ of (quite limited) integer values, namely $-1, 1, -1, 1, \ldots$, on and on. Thus we have:

$$\begin{aligned} x(1) &= (-1)^1 = -1 \\ x(2) &= (-1)^2 = 1 \\ x(3) &= (-1)^3 = -1 \\ x(4) &= (-1)^4 = 1 \end{aligned}$$

etc. Note, however, that it is more common to write $x_1 = -1$, $x_2 = 1$, $x_3 = -1$, $x_4 = 1$, etc. ■

Because, in the case of sequences, the domain is already understood to be the set of natural numbers, we often give more prominence to the terms (the sequence values as a function), and in fact, it is quite common to see the sequence defined by specifying the general nth term x. This style of denotation is illustrated in the two positive integer sequences $x: N \to N$ that follow in Examples 1.30 and 1.31.

Example 1.30 The sequence

$$x_n = 2n$$

defines (as its set of values) the even positive integers, 2, 4, 6, 8, 10, etc. ■

Example 1.31 By writing

$$x_n = \text{the } n\text{th prime number}$$

we have introduced the primes 2, 3, 5, 7, 11, etc. as a sequence. ■

In many of our discussions, the notion of a **finite sequence** is quite important, and in such cases, we are considering a function

$$x: \mathbf{n} \to X$$

with some set **n** (see Example 1.4) as our domain, rather than N. The terms of such a sequence would be denoted $x_1, x_2, \ldots, x_n$, by analogy with Example 1.29. Often the terms of the sequence are enclosed in angle brackets; for example, $\langle x_1, x_2, \ldots, x_n \rangle$.

If the range of a finite sequence is an alphabet set A (see Section 1.1) rather than an arbitrary set X, then we are considering a function $a: \mathbf{n} \to A$, and instead of the notation just used, we often omit the commas, writing $a_1 a_2 \ldots a_n$ and calling the resulting sequence a **word** (*of length* n) on the alphabet A. The reader will see from Example 1.32 that this terminology, by and large, is well chosen. It is customary to use the identifier A^* to refer to the set of all words on an alphabet A; that is,

$$A^* = \{w: w = a_1 a_2 \ldots a_n \text{ with each } a_i \in A\}$$

Here, we include words of all lengths $n \geq 0$. For technical reasons we even include a word of length zero—so short that it can't be seen! We call it the **null word** and use a special symbol λ (the Greek lambda) to refer to this artificial word.

Example 1.32 If A is the ordinary lowercase alphabet

$$A = \{a, b, c, \ldots, z\}$$

of Example 1.5, then the following are words:

cat finite
apple circle
robot halcyon

but so are the following:

mimsy	freind
zrg	nurd

showing that the notion of a word on an alphabet, though agreeing with our intuition to some extent, goes somewhat beyond the obvious meaning in its full range of application. Thus we have an infinite collection

$$A^* = \{\lambda, a, b, \ldots, z, \ldots, \text{cat}, \ldots, \text{zrg}, \ldots\}$$

including the null word, all words of length one, all words of length two, all words of length three, etc. ■

Example 1.33 If **P** is the Pascal alphabet (of Example 1.5), then the following string of symbols is considered to be a word on the alphabet **P**.

$$\text{if } x < 3 \text{ then } y := y + 1;$$

(The spaces are blank symbols.) In fact, every Pascal program is a word on the alphabet **P**, that is, a member of the set **P***. As in the previous example, however, it must be remembered that nonsensical combinations of these same symbols would still be considered as members of **P***. In fact, one of the jobs of a computer language compiler is to recognize a meaningful sequence of characters from an arbitrary sequence. ■

Properties of Functions

As we have seen, functions (or **mappings** as they are also called) simply transform members of one set into members of another. Functions are the main objects of study in calculus. Examples abound in computer science as well. A compiler transforms programs written in some high-level language such as Pascal or FORTRAN into *machine language*. Codes or encodings transform numbers or alphanumeric data into binary information. In this instance we would probably want the encoding process to be reversible, and we might then raise the question of the existence or nonexistence of an *inverse* for the function at hand. This is one of the notions that we would like to investigate in our survey of functional properties, a study that spans the next two subsections and beyond.

For this purpose, let us return to the general discussion of an arbitrary function $f: S \to T$ from a set S to a set T. We recall that f associates with each element x in the domain S one and only one element $f(x)$ in the range T. Thus functions are by nature single valued. Said another way, they should be **well-defined** in the sense that

$$x = y \quad \text{should imply} \quad f(x) = f(y)$$

By this we mean that the value of the function at x should not depend in any way on the representation that is chosen for x. On the other hand, if the converse also holds, that is,

$$f(x) = f(y) \quad \text{implies} \quad x = y$$

then the function f is said to be **one-to-one** (abbreviated 1–1). In what seems to be a different matter altogether, the function $f: S \to T$ is said to be **onto** T (or simply, **onto**) if every element t in the range T can be written, uniquely or not, as $t = f(s)$ for some s in S; that is, if every element of the range is the image of something in the domain. We will see later (Example 1.73) that for finite sets S and T of the same cardinality, these two notions coincide. In general, however, a function that is both one-to-one and onto is said to be a **correspondence** (or more specifically, a **one-to-one correspondence**). Obviously, these mappings are quite special (see the last diagram in Figure 1.7).

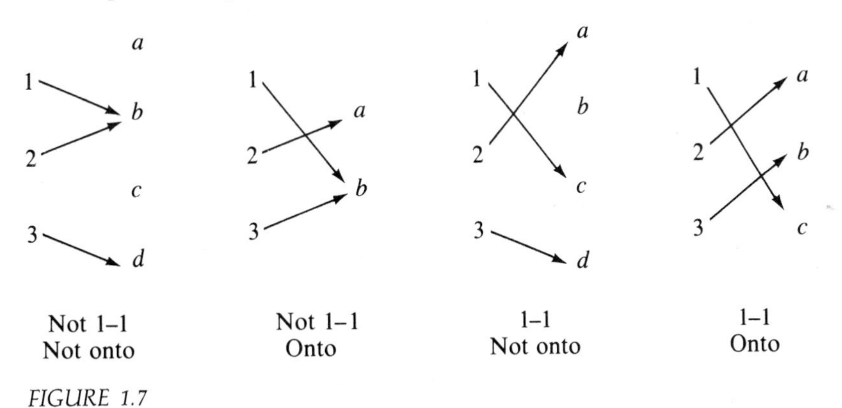

FIGURE 1.7

Example 1.34 It is instructive to see that, even for finite sets, it is possible for all four combinations (1–1 or not, onto or not) to arise (see Figure 1.7). In the first case, the function (call it f) is not one-to-one because $f(1) = f(2)$, whereas $1 \neq 2$; that is, the general condition

$$f(x) = f(y) \quad \text{implies} \quad x = y$$

does not hold. Neither is f onto, since a (and also c) do not appear as images. The second instance is onto, however, since we have

$$a = f(2) \qquad \text{and} \qquad b = f(1) \quad [\text{or } f(3)]$$

but again, this mapping is not one-to-one because $f(1) = f(3)$, whereas $1 \neq 3$. Note that the fourth illustration is an instance of a one-to-one correspondence. ■

Example 1.35 Consider the mapping $g: Z \to Z$ given by the equation

$$g(x) = x + 3$$

Because we can argue that

$$g(x) = g(y) \quad \text{implies} \quad x + 3 = y + 3, \quad \text{which implies} \quad x = y$$

we conclude that g is one-to-one. Moreover, any member y of the range (y an integer) can be written:

$$g(y - 3) = (y - 3) + 3 = y$$

and because $y - 3$ is certainly in the domain (of integers), we conclude that g is onto as well. Thus we have a one-to-one correspondence of the integers with themselves (see Figure 1.8).

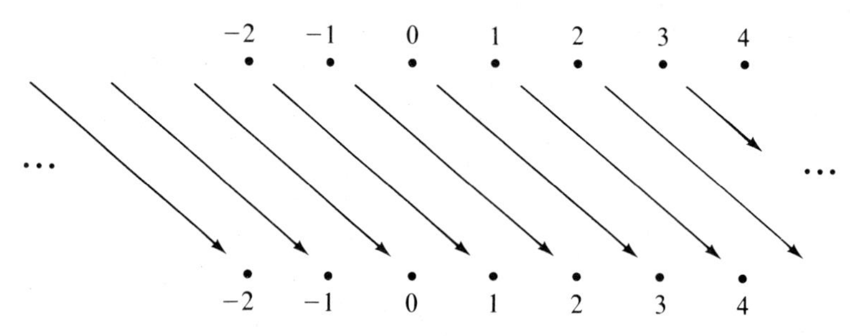

FIGURE 1.8

Composition of Functions

The notion of a **composite function** (a function of a function, as it is sometimes called) is central to the treatment of algebra, trigonometry, calculus, and discrete mathematics as well. Formally, we can discuss all of these situations in the following general setting. We suppose that we are given two functions $g: S \to T$ and $f: T \to U$ (note that the range of g is required to coincide with the domain of f). We may then define a new function (giving it the name $f \circ g$)

$$f \circ g: S \to U$$

by setting

$$f \circ g(x) = f(g(x))$$

Note that this is a genuine rule, provided that g and f are themselves given by rules. That is, to compute $f \circ g(x)$ we first compute $g(x)$, and we then use this value as an argument to f, computing $f(g(x))$. The result is a value in U, as required, whenever x is in S. (See Figure 1.9.) The new mapping $f \circ g$ is called the **composition** of f and g.

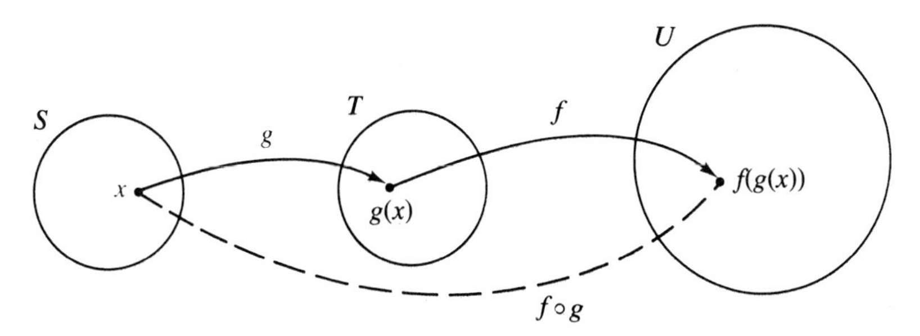

FIGURE 1.9

Example 1.36 If we have the two functions

$$g(x) = x^2 - 2x + 3 \qquad f(y) = 3y + 1$$

(suppose both have Z as domain and range, say), then the composition $f \circ g$ produces the function

$$f \circ g(x) = f(g(x)) = 3(x^2 - 2x + 3) + 1$$

again a function from Z to Z. ■

Example 1.37 In order to see that the same ideas are applicable to functions in a finite setting, consider g and f as pictured in Figure 1.10(a) and 1.10(b), respectively. Then $f \circ g$ is the composite function shown in Figure 1.10(c); that is,

$$\begin{aligned} f \circ g(a) &= f(g(a)) = f(3) = y \\ f \circ g(b) &= f(g(b)) = f(1) = z \\ f \circ g(c) &= f(g(c)) = f(2) = x \\ f \circ g(d) &= f(g(d)) = f(3) = y \end{aligned}$$

all in accordance with the definition of a composite function given above. ■

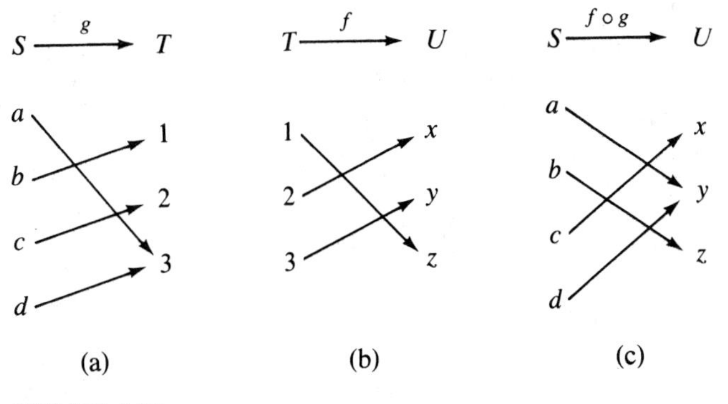

FIGURE 1.10

The notion of a composite function is useful to describe many different phenomena in mathematics. In particular we may now return to a proper discussion of the invertibility question. Suppose, to be completely general, that an arbitrary function $g: S \to T$ is given. We may then say that g is **invertible** if there is a function $f: T \to S$ (note, that it goes back the other way, as in Figure 1.11), such that

$$f \circ g(x) = x \quad \text{and} \quad g \circ f(y) = y$$

for every $x \in S$ and $y \in T$, respectively. Recalling that $f \circ g(x) = f(g(x))$, we see that such a function f (an **inverse** for g) recovers the argument x, given $g(x)$. Note that this will not always be possible. Consider, for example, the function g in Example 1.37. Since $g(a) = g(d) = 3$, there is no way that we can recover the argument x if we are only given that $g(x) = 3$. It could be a or it could be d. Noting that this particular function is not one-to-one, we can begin to appreciate the significance of the following important characterization of the *invertible* functions, that is, those that have an inverse.

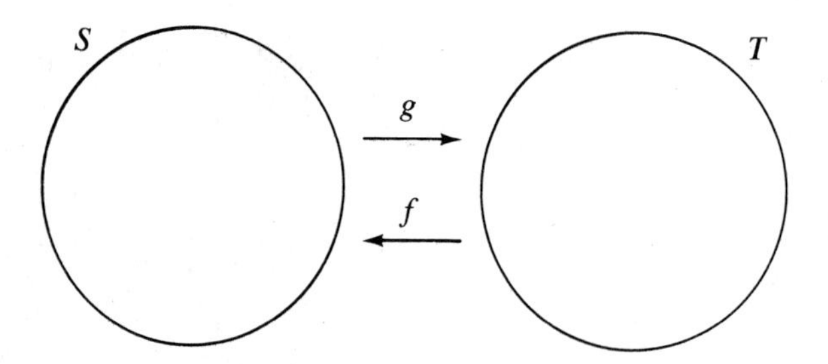

FIGURE 1.11

Characterization of Invertible Functions A function is invertible just in case it is one-to-one and onto (that is, a one-to-one correspondence).

We have already seen the reason that an inverse function cannot exist if the original function g is not one-to-one. It is equally impossible if g is not onto. For suppose y is not an image for g, so that there is no x with $g(x) = y$. Then $g \circ f(y) = g(f(y)) \neq y$, by our choice for y. Therefore, both onto and one-to-one are necessary conditions. On the other hand, if $g: S \to T$ is both one-to-one and onto, then we may define $f: T \to S$ as follows:

$$f(y) = \text{the unique } x \text{ with } g(x) = y$$

We then have:

$$f \circ g(x) = f(g(x)) = x \quad \text{and} \quad g \circ f(y) = g(f(y)) = y$$

as required. Note that the uniqueness of x follows from the fact that g is one-to-one.

Example 1.38 We have seen that the function $g: Z \to Z$ given in Example 1.35 is both one-to-one and onto. According to the preceding characterization, such a function should have an inverse. Realizing that g is simply addition of 3, it is reasonable to expect that its inverse would amount to a subtraction of 3 from its argument. Suppose, however, that we just try blindly to discover this from the prescription for the inverse:

$$f(y) = \text{the unique } x \text{ with } g(x) = y$$

as given above. We have to solve for x in the equation

$$x + 3 = g(x) = y$$

giving $x = y - 3$, a subtraction of 3, as expected. Note that we then obtain

$$f \circ g(x) = f(g(x)) = f(x + 3) = (x + 3) - 3 = x$$

and

$$g \circ f(y) = g(f(y)) = g(y - 3) = (y - 3) + 3 = y$$

as required in the definition of the inverse function. ■

Example 1.39 The function in the last part of Figure 1.7 is one-to-one and onto (see Example 1.34). Its inverse is obtained by reversing the arrows! ■

Finally, we introduce one further bit of notation for the notion of an inverse function. When f is an inverse for the function $g: S \to T$, it is useful to denote f as g^{-1}; then our original defining equations might be written:

$$g^{-1} \circ g(x) = x \quad \text{and} \quad g \circ g^{-1}(y) = y$$

thus further reinforcing the fact that g^{-1} (the inverse of g) serves to undo whatever it was that g accomplished.

Set Products

Before continuing in our discussion of functions, and particularly in our contrasting of functions with the more general relations between sets, we need to introduce one more set construct, the *product* operation. Products again build new sets from old ones, but unlike the union and intersection of sets, the product of two sets falls outside the system of subsets of a given universe U, even though the two sets may be subsets of U. Given any two sets S and T, their **product** $S \times T$ is defined as the collection of all (ordered) pairs (s, t), with $s \in S$ and $t \in T$. Using the set-building notation, we would write:

$$S \times T = \{(s, t): s \in S \quad \text{and} \quad t \in T\}$$

noting that (t, s) is not the same as (s, t). That is why we spoke of ordered pairs. Said another way, (s, t) and (s', t') are regarded as equal only if $s = s'$ and $t = t'$.

Example 1.40 If $S = \{a, b, c\}$ and $T = \{1, 2\}$ then

$$S \times T = \{(a,1),(a,2),(b,1),(b,2),(c,1),(c,2)\}$$

a set of cardinality 6. ■

More generally, for sets $S_1, S_2, \ldots, S_n$, with $n \geq 2$, we may define

$$S_1 \times S_2 \times \cdots \times S_n = \{(s_1, s_2, \ldots, s_n): s_i \in S_i \text{ for each } i\}$$

In the special case where all the S_i are the same, say $S_1 = S_2 = \ldots = S_n = S$, we let S^n denote this generalized product set. Thus for any set S we have

$$S^n = \{(s_1, s_2, \ldots, s_n): s_i \in S \text{ for all } i\}$$

and the elements of such a product set are called **n-tuples** from the set S. (Note that this term is just a generalization from the words doubles (pairs), triples, quadruples, etc.)

Example 1.41 If $B = \{0, 1\}$ (as in Example 1.4), then

$$B^n = \{(b_1, b_2, \ldots, b_n): b_i = 0 \text{ or } 1 \ \ (1 \leq i \leq n)\}$$

In words, B^n consists of all the n-tuples or *vectors*, each of whose coordinates is either 0 or 1. Thus in the special case of $n = 3$, we would have

$$B^3 = \{(0,0,0),(0,0,1),(0,1,0),(0,1,1),(1,0,0),(1,0,1),(1,1,0),(1,1,1)\}$$

a set of cardinality 8. ■

The Notion of a Relation

In order to introduce the notion of a relation, let us first establish its connection to the idea of a product set. Suppose $T = \{t_1, t_2, \ldots, t_m\}$ is a collection of unfilled jobs and $S = \{s_1, s_2, \ldots, s_n\}$ is a set of applicants. We might then let the ordered pair (s_i, t_j) signify that the person s_i is qualified for the job t_j. We may consider this ordered pair as an element of the product set $S \times T$. However, unless every applicant is qualified for every job, it is not $S \times T$ but a subset of $S \times T$ that we are dealing with in the personnel matching problem, and in the study of relations in general.

In more abstract terms, a **relation** R from a set S to a set T is simply a subset $R \subseteq S \times T$. We say that s **is related to** t (by the relation R) if $(s, t) \in R$. The notation $s\,R\,t$ is used to express this relation. A graphical representation can be given for describing relations from one finite set to another. Thus if

$$S = \{s_1, s_2, \ldots, s_n\} \quad \text{and} \quad T = \{t_1, t_2, \ldots, t_m\}$$

as before, we can draw a figure (reminiscent of our representation of finite functions) with elements of S listed at the left and those of T at the right. Then we simply draw an arrow from the point s_i to the point t_j in the event that $s_i \, R \, t_j$.

Example 1.42 For the two finite sets

$$S = \{s_1, s_2, s_3, s_4\} \quad \text{and} \quad T = \{t_1, t_2, t_3, t_4, t_5\}$$

we can introduce a relation R from S to T as shown in the graphical representation of Figure 1.12.

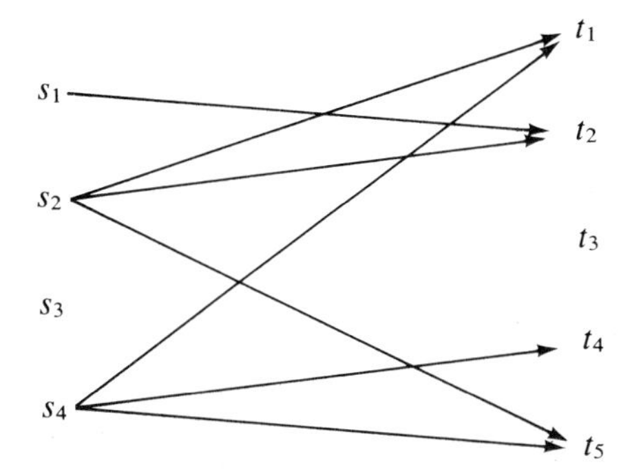

FIGURE 1.12

Then we may write

$$\begin{array}{ll} (s_1, t_2) \in R & (s_1, t_4) \notin R \\ (s_4, t_1) \in R & (s_1, t_3) \notin R \end{array}$$

etc., or alternatively, $s_1 \, R \, t_2, s_1 \not\!R \, t_4$, etc. Moreover, we have

$$R = \{(s_1, t_2), (s_2, t_1), (s_2, t_2), (s_2, t_5), (s_4, t_1), (s_4, t_4), (s_4, t_5)\}$$

as a subset of $S \times T$, with one member for each arrow in the figure. ■

With this one example, already the distinction between functions and relations is clear. A function associates with each object in its domain, one and only one object in its range. A relation may relate a given object to several objects, or to none. In Example 1.42, s_2 is related to t_1, t_2, and t_5, whereas s_3 is not related to any of the objects in T. And yet, it is important to understand that a function

$$f: S \to T$$

does represent a relation (from S to T). We have only to say that each $x \in S$ is related to the (one and only one) element $f(x) \in T$. In summary, we can say that every function gives rise to a relation, but not conversely. The notion of a relation is more general.

EXERCISES 1.3

1 Assuming the function $f: Z \to Z$ is defined by the equation

$$f(x) = x^3 - 2x + 1$$

compute the integer values that f associates with the arguments:

(a) 0 **(b)** 1 **(c)** -1 **(d)** 2

2 Repeat Exercise 1 for

$$f(x) = x^3 - 2|x| + 1$$

3 Compute the sequence values $x(7) = x_7$ assuming $x(n) = x_n$ is the sequence given in

(a) Example 1.29 **(b)** Example 1.30 **(c)** Example 1.31

4 For each of the following, decide whether the function $f: Z \to Z$ is one-to-one; onto. Verify your answers.

(a) $f(x) = x + 1$

(b) $f(x) = |x| + 1$

(c) $f(x) = ax + b \quad (a, b \in Z)$

(d) $f(x) = x^2$

(e) $f(x) = 2x$

(f) $f(x) =$ number of prime factors in x (if $|x| > 1$, otherwise $f(x) = 0$)

5 For each of the functions of Figure 1.13, decide whether they are one-to-one; onto. Verify your answers.

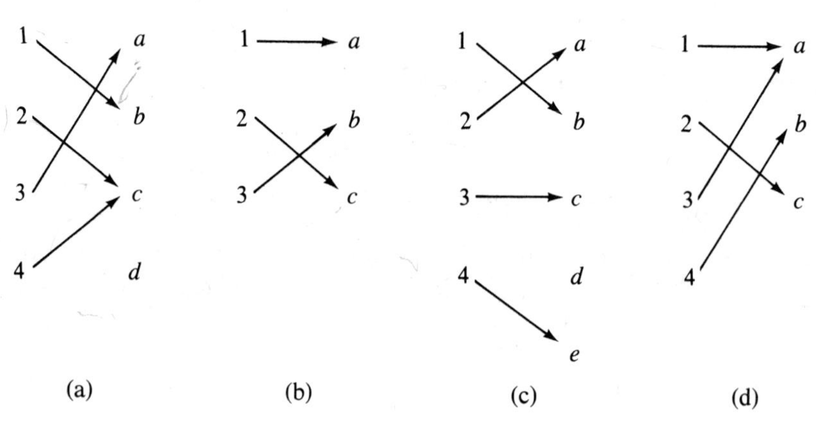

FIGURE 1.13

6 Describe the inverse function (one in each case) for the invertible functions of

(a) Exercise 4 **(b)** Exercise 5

7 Describe the function $h: Z \to Z$ given by

$$h(x) = 2(x^3 - 1) + 7$$

as a composite function $h = f \circ g$; that is, describe f and g.

8 A function $f: S \to T$ represents a relation (from S to T), and in this way, the function can be interpreted as a subset $F \subseteq S \times T$, namely, as the set

$$F = \{(s, f(s)): s \in S\}$$

Similarly, $g: S \to T$ can be viewed as a subset $G \subseteq S \times T$. We say that the two functions are *equal* (and we write $f = g$) if $F = G$ as sets. Show that

$$f = g \quad \text{means} \quad f(s) = g(s) \quad \text{for all } s \in S$$

9 Show that composition of functions is *associative*; that is, show that

$$f \circ (g \circ h) = (f \circ g) \circ h$$

whenever $h: R \to S$, $g: S \to T$, $f: T \to U$. [*Note:* This requires that we first understand what it means to say that two functions (with the same domain and range) are equal. (See Exercise 8.)]

10 Compute, that is, list the product set $S \times T$, assuming S and T are given as follows:

(a) $S = \{1, 2\}, \quad T = \{a, b, c\}$ **(b)** $S = \{1\}, \quad T = \{a, b, c, d\}$

(c) $S = \{\text{boy, girl}\}, \quad T = \{x, y\}$ **(d)** $S = \{a, b, c\}, \quad T = \{\varnothing\}$

11 Compute each of the following product sets and note their cardinality.

(a) B^4 **(b)** $\{1, 2, 3\}^3$

(c) $\{a, b, c, d\}^2$ **(d)** $\{a, b\} \times \{1, 2, 3\} \times B^2$

12 For arbitrary sets A, B, C, D show that

(a) $(A \cap B) \times (C \cap D) = (A \times C) \cap (B \times D)$

(b) $A \times (B \cup C) = (A \times B) \cup (A \times C)$

13 Interpret each of the functions of Exercise 5 as a relation.

14 Show that the composition of two arbitrary one-to-one functions is again one-to-one.

15 Show that the composition of two arbitrary onto functions is again onto.

16 What conclusion may be drawn from Exercise 14 and Exercise 15 regarding the composition of two one-to-one correspondences?

17 Give a graphical representation of the relation R from $\{1, 2, 3\}$ to $\{a, b, c, d\}$ when R is given as follows:

(a) $R = \{(1, a), (1, c), (2, b), (2, d), (3, a), (3, c)\}$

(b) $R = \{(1, d), (2, a), (2, c), (2, d)\}$

(c) $R = \{(1, b), (2, a), (3, d)\}$

(d) $R = \{(1, a), (1, b), (2, a), (2, d), (3, c), (3, d)\}$

18 Which of the relations of Exercise 17 represents a function?

Sec. 1.4 EQUIVALENCE RELATIONS AND PARTITIONS

The idea of an equivalence relation on a set and the notion of a partition of a set are really one and the same.

For many purposes, we may wish to ignore certain distinctions between the various objects of a set so that we may regard them as equivalent in some sense. We may not want to distinguish among congruent triangles or between persons of the same age group. In these settings, we are considering a relation among the members of a set, but a relation of a very special sort; we call it an *equivalence relation.* We shall find that such relations determine a partition of the set into nonoverlapping subsets. In fact, it turns out that these two ideas, partition and equivalence, simply represent two different ways of saying the same thing. This duality of perspective is a most important principle to learn.

Goals

After studying this section, you should be able to:

- Give examples of relations on a set, for example, divisibility, congruence, etc.
- Name and state the three defining properties of an *equivalence relation.*
- Determine whether or not a given relation is an equivalence relation.
- Identify the *equivalence classes* of a given equivalence relation.
- Discuss the characterization of equivalence relations as *partitions.*
- Show how a partition of a set gives rise to an equivalence relation.

Relations on a Set

As with functions, we will often find relations $R \subseteq S \times T$ where S and T are the same. Such a relation R from S to S, say, is then said to be a **relation on** S. Actually, these are perhaps the most commonly encountered relations. The relations $<$, $\leq$, etc., that one finds for integer arithmetic would qualify here, as relations on Z. In writing $6 < 11$, $7 < 148$, $7 \nless 6$, etc., we soon realize that indeed we are dealing with a relation on Z. That is, the set of all pairs (x, y) with $x < y$ is a subset, in fact, an infinite subset of $Z \times Z$. But this is only a start. We will soon discover that the idea of a relation on a set is extremely general, an idea that the student will encounter many times over.

Example 1.43 In a somewhat different vein, but still representing a relation on a set within the realm of integer arithmetic, consider the relation $|$ of (even) divisibility of one integer into another. Thus on the set N, say, we may agree to write:

$$n \mid m \quad \text{whenever } n \text{ divides } m \text{ (evenly)}$$

(Note that we have given this relation a special "reserved" name or symbol.) It then follows that

$$3 \mid 12 \qquad 5 \nmid 12$$
$$4 \mid 20 \qquad 5 \mid 20$$

etc. The reader should be cautioned, that whereas this example does appear within a section devoted to equivalence relations, it is definitely not an example of such a relation (see Exercise 1). But in this case again, the relation is infinite when considered as a subset of $N \times N$. And we have no difficulty in deciding, given arbitrary positive integers n, m, whether or not $n \mid m$. We may simply perform a long division of n into m:

$$\begin{array}{r|l} & q \\ n & m \\ & \underline{\quad} \\ & \underline{\quad} \\ & \ddots \\ & \overline{\ r\ } \end{array}$$

to obtain a quotient q and a remainder r. If $r = 0$ then $n \mid m$, otherwise not. As we have said, it's simply a matter of arithmetic. ■

Example 1.44 As a somewhat simpler and totally finite example, let us suppose that

$$S = \{s_1, s_2, s_3, s_4, s_5, s_6\}$$

and let us then consider the relation on S (from S to S) as depicted in Figure 1.14(a).

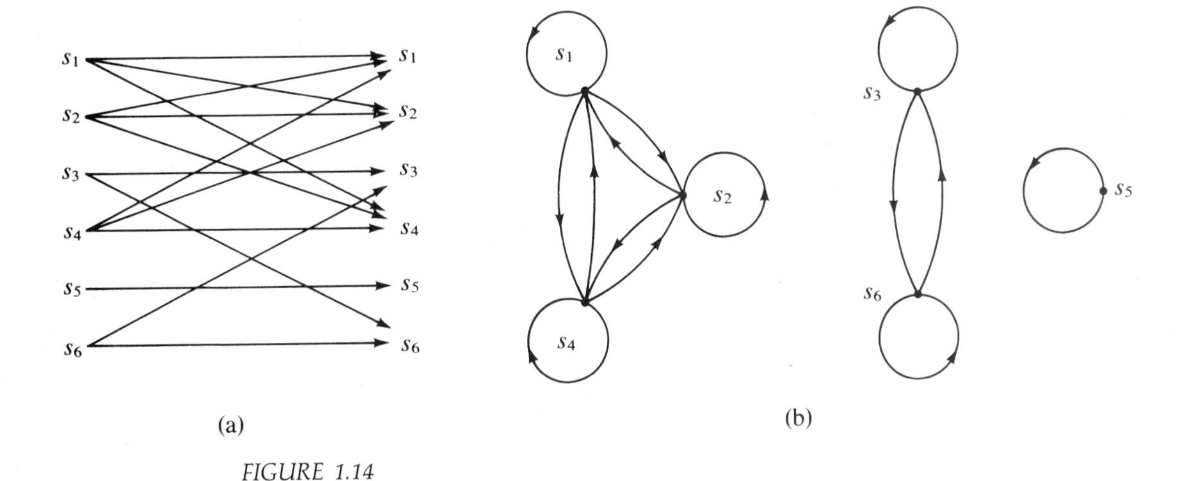

FIGURE 1.14

Simply in reading the arrows, we can decide that $s_1 \; R \; s_2$, $s_1 \not R \; s_3$, etc., and seemingly, there is nothing very profound going on here. But in figures such as this, it is helpful to collapse the two copies of the points s_i into one, as shown at the right in Figure 1.14(b). We then discover that somehow the set is partitioned by our relation into three nonoverlapping subsets. Within each of these subsets, all objects are related, one to another, but there are no relationships between objects of one subset and another. So indeed, there seems to be something quite special about this relation. What it is, we will soon discover. ■

Equivalence Relations

It is well that we begin with a simple but typical example of the notion of equivalence. We consider the set Z of integers and for a fixed $n > 1$, we introduce the relation on Z specified as follows:

$$a \underset{n}{\sim} b \quad \text{should mean} \quad n \mid (a - b)$$

and we then say that two such integers a, b are **congruent** (*modulo n*). (The military uses such a system of congruence (modulo 12) to tell time; 1300 hours (13 o'clock) is equivalent to 1 o'clock.) Then it is clear that for any $a \in Z$ we will have infinitely many integers related to a, namely those of the subset:

$$\{a, a \pm n, a \pm 2n, \ldots\}$$

for surely the difference between any two of these members is evenly divisible by n. In fact, taking $a = 0$, then $a = 1$, etc., we obtain the subsets:

$$A_0 = \{\ldots, -2n, -n, 0, n, 2n, \ldots\}$$
$$A_1 = \{\ldots, -2n + 1, -n + 1, 1, n + 1, 2n + 1, \ldots\}$$

and so on, and finally, for $a = n - 1$ we obtain

$$A_{n-1} = \{\ldots, -n - 1, -1, n - 1, 2n - 1, 3n - 1, \ldots\}$$

[If we were to continue with $a = n$, we would obtain A_0 all over again!] The reader is invited to conclude that the collection:

$$\mathscr{A} = \{A_0, A_1, \ldots, A_{n-1}\}$$

is a partition of Z into n disjoint infinite subsets.

Example 1.45 If $n = 4$, our relation becomes

$$a \underset{4}{\sim} b \qquad [\text{meaning} \;\; 4 \mid (a - b)]$$

and we can then cite the following:

$$12 \underset{4}{\sim} 4 \qquad 13 \underset{4}{\nsim} 4$$

$$6 \underset{4}{\sim} 30 \qquad 5 \underset{4}{\nsim} 7$$

as illustrations of related and unrelated pairs. This is because we are referring to divisibility of the difference (by 4), so that whereas

$$12 - 4 = 8 \qquad \text{(evenly divisible by 4)}$$

we have

$$13 - 4 = 9$$

and the latter is not evenly divisible by 4. In this instance, our partition of Z becomes

$$Z = \{A_0, A_1, A_2, A_3\}$$

where

$$\begin{aligned} A_0 &= \{\ldots, -8, -4, 0, 4, 8, \ldots\} \\ A_1 &= \{\ldots, -7, -3, 1, 5, 9, \ldots\} \\ A_2 &= \{\ldots, -6, -2, 2, 6, 10, \ldots\} \\ A_3 &= \{\ldots, -5, -1, 3, 7, 11, \ldots\} \end{aligned}$$

as described previously ■

By way of generalization, let S be set and let the abstract symbol $\sim$ denote a relation on S. The relation is said to be an **equivalence relation** if it satisfies the following three properties for all choices of $a, b, c \in S$.

E1. Reflexive: $a \sim a$

E2. Symmetric: if $a \sim b$ then $b \sim a$

E3. Transitive: if $a \sim b$ and $b \sim c$ then $a \sim c$

In high school algebra, we learn that equality is an equivalence relation; in fact, we often hear the transitive condition expressed as "Quantities equal to the same thing are equal to each other." And indeed, if we truly have a generalization here, then congruence (modulo n) should be another example of an equivalence relation (see Example 1.46). What we soon discover, however, is that the abstract notion of an equivalence relation quite simply abounds, throughout mathematics generally and in all of its fields of application.

Example 1.46 Consider the definition of congruence (modulo n), that is,

$$a \underset{n}{\sim} b \qquad [\text{meaning } a - b \text{ is a multiple of } n]$$

on the set of integers, as given previously. But note the equivalent rephrasing. We will show that such relations are in fact equivalence relations.

E1. $a \underset{n}{\sim} a$ because $a - a = 0 = 0 \cdot n$ (a multiple of n)

E2. If we suppose that $a \underset{n}{\sim} b$, that is, that $a - b = pn$ (some multiple of n), then this, in turn, implies that

$$b - a = -(a - b) = -pn \qquad \text{(again a multiple of } n)$$

yielding $b \underset{n}{\sim} a$.

E3. If $a \underset{n}{\sim} b$ and $b \underset{n}{\sim} c$, so that $a - b = pn$ and $b - c = qn$, then we have

$$\begin{aligned} a - c &= a - b + b - c \\ &= pn + qn = (p + q)n \qquad \text{(again a multiple of } n) \end{aligned}$$

so that $a \underset{n}{\sim} c$, as required. ■

Example 1.47 This time, let us take A^* as our set (refer to Section 1.3) for some alphabet A, and suppose we let a relation $\sim$ be defined on this set as follows:

$$x \sim y \quad \text{should mean } x \text{ and } y \text{ are words of equal length}$$

Note that we use the same abstract symbol $\sim$ with different meanings from one example to the next. Then surely (E1) holds—x is the same length as itself. Equally trivial is the verification of (E2), namely that if x and y are of equal length, then so are y and x. Finally, considering (E3), if x and y are of the same length and y and z are of the same length, then clearly

$$\text{length}(x) = \text{length}(y) = \text{length}(z)$$

that is, x and z are of the same length. (Note the use of transitivity of the equality relation on the set of integers.) It follows that we again have an equivalence relation, but of quite a different kind than any we have previously seen. ■

Example 1.48 Consider a set of functions (from Z to Z, say), and suppose that among these functions we take

$$f \sim g \quad \text{to mean} \quad f - g = \text{constant}$$

as in the case of $f(x) = 2x^2 - 3x + 5$ and $g(x) = 2x^2 - 3x - 1$. Then again we may check the reflexive, symmetric, and transitive properties, this time in an abbreviated form:

E1. $f - f = 0$ (a constant)

E2. If $f - g = c$ (some constant), then

$$g - f = -(f - g) = -c \qquad \text{(another constant)}$$

E3. If $f - g = c$ and $g - h = d$ (two constants c and d), then

$$f - h = f - g + g - h = c + d \qquad \text{(another constant)}$$

which shows transitivity.

Note the similarity to the argument of Example 1.46. Altogether, our analysis shows that we have yet another example of an equivalence relation here, this time an equivalence among functions. ∎

Having examined, in our last several examples, instances of equivalence relations on an abstract set, on the set of integers, on a set of words, and finally on a set of functions, we see indeed that the notion of equivalence is quite a general one.

In surveying these examples, the student must bear in mind that *equivalence* is usually quite different from equality. In Example 1.48, the two particular functions cited are definitely not equal to one another. They are, however, equivalent as long as we do not distinguish between functions that differ only by a constant (as was our intention there). But it should be remembered that our criteria for indistinguishability will vary from one example to another.

Equivalence Classes

Whatever our set and whatever our criteria for equivalence (as long as it is a true equivalence relation as defined by (E1)–(E3)), we will find that our discussion is greatly simplified by lumping together all of the mutually equivalent members of our set into a class of indistinguishable elements. In effect, this was the sense of the subsets $A_0, A_1, \ldots, A_{n-1}$ used in our discussion of congruence relations. In general, if $\sim$ is an equivalence relation on the set S, and a is any member of S, we let

$$[a] = \{x \in S : x \sim a\}$$

denote the **equivalence class** (determined by a). Thus, as explained above, $[a]$ collects together all of the members of S that are equivalent to a.

Example 1.49 In Example 1.44, there are just three equivalence classes. Thus we may compute

$$\begin{aligned}
[s_1] &= \{s_1, s_2, s_4\} \\
[s_2] &= [s_1] \qquad \text{(because } s_2 \; R \; s_1\text{)} \\
[s_3] &= \{s_3, s_6\} \\
[s_4] &= [s_1] \qquad \text{(because } s_4 \; R \; s_1\text{)} \\
[s_5] &= \{s_5\} \\
[s_6] &= [s_3] \qquad \text{(because } s_6 \; R \; s_3\text{)}
\end{aligned}$$

and we find that in choosing related members, we obtain identical equivalence classes. Indeed, we will discover that this is a general phenomenon, whatever equivalence relation we are considering. ■

Example 1.50 In Example 1.45, there are four equivalence classes, that is, we have

$$\begin{aligned}
[0] &= A_0 = \{\ldots, -8, -4, 0, 4, 8, \ldots\} \\
[1] &= A_1 = \{\ldots, -7, -3, 1, 5, 9, \ldots\} \\
[2] &= A_2 = \{\ldots, -6, -2, 2, 6, 10, \ldots\} \\
[3] &= A_3 = \{\ldots, -5, -1, 3, 7, 11, \ldots\}
\end{aligned}$$

identifying our new notation with the old. Once again, if we attempt to compute [4], we find that it is the same subset as [0], owing to the fact that $4 \underset{4}{\sim} 0$. Similarly, $[5] = [1]$, $[6] = [2]$, etc. ■

Example 1.51 In Example 1.47, let us suppose that our alphabet is

$$A = \{a, b, c, \ldots, z\}$$

for definiteness. Then the equivalence classes become:

$$\begin{aligned}
[\lambda] &= \{\lambda\} \qquad \text{(the null word is the only word of length zero)} \\
[a] &= \{a, b, c, \ldots, z\} = A \qquad \text{(all the one-letter words)} \\
[aa] &= \{aa, ab, ac, \ldots, zz\} = A^2 \qquad \text{(all two-letter words)} \\
[aaa] &= \{aaa, aab, aac, \ldots, zzz\} = A^3 \qquad \text{(all three-letter words)}
\end{aligned}$$

etc. (*Note:* We use this notation A, A^2, A^3, etc., quite generally in the sequel.) In anticipation of the important characterization of equivalence relations that will follow, we note that our set A^* is partitioned by the relation of Example 1.47 into a union

$$A^* = \{\lambda\} \cup A \cup A^2 \cup A^3 \cup \ldots$$

of nonoverlapping subsets—the set consisting of the null word alone, the set of all the one-letter words (A itself), the set of all two-letter words, the set of all three letter words, and so on. ■

We have seen that, in all of our examples, whenever we have an equivalence relation $\sim$ on a set S and we choose two related elements $a \sim b$, it follows that the equivalence classes $[a]$ and $[b]$ coincide. If, on the other hand, we have $a \nsim b$, then $[a]$ and $[b]$ cannot have any elements in common. For if we suppose $x \in [a]$ and $x \in [b]$, we would have $x \sim a$ and $x \sim b$. By symmetry and transitivity, we would conclude that $a \sim b$ (contrary to our assumption). Thus it appears that our equivalence classes are "just what the doctor ordered," toward the attempt to capture the partitioning of a set that seems to characterize each equivalence relation. But this has to be precisely stated.

Characterization of Equivalence Relations

Whenever we have observed an instance of an equivalence relation on a set, then have identified carefully the equivalence classes, it has become clear that we were looking at a partition of the set (by its classes) into nonoverlapping subsets. It always seems to happen. Nevertheless, such observations should be captured once and for all, in a precise wording, so that we will have the general situation clear in our mind. This precise wording follows. In preparation, the reader should review the definition of a partition as given in Section 1.2.

Characterization of Equivalence Relations Whenever $\sim$ is an equivalence relation on a set S, the set

$$\mathscr{A} = \{[a_1], [a_2], \ldots\}$$

of equivalence classes $[a]$ partitions S into a union of nonoverlapping subsets. Conversely, any partition

$$\mathscr{A} = \{A_1, A_2, \ldots\}$$

of a set S determines an equivalence relation on that set, simply by our agreement that

$$x \sim y \quad \text{should mean} \quad \text{“}x \text{ and } y \text{ are } \textit{together} \text{ in } \mathscr{A}\text{”}$$

By now, the meaning of the opening statement is clear. For the converse we should first make certain that we know what is meant by *together*. It means simply that x and y are found in one subset A of the partition. We then leave to the reader the verification that this relation "together" indeed amounts to an equivalence relation on S.

Example 1.52 Suppose we take the set

$$S = \{s_1, s_2, s_3, s_4, s_5, s_6\}$$

and consider the partition into three nonoverlapping subsets as follows:

$$A_1 = \{s_1, s_2, s_4\} \qquad A_2 = \{s_3, s_6\} \qquad A_3 = \{s_5\}$$

Then the relation "together" (call it $\sim$) yields $s_1 \sim s_2$, $s_1 \nsim s_3$, etc.; in fact, we are led to the same relation that was originally given in Example 1.44, an equivalence relation for sure. Indeed, we have come full circle. ■

EXERCISES 1.4

1 Explain why it is that the relation of divisibility (Example 1.43) is not an equivalence relation on N. Which of the three required properties (E1), (E2), (E3) fails to hold and why?

2 Which of the relations of Figure 1.15 is reflexive? Symmetric? Transitive? Provide specific instances of the failure whenever one of these properties fails to hold.

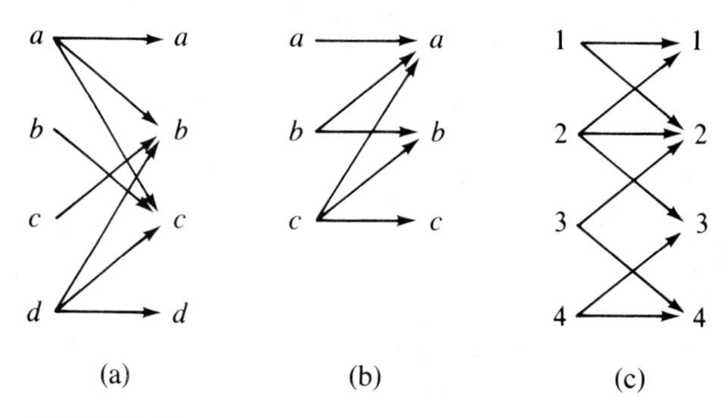

FIGURE 1.15

3 In relation to the discussion of congruence (modulo n), show precisely why it is that $A_n = A_0$, $A_{n+1} = A_1$, etc.

4 Show, in general, that when $a \sim b$ then $[a] = [b]$.

5 A relation R on a set is said to be *circular* if $x\ R\ y$ and $y\ R\ z$ implies $z\ R\ x$. Show that any reflexive and circular relation is an equivalence relation.

6 Given the set

$$S = \{x: x \text{ is a student at this college/university}\}$$

which of the following are equivalence relations on S?

(a) $x\ R\ y$ means x and y have the same major

(b) $x\ R\ y$ means x and y are siblings

(c) $x\ R\ y$ means x and y are enrolled in a course together

(d) $x\ R\ y$ means x and y have the same initials

(e) $x\ R\ y$ means x and y are friends

7 Let R be any relation on a set S. Its *transitive closure* R' is defined as the following relation:

$$x\ R'\ y \quad \text{means} \quad \begin{Bmatrix} \text{there is a finite sequence } x_1, x_2, \ldots, x_k \text{ in } S \\ x = x_1 \quad \text{and} \quad x_k = y \\ x_1\ R\ x_2, x_2\ R\ x_3, \ldots, x_{k-1}\ R\ x_k \end{Bmatrix}$$

Show that R' is transitive. Show that if R is reflexive and symmetric, then R' is an equivalence relation.

8 Verify that the relation "together" in the characterization of equivalence relations is a true equivalence relation.

9 Let the natural numbers in the range

$$10(i-1) < n \leq 10i \qquad (i \geq 1)$$

be called a *decade*. Show that the partition

$$\mathscr{A} = \{A_1, A_2, \ldots\}$$

of Example 1.25 determines the equivalence relation

$$n \sim m \qquad [\text{meaning } n \text{ and } m \text{ are in the same decade}]$$

10 Show that each of the following is an equivalence relation on the accompanying set.

(a) "Living in the same house" among people of the United States

(b) "Parallelism" of lines in the plane

(c) "Similarity" of triangles in the plane

(d) "Equicardinality" among sets in a collection

(e) $(n, m) \sim (r, s)$, meaning $ns = mr$, among ordered pairs of integers (n, m) with $m \neq 0$

Sec. 1.5 ELEMENTARY COUNTING TECHNIQUES

Let me count the ways.

For the design and analysis of algorithms and for other purposes as well, we need to have techniques for counting finite sets. In a probabilistic analysis we may need to determine the number of ways in which an event can take place, not wishing to enumerate each and every case. These counting techniques will then be indispensable. In another context these same tools may be helpful in estimating the running time of a computer program or in determining its storage requirements. So, these enumeration techniques, though properly belonging to the subject called *combinatorial mathematics*, are becoming increasingly important in their application to computer science, probability theory, operations research, and a number of other fields. We will state several of our counting techniques as if they were rules of calculation, although we do provide brief justification for these rules as they are encountered. The arguments are quite elementary and the student should not experience any particular difficulty, either in the understanding of these principles or in their use.

Goals

After studying this section, you should be able to:

- Define the concepts *permutation* and *combination.*
- Reduce the counting of a complex set to that of the simpler sets of which it is composed.
- Give examples of *Inclusion-Exclusion* arguments.
- State the rules of product and sum and give their justification.
- Give the rules for counting permutations and combinations and justify them.
- Compute binomial coefficients and explain their use in the binomial theorem.
- State the *pigeonhole principle* and provide examples of its use.
- Derive the *recurrence equation* for the binomial coefficients and illustrate its use in developing *Pascal's triangle.*
- Explain the notion of a *power set and* show how power sets may be counted.
- State the characterization of *equicardinal sets* and illustrate its application.

Rules of Product and Sum

As we have indicated, it will be assumed throughout this section that all of our sets are finite, even though it is likely that some of our results would be applicable in more general settings. Then, as we will see, the counting of a set is often greatly facilitated by our knowledge of how the set has been constructed. Is it a union of two smaller sets? Is it a product set? These questions are typical of those we want to ask. For it is definitely the case that a knowledge of the structure of a set, the manner in which it was formed, the arrangement of its elements, etc. will put us in a better position to carry out the enumeration.

The development of the intuitive set theory, beginning in our first two sections and continuing to the present point, has had one central theme—the building of more and more complex sets from simpler ones. In order to count the members in the more complex sets, rules of reduction are required to reduce the enumeration to the case of the simpler sets. That is essentially the purpose of our first two rules of enumeration.

Rule 1 **Rule of the Product**

$$|A \times B| = |A| \cdot |B|$$

Justification: There are $|A|$ ways to choose the first coordinate of each ordered pair $(a, b) \in A \times B$. The choice of b may then be made independently, there being $|B|$ choices for each a.

Corollary

For any finite set A we have

$$|A^n| = |A|^n$$

Example 1.53 In Example 1.40, where $S = \{a, b, c\}$ and $T = \{1, 2\}$, we should expect $|S \times T| = |S| \cdot |T| = 3 \cdot 2 = 6$ members in the product set $S \times T$. Of course, this is just the way it turned out. ■

Example 1.54 In Example 1.41 we introduced the product set

$$B^n = \{(b_1, b_2, \ldots, b_n): b_i \in B\}$$

where $B = \{0, 1\}$. According to the corollary, we should have

$$|B^n| = |B|^n = 2^n$$

Thus when $n = 3$, the set $B^n = B^3$ should have $2^3 = 8$ elements. Indeed, we were able to list just 8 members for B^3 in Example 1.41. ■

Example 1.55 If A is an n-symbol alphabet and if we let A^k denote the set of all words of length k on the alphabet A (as in Example 1.51), then again according to the corollary to Rule 1,

$$|A^k| = |A|^k = n^k$$

Thus if A is the standard alphabet of Example 1.5 (with the usual 26 letters), we should be able to compose $26^4 = 456{,}976$ four-letter words. ■

Rule 2

Rule of the Sum

$$|A \cup B| = |A| + |B| - |A \cap B|$$

Justification: When the number of elements in A and the number of elements in B are added together, the members of the intersection will have been counted twice. So we must compensate by subtracting $|A \cap B|$.

Example 1.56 There are 10 listed faculty members in the mathematics department and 7 in computer science at a certain university. On closer inspection, however, we find that 3 faculty members have joint appointments and are listed by both departments. Using the rule of the sum, we conclude that the number of persons attending a joint departmental faculty meeting should be $10 + 7 - 3 = 14$. ■

Example 1.57 The rule of the sum can be generalized to more than two sets in a union; then it is called the **Principle of Inclusion and Exclusion**. To illustrate the situation with three sets in a union, consider Figure 1.16. This figure indicates the number of tallies when we add the cardinality of the three separate sets A_1, A_2, and A_3. In computing $|A_1 \cup A_2 \cup A_3|$ by first adding $|A_1| + |A_2| + |A_3|$, we must then compensate by subtracting $|A_1 \cap A_2|$, $|A_1 \cap A_3|$, and $|A_2 \cap A_3|$, for they have been counted twice (see Figure 1.16). But then we have counted the multiple intersection a total of $3 - 3 = 0$ times, so we must add it back into our sum. Altogether we obtain the Inclusion-Exclusion formula:

$$|A_1 \cup A_2 \cup A_3| = (|A_1| + |A_2| + |A_3|) - (|A_1 \cap A_2| + |A_1 \cap A_3| + |A_2 \cap A_3|) + |A_1 \cap A_2 \cap A_3|$$

Indeed, the reader should begin to see a general principle emerging here. ■

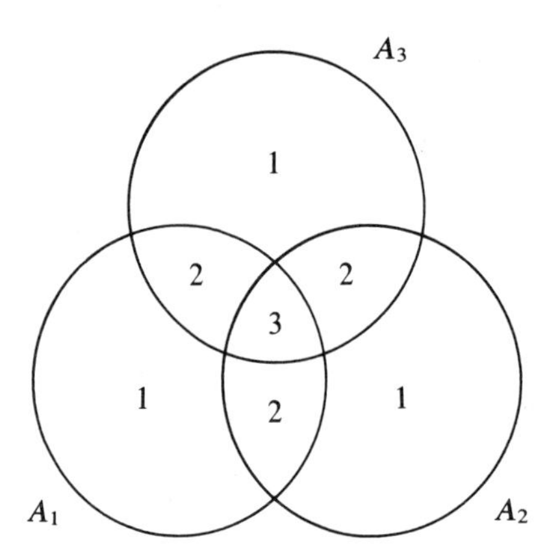

FIGURE 1.16

Example 1.58 At a hamburger stand, it is observed that

7% of the patrons use no condiments whatsoever

58% use mustard	29% use mustard and ketchup
58% use ketchup	30% use mustard and onions
47% use onions	25% use ketchup and onions

What percentage of the patrons order hamburgers with everything? (*Note:* When a patron is observed to have used mustard, she may have also used something else, and similarly with our other statistics.) We may use the Inclusion-Exclusion formula of Example 1.57 to compute (with the multiple intersection as the unknown x):

$$\begin{aligned} 100 - 7 &= (58 + 58 + 47) - (29 + 30 + 25) + x \\ 93 &= 163 - 84 + x \\ x &= 14 \end{aligned}$$

Basing all of the calculations on 100 patrons, this means that 14% order a hamburger with everything. ■

Permutations and Combinations

Undoubtedly the student has learned something about permutations and combinations in high school mathematics. As usually presented, these ideas are used to answer questions of the following sort: In how many ways can we choose such and such from this and that ...? The decision as to whether permutations or combinations are involved will generally depend on whether or not the order of the choices is significant. If yes, then we deal with permutations; if not, then it is probably the idea of a combination that is required.

In order to focus the discussion, suppose A is a set with n elements (an **n-set** as it is commonly called), say

$$A = \{a_1, a_2, \ldots, a_n\}$$

The decision to now call such a set an n-set rather than an "alphabet" has to do with our interest in focusing on the set's cardinality. Borrowing the notation from Examples 1.51 and 1.55, we will have occasion to refer to the sets A^k (consisting of the *k-letter words* on the set A). Recall that in Example 1.55 we were able to count these sets A^k; that is, $|A^k| = n^k$. But notice that the words of A^k may contain duplications of their symbols. If instead, we insist that we choose a sequence of length k from A without duplication, the resulting choice is called a **k-permutation** of the members of A. Note that we cannot choose more than n elements from an n-set without a duplication, and thus it must be assumed that $k \leq n$ in this discussion. When $k = n$, the k-permutations are simply called **permutations** of the set A, without further qualification.

Example 1.59 If $A = \{a, b, c, d\}$ $(n = 4)$, then we have the following 3-letter words:

aaa	*aca*	*baa*	*bca*	*caa*	*cca*	*daa*	*dca*
aab	*acb*	*bab*	*bcb*	*cab*	*ccb*	*dab*	*dcb*
aac	*acc*	*bac*	*bcc*	*cac*	*ccc*	*dac*	*dcc*
aad	*acd*	*bad*	*bcd*	*cad*	*ccd*	*dad*	*dcd*
aba	*ada*	*bba*	*bda*	*cba*	*cda*	*dba*	*dda*
abb	*adb*	*bbb*	*bdb*	*cbb*	*cdb*	*dbb*	*ddb*
abc	*adc*	*bbc*	*bdc*	*cbc*	*cdc*	*dbc*	*ddc*
abd	*add*	*bbd*	*bdd*	*cbd*	*cdd*	*dbd*	*ddd*

$64 = 4^3$ in number, according to the corollary to Rule 1 (and the discussion of Example 1.55). Of these, only the following are 3-permutations:

$$\begin{array}{cccccccc} abc & acd & bac & bcd & cab & cbd & dab & dbc \\ abd & adb & bad & bda & cad & cda & dac & dca \\ acb & adc & bca & bdc & cba & cdb & dba & dcb \end{array}$$

24 of them in all. ■

Example 1.60 The 52-set

$$\begin{aligned} A = \{&1H, 2H, \ldots, 9H, 0H, JH, QH, KH, 1S, 2S, \ldots, 9S, 0S, JS, QS, KS, \\ &1D, 2D, \ldots, 9D, 0D, JD, QD, KD, 1C, 2C, \ldots, 9C, 0C, JC, QC, KC\} \end{aligned}$$

can serve as a deck of playing cards (H = hearts, S = spades, D = diamonds, C = clubs). But there are 5-letter words, for example,

2H 9C 2H JS QS   2H 2H 2H JS JD

that cannot be interpreted as poker hands because of the duplications. If we are to count the number of distinct poker hands, the concept of a 5-permutation will not be appropriate either, because we would not want to distinguish between

2H 9C 3D JS QS   and   9C JS 2H QS 3D

say. For this purpose, we will need the idea of a 5-combination (a subset of cardinality 5), where the order of occurrence of the symbols is immaterial. ■

So, we have seen the need for an unordered counterpart to the notion of a permutation. Again we consider an n-set

$$A = \{a_1, a_2, \ldots, a_n\}$$

If we choose k distinct members from A without regard to the order of their selection, we arrive at the notion of a **k-combination** (or **k-subset**) of the members of A. As before, we should only consider $k \leq n$ here. Note, as indicated above, that except for the notation, a k-combination is the same thing as a k-element subset of A.

Example 1.61 In examining the 3-permutations of the 4-set A in Example 1.59, we would only recognize four of them:

$$abc \quad abd \quad acd \quad bcd$$

as distinct 3-combinations. The others are simply rearrangements of these. Note as well that these four 3-combinations correspond to the 3-element subsets:

$$\{a, b, c\} \quad \{a, b, d\} \quad \{a, c, d\} \quad \{b, c, d\}$$

respectively. ■

It really is quite easy to count the total number of distinct k-permutations or k-combinations of an n-set. For this purpose, we introduce the notation $P(n, k)$ and $C(n, k)$, respectively, to refer to these totals in the rules that follow. Moreover, we will find that it is convenient to express both of these rules in terms of the *factorial notation*

$$n! = n(n-1)\cdots(2)(1)$$

with which the reader is no doubt already familiar.

Rule 3 The number of k-permutations of an n-set is

$$P(n, k) = \frac{n!}{(n-k)!}$$

Justification: For each choice (among n) for the first member of a k-permutation, there remain only $n - 1$ choices for the second member, etc., so that

$$P(n, k) = n(n-1)\cdots(n-k+1) = \frac{n!}{(n-k)!}$$

as claimed.

Corollary There are $n!$ permutations of an n-set.

Example 1.62 If $n = 4$ and $k = 3$, as in Example 1.59, then we would expect

$$P(n, k) = \frac{n!}{(n-k)!} = \frac{4!}{(4-3)!} = 24$$

3-permutations. And indeed, that was the number we were able to list. ■

Example 1.63 The number of possible batting orders for a 12-member baseball team is $P(12, 9) = 12!/3! = 79{,}833{,}600$. ■

Suppose our n-set is simply $\mathbf{n} = \{1, 2, \ldots, n\}$. As we remarked previously a permutation (though it could rightly be called an n-permutation) of $\mathbf{n}$ is then a sequence

$$s = s_1, s_2, \ldots, s_n$$

of nonduplicated members from n. Because we do not permit duplications, the sequence can be regarded as a rearrangement of the symbols $1, 2, \ldots, n$. The corollary simply asserts that this can be done in $n!$ different ways.

Example 1.64 If $n = 4$, then $n! = 24$ and indeed we can find precisely 24 rearrangements (that is, permutations) of the symbols $1, 2, 3, 4$. Disregarding commas, we write:

1234	2134	3124	4123
1243	2143	3142	4132
1324	2314	3214	4213
1342	2341	3241	4231
1423	2413	3412	4312
1432	2431	3421	4321

and that is all. Note, however, that the (non)rearrangement $1, 2, 3, 4$ is regarded as a valid permutation. ■

In certain applications, particularly in dealing with problems in linear algebra, it is important to distinguish permutations as having an odd or even **parity**; that is, the *number of interchanges* required to transform the given rearrangement $s = s_1, s_2, \ldots, s_n$ into the standard order $1, 2, \ldots, n$ can be even or odd. It turns out that there is quite a simple way to determine the parity of a permutation. Beginning with the member 2, we first interchange 2 successively with all of the members of the sequence that are smaller, but to the right (only the member 1 could be a possibility here). We continue to execute this same rule: interchange j successively with all of the members of the sequence that are smaller, but to the right of j. When this process is completed, from 2 up to n, the sequence s will have been transformed into the standard order. Now suppose we say that an *inversion* occurs in a permutation $s = s_1, s_2, \ldots, s_n$ whenever a larger number precedes a smaller one. And suppose we count the total number of inversions by examining each member of the sequence (except the member 1), determining in each case the number of smaller members that follow it, and then adding them up. Then according to the above analysis, the oddness or evenness of this total number of inversions is the parity of the permutation s.

Example 1.65 Consider the permutation $5, 6, 3, 4, 2, 1$ of the numbers from 1 to 6. Then we note that

2 contributes 1 inversion

3 contributes 2 inversions

4 contributes 2 inversions

5 contributes 4 inversions

6 contributes 4 inversions

and the total number of inversions is 13—an odd parity. We thus regard 5, 6, 3, 4, 2, 1 as an odd permutation. Note that in performing the interchanges just discussed,

$$\begin{aligned} 563421 &\longrightarrow 563412 \longrightarrow 561432 \longrightarrow 561423 \longrightarrow 561243 \\ &\longrightarrow 561234 \longrightarrow 165234 \longrightarrow 162534 \longrightarrow 162354 \\ &\longrightarrow 162345 \longrightarrow 126345 \longrightarrow 123645 \longrightarrow 123465 \\ &\longrightarrow 123456 \end{aligned}$$

Thus we indeed transform 5, 6, 3, 4, 2, 1 into 1, 2, 3, 4, 5, 6 with 13 interchanges. ■

It happens that in considering all of the permutations of the symbols $1, 2, \ldots, n$ (for a fixed n), half of them will be odd and half of them will be even. This is simply because an interchange will transform an odd permutation into an even permutation, and vice versa.

Example 1.66 In reviewing the 24 permutations of 1, 2, 3, 4 as listed in Example 1.64, we obtain the classification

1234	even	2134	odd	3124	even	4123	odd
1243	odd	2143	even	3142	odd	4132	even
1324	odd	2314	even	3214	odd	4213	even
1342	even	2341	odd	3241	even	4231	odd
1423	even	2413	odd	3412	even	4312	odd
1432	odd	2431	even	3421	odd	4321	even

and, indeed, half of the permutations are odd and half of them are even. ■

Rule 4

The number of k-combinations of an n-set is

$$C(n, k) = \frac{n!}{k!(n-k)!}$$

Justification: According to the corollary to Rule 3, there are $k!$ permutations of each k-combination. Since we obtain all $k!$ permutations exactly once in this way, we may simply divide $P(n, k)$ by $k!$ to obtain

$$C(n, k) = \frac{P(n, k)}{k!} = \frac{n!}{k!(n-k)!}$$

as claimed.

Example 1.67 In Example 1.61, we were listing the 3-combinations of the 4-set $A = \{a, b, c, d\}$. We could only list four such combinations, because

$$C(4, 3) = \frac{4!}{3!(4-3)!} = 4$$

according to Rule 4. ■

Example 1.68 The number of distinct poker hands (see Example 1.60) is

$$C(52, 5) = \frac{52!}{5!47!} = 2{,}598{,}960$$

rather a large number. Certainly we would not want to list all of them. ■

Binomial Coefficients

The numbers $C(n, k)$ are also denoted $\binom{n}{k}$ and they are often called **binomial coefficients**. This terminology derives from the binomial theorem of elementary algebra, used in multiplying out the expression $(a + b)^n$. Because $(a + b)^n$ means

$$(a + b)(a + b) \cdots (a + b)$$

with n factors, the coefficient of $a^{n-k}b^k$ in the product is easily determined. It is the number of ways of choosing k of the n parentheses from which to take b's. But because this is the number $C(n, k)$ or $\binom{n}{k}$, the process of multiplying out yields the **binomial theorem**:

$$(a + b)^n = a^n + \binom{n}{1}a^{n-1}b + \binom{n}{2}a^{n-2}b^2 + \cdots + b^n$$

The reader is perhaps familiar with the Pascal triangle and its use in computing the binomial coefficients $C(n, k)$. Suppose we think of $C(n, k)$ as the number of k-element subsets S of an n-set A. Choosing a fixed element $a \in A$, we can argue that there are only two possibilities for such a subset S. Either $a \in S$ or $a \notin S$. In the first case, $S = T \cup \{a\}$, where T is a $(k - 1)$-subset of the $(n - 1)$-set $A \sim \{a\}$. In the second case, S is a k-subset of $A \sim \{a\}$. If we account for the number in both possibilities using a summation, then we arrive at the important **recurrence equation** for the binomial coefficients:

$$C(n, k) = C(n - 1, k - 1) + C(n - 1, k)$$

Such recurrence equations occupy an important position in combinatorial investigations. They are especially valuable when a closed or non-recursive formula is not available for the enumeration in question.

We use the preceding recurrence equation to compute the binomial coefficients, beginning with a listing of the obvious *boundary conditions*:

$$C(n, 0) = C(n, n) = 1$$

The list gives rise finally to the *Pascal triangle* (see Figure 1.17).

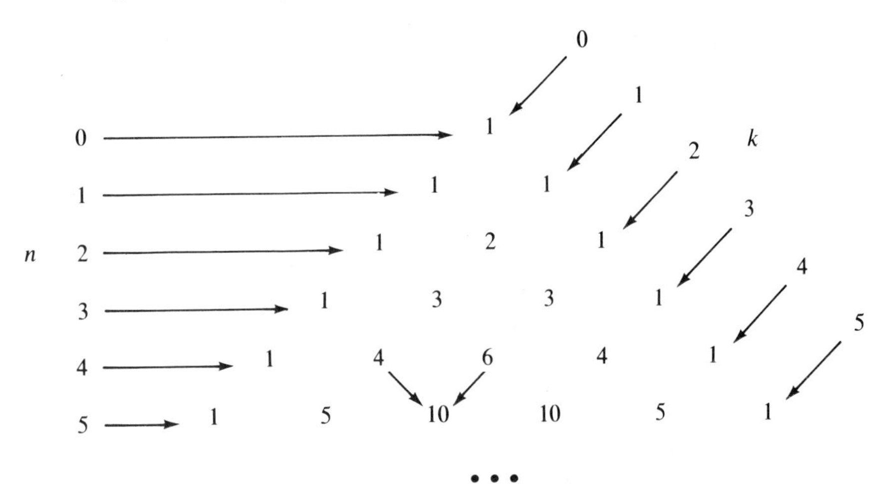

FIGURE 1.17

Example 1.69 Assuming that the Pascal triangle has already been completed through $n = 4$, we may compute

$$C(5, 2) = C(4, 1) + C(4, 2) = 4 + 6 = 10$$

(see Figure 1.17). Of course, the same result could have been obtained directly from the formula of Rule 4. However, this would not have provided any insight into the recursive computation of the coefficients. ■

Example 1.70 The row of entries (again in Figure 1.17) for $n = 5$ can be used to write down immediately the expression obtained in multiplying out $(a + b)^5$; that is,

$$(a + b)^5 = a^5 + 5a^4b + 10a^3b^2 + 10a^2b^3 + 5ab^4 + b^5$$

Similarly, we may use the recurrence equation to calculate yet another row—1, 6, 15, 20, 15, 6, 1, so as to compute

$$(a + b)^6 = a^6 + 6a^5b + 15a^4b^2 + 20a^3b^3 + 15a^2b^4 + 6ab^5 + b^6$$

in the same fashion. ■

Sets of the Same Cardinality

Counting is a fundamental activity. We may think that it always requires a mathematical sophistication, but an illustration can be given to show that this may not be the case. Consider a boatload of sailors arriving on a remote island, where the native inhabitants are unsophisticated in the mathematical skills, perhaps not even possessing the ability to count or having even the most elementary appreciation for our number system. The leader of the native tribe would like to present each of the arriving sailors with a gift, a coconut let us say. If he were dealing with a developed society, the captain of the ship could say, "Well, we have seventy-two sailors in all, so I guess your gift should amount to seventy-two coconuts." But this information would not be helpful to our tribe leader. Note, however, that if the leader presents one coconut to each of the sailors until all of the sailors have been accommodated, the effect is the same—without any counting whatsoever.

There is a lesson in this illustration. If we were to state the principle in its most abstract form, it would amount to the following characterization.

Characterization of Equicardinality Two sets are of the same cardinality in the event that there is a one-to-one correspondence of the one set onto the other.

This principle, trivial though it may appear, finds many useful applications in those situations in which we are unable to count a set according to its given arrangement, but in which we are able to establish a one-to-one correspondence with another set whose cardinality is known or is easily established. The following example illustrates this point. The example deals with the problem of counting all of the subsets of an n-set.

The Powerset Construction

In general, for any set S, we let

$$\mathscr{P}(s) = \{A : A \subseteq S\}$$

denote the set of all subsets of S. The script $\mathscr{P}$ is used because this set of sets is called the **power set** of S. Thus, if $S = \{a, b, c\}$, then

$$\mathscr{P}(S) = \{\varnothing, \{a\}, \{b\}, \{c\}, \{a, b\}, \{a, c\}, \{b, c\}, S\}$$

Recall (see Example 1.15) that $\varnothing$ and S are subsets of S in good standing. It would appear that as the cardinality of S grows and grows, that of $\mathscr{P}(S)$ becomes quite large. Note that there are $8 = 2^3$ subsets here, S being a 3-set. We would like to be able to generalize from this observation, and that is the point of Example 1.71.

Example 1.71 When S is an n-set, say

$$S = \{a_1, a_2, \ldots, a_n\}$$

and we want to count $\mathscr{P}(S)$, it happens that we can use our characterization of equicardinality to first establish a one-to-one correspondence of $\mathscr{P}(S)$ with the set B^n (Example 1.41). Then, because we have already counted B^n (Example 1.54), we will have counted $\mathscr{P}(S)$ indirectly.

Consider the function

$$f: \mathscr{P}(S) \to B^n$$

given by writing:

$$f(A) = (b_1, b_2, \ldots, b_n)$$

with $b_i = 1$ if $a_i \in A$ and $b_i = 0$ otherwise. If $f(A) = f(B)$ then $a_i \in A$ whenever $a_i \in B$ and conversely; that is, we would have $A = B$, showing that our mapping f is one-to-one. Moreover, if $(b_1, b_2, \ldots, b_n)$ is any n-tuple of B^n, we can find a set A with

$$f(A) = (b_1, b_2, \ldots, b_n)$$

simply by taking

$$a_i \in A \quad \text{just in case} \quad b_i = 1$$

and this shows that f is onto as well.

As we have said, we have already counted B^n in Example 1.54, that is,

$$|B^n| = 2^n$$

and according to what we have just established, we must have

$$|\mathscr{P}(S)| = 2^n$$

for the n-set S as well, all on account of our characterization of equicardinality. ■

The formula just established is so important that we will find it helpful to identify it as one of our rules.

Rule 5 ***Powerset Rule***

$$|\mathscr{P}(S)| = 2^n \qquad (S \text{ an } n\text{-set})$$

An Important Combinatorial Principle

It is interesting to see how a seemingly trivial observation can lead to significant applications, as just illustrated. Similar behavior is known to be associated with the use of the innocuous *pigeonhole principle*, so named because "You can't put n pigeons in fewer than n pigeonholes without two pigeons occupying the same hole." In a more precise wording, the principle can be phrased as follows:

Pigeonhole Principle If the distinct images of $f: S \to T$ are fewer than $|S|$ in number, then f is not one-to-one.

Example 1.72 Show that two people in San Francisco have the same birthdate. We will make the assumption that no one in San Francisco is over 100 years of age. The total number of possible birthdates is then 100(365) = 36,500—not counting February 29ths. The population of San Francisco is about 720,000. Mapping the population onto their birthdates, the pigeonhole principle shows that two people must have the same birthdate. Note that our approximations could not affect the conclusion. ■

Example 1.73 Show that for finite sets of the same cardinality, the notions of one-to-one and onto must coincide. If a function $f: S \to T$ is not onto (for finite sets S and T of the same cardinality), then there are fewer than $|T| = |S|$ images for the function. According to the pigeonhole principle, f cannot be one-to-one either. Conversely, if f is onto, it is easy to see that it must also be one-to-one. We leave this verification as an exercise for the student (see Exercise 13). ■

EXERCISES 1.5

1 A survey of 100 students in a certain dormitory reveals that:

47 students subscribe to Playboy
44 students subscribe to Newsweek
32 students subscribe to Time
11 students subscribe to Playboy and Newsweek
12 students subscribe to Playboy and Time
12 students subscribe to Newsweek and Time
3 students subscribe to all three magazines

How many students subscribe to none of these magazines?

2 Two classes, one in mathematics and one in computer science, meet at successive hours. There are 24 students in the mathematics course and 27 in the computer science course. How many students are represented here if it is known that 8 students are enrolled in both of these courses?

3 Compute the number of initials (first name, last name) of a person's name.

4 Compute the number of possible social security numbers.

5 Compute the number of possible Colorado license plates (two letters followed by four digits or three letters followed by three digits).

6 Obtain a formula for the number of different relations on an n-set.

7 Give an algebraic proof of the identity:

$$\binom{n}{k} = \binom{n-1}{k-1} + \binom{n-1}{k}$$

[*Hint:* Express the summands (on the right) over a common denominator.]

8 Compute the number of

(a) Four-digit decimal numbers

(b) Bridge hands

(c) Baseball lineups for a 15-member team

(d) Arrangements of 20 books on a shelf

(e) Fingernail colorings from 16 shades of nail polish

(f) Solutions to a 20-question multiple-choice (A, B, C, D, E) exam

(g) Subset of size 3 from a set with 7 members

(h) Subsets of less than half the size of a 7-member set

9 Compute the row for $n = 7$ in the Pascal triangle.

10 In FORTRAN, an identifier (for variables, arrays, functions, etc.) is a string of from one to six characters. The first character must be alphabetic (uppercase), whereas the others may be alphabetic or numeric (digits from 0 to 9). Compute the number of FORTRAN identifiers.

11 Compute the following:

(a) $C(7, 4)$ **(b)** $C(9, 7)$ **(c)** $C(8, 5)$

(d) $P(7, 4)$ **(e)** $P(9, 7)$ **(f)** $P(8, 5)$

12 Show that the sum of the entries in row n of the Pascal triangle is 2^n. (*Hint:* Choose $a = b = 1$ in the binomial theorem.)

13 If S and T are finite sets of the same cardinality, show that if $f: S \to T$ is onto, then it is necessarily one-to-one.

14 Determine the number of inversions and the parity of each of the following permutations:

(a) 5, 3, 6, 4, 1, 2 **(b)** 2, 5, 6, 4, 3, 1 **(c)** 6, 5, 4, 3, 2, 1

(d) 6, 4, 2, 3, 5, 1 **(e)** 1, 4, 3, 6, 5, 2 **(f)** 4, 6, 2, 1, 5, 3

15 Use interchanges to transform the permutations of Exercise 14 into the standard order.

16 Determine all of the permutations of the symbols 1, 2, 3 and classify them as to their parity.

Chapter Two

DEDUCTIVE MATHEMATICAL LOGIC

Contents

2.1 Axiomatic Mathematical Systems
2.2 Proof Techniques
2.3 Propositional Rules
2.4 Deductive Logic
2.5 Derived Logical Rules
2.6 Constructive vs. Classical Rules
2.7 The Statement Calculus

L. E. J. Brouwer (1881–1966)

Luitzen Brouwer was a Dutch mathematician who received his doctorate in 1907 at the University of Amsterdam. His subtlety of thought and his almost mystical insight into the foundations of mathematics are unrivaled to this day. He made numerous important contributions to topology in particular, but he is best known as the founder of the "intuitionist school" of mathematical thought. In this pursuit, he followed in the footsteps of Leopold Kronecker (see Chapter 3) and became, unknowingly perhaps, the precurser of generations of constructive mathematicians that followed. Some have called his treatment of the law of excluded middle the most significant advance in logical thought since Aristotle. Always controversial, Brouwer and his work stand as monuments to the creative intellect, a source of ideas and inspiration for years to come.

Writings

People try by means of sounds and symbols to originate in other people copies of mathematical constructions and reasonings which they have made themselves; by the same means they try to aid their own memory. In this way, the mathematical language *comes into being, and as its special case, the* language of logical reasoning.

Collected Works (Brouwer)

The first to systematically re-examine the language of mathematics, to adapt it to the constructive point of view, seems to have been Brouwer. He remarked that the meanings customarily assigned to the terms "and," "or," "not" and "implies" are not entirely appropriate to the constructive point of view, and he introduced more appropriate meanings where necessary.

The connective "and" causes no trouble. To prove "p and q," we must prove p and also prove q, as in classical mathematics. To prove "p or q" we must give a finite, purely routine method which after a finite number of steps either leads to a proof of p or to a proof of q. The classical meaning of "or" is too vague to be of constructive use. We define "not p" to mean that p is contradictory. By this we mean that it is inconceivable that a proof of p will ever be given. (Finally, note that) the connective "implies" is defined classically by taking "p implies q" to mean "not p or q." This definition would not be of much value constructively. Brouwer therefore defined "p implies q" to mean that one can give an argument which shows how to convert an arbitrary proof of p into a proof of q. Isn't this a more natural and intuitive definition anyway?

Having changed the meaning of the connectives, we should not be surprised to find that certain accepted modes of inference are no longer correct. The most important of these is the principle of the excluded middle— "p or not p." Constructively, this principle would mean that we had a method which in finitely many, purely routine steps would lead to a proof or disproof of an arbitrary mathematical assertion p. Of course we have no such method, and nobody has the least hope that we ever shall.

Aspects of Constructivism (E. Bishop)

Fortunately, the classical algebra of logic has its merits quite apart from the question of its applicability of mathematics. Not only as a formal image of the technique of commonsensical thinking has it reached a high degree of perfection, but also in itself, as an edifice of thought, it is a thing of exceptional harmony and beauty.

Collected Works (Brouwer)

Sec. 2.1 AXIOMATIC MATHEMATICAL SYSTEMS

The beauty of mathematics lies in its axiomatic method.

The axiomatic method is acknowledged as a fundamental system of reasoning, in mathematics and virtually every branch of scientific endeavor. The technique can be traced back at least as far as the work of Euclid (about 300 B.C.) in the treatment of geometry that bears his name. Euclid's *Elements* has had a significant and long-lasting influence on all of scientific thought. Modern high-school geometry is still modeled after this famous work. Indeed, some acquaintance with the axiomatic method is considered a mark of the educated person. We are fortunate therefore, that in this brief exposition of the technique, we will be able to appeal from time to time to the student's preliminary exposure to these ideas, as presented in the basic high-school geometry course.

Goals

After studying this section, you should be able to:

- Distinguish between *common notions* and *axioms*, providing examples of each.
- Identify the *primitive terms* in familiar axiomatic systems.
- State the logical *rule of substitution.*
- Express everyday implications and mathematical implications in the form $P \Rightarrow Q$.
- State the *law of contraposition* and provide illustrations of its use.
- Identify the antecedent and consequent of implications found in everyday discourse and in mathematical text.
- Provide examples of extensions of a mathematical language (theory) through the use of definitions.
- Give an axiomatic definition for (a fragment of) Euclidean geometry.
- Cite several elementary theorems of Euclidean geometry and provide an outline of their proof.

Primitive Terms

Every deductive science must begin from a set of indemonstrable principles—those that we do not try to prove. Otherwise there would be an endless sequence of steps in any subsequent demonstration. Among these basic principles, some are universal, that is, common to all of the sciences, whereas others are axiomatic within the particular science under study. Euclid made this distinction by referring to the *common notions* and the *postulates*, respectively. Examples of such common or universal notions are the statement "Quantities equal to the same thing are equal to each other," together with the implicit logical apparatus that one employs in deductive reasoning. However, we note that it is precisely this logical system, dating back to the work of Aristotle, on which our attention will ultimately focus in the studies of this chapter.

There is considerable confusion throughout mathematical history on the use of the terms *axiom* and *postulate*. In some early works, the common notions were called axioms, that is, self-evident statements so simple and obvious that their validity could be assumed at the start, as in asserting that "The whole is equal to the sum of its parts." In contrast, a postulate (from geometry) might have asserted that "Through two distinct points, passes one and only one line." The distinction here is again that of the universal versus the specific. In modern writing, it is more common to find such postulates called **axioms**, with the common notions left as implicitly understood, standing in the background. For the time being, at least, this will be our point of view.

In an axiomatic approach to a mathematical subject area, various terms that we would like to discuss are given a precise definition, using a language of terms that are already defined and understood. Yet, such an approach must start somewhere with a set of **primitive terms**, but they are left undefined. For example, in geometry the primitive terms might be "point" and "line," however, these primitive notions are not defined. Presumably, your intuition about the subject will then carry the burden of understanding.

Axioms must appear as reasonable to our intuition—otherwise they are not accepted as a basis for the study of a mathematical subject area. The axioms would then be left unproven in our system, and the axiomatic approach then becomes one of determining just which statements (theorems) follow as a necessary logical consequence (from the axioms).

Axioms of Order

To be more explicit, suppose we proceed as though we intend to apply the axiomatic approach in a simple situation, one where the student will have little difficulty in supplying the necessary intuition. We refer to the notion of **linear order** (or *total order*, as it is sometimes called). Presumably, it is not necessary to convince the reader of the fact that "ordered" collections abound in mathematics. The natural numbers, for example, are ordered $1, 2, 3, \ldots$, and we use the relation symbol $<$ in describing this order. The names in a phone book are ordered lexicographically. And so on.

If we want to study all such ordered collections axiomatically, we must first discover their commonality in abstract terms. It seems that of two distinct members x and y in such a collection, we can say that either x precedes y or y precedes x. Note that we choose a neutral terminology, "precedes," so that we can speak equally well of one number preceding (being less than) another, one name preceding another in a phone book, etc. We have adopted *precedes* as an undefined term that describes a relation among the members, or *points*, of our collections (the term "point" also is taken as primitive, and hence undefined).

To develop a system of axioms, we may as well start with the property we have just observed, that "If x and y are distinct points, then either x precedes y or y precedes x." Will we want further axioms? Suppose x and y are the same. Should we then say that x precedes y? Should y precede x? Neither would make much sense in the examples we are considering. (Note that our intuition is our guide.) So perhaps it is best to rule out either possibility with another axiom: "If x precedes y, then x and y are distinct points." Is this enough? Apparently not; consider the hours of time as depicted on a clock face. And let "x precedes y" be interpreted to mean that the arc from x to y (clockwise) is greater than zero and no more than 180 degrees. Then, for example, 3 precedes 5 and 5 precedes 10. Also, notice that this system would satisfy our two original axioms. If we want to limit our attention to *linear* (and specifically exclude *cyclical*) order, we will need yet another axiom: "If x precedes y and y precedes z, then x precedes z." Note that this is a "transitivity" requirement that we have seen before (Section 1.4) in our discussion of equivalence relations. Note further, in terms of the above illustration, that 3 precedes 5 and 5 precedes 10 should have "3 precedes 10" as a consequence (if our new axiom were to be satisfied). However, this would contradict the clock interpretation of "precedes." So, indeed, such cyclical notions of order have apparently been ruled out by the addition of this transitivity axiom.

To summarize, we axiomatize the notion of an *ordered set* X as a collection of points, together with a relation on X that we write as $<$ (but pronounce "precedes"), subject to the following axioms.

O1. If x and y are distinct points, then $x < y$ or $y < x$.

O2. If $x < y$, then x and y are distinct.

O3. If $x < y$ and $y < z$, then $x < z$.

We use the symbol $<$ in a generic sense, just as $\sim$ was used in our earlier axiomatization of equivalence relations. The fact that this symbol means "less than" in certain contexts should not prejudice the discussion unduly; it is often the case that we will choose our terminology in such a suggestive way, borrowing names and/or symbols from one specific example of an axiomatic theory, without meaning to imply that such typical interpretations are to be made universally.

Having introduced an axiomatic system, we may then proceed to use the axioms and our underlying logical apparatus (for now, just good common sense) to prove those theorems that follow as a necessary consequence.

Example 2.1 In the axiomatic system (O1)–(O3), we can prove the theorem:

$$x < x \text{ is impossible}$$

because, by substituting x for y in (O2), we obtain the statement

If $x < x$, then x and x are distinct.

However, it is utterly absurd to say "x and x are distinct." So it must be that $x < x$ cannot be valid. ■

We will find that logic is involved in this proof (Example 2.1) in at least two ways. First of all, **substitution** is an accepted technique in logic.

Rule of Substitution Any meaningful formula may be substituted for *all* the occurrences of a propositional variable in a valid statement, for example, in an axiom, a law, etc.

Example 2.2 When we write $x^2 - y^2 = (x + y)(x - y)$ as a law in algebra, we may substitute to obtain:

$$
\begin{aligned}
u^2 - v^2 &= (u + v)(u - v) \\
u^2 - 4 &= (u + 2)(u - 2) \\
(u + w)^2 - v^2 &= (u + w + v)(u + w - v) \\
x^2 - x^2 &= (x + x)(x - x)
\end{aligned}
$$

etc., as the student has learned to do in high school. ■

Similarly, in Example 2.1 the substitution of x for y in (O2) is entirely permissible. The rule of substitution stated above is so familiar to the student that we scarcely feel the need to call attention to its use.

Logical Implication

The other uses of logic in Example 2.1 are more subtle, and, to discuss these more fully, we will digress for a time to become a bit more precise in our description of logical implication.

In our everyday conversation, we often express "cause and effect" relationships of events with the phrase: "*If* such and such happens, *then* so and so will be the result." There is a fundamental understanding that the occurrence of the one event *implies* that of the other. So it is in mathematical reasoning. Many, if not most, of the theorems that we encounter, whatever the mathematical subject area, will assert that from a given statement P, another statement Q follows as a necessary consequence. We say "*if* P *then* Q" to express this relationship most succinctly. The first part of the statement, P, is called the **antecedent**, whereas Q is called the **consequent**. The reader should be warned that mathematicians will use a variety of phrases, all of which amount to the same **logical implication**, whatever the wording, just to avoid sterility of the prose. We may say "P implies Q" or "whenever

P holds, Q holds," or some similar phrase. Symbolically, we usually write $P \Rightarrow Q$ to express an implication, again, just a short-hand notation for "if P then Q."

Many expressions in everyday language are actually logical implications, and it is useful to become adept at recognizing them as such, recasting them in the form $P \Rightarrow Q$. The same is true for the various logical implications that are encountered frequently in mathematical texts. Often a bit of rephrasing is necessary to express such implications in a most succinct form, and the student should try to develop this facility to the greatest degree possible, as illustrated in the examples that follow.

Example 2.3 Suppose it is asserted that "If it is past seven o'clock, the coffee shop is open." This is an everyday use of implication, with

P: It is past seven o'clock

Q: The coffee shop is open

as antecedent and consequent, respectively. Note that the word "then" does not appear in the given phrase, and yet, we nevertheless interpret the statement as saying "if P then Q," writing in a more abbreviated form:

Past seven o'clock $\Rightarrow$ coffee shop open

to clearly delineate the two constituents of the implication. ■

Example 2.4 We have noted that the word "then" may be missing from the logical implications encountered in everyday discourse. For example, "If dreams come true, I'll be with you." Equally likely is the possibility that another word or phrase will take its place, for example, "therefore," "it follows that," "evidently," "consequently," etc., as in the following:

If the earth is flat, it follows that it has an edge.

By the same token, the word "if" may be replaced by "because," "granted that," or some similar phrase, as in writing:

Granted that the insured becomes disabled, premiums are waived.

Nevertheless, such illustrations are still to be understood as implications. ■

Example 2.5 In mathematical writing the wording is often equally cryptic, and a careful eye is often necessary to discover that an implication is intended. If we say

"Every prime greater than two is odd"

then in fact we have stated an implication, one perhaps best rewritten in the form:

$$n \text{ prime and } n > 2 \Rightarrow n \text{ odd}$$

using the variable n to convey the fact that it is every natural number for which such an implication is being claimed. ■

Law of Contraposition

In Example 2.1, we are making use of the so-called **law of contraposition**, namely that

if P then Q—and not Q

yields "not P" as its conclusion. For example, we would argue from the premise

If it is past seven o'clock, then the coffee shop is open

and

The coffee shop is not open

to the conclusion: "It is not past seven o'clock." Or in Example 2.1, we have the assumption

If $x < x$, then x and x are distinct

and

x and x are distinct is absurd

and we conclude "It cannot be the case that $x < x$."

Formally, we may symbolize this general principle of the *law of contraposition* (CON) in the form of a rule:

$$\frac{P \Rightarrow Q, \neg Q}{\neg P} \quad \text{(CON)}$$

Note the use of the implication symbol $\Rightarrow$ as discussed earlier; $\neg$ symbolizes negation. Negated statements such as $\neg Q$ should be read "not Q," "it is not the case that Q," or in words to this effect. Thus $\neg(x < x)$ would be read "x does not precede x." Note in (CON) that there is no division going on. We simply agree to use a dividing line to separate the ***premise(s)*** of a rule from its ***conclusion(s)***. It should be pointed out that we are using a more elaborate form of substitution here. The components, P, Q, etc., in any of our rules may be replaced with statements such as "$x < x$," "the coffee shop is open," etc., and the deduction from premise to conclusion is assumed to be valid in all such instances.

It is extremely important that the reader see such principles at work in these elementary settings before they are employed in the more complex situations that arise later. By way of illustration, we first return to our earlier proof to show the further understanding that is imparted by the new rule of logic (CON).

Example 2.6 The proof in Example 2.1 that $x < x$ is impossible in the axiomatic system (O1)–(O3) may be rephrased in three entries:

$$(x < x) \Rightarrow x \neq x \qquad \text{(O2)}$$
$$\neg(x \neq x) \qquad \text{(E1)}$$
$$\neg(x < x) \qquad \text{(CON)}$$

with annotated justifications provided at the right. The first entry comes about by way of substitution (of x for y) in axiom (O2). The second entry is justified by a use of the reflexive law (E1) for the equality relation (See Section 1.4)—considered as a "common notion" relative to the axiomatic system under study. Finally, and most importantly, with

$$P: x < x$$
$$Q: x \neq x$$

we have satisfied both of the premises $P \Rightarrow Q, \neg Q$ for the law of contraposition. Consequently, we are able to assert $\neg P$, that is, $\neg(x < x)$. ■

Example 2.7 Supposing the two assertions:

If the earth is flat, then it has an edge

and

The earth does not have an edge

are valid, then again using (CON), we may conclude

The earth is not flat

in agreement with Columbus. ■

Extension by Definition

We have seen that an axiomatic system will generally consist of certain undefined primitive terms and a list of axioms specifying certain relationships imposed on these terms, all consistent with out intuition about the subject under study. From there we may proceed to derive theorems that will further elucidate the intricacies of the subject area, using commonly accepted modes of logical deduction—a subject barely touched upon up to this point. As we have indicated, the further formulation of these accepted logical rules is the main object of concern in this chapter.

To describe the resulting axiomatic theory in a more convenient form, it is advisable, if not necessary, to introduce additional terminology, to extend the language of discourse by the definition of new concepts in terms of the primitive ones.

As a matter of fact, this was already done in our brief development of the axioms of order.

Example 2.8 To say that x and y are **distinct** is to say $\neg(x = y)$, that is, $x \neq y$, assuming that equality is the more primitive notion. But of course, this is a rather trivial instance of what we have in mind. ■

Example 2.9 A more substantive illustration may be taken from geometry. We may say that two lines are *parallel* if they have no points in common (and we write $l \| m$ for two such lines l and m). If lines and points are taken as primitive terms (and if the notion of a point *lying on* a line is a primitive relation), then we are defining "parallel" in terms of our primitive notions. ■

Example 2.10 In number theory (arithmetic, if you will), certain primitive operations are introduced (for example, addition, multiplication, etc.) on the numbers—themselves a primitive notion of the theory. Then such definitions as the following may be introduced: A number m *divides* a number n (evenly) if there is a number k with $n = km$—and we then write $m | n$. A number is *prime* if it is (different from one and) evenly divisible only by itself and one. Note how the definition of "prime" depends on our first having defined the notion of "divisibility." ■

In general, the development of a complex theory, such as Euclidean geometry or number thoery, will involve the definition of more and more concepts in layers, each further and further removed from the primitive notions appearing in the axiomatization. In principle, it would be possible to reword all of the theory in primitive terms alone. Clearly, however, this becomes unmanageable as the concepts become more and more complex. The extension of the language by means of definitions is the only way that we can ensure readability and understandability as the theory expands.

Euclidean Geometry

A more substantive illustration of the axiomatic technique can be given in reference to geometry, specifically Euclidean geometry as it is taught or could be taught in our high school mathematics programs. We say "could be taught" since there is a good chance that the ingredients of the axiomatic method are not ordinarily brought out so clearly as we intend to do here. It is more likely, for example, that the typical high school geometry student has gained the impression that points are real points of space, that the lines are perhaps drawn with chalk on a blackboard, and that all of the statements refer to a reality common to our experience.

It is true that the word *geometry* comes from two Greek words, *ge* meaning "earth" and *metrei* meaning "to measure." However, we find that the Greeks, up to and including Euclid, organized and studied geometry for their own mental enjoyment. They were not primarily concerned with its application. In fact, an educated person of those times would have regarded the applications of geometry (to surveying, for example) as beneath his dignity. It was the deductive reasoning—the construction of proofs from certain simple assumptions (axioms) that excited their interest—a truly remarkable and thoroughly Greek contribution to civilization.

Though Euclid and his contemporaries held a high regard for the necessity of an axiomatic basis for their work, recognizing the importance of the unproven statements of a theory, they were not always equally well in tune with the need for primitive undefined terms in the theory. Thus we often see in the original work attempts to define point as that which has no part, or line as a breadthless length. It is only in recent times that it was fully realized that such an approach leads to circularity in reasoning, to the same degree as if we were to attempt to prove all of our statements.

In modern treatments of geometry from a more sophisticated point of view, the notions of *point* and *line* are left as undefined primitive terms. So too is the relationship *on*, in reference to a point "lying on" a line. No attempt is made to interpret this terminology. Nevertheless, the student is encouraged to draw appropriate figures as usual, in trying to visualize the relationships being discussed in a particular argument. From just these three primitive notions, all further terminology can be developed as necessary, using the technique of "extension by definition," as before.

Example 2.11 Among the definitions that we might want to make are the following: Two lines *intersect* if they have a point in common, that is, if there is a point that lies on each of the lines. Two lines are *parallel* if they do not intersect. (Note that in this way, the definition of parallel is made to depend on the definition of intersect.) Two lines are considered *equal* if they consist of the same set of points, that is, if every point that is on either one of the lines is also on the other. A line *passes through* a given point if that point lies on the line. (Evidently, this is just another way of saying the same thing.) Two (or more) points are said to be *colinear* if there is a line that passes through all of the points. ■

In this brief review of Euclidean geometry, it is neither necessary nor possible for us to make any pretense toward completeness in our presentation. We are only interested in capturing some of the flavor of the subject in a somewhat more axiomatic treatment than perhaps was originally seen by the reader. As we mentioned earlier in this section, Euclid listed two groups of axioms (*common notions* and *postulates* (which we call *axioms*) and these served as the foundation for the study of geometry for centuries. More recently, others have revised Euclid's original system, bringing the language more up-to-date, and recasting the axiomatization in a more elegant form. In fact, it is these revised systems that have recently found their way into the high school geometry presentations. And we give only an excerpt from such an axiomatization here.

As we have said, we take *point*, *line*, and *on* as our primitive terms. We may then consider the following **axioms**, recognizing that these are only a portion of those that would be found in a complete axiomatization.

G1. There are at least two distinct points.

G2. Through any two distinct points there passes one and only one line.

G3. For each line, there is a point not on the line.

G4. Through a given point not on a given line, there is one and only one line parallel to the given line.

See Figures 2.1 through 2.4. Note that, as mentioned in Example 2.11, to say that a line passes through a point is just to say that the point lies on the line. Note further the implicit assumption that the notion of parallel lines has already been introduced.

FIGURE 2.1

FIGURE 2.2

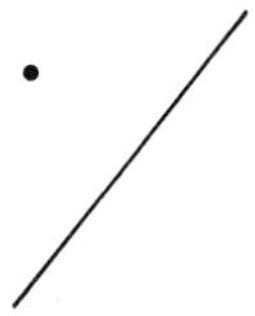

FIGURE 2.3

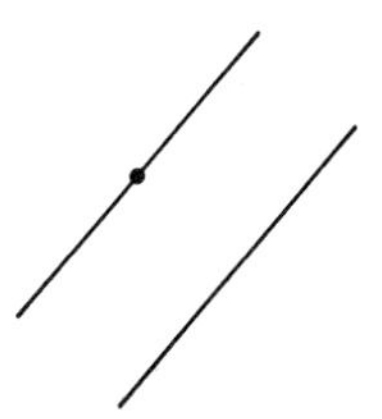

FIGURE 2.4

With little difficulty, we are then able to prove such propositions (theorems, actually) as the four that follow. Their proofs are presented in a prose-like style as opposed to the formal two-column demonstrations commonly seen in high school texts. The reader should examine our proofs, but *should not* study them in any detail at this time. They are offered here as arguments for later study, when we have the necessary logical apparatus for presenting a more formal analysis of the proof techniques involved.

Proposition 1 Two lines cannot intersect in more than one point.

Proof Suppose instead that two lines l, m intersect in points $A, B, \ldots$. Then through points A, B pass the lines l, m according to the definition of *intersect* (see Example 2.11 and Figure 2.5). However, through A, B there should pass not more than one line, by (G2). Since our assumption, "Suppose..." led to a contradiction, it must be that two lines *cannot* intersect in more than one point. □

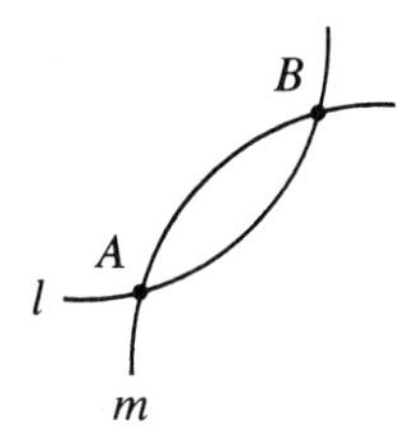

FIGURE 2.5

Proposition 2 There are (at least) three noncolinear points.

Proof Left as an exercise. □

Proposition 3 Every point is on at least two lines.

Proof Consider any point O. By (G1), there is another point $A \neq O$. By (G2), there is a line l passing through O, A. (See Figure 2.6.) Using (G3), there is

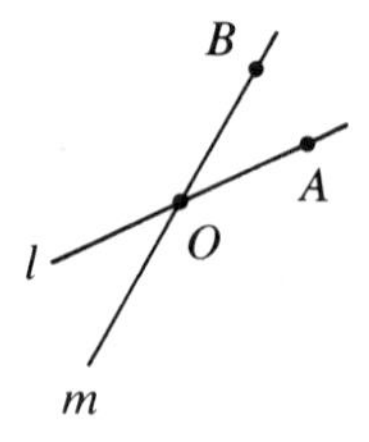

FIGURE 2.6

a point B not on l. By (G2), again, there is a line m through O, B. The lines l, m are distinct since B is on m but not on l. □

Proposition 4 Every line contains at least two points.

Proof Consider any line l. By (G3), there is a point A not on l. By Proposition 3, A is on two lines l_1 and l_2. (See Figure 2.7.) If l had no points, then by the definition of parallel, we would have $l_1 \parallel l$ and $l_2 \parallel l$, contradicting (G4). So l has at least one point B. Again by Proposition 3, there is another line k through B (see Figure 2.8). Either k or l must contain a point $C \neq B$, for otherwise they would be the same collection of points. If C is on l, we are done. If C is on k, then by (G3), there is a point D not on k (again, see Figure 2.8). If D is on l, then again we are done. Otherwise, by (G4), there is a line m through D parallel to k. Lines l and m intersect (at E, say), for otherwise k and l are two distinct lines parallel to m through B (contradicting (G4)). Since B is not on m, E is a second point on l. □

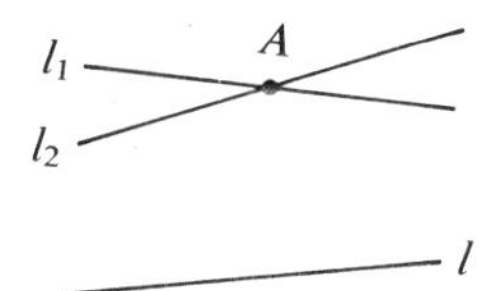

FIGURE 2.7

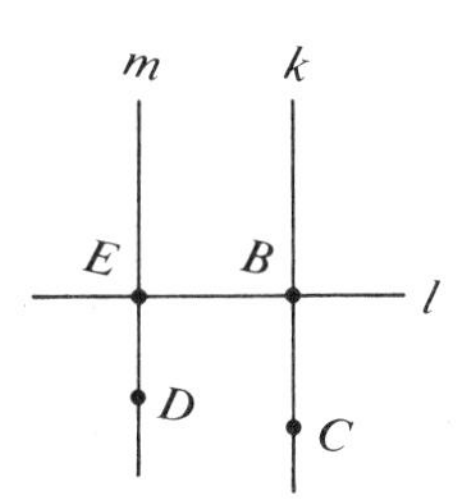

FIGURE 2.8

Note again that we have written the proofs in a kind of prose style. As a matter of fact, it is this prose style that is most commonly found in modern mathematical texts. There, it is the reader's responsibility to keep track of the logical structure of the proof as it is read. This is often quite a difficult task for beginners. So, we will devote some of our later effort toward an analysis of proof structure relating to these specific examples and others as well.

These few examples of proofs from Euclidean geometry will serve as additional motivation for our further studies in mathematical logic. It may very well be that the arguments in our proofs are entirely convincing to the reader (what could be better?). Yet, there is perhaps still the feeling that it is all akin to witchcraft, somehow. It is only through a more detailed study of logic that such a feeling can be dispelled.

EXERCISES 2.1

1 Provide an axiomatization for the notion of an equivalence relation.

2 Given the following interpretations show that Axioms (G1)–(G4) are satisfied.

point = one of the digits from the set $\{1, 2, 3, 4\}$

line = a pair of these digits, i.e., one of the following sets—$\{1, 2\}, \{1, 3\}, \{1, 4\}, \{2, 3\}, \{2, 4\}, \{3, 4\}$

A on l for point A, line l means $A \in l$

3 Given the following interpretations determine which axioms from among (O1)–(O3) are valid and which are not. Verify your assertions.

point = subset of $\{a, b, c\}$

A precedes B for points A, B means $A \subseteq B$

4 Given the following interpretations determine which axioms from among (O1)–(O3) are valid and which are not. Verify your assertions.

point = person, living or dead;

A precedes B for points A, B means A is an ancestor of B

5 Provide a proof of Proposition 2.

6 Let *line*, *point*, and *on* be primitive terms of a geometry with the following set of axioms:

H1. There is at least one line.

H2. Through any two distinct points, there passes one and only one line.

H3. All the points do not lie on the same line.

H4. Any two lines intersect.

In this geometry, provide proofs for the following propositions:

(a) Any two lines intersect in just one point.

(b) There is at least one point.

(c) No point lies on all of the lines.

7 Given the following interpretations:

point = one of the digits from the set $\{1, 2, 3, 4, 5, 6, 7\}$

line = any of the following triples of digits—$\{1, 2, 4\}$, $\{1, 3, 7\}, \{1, 5, 6\}, \{2, 3, 5\}, \{2, 6, 7\}, \{3, 4, 6\}, \{4, 5, 7\}$

A on l for point A, line l means $A \in l$

Show that the Axioms (H1)–(H4) of Exercise 6 are satisfied. Illustrate this geometry with a sketch.

8 Identify the antecedent P and the consequent Q in the following implications:

(a) If dreams come true, I'll be with you.

(b) I'll get wet if it rains.

(c) $a \sim b$ and $b \sim c$ implies $a \sim c$.

(d) Whether or not it rains, I'm going to play golf.

(e) Every point is on at least two lines.

(f) A rolling stone gathers no moss.

(g) $f(x) = f(y)$ only if $x = y$.

(h) Any set C satisfying $A \subseteq C$ and $B \subseteq C$ will contain $A \cup B$.

9 What conclusion may be drawn from the following pairs of statements?

(a) If dreams come true, I'll be with you

and

I'll not be with you

(b) If this is a sunny day, then I'm a monkey's uncle

and

I'm not a monkey's uncle

(c) If it is past seven o'clock, the coffee shop is open

and

It is past seven o'clock

(d) $a \sim b$ and $b \sim c$ imply $a \sim c$

and

$a \nsim c$

(e) If $x^2 > 4$ then $|x| > 2$

and

$|x| \leq 2$

(f) If the light was red, the driver is guilty

and

The driver is not guilty

Sec. 2.2 PROOF TECHNIQUES

The proof is in the pudding.

In this section, we continue our preparation for the study of mathematical logic. We have seen that in the mathematical systems that we might encounter, deductive arguments often present a blend of the use of axioms (and definitions) of a system and the use of pure logical rules that underlie all of our thought processes. So far, we have concentrated mainly on axioms, as illustrated in fairly elementary settings (axiomatizations of order, equivalence, geometry, etc.) that are quite close to the student's previous experience. Now we consider further logic itself. Fortunately, we will be able to draw on the student's good common sense (and past experience) to a large extent as we pave the way for the more detailed logical system to be presented later in this chapter.

As our primary focus here, we look at a few of the more common proof techniques that are typical of a valid argument. Little that we have to say will be new to the reader, at least in its overall thrust. As we have said, it's all "just good common sense." However, we will eventually formalize these techniques in a new way, helping the student to better understand how these logical thought patterns may be organized into an effective proof strategy. We discuss only a few of the ingredients of such a strategy here, those that we have previously encountered informally or those that are easily introduced through the examples we have studied. These in turn become instances of the broader rules of deduction that we introduce in the subsequent sections of this chapter.

Goals

After studying this section, you should be able to:

- Express everyday equivalences and mathematical equivalences in the form $P \Leftrightarrow Q$.
- Identify the statements P and Q for instances of equivalence $P \Leftrightarrow Q$ found in everyday discourse and in mathematical text.
- Verify implications by *forward-backward* reasoning.
- State the *law of detachment* and provide examples of its use.
- Discuss the meaning of the logical connectives "and," "or," and "not."
- Describe the technique of *proof by cases*, and provide examples of its use.

Implication and Equivalence

Ordinarily the statement Q is weaker than the statement P when we encounter an implication $P \Longrightarrow Q$, if indeed such an implication is valid. Thus $n < m \Longrightarrow n \leq m$ because to say that n is less than or equal to m is a weakening of the less-than relation. For this reason, the *converse* implication $Q \Longrightarrow P$ need not hold (when

$P \Rightarrow Q$ is valid). For instance, $3 \leq 3$ but $3 < 3$ does not hold. In those special instances in which both $P \Rightarrow Q$ and $Q \Rightarrow P$ are valid, we say that P and Q are **logically equivalent** statements, symbolized by writing $P \Leftrightarrow Q$. The reason for this terminology is clear. If $P \Leftrightarrow Q$, then the validity of P ensures that of Q; and conversely, from the statement Q the statement P follows as a consequence. There is then no logical distinction that can be drawn between the two statements P and Q. To say the one is to say the other.

As in the case of the implication relation, we find that various phrases are used to assert that $P \Leftrightarrow Q$, and we have already used a few of them in Chapter 1. We can say "*P if and only if Q*" (abbreviated P iff Q), "P and Q are *logically equivalent*," "P holds just in case Q holds," "P implies Q and conversely," or some variation of these phrasings. The student must become familiar with all of these so that a logical equivalence can be recognized as such.

Note that in verifying a logical equivalence $P \Leftrightarrow Q$, the proof will ordinarily split in two. We generally provide separate arguments that $P \Rightarrow Q$ and $Q \Rightarrow P$ are each valid implications in themselves. For this reason, the implication proofs are the more fundamental, and thus deserving of our main consideration in the sequel. Note that our various characterizations in Chapter 1 are instances of logical equivalence, because to *characterize* means to restate in a logically equivalent but different form. Moreover, a *definition* always represents an implicit instance of a logical equivalence, for we are then agreeing that one (new) term is to replace (be logically equivalent to) some other way of saying the same thing. However, here, the reader must be cautioned. It is customary to use the word "if" in stating definitions, even though "iff" is what is to be understood. We say that two lines are *parallel if* they do not intersect, but it is understood as well that if we say that they are parallel, then it is indeed the case that they do not intersect. This is a subtle point, to be sure. And perhaps the student has already caught on to this quirk in the mathematician's language. If not, now is the time to make this point clear.

Example 2.12 The characterization of invertibility for functions (Section 1.3) is an instance of a logical equivalence (as are all mathematical characterizations). Thus, the statements

P: f is invertible

Q: f is one-to-one and onto

are logically equivalent and we may write:

$$f \text{ is invertible} \Leftrightarrow f \text{ is one-to-one and onto}$$

Note that the argument surrounding the discussion of this characterization earlier was indeed designed to establish that $P \Rightarrow Q$ and $Q \Rightarrow P$. ■

Example 2.13 If l and m are two lines of Euclidean geometry, then we may write:

$$l \parallel m \Leftrightarrow l \text{ and } m \text{ have no points in common}$$

simply because of the fact that the word "parallel" has been defined to have this (latter) meaning. ■

Example 2.14 Suppose that an instructor says that he will give an A in the course to every student scoring over 90% on the final exam. Symbolically, the implication here is as follows:

$$\text{Final exam score} > 90\% \Rightarrow \text{course grade} = \text{A}$$

Now it may be that the instructor will also give A's to certain students scoring 90% or less on the final, perhaps because of their superior performance in earlier exams. But this does not mean that the instructor has gone back on his word. In order to have ensured a genuine logical equivalence:

$$\text{Final exam score} > 90\% \Leftrightarrow \text{course grade} = \text{A}$$

the instructor would have to have said, "I will give an A in the course to those students scoring over 90% on the final exam, *and only to those students*," or words to this effect. ■

In a general mathematical system, a statement is considered to be **valid** if it can be shown to follow from the axioms of the system (whatever they may be) according to accepted laws of logical thought. These are the laws, rules, and principles that we begin to examine in some detail in the next section. In the present discussions, we are content to describe some of these principles from an informal point of view. And we would hope that in thus providing a more gradual introduction to the theory, the reader will be in a better position to grasp these important logical concepts as they are encountered. Our general discussion of *forward and backward* reasoning is typical of those informal presentations of proof techniques that are helpful in preparation for this later work.

Forward and Backward

For the reasons given above, it is only natural that we will want to direct the bulk of our attention to the problem of establishing the validity of logical implications $P \Rightarrow Q$ for given statements P and Q. As we have said, the problem of verifying a logical equivalence generally reduces to that of verifying two implications. Once it is suspected that we are required to prove an implication, the next step entails our isolating the statements P and Q, the antecedent and the consequent of the implication, respectively. Thus in the expression

$$\text{(antecedent) } P \Rightarrow Q \text{ (consequent)}$$

the antecedent P is *the given*, the assumption(s) that we are to use in deriving Q, *the result* of the implication.

Example 2.15 Recalling Example 2.5 where it is claimed that

Every prime greater than 2 is odd

we were able to isolate

$$P: n \text{ prime and } n > 2$$
$$Q: n \text{ odd}$$

as antecedent and consequent, respectively, in the supposed implication $P \Rightarrow Q$. To be convinced that this implication is indeed valid, it is then necessary for us to construct an appropriate logical argument showing that Q necessarily follows from the assumption P. ∎

Ordinarily, in our attempting to verify an implication $P \Rightarrow Q$, we are not able to establish a direct link from P to Q. More often, we will require a series of intermediate statements, each following by some elementary line of reasoning from the preceding, but leading ultimately to an unbreakable chain of implications from P to Q (see Figure 2.9).

FIGURE 2.9

However, we do not always discover these intermediate statements in sequence. Often we are well advised to work both ends against the middle, so to speak. We refer to the **forward and backward** method of proof. It may be best, in fact, to start with the consequent Q, the statement we wish to prove valid (if P is valid). Perhaps we will be able to derive a new statement Q_1 with the property "if Q_1 is valid, then so is Q." All of our effort can then be directed to establishing Q_1.

Example 2.16 Considering Example 2.15 and looking at the consequent

$$Q: n \text{ odd}$$

we may think to write down a new statement

$$Q_1: 2 \nmid n$$

for certainly it is the case that $Q_1 \Rightarrow Q$. (In fact, $Q_1 \Leftrightarrow Q$ since $2 \mid n$ is the definition of even, and, of course, odd means not even.) Now our attention is directed to Q_1 rather than Q, and we hope to establish that $P \Rightarrow Q_1$. ∎

We may not be able to see that $P \Rightarrow Q_1$, but perhaps we will think of another statement Q_2 with the property "if Q_2 is valid, then so is Q_1." Continuing in this so-called *backward* mode of reasoning, it may be that ultimately we are led to P, so that we know:

If P is valid, then so is Q_k.
If Q_k is valid, then so is Q_{k-1}.
$\vdots$
If Q_2 is valid, then so is Q_1.
If Q_1 is valid, then so is Q.

We will then have definitely established that $P \Rightarrow Q$. Alternatively, we may be unsuccessful at arriving at P. Then it is perhaps best to counter with a so-called *forward* mode of reasoning. We may derive a sequence of intermediate statements P_1, P_2, etc., with the property that:

If P is valid, then so is P_1.
If P_1 is valid, then so is P_2.
$\vdots$

The goal of this forward reasoning is to ultimately reach the last statement obtained in the backward reasoning (even if this is Q itself). (Again, see Figure 2.9.) If this is accomplished, we will again have established that $P \Rightarrow Q$. Once this has been achieved, however, it is customary to write down the proof in the forward direction entirely, for ease of readability. In fact, however, this is perhaps one of the reasons that "textbook proofs" are so often baffling to beginning students. The thought processes that go into the construction of the proof are often obscured by this reorganization of the steps.

Example 2.17 Returning to the continuing discussion of Examples 2.15 and 2.16, we may next attempt a forward step. What can we derive from the assumptions

$$n \text{ prime} \quad \text{and} \quad n > 2$$

that might be helpful? Perhaps we will want to explore the definition of prime, recalling its meaning:

$$n \text{ is divisible only by itself and } 1$$

so that we may write $P \Rightarrow P_1$ where P_1 is the statement:

$$P_1 : k \nmid n \quad \text{unless } k = 1, n \quad \text{and} \quad n > 2$$

Recalling Q_1, we may think to ask if $2 \mid n$. Since $n > 2$ in the assumption P_1, evidently $2 \nmid n$ as in the statement Q_1. Altogether, we have established a chain of implications:

$$P \Rightarrow P_1 \Rightarrow Q_1 \Rightarrow Q$$

If we then write down the proof in the forward direction, we have:

n prime and $n > 2$

$k \nmid n$ unless $k = 1, n$ and $n > 2$

$2 \nmid n$

n odd

and we note how the thought processes have been obscured, particularly the jump to the statement $2 \nmid n$. Had we not already derived this statement by backward reasoning from Q, it is doubtful that this jump would have been made. ■

The student should be cautioned that at each forward or backward step, there may be many directions in which to go, many blind alleys, so to speak, perhaps leading us completely astray. The best advice we can give is to look at the other end, to see where we are heading—to Q_k (or Q) in the forward mode, to P_j (or P) in the backward mode—and to try to derive intermediate steps that take us nearer to our goal.

Example 2.18 In Example 2.16 we could have chosen to write

$$Q_1: n = 2k + 1 \quad \text{for some integer } k \text{ (again } Q_1 \Rightarrow Q\text{)}$$

but that would not have been nearly so helpful. Our first choice ($2 \nmid n$) is definitely the better of the two. Why did we choose it? Because we knew that prime is concerned with the notion of divisibility, and therefore, looking back at P, this seemed to us the better alternative. ■

Example 2.19 A similar multitude of choices will often be available in the forward direction of reasoning. Consider the proof of Proposition 3 in Section 2.1. After constructing the line l (through O, A) and establishing the existence of a point B not on l, we have the situation in Figure 2.10. We could then have chosen from among the following options, to establish that

1. There is a line m through B parallel to l
2. There is a line m through A, B
3. There is a line m through O, B

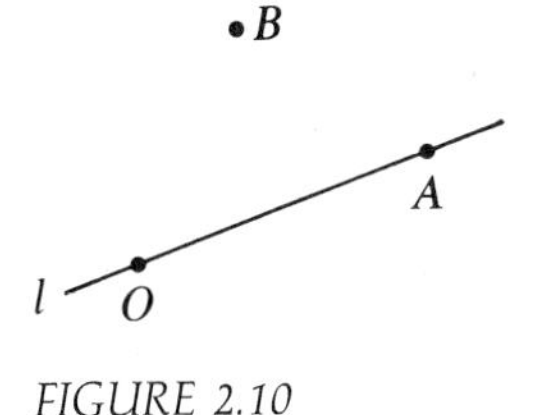

FIGURE 2.10

and we chose the latter. The others could have led us astray. Remember, however, that we were trying to show that two lines pass through O, and that undoubtedly influenced our choice. ■

Law of Detachment

Because so many theorems of mathematics are of the form $P \Rightarrow Q$, we need some way of breaking such implication statements apart. The law of contraposition (Section 2.1) is one of these. Often, however, we may know the validity of a statement P, and we may know that the implication $P \Rightarrow Q$ is valid. From this knowledge, we are permitted to detach Q from the implication, concluding that Q is valid as well. This rule of deduction, appropriately named the **law of detachment** may be symbolized:

$$\frac{P \Rightarrow Q, P}{Q} \qquad \textbf{(DET)}$$

and as we see, it permits us to do just that—to detach Q from the implication. For instance, if we consider the following meanings of P, Q, etc.,

$P \Rightarrow Q$: if n is an integer, then $2n + 1$ is an odd integer
P: n is an integer
Q: $2n + 1$ is an odd integer

then surely the law of detachment applies. In more everyday terms, suppose we consider the following interpretations of P, Q, etc.

$P \Rightarrow Q$: If it is past seven o'clock, then the coffee shop is open
P: It is past seven o'clock
Q: The coffee shop is open

Then once again, our common sense is supported by the application of the law of detachment. Surely we are entitled to conclude that the coffee shop is open, if the two given premises, P and $P \Rightarrow Q$, are valid.

The reader should notice a strong reciprocal relationship between the (DET) and (CON) laws, the latter as introduced in Section 2.1. (See the coffee shop example (Example 2.3) discussed there.) As a matter of fact, these two laws represent two different ways of breaking an implication apart whenever we know something more about its antecedent (DET) or its consequent (CON), respectively. When possible, we prefer to argue in a positive way (avoiding unnecessary negation in our thinking). In general, this leads to a clearer presentation of a logical argument. Thus the law of detachment becomes a principal part of the formal deductive logic system that we soon introduce, whereas the law of contraposition appears only later, as a derived law in a subsequent discussion. Consider the use of (DET) in analyzing the logical significance of a corollary to a theorem. In this case we should think of P as a theorem that has already been proved, and we should regard Q as the statement of a corollary to that theorem. If we can show $P \Longrightarrow Q$, that is, that the validity of P ensures that of Q, then the proof of the corollary is complete—using the law of detachment. Note the clarity of this line of reasoning, rather than our supposing that the corollary is absurd (as we must argue if we use the law of contraposition).

Example 2.20 The proof of Proposition 4 in Section 2.1 can be viewed as an application of the law of detachment. The latter part of that proof established the following:

If l (the given line) has one point, then it has two points.

that is, $P \Longrightarrow Q$, where the first half of the proof established that indeed

P: l has one point

By the law of detachment, we were able to conclude that

Q: l has two points

as required. ■

Logical Connectives

We have already discussed the way that a given statement P can be negated (we then write $\neg P$) to express "P is not the case," "P does not hold," or whatever phrasing we choose. This is only the most elementary of the logical connectives available in building up complex logical statements. We have yet to formally introduce the disjunction (*or*), and the conjunction (*and*), which together with the negation (*not*) lead to a powerful logical statement calculus, as we see later.

For the most part, the student will be able to get along with only an understanding that is based on the everyday usage of the words "or" and "and." However, there is one important word of caution that we must voice at the very outset. It is implicitly understood that "or" is to be used in the *inclusive* sense, and this convention sometimes departs from everyday usage. When I enter a restaurant and the waitress asks me what kind of pie I would like to have for dessert, I might say "apple or blueberry." But I would then be surprised if she brought me both kinds. On the other hand, this would be consistent with the mathematician's use of "or," to include the possibility of both alternatives. That is, "one *or* the other *or* both" is the proper interpretation.

Symbolically, we construct a **disjunction** of two statements P, Q by the use of the connective $\vee$, writing

$P \vee Q$ (read "P or Q")

and similarly, their **conjunction** is written

$$P \wedge Q \qquad \text{(read "}P\text{ and }Q\text{")}$$

When taken together with the use of the **negation** operation, as in writing

$$\neg P \qquad \text{(read "not }P\text{")}$$

we have the possibility of constructing arbitrarily complex logical statements. Note that our choice of symbols, $\vee$, $\wedge$, and $\neg$, makes for an easy comparison with the symbols used in our calculus of sets—that is, $\cup$, $\cap$, and $\sim$ (we need only to straighten out the curves!). As a matter of fact, the similarity goes a bit further than this, as we see later in this chapter.

Proof by Cases

It often happens that a statement is naturally partitioned into a disjunction $P \vee Q$, and whereas we are unable to argue directly that the given statement implies another statement R, we may be able to establish that $P \Rightarrow R$ and that $Q \Rightarrow R$. If the given disjunction is known to be valid, then we conclude that R is valid as well.

Suppose I am going to order either a hamburger or a grilled cheese sandwich. I notice that hamburgers cost $1.50, but that grilled cheese sandwiches cost $1.50, as well. I can conclude that my lunch will cost $1.50, whichever it is. This kind of reasoning is quite common, whether in everyday circumstances such as this, or in a more technical setting.

Example 2.21 In our proof of Proposition 4 of the Euclidean geometry discussion in Section 2.1, we came to a point where we knew that of the two lines k and l, either

$$P\text{: } k \text{ contains a point } C \neq B$$

or

$$Q\text{: } l \text{ contains a point } C \neq B$$

In the case of P (see Figure 2.8), our subsequent analysis led to the statement

$$R\text{: the existence of a second point } (E) \text{ on } l$$

whereas in the case of Q, the point C is already a second point on l. Thus we conclude that in either case l contains two points. ■

In order to capture this important proof technique in symbols, we simply write:

$$\frac{P \vee Q,\ P \Rightarrow R,\ Q \Rightarrow R}{R} \qquad \textbf{(CAS)}$$

and the meaning is immediately understood. We label the technique (CAS) as a three-letter mnemonic, an abbreviation for **proof by CASes.** Actually, the technique is more general. Whenever we are able to partition a hypothesis into two *or more* exhaustive cases and our desired conclusion is implied by each case, then the argument leading to this conclusion is certainly sound. Note the requirement that the cases *exhaust* all of the possibilities, however. Often it is a case of one or the other, as above, where we are certainly on safe ground.

EXERCISES 2.2

1 Identify the statements P and Q in the following logical equivalences $P \Leftrightarrow Q$.

(a) $x \in P$ just in case $p(x)$ holds.

(b) A function f is invertible just in case it is one-to-one and onto.

(c) $a \underset{n}{\sim} b$ should mean $n \mid a - b$.

(d) No man is an island.

(e) $a \sim b$ implies $b \sim a$, and conversely.

(f) I forgive you if and only if you apologize.

(g) The instructor will give an F in the course to those students scoring less than 50% on the final, but only to those students.

(h) Two sets are of the same cardinality just in case there is a one-to-one correspondence of the one set onto the other.

2 What conclusion can be drawn from the following pairs of statements? In each case, state the logical rule that led you to your conclusion.

(a) If dreams come true, I'll be with you

and

Dreams do come true

(b) When it rains, it pours

and

It is raining

(c) n prime and $n > 2$ implies n is odd

and

n is prime and $n > 100$

(d) Long distance phoning is expensive

and

My phone call was not expensive

(e) $x^2 - y^2 = 0$ implies $x = \pm y$

and

$x^2 = y^2$

(f) If it is raining and the sun shines, expect a rainbow

and

It is raining and the sun is shining

(g) Jones plays golf on Monday, Wednesday, and Friday

and

Jones plays golf on Tuesday, Thursday, and weekends

(h) Roses are red and violets are red

and

This flower is either a rose or a violet

3 Identify and describe the use of "proof by cases" (CAS) in

(a) Example 1.19

(b) Example 1.20

(c) Example 1.22

4 Reconstruct the argument, using forward-backward reasoning, to establish the property (E3) in

(a) Example 1.46

(b) Example 1.48

5 Reconstruct the argument in the proof of Proposition 4 using forward-backward reasoning.

6 Using forward-backward reasoning, prove that if a right triangle ABC (with $c =$ hypotenuse $=$ side opposite C) has area $c^2/4$, then it is isosceles (Figure 2.11). (*Hint:* Try to make use of the Pythagorean Theorem: $c^2 = a^2 + b^2$.)

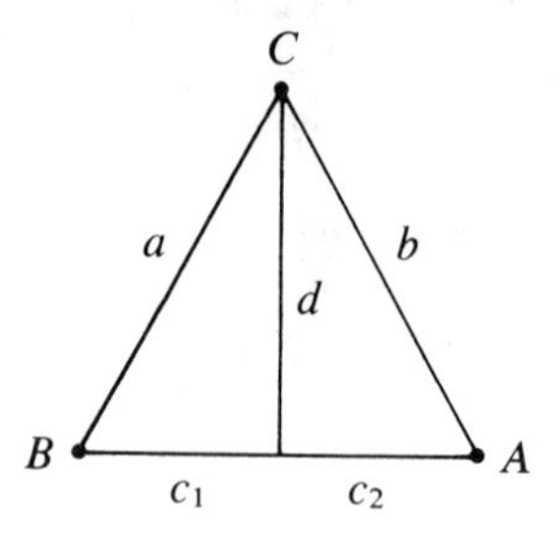

FIGURE 2.11

7 Using forward-backward reasoning, show that if d is a perpendicular bisector of c into segments c_1 and c_2, and if $a = 2c_1$, then the triangle of Figure 2.11 is equilateral.

8 Use forward-backward reasoning to establish that any integer satisfying the equation $-3n^2 + 2n + 8 = 0$ will satisfy $2n^2 - 3n = 2$.

9 List as many statements as you can (at least two) that are a result of applying forward reasoning for just one step to the following:

(a) Triangle ABC is equilateral.

(b) Circle O has the equation $(x - 3)^2 + (y - 2)^2 = 25$.

(c) Quadrilateral $ABCD$ is a square.

10 List as many statements as you can (at least two) that are a result of applying backward reasoning for just one step to the following:

(a) Triangle ABC and $A'B'C'$ are congruent.

(b) Lines l and m are parallel.

(c) Quadrilateral $ABCD$ is a rectangle.

11 Let A be the two-letter alphabet $A = \{a, b\}$ and consider the following *actions* for creating new words (in A^*) from old words:

A1. $x \rightarrow xx$ (duplicate the old word)

A2. $xbby \rightarrow xy$ (erase consecutive pairs of b's)

A3. $xaaay \rightarrow xby$ (replace a consecutive triple of a's by b)

A4. $xa \rightarrow xab$ (adjoin b to a word ending in a)

With these four possible actions applied in any order

(a) Use forward reasoning to derive all words that can be obtained in three steps from the initial word a.

(b) Use backward reasoning to derive all words that can lead to the terminal word bab in one step.

(c) Show that bab may be obtained from a.

(d) Show that $bbab$ may be obtained from a.

Sec. 2.3 PROPOSITIONAL RULES

The introduction and elimination rules of deduction can be used to develop a formalized proof technique.

We have embarked on a program to formalize our logical thought processes to cast our proof technique in a sharper focus. We will see that there is a straightforward underlying basis for such a technique, one that is quite easy to master, once we learn a few simple rules of deduction. For the most part, these deduction rules will seem to grow out of our informal studies thus far, in quite a natural way.

Certainly they will appear to be eminently logical to any reasonable person, and we will be left with no doubt as to their validity for expressing the way of correct reasoning. There are four pairs of rules to be discussed here, relating to implication, conjunction (and), disjunction (or), negation (not). A few others will follow, but they are more of an abbreviation or shorthand technique for extending our power of expression beyond the possibilities available with the original set of rules. Once these rules have been mastered, the student will have greatly enhanced his or her ability to read, to understand, and to construct a sound logical argument.

Goals

After studying this section, you should be able to:

- State the *introduction* and *elimination* rules for implication, disjunction, conjunction, negation, and equivalence, and illustrate their use in simple deductive proofs.
- Relate the deduction rules to the familiar proof techniques.
- Begin to appreciate the nature of a deductive proof.
- Discuss the meaning of the logical connectives "and," "or," and "not," in relation to their deduction rules.
- State the laws of the logical calculus and illustrate their derivation and their use.

Rules of Implication

We recall that the forward-backward style of reasoning may establish an implication $A \Longrightarrow B$ between two (usually mathematical) statements A and B. As we have said, such a relation may be read "A implies B," "B is a logical consequence of A," etc. As the wording signifies, we will ordinarily have determined that the result B follows from the assumption(s) A, by some correct mode of reasoning.

One of the goals of these next few sections is that of adding to the range of tools that can be used in making such inferences $A \Longrightarrow B$. Ordinarily, this notation is reserved, as we have indicated, for expressing a definite relationship *between two statements*. Moreover, the symbol $\Longrightarrow$ is often used informally, as in our earlier discussions, so as to be inappropriate for certain of the presentations that follow. For both of these reasons, it is best that we now introduce a separate symbol $\longrightarrow$ as a syntactic ingredient to our language, signifying logical implication. It is best to think of this new symbol as a logical connective placed between two statements A and B to form *a new statement*, written $A \longrightarrow B$, one that may or may not be valid, depending on the nature of A and B. Fortunately (or not), the two symbols have a similar appearance, so that one will scarcely notice any difference in the reading.

From the foregoing discussion, however, we can already anticipate the first **rule of implication**:

$$\frac{A \Longrightarrow B}{A \longrightarrow B}$$

which says that if we have established that A implies B by correct reasoning, then $A \longrightarrow B$ is valid. As long as we are splitting hairs, we might point out that we are now moving to a more formal logical system where it is important to make the distinction that B is a logical consequence of A, not from the broadest possible system of reasoning, but within the narrow confines of the proof system being developed, that is, that which uses only a certain set of propositional rules, together with a definite proof technique, all in the state of its early development. This is a subtle but important distinction, and our new use of symbolism should be sufficient to bring this out.

Our second **rule of implication** is as transparent as the first, particularly if we recall the law of detachment (Section 2.2). Thus if the statement A is valid and we have also determined that $A \longrightarrow B$ (most likely through our having established that $A \Longrightarrow B$), then we can affirm the validity of B, that is,

$$\frac{A, A \longrightarrow B}{B}$$

Taken as a pair, the first rule can be viewed as a means for the *introduction* of a logical implication into an argument, whereas the second rule accomplishes the *elimination* of an implication. With I and E as mnemonic abbreviations for introduction and elimination, respectively, we may summarize in writing the **rules of implication:**

$$[\to\text{I}] \quad \frac{A \Longrightarrow B}{A \longrightarrow B} \qquad [\to\text{E}] \quad \frac{A, A \longrightarrow B}{B}$$

Recall from Section 2.1 that there is no division involved here whatsoever. The horizontal line only serves as a delimiter:

$$\frac{\text{Premise(s)}}{\text{Conclusion(s)}}$$

to separate the *premise* (an assumption) of a rule from its *conclusion* (the result). All of our subsequent propositional rules will have this same structure, and for each of the other logical connectives (and, or, not, etc.), the same dichotomy between introduction and elimination of the connective will be observed. Furthermore, it must be remembered that for all of these rules, the terms A, B, etc. are to be viewed as placeholders in which particular *instances* ($p, q, p \vee q$, etc.) of logical statements may be substituted.

Conjunction and Disjunction Rules

The two pairs of rules that we are about to present can be viewed as constituting and elaboration on the meaning of the connectives *and* and *or*, beyond that provided by the initial discussion in Section 2.2. The introduction and elimination **rules for**

conjunction are symmetric, simply expressing the fact that the compound statement $A \wedge B$ is valid in the event that A is valid and B is valid; thus

$$[\wedge \mathrm{I}] \quad \frac{A, B}{A \wedge B} \qquad [\wedge \mathrm{E}] \quad \frac{A \wedge B}{A, B}$$

In the latter, we must realize that we often use only half of the conclusion; that is, we sometimes wish to conclude that A is valid, say, from the assumption $A \wedge B$. A similar remark is appropriate in reference to the first of the **rules for disjunction**:

$$[\vee \mathrm{I}] \quad \frac{A}{A \vee B, B \vee A} \qquad [\vee \mathrm{E}] \quad \frac{A \vee B, A \longrightarrow C, B \longrightarrow C}{C}$$

that is, we may only wish to conclude $B \vee A$, say, from the knowledge that A is valid. Certainly this is a reasonable conclusion. For example, if it is raining, we may surely say "it is snowing or it is raining." Of course, we may also turn the statement around, claiming "it is raining or it is snowing," if indeed it is raining. As in the present example, it all depends on what it is that we want to say, but either of the two conclusions should be considered legitimate.

The rule of disjunctive elimination $[\vee \mathrm{E}]$ deserves some discussion. It can be viewed as the embodiment of the technique of proof by cases (see the end of Section 2.2). Essentially, we wish to deduce a certain conclusion C from the assumption that $A \vee B$ is valid. Thinking of the mathematical use of the word "or," and not knowing whether it is A or B that is valid (or both), it is sensible to think that we should check both cases, to see that $A \longrightarrow C$ and that $B \longrightarrow C$, so that we are covered either way. In this sense, the $[\vee \mathrm{E}]$ rule is rather like a double-barreled $[\rightarrow \mathrm{E}]$ rule. Note however, as in the case of $[\rightarrow \mathrm{E}]$, that we often determine that $A \Longrightarrow C$ and $B \Longrightarrow C$, making implicit use of $[\rightarrow \mathrm{I}]$ in establishing the clauses $A \longrightarrow C$ and $B \longrightarrow C$ of the premise to $[\vee \mathrm{E}]$.

Example 2.22 To anticipate the discussions of Section 2.4, let us illustrate the use of some of our deduction rules by establishing the principle:

$$A \longrightarrow A \vee B$$

This is our first in a series of examples of formal proofs, and it is therefore imperative that the student make every effort to appreciate each detail. The style is as important as the substance, and this should be kept firmly in mind. We first assert a statement that represents a general instance of the principle; then in a series of indented entries, we construct the proof of this statement:

$p \rightarrow p \vee q$	[the statement to be proved]
$\quad p \Rightarrow p \vee q$	[the implication, validated as indented below]
$\qquad p$	[the assumption]
$\qquad p \vee q$	$[\vee \mathrm{I}]$
$\quad p \rightarrow p \vee q$	$[\rightarrow \mathrm{I}]$

for arbitrary logical expressions p, q (playing the same role as A, B above). In proving the implication $p \Rightarrow p \vee q$, the assumption p is first written down as a given. Then we apply the disjunctive introduction rule [$\vee$I] to assert $p \vee q$. In these two lines, however, we have established $p \Rightarrow p \vee q$, and consequently, we may apply the implication introduction rule [$\rightarrow$I] to conclude $p \rightarrow p \vee q$, the statement we set out to prove. ■

Example 2.23 In a completely analogous manner, we establish

$$A \wedge B \longrightarrow A$$

this time writing down an instance of the principle and its (indented) proof:

$$\begin{array}{lll} p \wedge q \rightarrow p & & \\ \quad p \wedge q \Rightarrow p & & \\ \quad\quad p \wedge q & & \\ \quad\quad p & & [\wedge\text{E}] \\ \quad p \wedge q \rightarrow p & & [\rightarrow\text{I}] \end{array}$$

without comment, except to note once again the derivation $p \wedge q \Rightarrow p$ necessary to invoke the [$\rightarrow$I] rule. ■

Rules of Negation

Unlike the preceding, the reader will find (in Section 2.5) that there are a number of propostional rules involving negation, and it is difficult to say which of these are the more fundamental. We have chosen the two that seem to lead to a more constructive style of proof, though we do show how to enlarge this view of the matter in our discussion of Section 2.6. To introduce negation into a logical argument, for example, in negating a statement A, we take the position that A should somehow lead to a contradiction. Such a contradiction, in general, will be expressed as a conjunction of a statement B and its opposite $\neg B$. This is our motivation for the first of the two **rules of negation**:

$$[\neg\text{I}] \quad \frac{A \longrightarrow B \wedge \neg B}{\neg A} \qquad\qquad [\neg\text{E}] \quad \frac{\neg A}{A \longrightarrow B}$$

certainly consistent with the position enunciated above.

For our rule of negation elimination, we are reminded of the argument in Example 1.15 where we concluded that the empty set is a subset of any given set [A: For each member of the empty set (there aren't any, i.e., $\neg A$) we can say anything we choose, e.g., B]. Or, if this analogy is not completely convincing, we can recall Bishop's remarks on Brouwer's interpretation of the meaning of the logical connectives, particularly implication. Thus if A is absurd, then any proof of A—there are none—can be converted automatically into a proof of B, using a do nothing procedure; that is, $A \longrightarrow B$.

Example 2.24 The proof of Proposition 1 in the discussion of Euclidean geometry of Section 2.1 made explicit use of the $[\neg \mathrm{I}]$ rule, with

p: Two lines intersect in two (or more) points

q: Through two points there passes exactly one line

and the argument that $p \to q \wedge \neg q$ was used to conclude that two lines cannot intersect in more than one point, that is, $\neg p$. ∎

Example 2.25 A closely related negation elimination rule

$$[\neg \mathrm{E}]' \qquad \frac{A, \neg A}{B}$$

has much the same flavor as $[\neg \mathrm{E}]$. In fact, it can be derived from $[\neg \mathrm{E}]$ as follows:

$$\begin{array}{ll} p & \\ \neg p & \\ p \to q & [\neg \mathrm{E}] \\ q & [\to \mathrm{E}] \end{array}$$

Note again that our proof is valid for any statements p, q (substituted for A, B, respectively). Furthermore, the style of this argument (where the instances of the premises p, $\neg p$ are listed first, then the following statements are justified by an appeal to a deduction rule) is one that we will repeat many times. ∎

Definition of Equivalence

In a certain sense, we could say that our list of propositional rules is now complete. However, it is convenient to introduce additional rules for the notion of *logical equivalence* $A \longleftrightarrow B$ as an abbreviation for those situations where $A \longrightarrow B$ and $B \longrightarrow A$ are both valid. From our preliminary discussion of proof techniques (Section 2.2), we recall that the relation $A \Leftrightarrow B$ holds between statements A and B whenever we are able to establish both $A \Longrightarrow B$ and $B \Longrightarrow A$. It thus appears that our symbol $\longleftrightarrow$ stands in the same relation to $\Leftrightarrow$ as did the implication symbol $\longrightarrow$ to $\Longrightarrow$. Thus we may view the equivalence symbol $\longleftrightarrow$ as a new logical connective for combining certain statements A and B to obtain a new statement $A \longleftrightarrow B$. And indeed, our reciprocal **rules of equivalence**:

$$[\leftrightarrow \mathrm{I}] \qquad \frac{A \longrightarrow B,\ B \longrightarrow A}{A \longleftrightarrow B} \qquad\qquad [\leftrightarrow \mathrm{E}] \qquad \frac{A \longleftrightarrow B}{A \longrightarrow B,\ B \longrightarrow A}$$

identify those circumstances where this can be done. Because the two rules are reciprocal, we reinforce the idea that equivalence only serves as an abbreviation for something that could be said at greater length.

Although the notion of equivalence is only an abbreviation in notation, it does serve quite a useful purpose in the manipulation of logical statements. We are able to state a collection of laws that permit the simplification of logical expressions, the derivation of equivalent statements that may be easier to read, to understand, or to analyze. In this sense the laws introduce a formal calculus for reasoning about logical statements. Such a calculus is indispensable in the study of computer programming language, for example, where the logical expressions are an essential ingredient of the syntax, used primarily to effect branching of a program or to terminate program loops.

Because of these applications, where it is a case of logical identity, we will often find it convenient to use a more appropriate symbol $\equiv$ rather than the two-way arrow $\longleftrightarrow$ introduced above. The reader should be willing to switch back and forth from either notation as the situation warrants. Moreover, at a later point we may become even less precise, using an ordinary equal sign to mean the same thing. The context should suffice to make the meaning clear.

The Logical Calculus

The laws of the logical calculus are completely analogous to those we have given previously for sets (Section 1.2). This is quite fortunate, for the reader then has nothing new to learn. A quick glance at the following list indeed reveals that this is so.

L1a.	$A \vee A \equiv A$	**L1b.**	$A \wedge A \equiv A$
L2a.	$A \vee B \equiv B \vee A$	**L2b.**	$A \wedge B \equiv B \wedge A$
L3a.	$A \vee (B \vee C) \equiv (A \vee B) \vee C$	**L3b.**	$A \wedge (B \wedge C) \equiv (A \wedge B) \wedge C$
L4a.	$A \vee (A \wedge B) \equiv A$	**L4b.**	$A \wedge (A \vee B) \equiv A$
L5a.	$A \vee (B \wedge C) \equiv (A \vee B) \wedge (A \vee C)$	**L5b.**	$A \wedge (B \vee C) \equiv (A \wedge B) \vee (A \wedge C)$

Indeed, these are our old friends, the *idempotent, commutative, associative, absorption,* and *distributive laws,* dressed up in new garb.

On the other hand, the reader should not be willing to accept these laws too readily, simply because they are so strongly reminiscent of something we have seen before. It must be realized that we are now working in an entirely new subject area. We are no longer speaking of sets. The operations are logical operations and the equalities are actually logical equivalences in the precise sense introduced in the previous subsection. This means that the whole set of laws (L1a)–(L5b) must be reexamined in this new context, quite apart from anything we may have learned previously.

It may be instructive, therefore, to provide a proof for one of those same identities [here called (L4a)] considered in Section 1.2 in reference to the calculus of sets (see Example 1.19). The student may find the argument a bit difficult to follow at this stage of our development. But that is only because certain formalities of style have yet to be discussed. Without undue attention to such details here, let us just say that we present this proof only in anticipation of the discussion of Section 2.4, in order that the reader will know what to expect.

Example 2.26 We first provide a statement of our theorem (the opening line). In the proof that follows the theorem it is perhaps unfortunate that we find occasion to make use of a derived rule $A \longrightarrow A$, one not actually encountered until Section 2.5 (under the name [→A]).

$p \vee (p \wedge q) \leftrightarrow p$

$p \rightarrow p \vee (p \wedge q)$	[Example 2.22]
$p \vee (p \wedge q) \Rightarrow p$	
$\quad p \vee (p \wedge q)$	
$\quad p \rightarrow p$	[→A]
$\quad p \wedge q \rightarrow p$	[Example 2.23]
$\quad p$	[∨E]
$p \vee (p \wedge q) \rightarrow p$	[→I]
$p \vee (p \wedge q) \leftrightarrow p$	[↔I]

But this trivial observation (that implication is a reflexive relation) should be rather obvious on the face of it. Nevertheless, the reader should note the use of two earlier principles, as established in Examples 2.22 and 2.23, which give some indication of the depth of the logical theory being developed. ■

Example 2.27 By way of application, suppose in a certain computer program, we encounter the alternative statement:

if $(p \vee (p \wedge q)) \wedge r$ then
  S
else
  T

where S and T are sequences of instructions to be executed in the one case or the other. Note that p, q, r may themselves be logical conditions, for example,

$$p\colon x \leq 3$$
$$q\colon y > 1$$
$$r\colon z = 0$$

involving the values of certain program variables x, y, z. Then, by virtue of (L4a), we can assert that the given logical expression is unnecessarily complex, that the simpler expression

```
if p ∧ r then
  S
else
  T
```

would have exactly the same effect. ■

EXERCISES 2.3

1 Establish the following logical principles:

(a) $B \longrightarrow A \vee B$ (b) $A \wedge B \longrightarrow B$

2 Establish the following logical principle:

$$A \wedge B \longrightarrow A \vee B$$

3 Discuss the meaning of the logical connectives "and" and "or" in relation to their rules of deduction.

4 What is meant by a logical contradiction, that is, what is the form of such a statement?

5 On what grounds do we accept the validity of an implication $p \longrightarrow q$ when p is known to be absurd?

6 Establish the following logical principles:

(a) $A \wedge B \longrightarrow B \wedge A$ (b) $A \vee B \longrightarrow B \vee A$

(c) $B \longrightarrow (A \longrightarrow B)$ (d) $A \wedge B \longrightarrow A \wedge (B \vee C)$

7 Establish the following rules of deduction:

(a) $\dfrac{B}{A \longrightarrow B}$ (b) $\dfrac{A \wedge B}{A \vee B}$

(c) $\dfrac{A \wedge \neg A}{B}$ (d) $\dfrac{A \longrightarrow \neg A}{A \longrightarrow B}$

8 Provide a proof for the following identities.

(a) Law (L1a) (b) Law (L1b) (c) Law (L2a) (d) Law (L2b)

(e) Law (L3a) (f) Law (L3b) (g) Law (L5a) (h) Law (L5b)

9 Provide a proof for the identity (L4b). (*Hint:* Make use of Example 2.23 in first establishing that $p \wedge (p \vee q) \rightarrow p$.)

10 Symbolize the following arguments, using the suggested notation. Then derive the conclusion from the assumption(s), justifying each step by citing the appropriate propositional rule(s).

(a) Water contains oxygen. So either water or milk contains oxygen. (w, m)

(b) If I study harder, I will be on the dean's list. I shall study harder, so therefore, I shall be on the dean's list. (s, d)

(c) If n is positive, n^2 is positive. If n is negative, n^2 is positive. Moreover, n is either positive or negative. Consequently n^2 is positive. (p, n, n^2)

(d) Line a is either longer than line b or shorter than b. If it is longer than b, then a and b are not the same length, and if it is shorter than b, then again, a and b are not the same length. Consequently, line a is not the same length as line b. (l, s, n)

(e) Jones is pitching tomorrow. Therefore, either Jones or Smith is pitching tomorrow. (j, s)

(f) Jones and Smith are going to Europe. Therefore, Jones is going to Europe. (j, s)

(g) Jones is going to Europe. Smith is going to Europe. Therefore, Jones and Smith are going to Europe. (j, s)

(h) The instructor is giving an F in the course to all students scoring less than 50% on the final, but only to those students. Thus students receiving an F in the course will have scored less than 50% on the final. (f, l)

Sec. 2.4 DEDUCTIVE LOGIC

A proof of a theorem in formal logic is a series of statements deduced from the developing proof by the application of logical rules.

The logical rules of the previous section lead to a formalized proof technique, that if studied quite carefully, can greatly enhance the student's ability to read, to understand, or to construct a sound logical argument. These skills are of the utmost importance in mathematical reasoning, whether in mathematics studies per se, or in computer science or any of a number of fields of application. These techniques will systematize the approach we have taken in discussing informal proofs thus far. Moreover, the style of proof will be patterned directly after the format the student has learned in high school mathematics courses, particularly geometry, where each step is justified by appeal to a definite rule. Furthermore, as we shall see, the rules have been chosen to enable the student to discover the plan of a proof from the initial statement of the problem. This approach then serves to eliminate much of the guesswork that is otherwise involved, developing your experience toward building a repertoire of techniques as you move from one problem to another.

Goals

After studying this section, you should be able to:

- State the complete set of introduction and elimination rules, and apply them to the development of formal proofs.

- Outline the format of a proof in deductive logic.
- Discuss the *scope rule* and give examples of its use.
- Outline the conventions on proof indentation and give reasons why such a technique should be followed.
- Derive formal proofs in deductive logic.
- Outline the general guide to the choice of deduction rules and make use of this guide in developing proofs.
- Discuss the analogies between programming and theorem proving.

Summary of Rules

Before describing the essential ingredients of the proof technique, it is a good idea to review first the logical rules of the preceding section. For the purposes we intend, it is best to have these rules summarized in one convenient point of reference, as shown in Figure 2.12. Note once again the way that the rules appear in pairs, one for the introduction [I] and one for the elimination [E] of a particular logical connective. It is this feature that helps to guide our thought processes later on as we learn to construct logical arguments step by step, from one statement to another.

Propositional Deduction Rules

[→I] $\dfrac{A \Rightarrow B}{A \longrightarrow B}$	[→E] $\dfrac{A, A \longrightarrow B}{B}$
[∧I] $\dfrac{A, B}{A \wedge B}$	[∧E] $\dfrac{A \wedge B}{A, B}$
[∨I] $\dfrac{A}{A \vee B, B \vee A}$	[∨E] $\dfrac{A \vee B, A \longrightarrow C, B \longrightarrow C}{C}$
[¬I] $\dfrac{A \longrightarrow B \wedge \neg B}{\neg A}$	[¬E] $\dfrac{\neg A}{A \longrightarrow B}$
[↔I] $\dfrac{A \longrightarrow B, B \longrightarrow A}{A \longleftrightarrow B}$	[↔E] $\dfrac{A \longleftrightarrow B}{A \longrightarrow B, B \longrightarrow A}$

FIGURE 2.12

Formalized Proof Technique

We have seen several examples of formal proofs in logic in the preceding section (among Examples 2.22 through 2.26). The reader should keep the format of such illustrations in mind in the discussion that follows. The **statement of a theorem** in our formal system will take on one or the other of the following two appearances:

$$q \Rightarrow p \quad \text{or} \quad p$$

In the first case, there is an **assumption** q, and it is claimed that the **result** p is a necessary consequence. By this we mean that one should be able to reason, using only our list of deduction rules, from q to p in a series of steps. In the second case, no assumptions are made, but still, we are expected to be able to develop a series of statements, again using only the deduction rules of Figure 2.12, leading eventually to the result p.

More precisely, the **proof of a theorem** $q \Longrightarrow p$ or p will consist of a finite number of entries following the statement of the theorem itself. When written out, the proof is indented (for visual reasons) and the last entry at this level of indentation is the result p. Each entry must have one of the following forms:

---

q—the assumption in $q \Longrightarrow p$.

r—$\dfrac{r_i}{r}$ is an instance of a deduction rule, and the premises r_i are earlier entries within the scope of the entry being made.

s—s is the complete proof of another theorem, one whose format must follow the same structure all over again.

Any prior entries in a proof are said to be *within the scope* of its later entries, or of the entries in any of its subproofs. The student who has had some previous exposure to blockstructured computer-programming languages, for example, Pascal, will have no difficulty in interpreting this scope rule. It is this same principle in programming languages, where *global* references can be made to variables occurring in a surrounding procedure.

Example 2.28 Consider again the theorem and proof given in Example 2.26. The statement of the theorem $p \vee (p \wedge q) \leftrightarrow p$ is given first. (Note that this whole statement plays the role of p in our foregoing discussion.) The proof is indented, and consists mainly of the verification of the two implications $p \rightarrow p \vee (p \wedge q)$ and $p \vee (p \wedge q) \rightarrow p$, as expected when we are treating an equivalence. The second of these requires our establishing $p \vee (p \wedge q) \Rightarrow p$ in a proof of its own, further indented beyond the initial level, as indicated previously where we say in reference to S: "whose format must follow the same structure all over again." This way, we may read the overall theorem and proof on one level:

$p \vee (p \wedge q) \leftrightarrow p$

$p \rightarrow p \vee (p \wedge q)$	[Example 2.22]
$p \vee (p \wedge q) \Rightarrow p$	[subproof, as indented]
$p \vee (p \wedge q) \rightarrow p$	[$\rightarrow$I]
$p \vee (p \wedge q) \leftrightarrow p$	[$\leftrightarrow$I]

and the interior proof of the second statement can be ignored if we wish. If later we would like to examine this subproof, it stands alone as a separate argument:

$$
\begin{array}{ll}
p \vee (p \wedge q) \Rightarrow p & \\
\quad p \vee (p \wedge q) & \\
\quad p \rightarrow p & [\rightarrow\text{A}] \\
\quad p \wedge q \rightarrow p & [\text{Example 2.23}] \\
\quad p & [\vee\text{E}]
\end{array}
$$

Note that the first line of this argument is the assumption in the implication to be proved. According to the above *qrs* trichotomy, such an entry q may stand alone as a line in the proof (of $q \Rightarrow p$), without justification. How else can we determine whether the implication holds, without first assuming the premise, to see what follows. Now, had we not already proved $A \wedge B \rightarrow A$ in Example 2.23, the proof of $p \wedge q \rightarrow p$ would have been indented at yet a third level:

$$
\begin{array}{ll}
p \wedge q \rightarrow p & \\
\quad p \wedge q \Rightarrow p & \\
\quad\quad p \wedge q & \\
\quad\quad p & [\wedge\text{E}] \\
\quad p \wedge q \rightarrow p & [\rightarrow\text{I}]
\end{array}
$$

and so on. Here, the first line of the proof is an s in the above *qrs* trichotomy. That is, it is the (heading of a) complete proof of another theorem, so it needs no rule as justification, but only its own (indented) proof. The first line of this indented proof is the assumption in the implication to be proved, so that it too requires no justification.

In this way, we see that proofs, even of elementary theorems such as this, can be *nested* rather deeply in their levels. Unless we have a systematic format to follow, the complexity could quite easily get out of hand. ■

We can, of course, be even more systematic. We could insist that the entries in a proof be numbered, and that all annotations [—] be augmented to include a list of the numbers of the lines where the annotating rule's premises have been satisfied. Some instructors and/or students may wish to adopt this more detailed system. But our preference is to compromise—provide only the name of the rule, and leave it up to the reader to search upward through the proof to verify that the rule's premises have indeed been met.

Conventions on Indentation

In our style of *indentation*, we again borrow from the techniques of (structured) programming. Whenever we begin an interior subproof, we indent the entries so as to enhance the readability of our overall proof. In this way, the levels of a proof are stratified in the same way as with our indenting of the nested constructs in a structured programming language such as Pascal. When we leave a subproof, we return the coding of our entries to the original level of indentation. Proofs, like programs, can become very complex; without such an indentation convention, they can be quite difficult to follow. On the other hand, if such a convention is strictly observed, the tendency toward complexity is mitigated considerably. We can read through a proof (at one particular level), ignoring for a time those detailed entries at a deeper level, and return later if we desire to a consideration of the subproofs. This is a distinct advantage.

Example 2.29 Here we present an example designed to more fully acquaint the reader with the use of the implication introduction and elimination rules, and again, to reinforce an understanding of the important indentation conventions.

$$
\begin{array}{lr}
p \to (r \to s) \Rightarrow r \to (p \to s) & \\
\quad p \to (r \to s) & \\
\quad r \Rightarrow (p \to s) & \\
\quad\quad r & \\
\quad\quad p \Rightarrow s & \\
\quad\quad\quad p & \\
\quad\quad\quad r \to s & [\to\text{E}] \\
\quad\quad\quad s & [\to\text{E}] \\
\quad\quad p \to s & [\to\text{I}] \\
\quad r \to (p \to s) & [\to\text{I}]
\end{array}
$$

Note that there are so many nested assumptions in the proof that we don't need any justifications until the last four lines. Moreover, we observe that we then detach results of the implications with the [→E] rule, after noting the earlier appearances of the required premises. ■

An Abbreviated Format

Throughout our examples, the reader will have noticed the recurring pattern:

$$
\begin{array}{lr}
q \Rightarrow p & \\
\quad q & \\
\quad \vdots & \\
\quad p & \\
q \to p & [\to\text{I}]
\end{array}
$$

with an indented subproof of $q \Rightarrow p$. So often does this occur that it is worth our while to introduce the **abbreviated format**:

$$
\begin{array}{ll}
q \to p & [\to\mathrm{I}] \\
\quad q & \\
\quad \vdots & \\
\quad p &
\end{array}
$$

meant to express the same thing. In most instances, the clarity is thereby enhanced, and we save one line of writing with each such abbreviation. For instance, the abbreviated rewriting of Example 2.26 would appear as follows:

$$
\begin{array}{ll}
p \vee (p \wedge q) \leftrightarrow p & \\
\quad p \to p \vee (p \wedge q) & [\text{Example 2.22}] \\
\quad p \vee (p \wedge q) \to p & [\to\mathrm{I}] \\
\qquad p \vee (p \wedge q) & \\
\qquad p \to p & [\to\mathrm{A}] \\
\qquad p \wedge q \to p & [\text{Example 2.23}] \\
\qquad p & [\vee\mathrm{E}] \\
\quad p \vee (p \wedge q) \leftrightarrow p & [\leftrightarrow\mathrm{I}]
\end{array}
$$

We hope our adoption of this new format will not cause unnecessary confusion for the reader.

Example 2.30 So far, the student has seen examples of the use of almost all of our basic introduction and elimination rules, except for $[\wedge\mathrm{I}]$ and $[\neg\mathrm{I}]$. We make up for this omission here in proving the theorem:

$$p \to q, \quad \neg q \Rightarrow \neg p$$

We first note the fact that our theorem structure has been extended to allow for the possibility of *more than one assumption* (separated by commas)—here, the two assumptions $p \to q$ and $\neg q$. It follows that both of these can be listed in our proof without rationale, as shown below:

$$
\begin{array}{ll}
p \to q, \neg q \Rightarrow \neg p & \\
\quad p \to q & \\
\quad \neg q & \\
\quad p \to q \wedge \neg q & [\to\mathrm{I}] \\
\qquad p & \\
\qquad q & [\to\mathrm{E}] \\
\qquad q \wedge \neg q & [\wedge\mathrm{I}] \\
\quad \neg p & [\neg\mathrm{I}]
\end{array}
$$

At the same time, we see that our new abbreviated format is being used in the verification of $p \to q \wedge \neg q$. Note that the statement $\neg q$ is within the scope of this subproof, where the conjunction introduction rule [$\wedge$I] is applied to conclude $q \wedge \neg q$. (See the continuing discussion in Example 2.32.) ■

Development of Proofs

Often the student looks at the proof of a theorem in wonder and awe. However, appearances can be quite deceiving. As we have seen, there is quite a difference between a well-crafted wording of a proof and the underlying logical skeleton of the argument. The former depends on one's skill with words, and certainly, that is a talent worth developing. On the other hand, we often need to perform the inverse operation, stripping away the verbiage to see what lies beneath it all. That is another skill we must learn. The earlier discussions of this chapter should have been helpful in this respect. We must also be able to construct a logical argument from scratch, which can be difficult for the beginning student.

The formal logical system just introduced, if studied quite carefully, can do wonders for our powers of understanding. But more importantly, the proper use of this system and its extensions can be an immeasurable aid in developing our proof technique on an individual basis. It happens that proofs are not constructed out of a vacuum, really. If we look more closely at our system, we will find that much of the mystery that beginners tend to associate with the proving process is unwarranted. We will discover that the process can be systematized to a large extent, as a goal-oriented activity.

Suppose that a theorem of the form $q \Rightarrow p$ is to be proved. Here, it must be remembered that p and q are themselves logical formulas of one kind or another, but unspecified for the present. From the idea of a formalized proof, as previously introduced, any (statement and) proof must take the form:

$$\begin{array}{l} q \Rightarrow p \\ \quad q \\ \quad \vdots \\ \quad p \end{array}$$

Thus, we state the theorem and indent its proof, beginning first of all with the assumption q (It is unlikely that the assumption will not be used!) and leading finally to the desired result p. The trick is to fill in the gap. Somehow we must develop a logical argument to confirm the last line p.

In this regard, there are (as expected, really) two ways to proceed. We can look at the assumption(s) q, hoping to derive something that looks similar to p, or at least closer to p than where we now stand. Alternatively, we can examine p itself, hoping that its structure will suggest a deduction rule that could yield p as its conclusion. (Note the connection to the forward-backward style of reasoning of Section 2.2.) In either of these approaches, we are to be guided by the propositional deduction rules themselves. There were only eight of them originally. (Unless an equivalence is a part of one of our assumptions q or of the desired result p, we can forget about the rules [$\leftrightarrow$I] and [$\leftrightarrow$E].) Fortunately, few of these will be applicable—either to q, by being a match with the form of a rule's premise, or to p, by having the form of a rule's conclusion.

Example 2.31 Suppose p has the form $r \to s$ for certain subexpressions r and s. Then, in attempting to match one of our rules' conclusions with p, the most likely candidates would be [$\to$I], [$\neg$E], and [$\leftrightarrow$E]. If it is not easy to see how an equivalence $r \leftrightarrow s$ might be derived from q, then we can discount [$\leftrightarrow$E] as a possibility. Similarly, if we do not see a chance to derive $\neg r$ from q, then [$\neg$E] could also be eliminated from consideration. Supposing that we are left with [$\to$I] as the best hope, we would then expand our developing proof as follows:

$$q \Rightarrow (r \to s)$$

$$\begin{array}{ll} q & \\ r \to s & [\to\text{I}] \\ \quad r & \\ \quad \vdots & \\ \quad s & \end{array}$$

(Note the abbreviated format and the annotation [$\to$I] as the rationale for our last step.) We then observe that we have reduced the original problem, show $q \Rightarrow p$, to the smaller problem, show $r \Rightarrow s$. We say smaller because r and s are subexpressions of p. The development then continues as before, and we treat the smaller problem as we did the original problem. ■

This example outlines the main plan of attack. That is, we examine our list of deduction rules to see which ones might be applicable to the given situation—looking first at the desired result, then at the available assumption(s), in turn, hoping to replace the current problem with a simpler one. Beyond these general words of advice, the only thing else we can say is "practice, practice, practice."

Experience will count for a lot. Therefore, the student should review our list of deduction rules quite carefully, together with the derived rules that we later develop, and learn to tell at a glance which rules are most appropriate to a given situation. As a general guide to this study, we can state the following principles:

---

1. The introduction rules, particularly [$\wedge$I] and [$\vee$I] tend to produce longer formulas from shorter formulas.
2. Since the conclusions of the five introduction rules have entirely different forms, most likely, only one of them will be useful to validate a desired result.

3. The elimination rules, particularly [$\rightarrow$E], [$\wedge$E], and [$\vee$E], have quite a general conclusion (for example, B, A, C, respectively).
4. Because of principle (3), almost any elimination rule will be applicable in attempting to substantiate a desired conclusion, and judgment becomes most important.
5. One should be guided with the choice in principle (4) by the entries that have previously been made in the developing proof, looking to see if a rule's premise(s) will have been satisfied.

But again, this guide can only be useful if the student makes a conscious effort to put the suggestions into practice on a continuing basis.

Example 2.32 Returning to Example 2.30, let us see if we can imagine how this proof might have been devised. After stating the two assumptions:

$$p \rightarrow q$$
$$\neg q$$

it becomes clear that we need to introduce somehow a negation relative to p. Thus suggesting the negation introduction rule [$\neg$I], we hope to be able to derive a contradiction from p itself. The assumptions suggest that $q \wedge \neg q$ will be the most likely contradiction—in fact, we begin to see immediately that this is the case (see Example 2.30). ■

EXERCISES 2.4

1 Provide proofs for each of the following.

(a) $p \Rightarrow p \vee \neg q$

(b) $p \leftrightarrow q \Rightarrow q \leftrightarrow p$

(c) $p \wedge \neg q \Rightarrow \neg q$

(d) $p \wedge (q \wedge r) \Rightarrow q \vee r$

(e) $q \vee s \rightarrow p \wedge q \Rightarrow q \vee s \leftrightarrow p \vee q$

2 Provide proofs for each of the following.

(a) $p \wedge q, \quad p \rightarrow r \Rightarrow r$

(b) $p, q \rightarrow r, \quad p \rightarrow r \Rightarrow p \wedge r$

(c) $p \wedge q, \quad p \rightarrow r \Rightarrow r$

(d) $p \leftrightarrow q, \quad q \Rightarrow p$

(*Note:* There are two or more assumptions here, as first encountered in Example 2.30.)

3 Provide proofs for each of the following.

(a) $p \rightarrow q \wedge r, \quad p \Rightarrow r$

(b) $p, q \wedge (p \rightarrow r) \Rightarrow q \wedge r$

(c) $p \rightarrow (q \wedge r), \quad p \Rightarrow r$

(d) $\neg p, \quad (\neg p \rightarrow q) \vee [p \wedge (r \rightarrow q)] \Rightarrow r \rightarrow q$

(*Note:* See the Note to Exercise 2.)

4 Provide proofs for each of the following.

(a) $p \vee (q \wedge r), \quad p \rightarrow s, \quad (q \wedge r) \rightarrow s \Rightarrow s \vee p$

(b) $p \wedge q, \quad p \rightarrow r \Rightarrow r \vee (q \rightarrow r)$

(c) $p, q \wedge (p \rightarrow s) \Rightarrow q \wedge s$

(d) $p \vee q, \quad p \vee r \rightarrow s, \quad q \vee t \rightarrow s \Rightarrow s$

(*Note:* See the Note to Exercise 2.)

5 Provide proofs for each of the following.

(a) $p \wedge q \rightarrow p \vee q$

(b) $p \leftrightarrow p \vee p$

(c) $q \rightarrow [r \rightarrow (q \wedge r)]$

(d) $q \leftrightarrow q \wedge q$

(e) $p \rightarrow [(r \vee s) \rightarrow p]$

(f) $(p \rightarrow q) \wedge \neg q \rightarrow \neg p$

6 Symbolize the following arguments using the suggested notation. Then derive the conclusion from the assumption(s) as a formal proof.

(a) If the pilot was conscious and knew the rate of descent of his airplane, and if the altimeter was accurate, then mechanical failure was responsible for the crash. Inspection of the wreckage shows that there was no mechanical failure and that the altimeter was accurate. Therefore, if the pilot was conscious, he did not know the rate of descent of his airplane. (c, k, a, m)

(b) The company will waive the premiums on the policy only if the insured becomes disabled. However, if the company has not waived the premiums, and said premiums are not paid when due, this policy will lapse, and the company will either refund its cash value to the insured or issue term insurance for a limited period. Therefore, if the insured is not disabled, and if the premiums are not paid when due, then this policy will lapse (w, d, p, l, r, i). (*Hint:* It may be convenient to apply the law of contraposition (CON) from Section 2.1.)

7 Provide proofs for each of the following.

(a) $p \vee (q \wedge r) \Rightarrow (p \vee q) \wedge (p \vee r)$

(b) $p \wedge q \Rightarrow \neg(p \wedge \neg q)$

(c) $p \wedge q \wedge (p \rightarrow r) \Rightarrow r \vee (q \rightarrow r)$

(d) $\neg q \Rightarrow q \rightarrow \neg p$

(e) $p \wedge \neg q \rightarrow r \wedge \neg r \Rightarrow p \rightarrow q$

8 Reconstruct the proof of Example 2.30 in the unabbreviated format.

9 Reconstruct the proof of Example 2.29 in the abbreviated format.

10 Reconstruct the following proofs using a systematic numbering system and providing annotations that include a list of the numbers of the lines where the annotating rule's premises have been satisfied.

(a) Example 2.29

(b) Example 2.30

11 Provide proofs for each of the following.

(a) $(p \wedge q) \wedge r \rightarrow p \wedge (q \wedge r)$

(b) $p \rightarrow (q \rightarrow p \wedge q)$

(c) $p \wedge q \rightarrow q$

12 In the context of the axiomatization (O1), (O2), (O3) of Section 2.1, provide a formal proof of the theorem

$$x < y \Rightarrow \neg(y < x)$$

(*Hint:* Make use of the theorem proved in Example 2.1.)

Sec. 2.5 DERIVED RULES

The instructor may view this section as being optional, although the derived rules can lead to a more concise proof.

For most purposes, the ten basic introduction and elimination rules of Section 2.3 provide an adequate foundation for understanding or formulating a logical argument. Certainly this system as extended in Section 2.6 will give the student a good basic repertoire of proof techniques, a system capable of treating almost any problem that he or she might encounter in further studies. Nevertheless, there are a number of important rules deriving from these basic ones that can greatly enhance your skill. These *derived rules* fall into two general categories: those that can be used as shortcuts in a formal proof and those that serve primarily to round out our appreciation and understanding of the nature of logic itself. We make no such distinction here, however, preferring instead to classify these new rules according to their relationship to the five logical operations: implication, conjunction, disjunction, negation, and equivalence. This scheme provides a better focus for our discussion and serves as well as a means to organize the rules into convenient units—three new derived rules for each logical operation.

Throughout this section, the term "derived" is used in the strictest sense, meaning that we are able to establish the validity of the new rules by appealing only to the application of the original introduction and elimination rules. The latter, we recall, were simply accepted through an appeal to the reader's good common sense and intuition. By now, they have become second nature. This makes it easier to add to our knowledge, having first assured ourselves that our foundation is secure.

Goals

After studying this section, you should be able to:

- Discuss the notion of a *derived logical rule* and illustrate the method of proof used to confirm such a rule.

- Begin to develop an enhanced facility with the derivation of deductive logical proofs.
- State the derived rules involving the basic logical connectives, illustrate their derivation, and begin to incorporate them in the overall proof technique.
- Recreate Euclid's argument, which claims that there can be no largest prime.

Rules Involving Implication

The first two of our three derived implication rules simply express the fact that *implication is a reflexive and transitive relation* (see Section 1.4) on the class of logical statements.

Derived Implication Rules

[→A] $\dfrac{}{A \longrightarrow A}$ [→B] $\dfrac{A \longrightarrow B,\ B \longrightarrow C}{A \longrightarrow C}$

[→C] $\dfrac{A \longrightarrow B}{\neg B \longrightarrow \neg A}$

(We use this same A-B-C naming scheme throughout this section.) Note that there is no premise to the first of these rules. As a result, we are able to insert a statement of the form $p \rightarrow p$ anywhere in a proof, without regard to any of the previous entries. In fact, we have done this already in Example 2.26. The second rule is one in complete agreement with our common sense. Certainly if I say, "When it is raining, I always carry an umbrella," and "If I carry an umbrella, I never get wet," then as a consequence, "When it is raining, I never get wet." The third rule is closely related to the earlier rule of contraposition (CON), Section 2.1, in asserting that A implies B establishes that not B implies not A. Referring to the above illustration this would be expressed, "If I am not carrying an umbrella, then it is not raining." Note, however, that common sense is not sufficient here. We aim to show that these new rules (in fact, all 15 of them) can be derived from the basic introduction and elimination rules of Section 2.3.

Example 2.33 Suppose we first verify rule [→A]. As shown in the following proof in *un*abbreviated form

$$p \Rightarrow p$$
$$p$$
$$p \rightarrow p \qquad [\rightarrow\text{I}]$$

we first establish that $p \Rightarrow p$ (where the assumption leads to the conclusion in zero steps!), then use the rule [→I] to obtain $p \rightarrow p$ as required. ■

Example 2.34 Leaving the verification of [→B] as an exercise for the reader, we consider [→C] instead. The derivation is as follows:

$p \to q$	
$\neg q \to \neg p$	[→I]
$\quad \neg q$	
$\quad p \to q \wedge \neg q$	[→I]
$\quad\quad p$	
$\quad\quad q$	[→E]
$\quad\quad q \wedge \neg q$	[∧I]
$\quad \neg p$	[¬I]

Note again, that only our original introduction and elimination rules are used to verify the steps of the proof. Note as well that our proof is valid for any logical formulas p, q whatsover. ■

The reader must remember to keep in mind that one of the main objectives in our developing these derived rules is that of providing shortcuts in our proof technique.

Example 2.35 To illustrate this aspect of our development, suppose we wish to establish the following implication:

$$\neg p \to \neg(p \wedge q)$$

Recognizing the similarity in this structure to that in the conclusion of [→C], we first establish that $p \wedge q \to p$ with an appeal to Example 2.23. Altogether, our proof is then only two lines long:

$p \wedge q \to p$	[Example 2.23]
$\neg p \to \neg(p \wedge q)$	[→C]

We use the newly derived rule [→C] in the second step. Were it not for the shortcut provided by [→C], all of the steps of Example 2.34 would have had to have been repeated here. ■

Rules Involving Conjunction

The first two of our derived rules involving conjunction are reciprocal, and in fact, together they could be used to establish a logical equivalence of the two expressions $A \wedge B \longrightarrow C$ and $A \longrightarrow (B \longrightarrow C)$. The usefulness of this equivalence should not be underestimated. Keep in mind that one or the other of these two forms may appear as a desired conclusion of an argument, whereas the other form may somehow be easier to derive from the premise(s) at hand.

Derived Conjunction Rules

$[\wedge A]\quad \dfrac{A \wedge B \longrightarrow C}{A \longrightarrow (B \longrightarrow C)}$ $\qquad [\wedge B]\quad \dfrac{A \longrightarrow (B \longrightarrow C)}{A \wedge B \longrightarrow C}$

$[\wedge C]\quad \dfrac{A \wedge \neg A}{B}$

The last of these three rules is quite interesting, for in fact it claims, that "from an absurdity, one can conclude anything!"

Example 2.36 The proof of $[\wedge C]$ is reminiscent of that for $[\neg E]'$ as given in Example 2.25. In fact, we can use the earlier result to derive $[\wedge C]$ for arbitrary logical expressions p, q as follows:

$p \wedge \neg p$	
p	$[\wedge E]$
$\neg p$	$[\wedge E]$
q	$[\neg E]'$ ■

Rules Involving Disjunction

Again, we see an interesting similarity in the case of the first two of our derived rules involving disjunction, in fact, a kind of contrapositive relationship between the two. Moreover, each of these is seen to be a sort of generalization of the rule $[\vee E]$ of proof by cases.

Derived Disjunction Rules

$[\vee A]\quad \dfrac{A \vee B,\ A \longrightarrow C,\ B \longrightarrow D}{C \vee D}$

$[\vee B]\quad \dfrac{\neg A \vee \neg B,\ C \longrightarrow A,\ D \longrightarrow B}{\neg C \vee \neg D}$ $\qquad [\vee C]\quad \dfrac{\neg A \vee B}{A \longrightarrow B}$

The last of these three rules can be considered to be a new way of establishing an implication (rather than appealing to the direct method $A \Longrightarrow B$, followed by the use of $[\rightarrow I]$).

Example 2.37 We provide a derivation of [∨A] as follows:

$p \vee q$	
$p \to r$	
$q \to s$	
$p \to r \vee s$	[→I]
$\quad p$	
$\quad r$	[→E]
$\quad r \vee s$	[∨I]
$q \to r \vee s$	[→I]
$\quad q$	
$\quad s$	[→E]
$\quad r \vee s$	[∨I]
$r \vee s$	[∨E]

and the student should be able to imitate this argument in providing a derivation of [∨B]. ■

Rules Involving Negation

Our three derived rules for negation are each of interest in their own right. The first of these [¬A] sometimes goes by the name "reduction to an absurdity," and it seems to recall the argument in Russell's Paradox (Section 1.1). In any case, should A lead to not A by any sound method of reasoning, then surely A itself is absurd. At least, that is the claim here. (In Russell's Paradox, both the statement R and its negation were shown to be impossible on these grounds.) The second of our derived negation rules has already been discussed (Section 2.1) under the name *law of contraposition.*

Derived Negation Rules

[¬A] $\dfrac{A \longrightarrow \neg A}{\neg A}$ [¬B] $\dfrac{A \longrightarrow B, \neg B}{\neg A}$

[¬C] $\dfrac{A \vee B, \neg A}{B}$

Considering the last of these three rules, suppose I tell you that I am going to do one of two things. Then I later inform you that I am not going to do the first of these. You are then obliged to conclude that I will definitely choose the second of my options. Again, it's just good common sense.

Example 2.38 As expected, the proof of [$\neg$A] can be based on an appeal to our basic negation introduction rule as follows:

$$
\begin{array}{ll}
p \to \neg p & \\
p \to p \wedge \neg p & [\to\text{I}] \\
\quad p & \\
\quad \neg p & [\to\text{E}] \\
\quad p \wedge \neg p & [\wedge\text{I}] \\
\neg p & [\neg\text{I}]
\end{array}
$$

Note that it is the contradiction $p \wedge \neg p$ that is used here. ∎

Example 2.39 Euclid used the line of reasoning expressed in [$\neg$A] in his proof "there is no largest prime number." Let p be the assertion:

p: There is a largest prime number (call it N)

We then form the number

$$N' = (2 \cdot 3 \cdot 5 \cdot 7 \cdot \cdots \cdot N) + 1$$

where the product in parentheses comprises every prime number up to and including N. Since each of these primes leaves a unit remainder on dividing into N', it must be that each prime factor of N' is larger than N. So the assumption that N was the largest prime leads to the conclusion that it is not, that is, $p \to \neg p$, and the hypothesis of [$\neg$A] is satisfied. We thus conclude with Euclid that there is no largest prime, after all. ∎

Example 2.40 We find that we can use our alternate negation elimination rule [$\neg$E]′ to derive the equally interesting [$\neg$C], as follows:

$$
\begin{array}{ll}
p \vee q & \\
\neg p & \\
p \to q & [\to\text{I}] \\
\quad p & \\
\quad q & [\neg\text{E}]' \\
q \to q & [\to\text{A}] \\
q & [\vee\text{E}]
\end{array}
$$

Note the use, once again, of the reflexive property [$\to$A] of the implication relation. ∎

Example 2.41 In studying these derived rules, the student should be careful not to come away with the impression that these are the only rules that could have been derived. In fact, it might be a useful exercise to attempt to fashion additional rules of your own, guided by common sense and your new found experience, and then to attempt to show that such rules indeed follow from those already presented. Such is the case with the rule

$$\frac{\neg(A \vee B)}{\neg A \wedge \neg B}$$

one that we mention because of its importance later on. We establish its validity as follows:

$$
\begin{array}{ll}
\neg(p \vee q) & \\
p \rightarrow (p \vee q) \wedge \neg(p \vee q) & [\rightarrow\text{I}] \\
\quad p & \\
\quad p \vee q & [\vee\text{I}] \\
\quad (p \vee q) \wedge \neg(p \vee q) & [\wedge\text{I}] \\
\neg p & [\neg\text{I}] \\
q \rightarrow (p \vee q) \wedge \neg(p \vee q) & [\rightarrow\text{I}] \\
\quad q & \\
\quad p \vee q & [\vee\text{I}] \\
\quad (p \vee q) \wedge \neg(p \vee q) & [\wedge\text{I}] \\
\neg q & [\neg\text{I}] \\
\neg p \wedge \neg q & [\wedge\text{I}]
\end{array}
$$

The reader should observe the duality in the two halves of the proof. ■

Rules Involving Equivalence

In our last group of derived rules, we simply state that logical equivalence is, in the sense of Section 1.4, a true equivalence relation—reflexive, symmetric, and transitive, respectively.

Derived Equivalence Rules

[↔A] $$\frac{}{A \leftrightarrow A}$$ [↔B] $$\frac{A \leftrightarrow B}{B \leftrightarrow A}$$

[↔C] $$\frac{A \leftrightarrow B, B \leftrightarrow C}{A \leftrightarrow C}$$

Surely these results come as no surprise to the reader. In fact, the derivations are quite straightforward if we simply recall the definition of logical equivalence as given in Section 2.3.

EXERCISES 2.5

1 Establish the validity of the derived rules:

(a) $\dfrac{A \longrightarrow C, B \longrightarrow C}{A \vee B \longrightarrow C}$ (b) $\dfrac{C \longrightarrow A, C \longrightarrow B}{C \longrightarrow A \wedge B}$

Taken together with the results of Examples 2.22, 2.23, and Exercise 1 of Section 2.3, discuss the analogy with the characterization of unions and intersections as presented in Section 1.2.

2 Discuss the use of the rule [$\neg$A] in the following proof that there can be no integer between 0 and 1. Suppose there is. Then there is a smallest one, call it n, so that $0 < n < 1$. Multiplying through by n we obtain $0 < n^2 < n < 1$. But n^2 is an integer and smaller than n between 0 and 1. Hence n is not the smallest such integer.

3 Symbolize the following arguments using the suggested notation. Then derive the conclusion from the assumption(s) as a formal proof.

(a) Hydrogen is the lightest of the elements, so it is true that hydrogen is or is not the lightest element. (l)

(b) If the earth were a flat disk atop the back of a tortoise, as some have asserted, then Magellan and others would never have been able to circumnavigate the globe. But Magellan and others have been able to circumnavigate the globe. Therefore, it cannot be true that the earth is a flat disk on the back of a tortoise. (f, c)

(c) Because the fire department was called, it is obviously not the case that it was not called. (c)

(d) If the current is flowing, then the light is on. So if the light is not on, then the current is not flowing. (c, l)

(e) Insofar as (if) the laws of mathematics refer to reality, they are not certain. Hence, insofar as they are certain, they do not refer to reality. (r, c)

(f) This statement must be either true or false, and it is not true. Therefore, it is false. (t, f)

(g) If the enemy is fully prepared, we had better increase our strength. To increase our strength, we must conserve our natural resources. Therefore, if the enemy is fully prepared, we must conserve our natural resources. (p, i, r)

(h) If n is even, then n^2 is even, and if n is odd, then n^2 is odd; n is either odd or even. Thus n^2 is either odd or even. (e, e^2, o, o^2)

4 Provide a verification of the derived rule [$\rightarrow$B].

5 Provide a verification of the following derived rules.

(a) Rule [$\wedge$A] (b) Rule [$\wedge$B] (c) Rule [$\vee$B] (d) Rule [$\vee$C]

6 Provide a verification of the derived rule [$\neg B$].

7 Provide a verification of the following derived rules.

(a) Rule [$\leftrightarrow$A] (b) Rule [$\leftrightarrow$B] (c) Rule [$\leftrightarrow$C]

8 Provide proofs for each of the following.

(a) $p \vee q \rightarrow \neg(p \vee q) \Rightarrow \neg p \wedge \neg q$

(b) $p \vee q \rightarrow r, \quad \neg r \Rightarrow \neg p \wedge \neg q$

(c) $(p \wedge q) \wedge (r \wedge s), \quad \neg(p \wedge q) \Rightarrow r$

(d) $p \vee (q \wedge r), \quad p \rightarrow r \Rightarrow r \vee q$

9 Provide proofs for each of the following.

(a) $\neg p \vee \neg(q \vee r), \quad s \rightarrow p \Rightarrow \neg s \vee \neg r$

(b) $\neg p \vee q, \quad q \rightarrow r \Rightarrow p \rightarrow r$

(c) $\neg p \vee q \Rightarrow \neg q \rightarrow \neg p$

(d) $p \rightarrow [q \rightarrow (r \wedge s)] \Rightarrow p \wedge q \rightarrow s$

10 Provide proofs for each of the following.

(a) $(p \vee q) \wedge \neg(p \vee q) \Rightarrow p \vee q$

(b) $p \rightarrow q \vee r, \quad p \rightarrow \neg(q \vee r) \Rightarrow \neg p \vee q$

(c) $p \rightarrow q \wedge \neg q, \quad r \rightarrow p \Rightarrow \neg r$

(d) $p \rightarrow q \wedge \neg q, \quad p \vee r \Rightarrow r$

Sec. 2.6 CONSTRUCTIVE VS. CLASSICAL RULES

In most instances, it is preferable to have developed a "constructive" proof.

The reader must certainly acknowledge that we are now in the possession of a broad and effective proof technique, based on the formal presentation of the last three sections. Yet, there are certain *classical* methods of reasoning that are still beyond our grasp. This gap is intentional. By design, we have carefully chosen a set of introduction and elimination rules (Section 2.3) that will ensure the formulation of *constructive* proofs. In so doing, an entire train of thought has been hidden from view. We seek to correct this omission here, and to use the opportunity to show how the classical and constructive approaches may be contrasted. Then the student will always be aware of the distinctions and will thus be able to exercise good judgment in building the clearest possible argument in support of a given thesis. At the same time, our presentation will develop a critical faculty in reasoning about proofs and in reading the logical arguments of others, so that it can be said, "This argument is more convincing or more informative than another." These are the skills, after all, that show a maturity of understanding and the working of a critical mind.

Goals

After studying this section, you should be able to:

- State the double negation introduction and elimination rules.
- State the *law of the excluded middle*, derive its proof, and discuss its significance.
- Contrast the nature of *constructive* vs. *nonconstructive* proofs.
- Discuss the notions of a *universal contradictory statement* and a *universal valid statement*.

- List the *extended classical laws*, give samples of their proofs, and provide examples of their use.
- Explain how it is that two logical operations suffice in the classical system.
- Identify the axioms for *Boolean algebra* and give examples of algebraic systems satisfying these axioms.
- Outline the controversy between constructive and classical mathematics in a historical context.

Rules of Double Negation

We begin our discussion with the presentation of two mutually reciprocal **introduction and elimination rules of double negation**. One of these [¬¬I] we show to be a derived rule in the sense of Section 2.5; the other [¬¬E] is entirely new, and in fact, *impossible to derive* in our existing system!

$$[\neg\neg\mathrm{I}]\quad \frac{A}{\neg\neg A} \qquad\qquad [\neg\neg\mathrm{E}]\quad \frac{\neg\neg A}{A}$$

Example 2.42 To show that [¬¬I] is indeed derivable within the existing (constructive) logical system, we argue as follows:

p	
$\neg p \to p \wedge \neg p$	[→I]
$\quad\neg p$	
$\quad p \wedge \neg p$	[∧I]
$\neg\neg p$	[¬I]

for any logical statement p. Note the use of the single negation introduction rule, which uses the fact that $\neg p$ leads to an absurdity (under the assumption p). ■

Indeed it would be difficult to show that the companion rule [¬¬E] *cannot be derived* within the existing system. That would mean that no matter how long and how hard we try, no proof of this rule could ever be found! Nevertheless, it is known that this is the case, tantalizingly simple though the rule may seem to be.

To get just an idea as to why this may be so, consider the following analysis based on the Bishop-Brouwer interpretation of negation:

$\neg A$ means "A is absurd"

$\neg(\neg A)$ means "It is absurd that A is absurd"

However, there is as yet no way to conclude (from this last line) that A is valid. That is, a demonstration of the impossibility of the impossibility of a statement A does not constitute a proof that A is valid. At least, so say the constructivists (more on this later). A **constructive** point of view would insist on a direct logical deduction of A *from the existing rules* of Section 2.3 and those derived from them.

On the other hand, there is a certain way of thinking—that A is either true or false, according to some higher authority. If it is *not* true that A is *false* [i.e., if $\neg(\neg A)$], then A must be true, and in this sense, we accept the rule [$\neg\neg$E]. Note however, that this places a somewhat different interpretation on the whole meaning of negation. So far, we have said nothing about truth or falsity, but only of validity in the sense of some conclusion being derivable from our rules of deduction. If we accept this new interpretation of logic (the **classical** view), then we may still speak of validity or not (or alternatively, truth or falsity), but we operate within the enlarged logical system that includes *both* rules of double negation, [$\neg\neg$I] *and* [$\neg\neg$E]. Our first task is then to see where this may lead.

Law of the Excluded Middle

This law has already been spoken by Hamlet: "To be or not to be, that is the question" (as if there were no other alternatives). Indeed, we will find that a powerful proof technique is available if only this is so, if for *any* statement p, we can assert: Either p or not p. (Informally, we have already made use of this line of reasoning, but only on rare occasions.) As a law, it is a general conclusion without premises, a tautological sentence, true of any statement p, as long as we accept the double negation elimination rule.

Law of the Excluded Middle

$$A \vee \neg A \qquad \textbf{(EXM)}$$

The proof of this law is rather intricate, even within the extended classical deduction system where both double negation laws (particularly that of elimination) are accepted.

Example 2.43 The statement and its proof may be given as follows:

$p \vee \neg p$	
$\quad \neg(p \vee \neg p) \rightarrow p \wedge \neg p$	[$\rightarrow$I]
$\quad\quad \neg(p \vee \neg p)$	
$\quad\quad \neg p \wedge \neg\neg p$	[Example 2.41]
$\quad\quad p \wedge \neg p$	[$\neg\neg$E]
$\quad \neg\neg(p \vee \neg p)$	[$\neg$I]
$\quad p \vee \neg p$	[$\neg\neg$E]

Note the use of the double negation elimination law in the last line of the proof. More importantly, however, the reader should observe that we have argued somewhat improperly in the earlier use of [$\neg\neg$E], by taking a shortcut in our reasoning. It would be instructive for the reader to identify this gap in the proof and to fill in the missing steps (see Exercise 1). ■

In practice, the law of the excluded middle finds its main application in certain proofs by contradiction or **indirect proofs** as they are sometimes called. Note that this latter name already suggests that perhaps there is something suspicious in the whole procedure. Typically, the argument runs as follows. We want to assert some statement p. We argue (using the law of the excluded middle) that either p or not p, one or the other must be the case. We then suppose it is the latter, say. If this should lead to a contradiction, then we conclude that p must hold—even though we have done nothing directly that would lead us to assert the statement p. Formally, we outline the strategy as follows:

$$
\begin{array}{ll}
p \vee \neg p & \text{(EXM)} \\
\neg p \rightarrow q \wedge \neg q & [\rightarrow\text{I}] \\
\quad \neg p & \\
\quad \vdots & \\
\quad q \wedge \neg q & \\
\neg\neg p & [\neg\text{I}] \\
p & [\neg\text{C}]
\end{array}
$$

Note how the last line of the proof follows from the derived rule [$\neg$C], together with an implicit use of the commutative law (L2a) for the disjunction operation. Alternatively, it might be said that this last line follows from an appeal to [$\neg\neg$E], but that is not surprising. In fact, the reader may wish to show, as suggested in Exercise 2, that the validity of either (EXM) or [$\neg\neg$E] will imply the validity of the other.

Constructive vs. Nonconstructive Proofs

Such indirect arguments as outlined above are particularly bothersome (at least to some persons) when we use their conclusion to affirm the existence of an object of a certain kind in an infinite set.

Example 2.44 Consider what is known as a *perfect number*, that is, a positive integer that is equal to the sum of its divisors, other than itself, for example, $6 = 1 + 2 + 3$. These are the numbers 6, 28, etc., but so far, only the first 15 or 20 of these are known. Presumably there are others, but that is not important here. What is interesting is that all of the known perfect numbers are even, and yet, nobody has proved that there cannot be odd perfect numbers.

Now consider the statement

p: There is an odd perfect number

and suppose, as above, that we were somehow able to show that $\neg p$ (there are no odd perfect numbers) somehow leads to a contradiction. Then by indirect reasoning, we would presumably have established p. And yet, we are no closer to finding an odd perfect number than we were before! ■

This argument would bother some people (the constructivists), those who would take the *existence* of an object to mean that a finite procedure had been given for constructing it. The lesson (for all of us) is this: examine the proof of a statement as closely as you would the statement itself. Make a distinction between constructive and nonconstructive proofs, those that are merely existential. There is often a difference in meaning.

Extended Classical Laws

When we move to the extended classical system, we find there are additional laws of the logical calculus, beyond those given in Section 2.3, (L1a)–(L5b). Primarily, these are the ones that mirror the extended laws (S6a)–(S8b) of the calculus of sets (Section 1.2). When taken as a whole, the extended logical calculus provides a most effective machinery for simplifying logical expressions.

In speaking of a logical contradiction, for example, $q \wedge \neg q$, it is clear that one contradiction is as good as another. In fact, the derived rule [$\wedge$C] shows that we may deduce anything we want from a given contradiction—in particular, another contradiction. Thus it is sensible to introduce a **universal contradictory statement** 0, to stand for all of these. Similarly, we introduce a **universal valid statement** 1 ($p \to p$ is suggested here, because it requires no premises), so that we can assert the *defining properties*:

$$0 \longrightarrow A \longrightarrow 1$$

for any logical statement p in the role of A. Note that $0 \longrightarrow A$ is just another way of asserting rule [$\wedge$C]. In contrast, $A \longrightarrow 1$ is justified by our thinking of 1 as some statement (for example, $p \to p$) that is valid without any particular assumptions, as in rule [$\to$A]. In any case, we will see that 0 and 1 play the same role in the logical calculus as did $\emptyset$ and U in the calculus of sets. This suspicion is in fact confirmed by comparing the following laws (L6a)–(L8b) with (S6a)–(S8b) of Section 1.2.

L6a. $A \vee 0 \equiv A$	**L6b.** $A \wedge 0 \equiv 0$
L7a. $A \vee 1 \equiv 1$	**L7b.** $A \wedge 1 \equiv A$
L8a. $A \vee \neg A \equiv 1$	**L8b.** $A \wedge \neg A \equiv 0$

As expected, these extended classical laws follow quite easily from the defining properties of 0 and 1, together with the usual proof techniques.

Example 2.45 We prove the identity (L6a) as follows:

$$
\begin{array}{ll}
p \vee 0 \rightarrow p & [\rightarrow\text{I}] \\
\quad p \vee 0 & \\
\quad p \rightarrow p & [\rightarrow\text{A}] \\
\quad 0 \rightarrow p & [\text{Definition of } 0] \\
\quad p & [\vee\text{E}] \\
p \rightarrow p \vee 0 & [\rightarrow\text{I}] \\
\quad p & \\
\quad p \vee 0 & [\vee\text{I}] \\
p \vee 0 \leftrightarrow p & [\leftrightarrow\text{I}]
\end{array}
$$

making use of the defining property of 0 statement. ■

Example 2.46 Similarly, we may prove (L8a):

$$
\begin{array}{ll}
p \vee \neg p \rightarrow 1 & [\text{Definition of } 1] \\
1 \rightarrow p \vee \neg p & [\rightarrow\text{I}] \\
\quad 1 & \\
\quad p \vee \neg p & (\text{EXM}) \\
p \vee \neg p \leftrightarrow 1 & [\leftrightarrow\text{I}]
\end{array}
$$

using the defining property of 1 and the law of excluded middle. ■

Example 2.47 We should not underestimate the use of these new laws in simplifying expressions. For instance, we have:

$$
\begin{array}{ll}
[(p \vee (p \wedge r)) \wedge \neg q) \wedge \neg p \vee (r \wedge (\neg p \vee r)] & \\
\quad = (p \wedge \neg q) \wedge \neg p \vee (r \wedge (r \vee \neg p)) & \text{(L4a), (L2a)} \\
\quad = (\neg q \wedge p) \wedge \neg p \vee r & \text{(L4b), (L2b)} \\
\quad = \neg q \wedge (p \wedge \neg p) \vee r & \text{(L3b)} \\
\quad = \neg q \wedge 0 \vee r & \text{(L8b)} \\
\quad = 0 \vee r & \text{(L6b)} \\
\quad = r & \text{(L2a), (L6a)}
\end{array}
$$

in which we use the complete set of laws developed so far. Of course, such simplifications should be appreciated in the context of the discussion in Example 2.27. ■

The classical logical system affords yet another important dual pair of laws beyond those available in the original system, those that bear the name of the logician, Augustus deMorgan. As will be evident from their form, these laws are most useful in negating a disjunction or a conjunction:

L9a. $\neg(A \vee B) \equiv \neg A \wedge \neg B$ **L9b.** $\neg(A \wedge B) \equiv \neg A \vee \neg B$

When they are added to the law

L10. $\neg\neg A \equiv A$

---

we have a rather complete set of laws for handling negation. Note again the comparisons with the corresponding laws in the calculus of sets.

Example 2.48 The classical rules admit an equivalent formulation of the implication operation in terms of disjunction and negation; that is, we have:

$$A \longrightarrow B \equiv \neg A \vee B$$

an identity that strengthens the derived rule [∨C] given in Section 2.5. We first derive a deduction in one direction:

$p \rightarrow q$	
$p \wedge \neg q \rightarrow q \wedge \neg q$	[→I]
$\quad p \wedge \neg q$	
$\quad p$	[∧E]
$\quad q$	[→E]
$\quad \neg q$	[∧E]
$\quad q \wedge \neg q$	[∧I]
$\neg(p \wedge \neg q)$	[¬I]

and then in the other:

$\neg p \vee q$	
$p \rightarrow q$	[→I]
$\quad p$	
$\quad \neg\neg p$	[¬¬I]
$\quad q$	[¬C]

Together, these deductions can be used to establish the string of implications:

$$(p \rightarrow q) \rightarrow \neg(p \wedge \neg q) \equiv \neg p \vee \neg\neg q \equiv \neg p \vee q \rightarrow (p \rightarrow q)$$

showing that indeed, the given identity must hold. Note that our second derivation is the argument we would use to establish [$\vee$C]. ■

As a somewhat surprising corollary to this example (and other results we have already derived), we can show that only two logical operations are required in the classical logical system, for example, conjunction and negation will suffice. Of course, the reduced notational system is nowhere near as expressive.

Example 2.49 We can use the identity of Example 2.48, (L10), and DeMorgan's Laws to express disjunction, implication, and equivalence in terms of conjunction and negation alone, as follows:

$$p \vee q \equiv \neg\neg(p \vee q) \equiv \neg(\neg p \wedge \neg q)$$
$$p \rightarrow q \equiv \neg p \vee q \equiv \neg(\neg\neg p \wedge \neg q) \equiv \neg(p \wedge \neg q)$$
$$p \leftrightarrow q \equiv p \rightarrow q \wedge q \rightarrow p \equiv \neg(p \wedge \neg q) \wedge \neg(q \wedge \neg p)$$

Note that the first identity in the last line is an easy consequence of the introduction and elimination rules for equivalence. ■

In a closing remark, it is appropriate to note that laws (L1)–(L10) may be taken as axioms in defining the notion of a *Boolean algebra* (after George Boole, an English logician). We speak of a set of (primitive) objects, together with primitive operations $\vee$, $\wedge$, and $\neg$ that satisfy (L1)–(L10). We may then interpret:

objects = logical statements
$\vee$ = disjunction
$\wedge$ = conjunction
$\neg$ = negation

or alternatively:

objects = sets in a universe
$\vee$ = union
$\wedge$ = intersection
$\neg$ = complementation

and according to the analysis of our two opening chapters, we have in each case an example, an instance of a Boolean algebra. As a matter of fact, this is often the way that axiomatic mathematical structures are introduced—through generalization of the characteristic properties found in studying a specific mathematical system.

Brief Historical Comment

The distinction between classical and constructive mathematics can be traced all the way back to Kronecker, and most certainly to Brouwer, whose criticism of classical set theory and its ramifications led to a philosophical stance called "Intuitionism." The kernel of this philosophy is the notion of mathematics as a constructive activity, strongly based on the intuitively given natural numbers. It recognizes as fundamental the ability of the individual mathematician to perform a series of mental constructions, leading from one structure to another, perhaps unendingly. However, it strongly rejects the *completed infinity*, and in particular, the transfinite reasoning brought out in the classical axiomatic set theory, a theory much in contrast to that introduced in Chapter 1.

We have seen that an unbridled excursion into set theoretical constructions can lead to the paradoxes of self-contradictory statements. The intuitionists would say that it is unreasonable that our traditional logic—going back to the Greeks—should be applicable to such transfinite theories. Indeed, the observation that $R \in R$ or $R \notin R$ seems not to hold in the case of Russell's set R can be seen as an instance of the failure of the law of the excluded middle. The intuitionists reject the law of the excluded middle (or, equivalently as we have seen, the double negation elimination rule), particularly in its application to infinite situations—where they would claim that it was never meant to apply. An instance of their style of argument runs as follows.

The statement, "There is an odd perfect number between 1 and a 1 billion" must be either true or false. For such a decision in the case of a finite set, however large, we can in principle "run through" all the numbers and discover for ourselves which is the case. However, consider the broader statement: "There is an odd perfect number." Suppose it is true. Then presumably, if we start to run through the odd positive integers, we will eventually come to one that is perfect. Now suppose it is false. Does this mean, that if we run through *all* the odd integers, we will *never* come to one that is perfect? Or does it mean only that there is an inherent contradiction in the notion of an odd perfect number?

Consider a mathematical researcher who seeks to answer this broader question, once and for all. Realistically, he or she has two possible plans of attack. Since he or she can't look at all the odd integers in trying to establish the falsity of the above statement, he or she may instead seek to show by some constructive method of reasoning that the existence of an odd perfect number would somehow lead to a contradiction. Alternatively, he or she may try to construct an odd perfect number, believing perhaps that the statement is true. But the two alternatives: either

an odd perfect number can be constructed

or

the notion odd perfect number is absurd

seem not to have exhausted all the possibilities! We may never succeed in devising a procedure that will construct such a number, *and* we may never succeed in the proof that the existence of such a number leads to a contradiction. And in fact, the constructivists, latter day intuitionists, would argue, in general, that to believe otherwise is to adhere to the overly idealistic notion that every mathematical problem is solvable.

Certainly this latter position is an untenable one. Generally speaking, it would mean that we had a method that in finitely many purely routine steps would lead to a proof or a disproof of any mathematical assertion. We have no such method and furthermore, nobody believes that we ever will. Yet, this is one possible interpretation that can be placed on the classical view of mathematics—that there is a *Supreme Intellect* for which no unsolvable problems can exist. To this Supreme Intellect, the difficulties pointed out by the intuitionists do not exist.

We need not question the truth of this thesis. Ours is not to defend or to expound one philosophy over another, but only to make the student aware of the subtle distinctions that can exist between these two points of view—the constructive vs. the classical. In discrete mathematics, however, the constructive approach is usually the more natural. Many of the structures with which we deal are finite, where the difficult philosophical questions of which we spoke do not arise. When we do treat infinite structures, we most often approach the material in an algorithmic spirit, where again a constructive point of view is a necessary component to our thinking.

EXERCISES 2.6

1 Identify the shortcut in the proof of Example 2.43 and provide an expanded proof that is strictly correct, filling in the missing steps.

2 Show that the double negation elimination rule $[\neg\neg E]$ is logically equivalent to the law of the excluded middle. [*Hint:* Show that $[\neg\neg E]$ is derivable from (EXM).]

3 Find the next perfect number after 6 and 28, as listed in Example 2.44. (*Hint:* A computer program may be helpful here.)

4 Provide proofs for each of the following identities:

(a) Law (L6b) **(b)** Law (L7a) **(c)** Law (L7b) **(d)** Law (L8b)

5 Simplify the following expressions, justifying each step as in Example 2.47.

(a) $(r \wedge p) \vee (r \wedge q) \vee [(\neg p \vee \neg q) \wedge r] \vee \neg p$

(b) $\neg p \wedge [(r \vee q) \vee \neg p] \vee (q \wedge r) \vee (\neg r \wedge q)$

6 Provide proofs for each of the following identities.

(a) Law (L9a) **(b)** Law (L9b) **(c)** Law (L10)

In particular, show that (L9a) is constructive, but (L9b) is not.

7 Symbolize the following arguments using the suggested notation. Then derive the conclusion from the assumption(s) as a formal proof.

(a) It is certainly not the case that John's statement was not true. Therefore, his statement was true. (t)

(b) If line a is perpendicular to line b and line c is perpendicular to line b, then a is parallel to line c. Thus, we can infer that if a is not parallel to c, then either a is not perpendicular to b or c is not perpendicular to b. (a, c, p)

(c) Line a is either longer than b or shorter than b. If it is longer than b, it is not the same length as b, and if it is shorter than b, it is not the same length as b. Therefore, line a is not the same length as line b. (l, s, e)

(d) Angle a is a right angle if it measures exactly $90°$, but if it is less than $90°$ it is acute, and if it is greater it is obtuse. If it is not exactly $90°$, then it is either less or greater. Hence, if angle a is not a right angle, it is either acute or obtuse. (r, e, l, g, a, o)

(e) If two gases are at the same temperature, then their molecules have the same average kinetic energy. Equal volumes of two gases contain the same number of molecules. The pressure of two gases is equal if the number of molecules and their kinetic energies are equal. Thus if two gases have the same temperature and volume, they must have the same pressure. (t, e, v, n, p)

(f) If ABC is a triangle, then $AC = BC$ if and only if angle $CAB =$ angle CBA. If ABC is a triangle and is isosceles, then $AC = BC$ if AB is the shortest side. Hence, if ABC is an isosceles triangle and AB is the shortest side, then $AC = BC$ and angle $CAB =$ angle CBA. (t, l, a, i, s)

(g) We have $ad = bc$ if and only if $a/b = c/d$. If it is not the case that $a/b > c/d$, then either $a/b < c/d$ or $a/b = c/d$. If $a/b = c/d$ then it is not the case that either $a/b > c/d$ or $a/b < c/d$. Therefore, $ad = bc$ if and only if it is not the case that $a/b > c/d$ or $a/b < c/d$. (p, e, g, l)

(h) If triangle ABC is isosceles, then side AC equals side BC if AB is the shortest side. If ABC is isosceles and if M is the midpoint of AB, angle ACM equals angle BCM. Triangle ACM is congruent with triangle BCM if and only if angle ACM equals angle BCM, AC equals BC, and CM equals CM. M is the midpoint of AB, and of course, CM equals CM. If triangle ACM is congruent with triangle BCM, then angle CAB equals angle CBA. So if triangle ABC is isosceles and AB is the shortest side, then angle CAB equals angle CBA. (i, q, s, m, a, c, l, e)

(i) If $ABCD$ is a quadrilateral, and if it has four right angles and four equal sides, then it is a rectangle and a square. If the four sides are not equal, then it is not a square, and if the four angles are not equal, then it is neither a square nor a rectangle. Hence, either $ABCD$ is not a quadrilateral, or else, if it is a rectangle and has four equal sides, then it is a square. (q, a, e, r, s)

(j) If the universe is infinite and not expanding, and if its matter is distributed uniformly, then the number of stars at any given distance from the earth increases as the square of that distance. Now the light from any particular star decreases as the square of the distance; and if the number of stars at any given distance increases as the square of the distance, while the light from each source decreases by the same factor, then the light reaching the earth from all of the sources at any given distance is equal to that reaching the earth from any other given stellar distance whatsoever. If the universe is infinite, then if the light from any distance is equal to that from any other, then the night sky is not dark, but is ablaze with light. But the night sky is dark. Thus, we must infer that if the universe is infinite, and if matter is distributed uniformly, then the universe is expanding. (i, e, m, s, l, o, d, a)

8 Provide proofs for each of the following.

(a) $\neg p \to q \wedge \neg q \Rightarrow p$ **(b)** $p \to r, \quad \neg p \to s \Rightarrow r \vee s$

(c) $\neg(\neg p \vee q) \Rightarrow p$ **(d)** $\neg p \to q, \quad \neg q \wedge r \Rightarrow p$

9 Provide proofs for each of the following.

(a) $\neg(\neg p \vee q) \Rightarrow p \wedge \neg q$ **(b)** $\neg(p \vee \neg q) \Rightarrow q$

(c) $\neg(p \wedge \neg q), \quad p \Rightarrow q$ **(d)** $p \Rightarrow q \vee \neg q$

Sec. 2.7 *THE STATEMENT CALCULUS*

Boolean algebra is said to embody our laws of thought.

It is trivial to remark that upon completing a deduction, the result is deducible in a definite sense from the given assumption(s). The student must be warned, however, that the converse does not necessarily hold. If we have failed to find a deduction of a certain desired conclusion from a given premise, that does not prove that such a deduction does not exist. We may just not have tried hard enough. When we are unsuccessful or unlucky in such an attempt, we may try indefinitely and our continued failure may have no significance at all. Thus whereas deducibility can definitely be asserted when we are successful in applying the rules, something else is needed for asserting nondeducibility. That "something else" is the main subject of this section.

Having accepted the classical rules of logic—either because of a philosophical conviction that they do indeed mirror our thought processes or because we recognize that they have generally served us well in the past—we are led to a simple and effective means for establishing the truth or falsity of logical propositions in every configuration of their constituent propositional variables. This is the method of *truth-table evaluation*, one that the student will find occasion to use not only for the purpose just mentioned, but also for a wide variety of situations, particularly in computer science. We will find that this method is consistent with the deduction techniques studied previously, and in a sense, we will see that either one may be considered as an alternative means of analysis to the other.

Our discussion begins with an introduction to the *well-formed formulas* of the logical calculus, those propositions that are syntactically correct. In effect, these are the expressions that we are able to form in the decision-making apparatus of a programming language. The connection with our earlier studies of this chapter is then established when we speak of those formulas that are *tautologies*, uniform truths in a definite sense. These are seen to coincide with the *theorems*, those propositions that admit a deductive proof. We then possess the means for establishing nondeducibility, and in fact, all of those matters that are difficult to affirm in the deductive approach. In this way, we will have come full circle, establishing a most cohesive bond between two greatly differing points of view, and this bond will solidify the student's understanding of the basic theory of logic.

Goals

After studying this section, you should be able to:

- Decide whether a given logical expression is well formed.
- Describe the complete hierarchical system for the logical calculus, and illustrate its use in resolving ambiguity.
- Derive the characteristic tables for the complete set of logical operations and discuss their significance.
- Compute truth tables of given logical statements.
- Apply the *truth-table method* to the analysis of alternative statements in programming languages and everyday semantic consequences.
- Define the notion of a *tautology* and explain why it is that every theorem is a tautology.
- Discuss the concepts of *adequacy* and *consistency* of mathematical logic.

Well-Formed Formulas

We now return to the setting originally envisioned by George Boole in his development of a calculus, a method of reasoning with statements or propositions in a logical manner. The statements are composed, as we have said, of certain elementary constituents called **propositional variables** p, q, r, etc., together with the fundamental logical connectives, $\vee$, $\wedge$, $\neg$, $\rightarrow$, $\leftrightarrow$, used in forming complex statements from simpler ones. Rather than seeking to derive proofs of these compound statements, we now accept these as unambiguous declarative sentences whose truth or falsity might be determined solely on the basis of the truth or falsity of the constituent variables.

The propositional variables can be used to denote most anything, depending on the nature of the field of investigation. Because we cannot anticipate the application in advance, it is best to treat these variables abstractly, giving little thought to their intended meaning. Ordinarily, however, they are assumed to be so elementary as to not be subject to further decomposition. Using a prescribed set of logical connectives as above, we then agree that a **well-formed formula** (abbreviated wff), i.e., a syntactically correct expression of our statement calculus, is either

1. One of the constants 0 or 1;
2. A propositional variable (standing alone); or

3. A statement for wff's a and b of the form
   (a) $a \vee b$
   (b) $a \wedge b$
   (c) $\neg a$
   (d) $a \rightarrow b$
   (e) $a \leftrightarrow b$
   (f) (a)

Example 2.50 By way of illustration, suppose we demonstrate that the expression

$$(p \wedge \neg q \vee r) \wedge p$$

is well formed. In a series of statements, we may assert:

p, q, and r are wff's	(2)
$\neg q$ is a wff	(3), part c
$p \wedge \neg q$ is a wff	(3), part b
$p \wedge \neg q \vee r$ is a wff	(3), part a
$(p \wedge \neg q \vee r)$ is a wff	(3), part f

and finally, the given expression is well formed, again, because of part b in (3). ■

Example 2.51 In (3), part f, the parentheses are used to remove potential ambiguities. For example, we wouldn't know whether to interpret.

$$p \wedge q \vee r$$

as $(p \wedge q) \vee r$ or as $p \wedge (q \vee r)$, and parentheses are best inserted for clarity. Alternatively, a hierarchical convention could be introduced that would settle such questions. For instance, we could insist, as it is normally understood, that $\wedge$ has a higher priority than $\vee$, so that the above sentence would necessarily be interpreted as $(p \wedge q) \vee r$. We would then have to insert parentheses in $p \wedge (q \vee r)$ if the second interpretation were intended. ■

More generally, we can introduce a complete hierarchical system that will rank all of our logical symbols according to their *priority*, once and for all. Such a system is described in Table 2.1. Then we use parentheses in the usual way, if we wish to override the priorities that this table imposes. Furthermore, we can agree that in the case of a tie, as could happen only for two identical symbols,

TABLE 2.1

Symbol	$\leftrightarrow$	$\rightarrow$	$\vee$	$\wedge$	$\neg$
Priority	1	2	3	4	5

the occurrence furthest to the left is to take precedence. In fact, this sort of tie-breaking rule is imposed in the parsing of the logical expressions of a computer-programming language. In any case, we will assume that such a hierarchical system has been introduced to ensure an unambiguous interpretation of all of our logical statements.

Example 2.52 We may check that the expression

$$p \wedge \neg q \vee r \wedge p$$

is well formed, as in Example 2.51. But how is it to be interpreted? According to our hierarchical system, the negation has the highest priority, so the given expression is understood as if it were written:

$$p \wedge (\neg q) \vee r \wedge p$$

but even this presents a problem. Among the three remaining operators, the conjunctions have a higher priority, and furthermore, the conjunction on the left takes precedence because of the tie-breaking rule. Thus our expression is interpreted as if it had been parenthesized in the form:

$$[p \wedge (\neg q)] \vee (r \wedge p)$$

Note the distinction between this form and that originally given in Example 2.50:

$$(p \wedge \neg q \vee r) \wedge p$$

There it is partially parenthesized, but the rest of the interpretation is left to the reader's good sense. ■

Characteristic Tables

Having provided a rather complete syntactical basis for the formation of logical statements, we can now begin to give a semantics, that is, *a meaning*, to associate with each well-formed formula. We can accomplish this step by step, basing the meaning on the form of the statement itself, and deriving it from elementary *characteristic tables* of the fundamental logical operations. The meaning—the truth or falsity of a statement—should depend entirely on the truth or falsity of the constituent variables.

Example 2.53 Consider the following propositional variables and their accompanying interpretations:

p: It is raining

q: The sun is shining

The complex statement $p \wedge q$, that is,

It is raining and the sun is shining

should be true only if p is true *and* q is true (and we should then expect a rainbow). ■

As a matter of fact, Example 2.53 is representative of a general observation, that the *conjunction operation* on the set $B = \{0, 1\}$ should have the characteristic behavior of Table 2.2(a), that is,

$$\begin{aligned} 0 \wedge 0 &= 0 \\ 0 \wedge 1 &= 0 \quad \text{(the upper right-hand entry)} \\ 1 \wedge 0 &= 0 \quad \text{(the lower left-hand entry)} \\ 1 \wedge 1 &= 1 \end{aligned}$$

TABLE 2.2

(a)

$\wedge$	0	1
0	0	0
1	0	1

(b)

	$\neg$
0	1
1	0

(c)

$\vee$	0	1
0	0	1
1	1	1

(d)

$\rightarrow$	0	1
0	1	1
1	0	1

(e)

$\leftrightarrow$	0	1
0	1	0
1	0	1

Here, the column on the left expresses possible arguments

$$\begin{aligned} 0 &= \text{false} \\ 1 &= \text{true} \end{aligned}$$

for p, and similarly, the row at the top represents possible arguments for q. The interior entries of Table 2.2(a) provide the resulting **truth value** (0 or 1) of $p \wedge q$, that is, they indicate whether $p \wedge q$ is false or true, respectively, for the various combinations of argument values for p and q. However, it can be shown that such observations are not merely consistent with our intuitive common sense. They are in fact derived from our earlier studies in this chapter. The same can be said of our expectation for the characteristic behavior of the unary *negation operation*, that is,

$$\begin{aligned} \neg 0 &= 1 \\ \neg 1 &= 0 \end{aligned}$$

See Table 2.2(b). As before, the meaning here is that $\neg p$ is true when p is false, and conversely. In fact, we will find that this behavior is a natural consequence of all that has preceded.

Example 2.54 Using the logical laws of Section 2.6, (L6a)–(L8b), we compute the entries in our characteristic table for conjunction:

$$\begin{aligned} 0 \wedge 0 &= 0 && \text{(L6b)} \\ 0 \wedge 1 &= 0 && \text{(L7b)} \\ 1 \wedge 0 &= 0 \wedge 1 = 0 && \text{(L2b)} \\ 1 \wedge 1 &= 1 && \text{(L7b)} \end{aligned}$$

as listed in Table 2.2(a). Then, in reference to the characteristic table for negation, we first note that

$$1 \wedge \neg 1 = 0 \qquad \text{(L8b)}$$

from which it follows from the above that $\neg 1 = 0$. Using this result, we may then compute:

$$\neg 0 = \neg\neg 1 = 1 \qquad \text{(L10)}$$

as listed in Table 2.2(b). ■

It is interesting to observe that a complete set of characteristic tables for all of the logical operations can be obtained from those for conjunction and negation alone. From Example 2.49 we recall the following identities:

$$\begin{aligned} p \vee q &= \neg(\neg p \wedge \neg q) \\ p \to q &= \neg(p \wedge \neg q) \\ p \leftrightarrow q &= \neg(p \wedge \neg q) \wedge \neg(q \wedge \neg p) \end{aligned}$$

We see easily that these can be used to derive the characteristic tables for the operations of *disjunction*, *implication*, and *equivalence*, as shown in Table 2.2(c), (d), (e). These three parts of the table can be expressed in words as follows:

- A disjunction is true when either (or both) of its arguments is true.
- An implication is always true, except when its antecedent is true and its consequent is false.
- An equivalence is true when its arguments agree in their truth value.

Note once again the fact that it is the *inclusive* or that we must acknowledge in the case of the disjunction operation. Finally, we note that the meaning of the implication operation as given here sometimes seems a bit unusual. When we say,

"If the moon is blue, then today is Friday," the antecedent is (presumably) false. Because the one condition under which an implication can be false is not fulfilled, however, we must regard this implication as true. Such behavior may seem a bit odd. However, what we ask of an implication, in general, is whether it is true that whenever the antecedent is true, then so is the consequent. Thus there is never a problem (of truth) when the antecedent is false.

Example 2.55 We use this interpretation in everyday speech, as in our saying:

If this is a sunny day, then I'm a monkey's uncle. ■

Tautology and Truth Table

As we have said, the structure of an arbitrary logical statement, built up from constants and propositional variables through the fundamental operations, will allow us to associate a meaning to each statement. We will see, however, that this meaning is merely a tabular representation of the truth or falsity of the given logical statement in all possible configurations of (the truth or falsity of) its constituent variables. If we assume a configuration of the propositional variables, taking on truth values 0 or 1, then we are able to use the characteristic tables of the logical operations to obtain a corresponding truth value for the statement as a whole, regardless of its complexity. This is the important method of **truth table evaluation**.

Example 2.56 To illustrate the truth table computation method, we derive the table for the expression:

$$\neg[(p \wedge \neg q \vee r) \wedge p] \vee p \wedge q$$

in a sequence of steps as outlined in Table 2.3 that are built up from the tables of the constituent subexpressions as shown. ■

TABLE 2.3

p	q	r	$\neg q$	$p \wedge \neg q$	$(p \wedge \neg q \vee r)$	$(\ ) \wedge p$	$\neg[(\ ) \wedge p]$	$p \wedge q$	*expression*
0	0	0	1	0	0	0	1	0	1
0	0	1	1	0	1	0	1	0	1
0	1	0	0	0	0	0	1	0	1
0	1	1	0	0	1	0	1	0	1
1	0	0	1	1	1	1	0	0	0
1	0	1	1	1	1	1	0	0	0
1	1	0	0	0	0	0	1	1	1
1	1	1	0	0	1	1	0	1	1

This example should be studied quite carefully. First of all, it should be observed that the number of rows in the truth table is governed by one of our fundamental counting results from Section 1.5, that is, the fact that $|B^n| = 2^n$. Moreover, the elements of the set B^n are the original row entries of the table, on which all subsequent computations are based. (Here, n is the number of propositional variables involved in the analysis.) Finally, the reader should observe that the column for the expression as a whole is determined by having first computed the columns to determine its constituent subexpressions, using the characteristic tables derived earlier.

Now we said we were going to talk about meaning. But what is the meaning of the statement in Example 2.56? Perhaps the best we can say is that the statement is sometimes true and sometimes false, depending on the truth or falsity of its propositional variables. That certainly seems sensible. However, it opens the door to an important special case. We say that a logical expression is a **tautology** if it happens that its truth table (as a column) consists entirely of ones. Obviously, this is quite unusual, but surely it happens. (Compute the truth table of $p \vee \neg p$!) Such expressions are then uniformly true, or true in all configurations of their constituent propositional variables. To fully appreciate the significance of tautologies, we return to a further discussion of the notion of a tautology later in this section.

An important application for the truth table method of analysis occurs in relation to the decision-making apparatus of a programming language. Typically, such a language will provide for *alternative statements* of the form (refer to Example 2.27):

if e then S else T

where one of the actions S or T is to be taken, respectively, depending on the truth value (1 or 0) of the logical expression e. This expression will ordinarily be composed of logical constituents, p, q, r, etc., and will be built up from these using the conventional logical operations, $\vee, \wedge, \neg$, etc. Then again, in order to know whether the action S or T will be taken in the course of a program's execution, it is necessary to know the truth value of e at the point where the alternative statement is encountered.

Example 2.57 In a programming language like Pascal, we are likely to encounter statements of the form:

```
if x ≠ 3 or y < 0 then
   S
else
   T
```

and if we identify p, q as the elementary statements:

$$p: x = 3$$
$$q: y < 0$$

then we are better able to analyze the given program statement if we first compute the truth table as shown in Table 2.4. We are then able to say that statement S will be executed in all cases except that where $x = 3$ and $y \geq 0$ (in which case T

TABLE 2.4

p	q	$\neg p$	$\neg p \vee q$
0	0	1	1
0	1	1	1
1	0	0	0
1	1	0	1

will be executed). Of course, the analysis is only useful if we know the values of x and y at the point in the program where our alternative statement is encountered. And again, the analysis is not all that difficult here. However, imagine if our logical expression had been that of Example 2.56. ■

Semantic Implication

One further argument for the usefulness of the truth table evaluation technique has yet to be mentioned. Suppose we have a collection of wff's $a_1, a_2, \ldots, a_n$, and the claim is made that a further statement b is a necessary **semantic consequence** of the a's. We can take this to mean that the compound statement

$$a_1 \wedge a_2 \wedge \cdots \wedge a_n \to b$$

is a tautology. Again, whether or not this is indeed the case can be checked by a purely mechanical procedure: Compute the truth table of $a_1 \wedge a_2 \wedge \cdots \wedge a_n \to b$ to see if it consists entirely of ones. If so, then we agree that b is a semantic consequence of the a's, otherwise not. Such a procedure can be used to analyze complex arguments, legal documents, political pronouncements, insurance policies, and the like.

Example 2.58 (Lewis Carroll) Below there are given three statements involving:

p = terriers
q = zodiac wanderers
r = comets
s = curly tailed objects

namely:

(a) No terriers wander among the signs of the zodiac.

(b) Nothing that does not wander among the signs of the zodiac is a comet.

(c) Nothing but a terrier has a curly tail.

and we are asked whether the statement:

(d) No comet has a curly tail

is a semantic consequence of these. Translating Lewis Carroll's language into the formal language of mathematical logic, we have:

$$\begin{array}{l} a = q \to \neg p \\ b = \neg q \to \neg r \\ c = s \to p \\ \hline d = s \to \neg r \end{array}$$

and according to our discussion above, d is a semantic consequence of a, b, c in the event that $(a \wedge b \wedge c) \to d$ is a tautology. Computing the truth table of this expression in stages, as before, we obtain Table 2.5. A column of ones is obtained in the end. Note that this last result depends solely on the fact that no pair of entries $(1, 0)$ is found in the $a \wedge b \wedge c$ and d columns, recalling the characteristic table of the implication operation. Because our implication here is true, the example shows the relevance of terriers to astronomy (or is evidence of Lewis Carroll's fine sense of humor). ■

TABLE 2.5

p	q	r	s	a	b	c	$a \wedge b \wedge c$	d	*expression*
0	0	0	0	1	1	1	1	1	1
0	0	0	1	1	1	0	0	1	1
0	0	1	0	1	0	1	0	1	1
0	0	1	1	1	0	0	0	0	1
0	1	0	0	1	1	1	1	1	1
0	1	0	1	1	1	0	0	1	1
0	1	1	0	1	1	1	1	1	1
0	1	1	1	1	1	0	0	0	1
1	0	0	0	1	1	1	1	1	1
1	0	0	1	1	1	1	1	1	1
1	0	1	0	1	0	1	0	1	1
1	0	1	1	1	0	1	0	0	1
1	1	0	0	0	1	1	0	1	1
1	1	0	1	0	1	1	0	1	1
1	1	1	0	0	1	1	0	1	1
1	1	1	1	0	1	1	0	0	1

Consistency and Adequacy

We have presented two separate methods for the logical analysis of propositional formulas f—the deductive method of proof and the truth table method of evaluation. Recall that we say that f is a theorem if its validity can be deduced within the (classical) system of rules. On the other hand, we say that f is a tautology if it yields a truth table consisting entirely of ones. It may come as a surprise to the reader to learn that these two notions coincide! In words,

---

Every theorem is a tautology, that is, is true

and

Every tautology is a theorem, that is, is provable

---

so that there is no difference, when all is said and done.

The first assertion, "every theorem is a tautology" is called the **claim of consistency** for the deductive theory. This is best appreciated by our understanding that the deductive rules transmit truth from their premises to their conclusion. The converse, "every tautology is a theorem" is called the **notion of adequacy** for the deductive theory; the deductive rules are adequate to prove any given tautological statement. This is more difficult to understand. In part, this is so because we are likely to overestimate the meaning of tautology, to think that we are saying something more. We are not referring here to all truths in some broad metaphysical sense, but only to those logical statements of our limited calculus that just happen to evaluate as all ones for each configuration of the constituent variables. Even so, the argument that shows that all such statements indeed admit a proof (whether we have found it or not) in our deductive system, is quite a difficult one, and we do not attempt such a presentation here.

On the other hand, the question naturally arises as to whether this fundamental result does not render our deductive apparatus obsolete. Because the truth table evaluation technique is a purely mechanical test for tautology (= theoremhood), why bother with the deduction rules at all? The answer is simply this. Although a truth table may indeed tell us *that* a certain conclusion may be derived from certain premises, it cannot tell us *how* to reach that conclusion. Unless we actually construct a proof, we will never see the connection or the line of reasoning that makes the conclusion inevitable.

The best position to take is that both of our methods of analysis have their place—for one purpose or another. Certainly there is no denying the value in the truth table technique for establishing the fact that no proof of a given formula f can exist. If we compute the truth table for f and we find that f is not a tautology, then by what we have just learned, f is not a theorem either. And we can then cease all our attempts to find a proof. We said earlier that something else (beyond our proof technique) is needed for asserting nondeducibility—and that "something" is definitely our truth table method of analysis.

Example 2.59 As we have said, every theorem is a tautology. To illustrate this concept, consider the theorem:

$$p \rightarrow q \leftrightarrow \neg p \vee q$$

derived from the result of Example 2.48. In a two-variable truth table, we compute successively the tables for the expressions $p \rightarrow q$, $\neg p$, $\neg p \vee q$, and finally, that for the statement of the theorem itself, as shown in Table 2.6. We note that the final column indeed represents a tautology. Note again how we must use the characteristic tables of the individual logical operations. ■

TABLE 2.6

p	q	$p \to q$	$\neg p$	$\neg p \vee q$	*statement*
0	0	1	1	1	1
0	1	1	1	1	1
1	0	0	0	0	1
1	1	1	0	1	1

Example 2.60 Conversely, we have also learned that every tautology is a theorem (that is, has a deductive proof in the classical logical system). To appreciate the meaning of this statement, the student should consider the tautology:

$$(q \to \neg p) \wedge (\neg q \to \neg r) \wedge (s \to p) \to (s \to \neg r)$$

derived in Example 2.58. Because it is a tautology, it should have a proof. Using the conjunction as a premise, we provide a skeleton of the proposed argument:

$$
\begin{array}{ll}
(q \to \neg p) \wedge (\neg q \to \neg r) \wedge (s \to p) & \\
q \to \neg p & [\wedge \mathrm{E}] \\
\neg q \to \neg r & [\wedge \mathrm{E}] \\
s \to p & [\wedge \mathrm{E}] \\
s \to \neg r & [\to \mathrm{I}] \\
\quad s & \\
\quad \vdots & \\
\quad \neg r &
\end{array}
$$

Note that we write $[\to\mathrm{I}]$ in the hope that we will be able to deduce that $s \Rightarrow \neg r$ as indicated. We leave the completion of the missing steps as an exercise for the student. (See Exercise 8.) ■

EXERCISES 2.7

1 Decide whether or not the following expressions are *well formed.*

(a) $p \vee q \wedge \neg r \neg p$

(b) $(r \wedge p) \vee (r \wedge q) \vee [(\neg p \vee \neg q) \wedge r] \vee \neg p$

(c) $\neg p \wedge [(r \vee q) \to \neg p] \vee (q \wedge r) \leftrightarrow (\neg r \wedge q)$

(d) $(r \wedge \neg q) \to \neg p) \vee 0 \wedge \neg 1$

In the case of wff's, show specifically how the formula is built up out of (1), (2), (3) in the text.

2 Parenthesize the following expressions according to the strict hierarchical system of the text.

(a) $r \wedge \neg q \leftrightarrow \neg p \vee 0 \wedge \neg 1$

(b) $p \wedge \neg q \vee r \wedge \neg q$

(c) $q \vee p \wedge \neg r \rightarrow s$

(d) $q \leftrightarrow q \wedge \neg r \vee s \rightarrow p$

3 Show in detail how the characteristic tables are determined:

(a) For disjunction, using $p \vee q = \neg(\neg p \wedge \neg q)$

(b) For implication, using $p \rightarrow q = \neg(p \wedge \neg q)$

(c) For equivalence, using $p \leftrightarrow q = \neg(p \wedge \neg q) \wedge \neg(q \wedge \neg p)$

4 Derive truth tables for the following logical expressions.

(a) $[(p \rightarrow q) \vee (p \rightarrow r)] \wedge (\neg q \vee r)$

(b) $(\neg q \vee p) \vee (q \wedge \neg r) \vee (\neg p \wedge r)$

(c) $[(q \rightarrow r) \wedge (q \rightarrow p)] \vee (r \wedge \neg q)$

(d) $[p \vee (q \wedge \neg r)] \wedge [(q \wedge p) \vee \neg r]$

(e) $[(p \wedge s \rightarrow q \vee r) \vee (q \wedge \neg r) \vee (r \wedge \neg s)] \wedge \neg p$

(f) $(r \vee s \leftrightarrow p \wedge \neg q) \wedge (r \wedge s \rightarrow \neg p \vee \neg s)$

Which (if any) are tautologies?

5 Derive truth tables for the following logical expressions.

(a) $(p \wedge q) \rightarrow (\neg p \vee r)$

(b) $(p \rightarrow \neg q) \leftrightarrow (r \rightarrow q \wedge \neg p)$

(c) $(p \vee r \rightarrow p) \wedge (q \rightarrow q \vee r)$

(d) $(p \wedge \neg q) \rightarrow (\neg\neg p \leftrightarrow q)$

(e) $(p \leftrightarrow q) \rightarrow \neg[\neg(\neg r \wedge p) \wedge (q \vee r)]$

(f) $(q \rightarrow r) \rightarrow [(p \rightarrow q) \rightarrow (p \rightarrow r)]$

Which (if any) are tautologies?

6 (Lewis Carroll) Show that from the three statements:

(a) Nobody who really appreciates Beethoven fails to keep silence while the Moonlight Sonata is being played.

(b) Guinea pigs are hopelessly ignorant of music.

(c) No one who is hopelessly ignorant of music ever keeps silence while the Moonlight Sonata is being played.

the further statement:

(d) Guinea pigs never really appreciate Beethoven.

follows as a semantic consequence.

7 Show that the expression obtained in Exercise 6 is a theorem by providing a formal proof of its validity.

8 Complete the argument of Example 2.60, filling in the missing steps.

9 Show that the following expressions are tautologies.

(a) $p \wedge q \rightarrow p \vee q$

(b) $p \wedge (q \wedge r) \rightarrow p \wedge q$

(c) $p \rightarrow (q \vee \neg q)$

(d) $(q \rightarrow r) \rightarrow [(p \rightarrow q) \rightarrow (p \rightarrow r)]$

10 Use truth tables to decide which of the following pairs of expressions are logically equivalent.

(a) $(p \wedge \neg q) \rightarrow (\neg\neg p \leftrightarrow q)$ and $p \vee \neg q$

(b) $r \rightarrow (\neg q \vee p)$ and $p \wedge \neg r$

(c) $(p \rightarrow \neg q) \leftrightarrow (r \rightarrow (\neg q \vee p))$ and $q \vee (\neg r \wedge \neg p)$

(d) $p \vee \neg q$ and $\neg p \wedge \neg q$

[*Hint:* If they are logically equivalent, shouldn't their truth tables be identical?]

Chapter Three

DISCRETE NUMBER SYSTEMS

Contents

3.1 The Natural Numbers
3.2 Induction and Recursion
3.3 Divisibility of Integers
3.4 Base Conversion
3.5 The Rational Numbers
3.6 Computer Arithmetic

Leopold Kronecker (1823–1891)

Born into a prosperous Prussian family, Leopold Kronecker received his doctorate at the University of Berlin in 1845. After spending several years in the banking business, he voluntarily assumed academic duties at Berlin, where he distinguished himself with his beautiful lectures and publications, the latter still regarded as an inexhaustible source of ideas. The climax of Kronecker's career was his prolonged philosophical disagreement with the leading mathematical analysts of his time. Kronecker was a born algebraist. He wanted to find the "natural way," that is, to use intuition and taste rather than a convenient scientific technique, and thus cause his mathematics to reflect his own unique philosophical inclination. In this pursuit, he led a revolt [culminating in the ideas of Brouwer (see Chapter 2)] whose full impact has perhaps yet to be felt.

Writings

All results of the profoundest mathematical investigation must ultimately be expressible in the simple form of properties of the integers.

Collected Works **(Kronecker)**

Kronecker was what is called an "algorist" in most of his works. He aimed to make concise, expressive formulas tell the story and automatically reveal the action from one step to the next so that, when the climax was reached, it was possible to glance back over the whole development and see the apparent inevitability of the conclusion from the premises. Details and accessory aids were ruthlessly pruned away until only the main trunk of the argument stood forth in naked strength and simplicity. In short, Kronecker was an artist who used mathematical formulas as his medium.

Much of Kronecker's work has a distinct arithmetical tinge, either of rational arithmetic or of the broader arithmetic of algebraic numbers. Indeed, if his mathematical activity had any guiding clue, it may be said to have been his desire, perhaps subconscious, to arithmetize *all of mathematics, from algebra to analysis. "God made the integers," he said, "all the rest is the work of man." Universal arithmetization may be too narrow an ideal for the luxuriance of modern mathematics, but at least it has the merit of greater clarity than is to be found in some others.*

In an atmosphere of confident belief in the soundness of analysis, Kronecker assumed the unpopular role of the philosophical doubter. (In fact,) the most original part of his work, in which he was a true pioneer, was a natural outgrowth of his philosophical inclinations. And when he contemplated mathematics from this questioning point of view he did not spare it because it happened to be the field of his own particular interest, but infused it with an acid, beneficial skepticism. Although but little of this found its way into print it annoyed some of his contemporaries intensely and it has survived. The doubter did not address himself to the living, but, as he said, "to those who shall come after me."

Kronecker, The Doubter **(E. T. Bell)**

Sec. 3.1 THE NATURAL NUMBERS

God made the integers; all the rest is the work of man.

Nowhere is this statement more clearly represented than in the derivation of an axiomatization of the natural numbers. Embedded in this axiomatization is the germ of the important principle of mathematical induction. This principle in turn is the foundation of all of the recursive techniques of analysis, so important for their applications to computer science, and in fact to all areas of the mathematical sciences. The student should study this material quite thoroughly, as the principles involved are of great importance in our later work.

Goals

After studying this section, you should be able to:

- Discuss the axioms for the natural numbers, with particular attention to the constructive aspect of the characterization.
- Describe the inductive feature of the axiomatization of the natural numbers.
- List the familiar laws of arithmetic, showing how they derive from the axiomatization.
- State the *Well-ordering Principle* for the integers and provide examples of its use.
- Discuss the *Archimedean Principle* of the integers.
- Provide an *inductive definition* of the operations of ordinary arithmetic.
- Describe the *extended number system* (including zero and the negative integers) and show that the usual arithmetic properties are valid in the extended system.

Axioms for the Natural Numbers

The object of our present concern—the natural numbers—offer a most appropriate chance to review the axiomatic method as discussed in Section 2.1. The discussion should be quite easy to follow, especially because we are familiar with the resulting system since the days of our earliest schooling. In fact, some may feel that we are all brought into this world with an intuitive appreciation for the number concept, quite apart from any knowledge we may have gained along the way. Certainly, if robins can count to three or four in keeping track of their eggs, it may be reasonably assumed that we might do as well with our own native abilities. So it must have seemed to Kronecker. And this assumption must underlie all of our subsequent discussion.

Moreover, we again refer to the constructive aspect of mathematics. In the treatment we present here, we do not conceive of the set of natural numbers:

$$N = \{1, 2, 3, \ldots\}$$

as a completed whole, but rather as a set continually in the process of formation, from one number to the next. In fact, the whole idea of the *successor* of a given number (moving from one number to the next) is taken as a primitive notion to the axiomatization—the notion of a number is itself a primitive term of the system. It is from this standpoint that our axioms for the natural numbers are framed. We write n^+ to denote the **successor** of n, but to facilitate understanding at the start, the reader may wish to (mentally) replace n^+ by $n + 1$ in his or her thinking; that is what it will amount to in the end. Now we may introduce five axioms as follows:

N1. $1 \in N$

N2. $n \in N \Rightarrow n^+ \in N$

N3. $n^+ \neq 1$

N4. $n^+ = m^+ \Rightarrow n = m$

N5. $X \subseteq N$ satisfying (N1), (N2) with respect to $X \Rightarrow X = N$

---

where in (N5) we mean that if the subset X satisfies

(N1) $1 \in X$

(N2) $n \in X \Rightarrow n^+ \in X$

then necessarily it is the whole set N.

The first axiom simply states that 1 is a natural number. In (N2) we assert that if n is a natural number, then so is its successor n^+ (think $n + 1$). In (N3), we insist that no number can have 1 as its successor—after all, we are starting at 1 and counting upwards. The fourth axiom states that successors are uniquely determined. It would indeed be unsettling if it were otherwise, say $n \neq m$, but $n^+ = m^+$ for some numbers n and m. Finally, there is the most important axiom (N5), stating that any subset X of the natural numbers that happens to contain 1, and moreover, contains with each member, also its successor, must coincide with N. Certainly this seems reasonable; that's what N is all about. However, let us look at the situation more carefully by way of an example.

Example 3.1 Suppose we were to introduce our own successor function

$$n \to n^+$$

in such a way that $n^+ = n + 2$. Then the natural numbers would be partitioned into odd and even numbers, that is, we would have the following:

$$1^+ = 3, \quad 3^+ = 5, \quad 5^+ = 7, \text{ etc.}$$

and

$$2^+ = 4, \quad 4^+ = 6, \quad 6^+ = 8, \text{ etc.}$$

and it would then be possible to have a subset, that is, $X = \{\text{odds}\}$ meeting the conditions

$$\textbf{(N1)} \quad 1 \in X$$
$$\textbf{(N2)} \quad n \in X \Rightarrow n^+ \in X$$

without having $X = N$. This whole situation would violate (N5). ■

Of course, this example simply illustrates that it is not the successor function we had in the back of our mind when we drew up the axioms. We were really thinking of $n^+ = n + 1$ when we listed (N1)–(N5), particularly (N5). And with this understanding, we may indeed expect that for any subset $X \subseteq N$, we have

$$\left.\begin{array}{r} 1 \in X \\ n \in X \Rightarrow n^+ \in X \end{array}\right\} \Rightarrow X = N$$

as expressed in (N5). However, note that *we don't define* $n^+ = n + 1$ as our successor function. It is left unspecified and only its desired properties are listed. Neither is it spelled out anywhere in (N1)–(N5) that what is called N is really the set

$$N = \{1, 2, 3, \ldots\}$$

as introduced in Section 1.1.

Example 3.2 Consider the set

$$N = \{1, 1^+, (1^+)^+, ((1^+)^+)^+, \ldots\}$$

without regard to the meaning of 1, or of + for that matter. We find that the required properties (N1)–(N5) are met by this interpretation. The reader should pause for a moment to see that indeed this is the case. Yet, the members of this version of N are merely algebraic expressions in the abstract symbols 1 and + (together with parentheses). Note that we are not prohibited from thinking of 1 in the usual way, of thinking that $1^+ = 2$, that $(1^+)^+ = 2^+ = 3$, etc., but this is merely a matter of convenience, and not essential to the understanding. ■

As this example clearly shows, we are really talking here about expressions that look and act like the natural numbers, quite apart from their representation. But indeed, that's one of the main purposes of an axiomatization—to capture the essence of an idea without getting bogged down in unnecessary detail.

Definition of Arithmetic Operations

Assuming that the axiomatization (N1)–(N5) of the natural numbers is consistent with our intuition, we may proceed to do arithmetic. As Kronecker said, "all the rest is the work of man." We will find that if we can count (that's what the successor function provides), then we can add, and if we can add, then we can multiply. That, at least, is the thrust of the argument we are about to present. The reader may have realized all of this as a child. If not, it is certainly brought home most clearly in the development that follows.

Let us suppose that the set N and its successor function $n \to n^+$ meet the conditions of (N1)–(N5). Then we may introduce an **addition operation** on N with the inductive definition:

A1. $m + 1 = m^+$

A2. $m + n^+ = (m + n)^+$

---

where the idea is as follows. Think of $m \in N$ as fixed. For any such m, the second argument to the addition operation must again be a natural number, that is, a member of N. In (A1), we prescribe the sum for the case in which this second argument is 1, noting that the whole expression then gives some credence to our thinking of n^+ as $n + 1$. In all other cases, this second argument is the successor n^+ of some [unique—by (N4)] natural number $n \in N$. Assuming that $m + n$ has already been determined, we then take $m + n^+$ to be its successor $(m + n)^+$ as noted.

Example 3.3 Disregarding parentheses in Example 3.2, we have:

$$N = \{1, 1^+, 1^{++}, 1^{+++}, \ldots\}$$

and our above prescription (with $m = 1^{++}$, say) yields:

$$\begin{aligned}
1^{++} + 1 &= 1^{+++} \\
1^{++} + 1^+ &= (1^{++} + 1)^+ = (1^{+++})^+ = 1^{++++} \\
1^{++} + 1^{++} &= (1^{++} + 1^+)^+ = (1^{++++})^+ = 1^{+++++} \\
1^{++} + 1^{+++} &= (1^{++} + 1^{++})^+ = (1^{+++++})^+ = 1^{++++++}
\end{aligned}$$

etc., all of which the reader may take to mean:

$$\begin{aligned}
3 + 1 &= 4 \\
3 + 2 &= 5 \\
3 + 3 &= 6 \\
3 + 4 &= 7
\end{aligned}$$

and so on. However, it should not be supposed that we are saying nothing more here than $3 + 4 = 7$, for example. The result is not important (except for accountants). It is the means of obtaining the result that should be noted—that is, the *inductive* character of the reasoning. Thus $3 + 4 = 3 + 3^+ = (3 + 3)^+ = 6^+ = 7$, showing that we are able to compute $3 + 4$ if we have already computed $3 + 3$. Similarly, we can compute $3 + 3$ if we know $3 + 2$, and $3 + 2$ if we know $3 + 1$. But the latter is $3 + 1 = 3^+ = 4$, so all is determined. ■

If our definition of addition truly represents an operation on the natural numbers, it should be applicable to all pairs $m, n \in N$. Consider any $m \in N$, and let X_m denote the set

$$X_m = \{n\colon m + n \text{ is defined}\}$$

where we mean defined by (A1), (A2). Then, certainly $1 \in X_m$ by virtue of (A1). Moreover, if we have $n \in X_m$ so that $m + n$ is defined, then according to (A2), we also have $n^+ \in X_m$, that is,

$$m + n^+ = (m + n)^+$$

as listed above. That's the whole point of an *inductive definition*: to base the current calculation on those that have preceded. As a consequence, we see that the set X_m is a subset of N meeting the requirements of (N5). It then follows from (N5) that each $X_m = N$, that is, that our addition is indeed defined for all pairs $m, n \in N$.

In a manner completely analogous to the above, we may proceed to introduce a **multiplication operation** on N:

M1. $m \cdot 1 = m$

M2. $m \cdot n^+ = m \cdot n + m$

---

noting that in (M2), we assume that addition has already been defined, as is indeed the case. Note again how we first prescribe the product in case the second argument is 1. Then we define $m \cdot n^+$ assuming that $m \cdot n$ has already been defined, just as we did for addition. We leave the reader the task of verifying again that (M1) and (M2) provide for an operation that is applicable to all pairs $m, n \in N$ (see Exercise 1).

Example 3.4 To compute $3 \cdot 5$, we use (M2) and write

$$3 \cdot 5 = 3 \cdot 4^+ = 3 \cdot 4 + 3 = 12 + 3 = 15$$

and note that we need to have computed $3 \cdot 4$ already. Note as well the application of the addition operation. ■

In short, we can multiply if we can add, and (as seen earlier), we can add if we can count. Indeed, that is the thrust of definitions (A1), (A2), (M1), and (M2).

Laws of Integer Arithmetic

The constructive approach to arithmetic, as presented here, offers an opportunity to appreciate the familiar *commutative, associative,* and *distributive laws* in a whole new light:

I1a.	$m + n = n + m$	**I1b.**	$m \cdot n = n \cdot m$
I2a.	$(m + n) + p = m + (n + p)$	**I2b.**	$(m \cdot n) \cdot p = m \cdot (n \cdot p)$
I3.	$m \cdot (n + p) = m \cdot n + m \cdot p$		

We memorize them in our earliest school days, but somehow they never take on much meaning. They seem to have been given to us without explanation, and we never quite understand their significance. By way of contrast, our constructive approach to these questions (of commutativity, associativity, distributivity, etc.) shows specifically how the characteristic behavior is transmitted, from one number to its successor—*by induction*, as we say.

Before we proceed to establish the validity of these various arithmetic laws, we first state and prove two lemmas that are important to the arguments that follow. Each of these lemmas is proved by induction, thus offering examples of the style of proof we use in much of what follows. Thus it is imperative that the student pay close attention here so that the arguments that follow will be easier to understand.

Lemma 1 $n^+ = 1 + n$ for every $n \in N$.

Proof We let the set

$$X = \{n : n^+ = 1 + n\}$$

denote the class of natural numbers for which our statement is true. Obviously, the idea is to show that $X = N$. We first try $n = 1$, that is, we compute:

$$1^+ = 1 + 1 \qquad \text{(by (A1))}$$

showing that $1 \in X$. Now assuming that $n \in X$ (for some $n \in N$) we have:

$$\begin{aligned}(n^+)^+ &= (1 + n)^+ &&\text{(by the \textit{inductive hypothesis})}\\ &= 1 + n^+ &&\text{(by (A2))}\end{aligned}$$

showing that it follows that $n^+ \in X$ as well. Thus we have established that X meets the two conditions of (N5), so $X = N$ as required. □

In proofs by induction, as indicated above, we first establish that the required condition (whatever it is) is met by the natural number 1. Then we show that the assumption that the condition holds for an arbitrary natural number n leads to the conclusion that this same condition holds for its successor n^+. Then an appeal to axiom (N5) establishes that the condition holds for all natural numbers. When we exhibit such proofs in abbreviated form, the assumption—that the condition holds for some $n \in N$—is called the **inductive hypothesis**, abbreviated (IH). Moreover, in such abbreviated proofs we omit any reference to the set X, thus leaving these details to be kept in mind by the reader.

Example 3.5 If we recreate the proof of the Lemma 1 in abbreviated form, it would read as follows:

$$\begin{aligned} 1^+ &= 1 + 1 && \text{(A1)} \\ (n^+)^+ &= (1 + n)^+ && \text{(IH)} \\ &= 1 + n^+ && \text{(A2)} \end{aligned}$$

no more, no less. ■

To appreciate more fully this abbreviated style of inductive proof, we use the same approach to establish the following lemma.

Lemma 2 $1 \cdot n = n$ for every $n \in N$.

Proof In abbreviated form

$$\begin{aligned} 1 \cdot 1 &= 1 && \text{(M1)} \\ 1 \cdot n^+ &= 1 \cdot n + 1 && \text{(M2)} \\ &= n + 1 && \text{(IH)} \\ &= n^+ && \text{(A1)} \end{aligned}$$

showing that indeed $1 \cdot n = n$ for every $n \in N$ (by induction). □

As further illustrations of the technique, we now provide several examples that establish the validity of the arithmetic laws listed earlier.

Example 3.6 First we establish the associative law for addition, (I2a), using induction on the variable p, as follows:

$$\begin{aligned} (m + n) + 1 &= (m + n)^+ && \text{(A1)} \\ &= m + n^+ && \text{(A2)} \\ &= m + (n + 1) && \text{(A1)} \\ (m + n) + p^+ &= ((m + n) + p)^+ && \text{(A2)} \\ &= (m + (n + p))^+ && \text{(IH)} \\ &= m + (n + p)^+ && \text{(A2)} \\ &= m + (n + p^+) && \text{(A2)} \end{aligned}$$

noting, in particular, the use of the inductive hypothesis that $(m + n) + p = m + (n + p)$ holds for some $p \in N$. ■

Example 3.7 Now we can establish the commutative law for addition, (I1a), as follows:

$$\begin{aligned} m + 1 &= m^+ && \text{(A1)} \\ &= 1 + m && \text{(Lemma 1)} \\ m + n^+ &= (m + n)^+ && \text{(A2)} \\ &= (n + m)^+ && \text{(IH)} \\ &= 1 + (n + m) && \text{(Lemma 1)} \\ &= (1 + n) + m && \text{(I2a)} \\ &= n^+ + m && \text{(Lemma 1)} \end{aligned}$$

and we note that we must have proved property (I2a) first. ■

Example 3.8 To establish the distributive law (I3), we argue by induction on p as follows:

$$\begin{aligned} m \cdot (n + 1) &= m \cdot n^+ && \text{(A1)} \\ &= m \cdot n + m && \text{(M2)} \\ &= m \cdot n + m \cdot 1 && \text{(M1)} \\ m \cdot (n + p^+) &= m \cdot (n + p)^+ && \text{(A2)} \\ &= m \cdot (n + p) + m && \text{(M2)} \\ &= (mn + mp) + m && \text{(IH)} \\ &= mn + (mp + m) && \text{(I2a)} \\ &= mn + mp^+ && \text{(M2)} \end{aligned}$$

Again we note the use of the associative property for addition (I2a). ■

Observe that we have yet to establish the commutativity property (I1b) for multiplication. Consequently, the dual distributive law

$$(n + p) \cdot m = n \cdot m + p \cdot m$$

is not yet a consequence of (I3). We leave the proof of this law, and also those of (I1b) and (I2b), as exercises for the student.

Extension of the Number System

As everyone knows, the integers consist of the (positive) natural numbers, zero, and the negative numbers: $-1, -2, -3$, etc. It is a simple matter to extend the arithmetic system N as just described, to obtain the arithmetic number system

$$Z = \{\ldots, -3, -2, -1, 0, 1, 2, 3, \ldots\}$$

as normally encountered in grade school studies.

We have first to define a new number called **zero** (represented by 0, as usual), satisfying the properties:

I4a. $n + 0 = n$ **I4b.** $n \cdot 0 = 0$

for all $n \in N$ (and later, for all $n \in Z$). Then the introduction of *negative numbers* is accomplished by associating with each $n \in N$, a new number (written $-n$), with the defining property:

I5. $n + (-n) = 0$

for every $n \in N$. We leave to the reader the task of verifying that the usual arithmetic properties of commutativity, associativity, and distributivity hold as well in the extended number system Z.

Ordering of the Integers

Returning momentarily to the restricted system of natural numbers, we may introduce the usual *less than* relation by agreeing to the definition:

$$n < m \Leftrightarrow m \text{ is among the successors } n^+, (n^+)^+, \ldots, \text{ of } n$$

(Of course, when we write $n \leq m$ we mean that $n < m$ or $n = m$.) The reader may verify that this definition indeed satisfies the properties (O1), (O2), (O3) expected of an order relation, as axiomatized in Section 2.1. In fact, we can say something more—that the natural numbers are **well ordered**, in the sense of the following statement:

Theorem 3.1 ***Well-ordering Principle*** Every nonempty subset of N has a least element.

Proof We want to show that the assumption that there is a nonempty subset of N that has no least element leads to a contradiction (see Exercise 11). □

The Well-ordering Principle can be used to establish the (rather obvious) fact that there is no integer between 0 and 1 (as shown in Exercise 2 of Section 2.5). Assuming that there is one, then there must be a least one, and this leads to a contradiction. This application, in which we have an intuitive idea of what should be happening, is typical, but the Well-ordering Principle makes everything precise. A similar comment is appropriate in reference to our next theorem, which dates back to the time of Archimedes:

Theorem 3.2 ***Archimedean Principle*** For $a, b \in N$ there exists an $n \in N$ with $nb > a$.

Proof Take $n = a + 1$. Then we have

$$(a + 1)b = ab + b > ab \geq a$$

as required (see Exercise 15). □

EXERCISES 3.1

1 Show that (M1), (M2) provide for an operation that is applicable to all pairs $m, n \in N$. (*Hint:* For each $m \in N$ let X_m denote the set

$$X_m = \{n : m \cdot n \text{ is defined}\}$$

and show that each $X_m = N$.)

2 Show explicitly how it is that:

(a) The computation of $3 + 5$ depends on our having computed $3 + 4$ (and on our knowledge of the successor function)

(b) The computation of $4 + 3$ depends on our having computed $4 + 2$ (and on our knowledge of the successor function)

(c) The computation of $3 \cdot 6$ depends on our having computed $3 \cdot 5$ (and on our knowledge of addition)

(d) The computation of $4 \cdot 3$ depends on our having computed $4 \cdot 2$ (and on our knowledge of addition).

3 Show that the set

$$N = \{1, 1^+, 1^{++}, 1^{+++}, \ldots\}$$

together with the successor function $n \to n^+$ meets the requirements of the axiomatization (N1)–(N5).

4 What partition of N results from choosing

$$n \to n^+ = n + 5$$

as a successor function? Show explicitly how axiom (N5) fails in this case.

5 Exhibit an inductive definition of the *exponentiation operation* m^n in a pair of equations (E1), (E2) patterned after (A1), (A2) and (M1), (M2). (*Hint:* In (E2), you need to assume that multiplication is already well understood.)

6 Use the definition of m^n in Exercise 5 to establish each of the following:

(a) $m^{n+k} = m^n m^k$ **(b)** $(mn)^k = m^k n^k$ **(c)** $(m^n)^k = m^{nk}$

7 Use induction to establish the dual distributive law

$$(n + p) \cdot m = n \cdot m + p \cdot m$$

for all $n, m, p \in N$.

8 Prove the following laws by induction.

(a) The commutative law (I1b) for multiplication

(b) The associative law (I2b) for multiplication

[*Hint:* It is best to try to establish part (b) before part (a).]

9 Establish the validity of these arithmetic laws in the extended number system Z:

(a) Law (I1a) **(b)** Law (I1b) **(c)** Law (I2a)

(d) Law (I2b) **(e)** Law (I3)

10 Show that the given definition $n < m$ for the ordering of the natural numbers satisfies the axioms (O1), (O2), (O3) given in Section 2.1.

11 Complete the proof of Theorem 3.1. (*Hint:* Assuming that A is a nonempty subset of N having no least element, let

$$X = \{n \in N : n \leq m \text{ for all } m \in A\}$$

and use (N5) to conclude that $X = N$.

12 Prove each of the following:

(a) $n < k \Longrightarrow m + n < m + k$ **(b)** $m + n = m + k \Longrightarrow n = k$

13 Prove each of the following:

(a) $m + k < n + k \Longrightarrow m < n$ **(b)** $mk < nk \Longrightarrow m < n$

14 Which logical deduction rule (from Chapter 2) is the basis for the proof of Theorem 3.1?

15 Show that for all $a, b \in N$,

(a) $a + b > a$

(b) $ab \geq a$ with equality only if $b = 1$

Sec. 3.2 INDUCTION AND RECURSION

Mathematical induction is a most powerful proof technique, and the closely related notion of recursion is likewise, a powerful problem-solving tool.

The inductive feature of our axiomatization of the natural numbers leads to a powerful new proof technique, suitable for establishing the validity of certain statements over the whole set of natural numbers. At first glance it would appear that no finite proof would be possible—there are infinitely many cases to examine. To the contrary we find that proofs by induction are quite easily handled once the technique has been learned. The time spent learning this procedure is indeed time well spent. There is perhaps no single idea that is of greater importance than induction (and the closely related notion of recursion), in particular as it applies to the general problem-solving activity in computer science and in all phases of applications in the mathematical sciences.

Goals

After studying this section, you should be able to:

- Describe the *Principle of Mathematical Induction* and show how it derives from the axiomatization of the natural numbers.
- Give inductive proofs for elementary mathematical propositions involving the natural numbers.
- Use *recursion* to define functions, sequences, and sets.
- Illustrate the use of recursion in problem-solving activities.

The Principle of Mathematical Induction

Suppose we are given a mathematical statement $P(n)$ concerning the natural number n, and we would like to assert that $P(n)$ is valid for every $n \in N$. Were we to check the validity of $P(n)$ for $n = 1, 2, 3$, etc., successively, we would only succeed in proving that the proposition holds for those values of n we have tested. No matter how many specific cases we examine, we could never generalize to say that $P(n)$ holds *for all n*.

Yet, there is a well-established technique for proving just such generalized statements. It is called the **Principle of Mathematical Induction**, symbolized as a deduction rule in the form:

$$\frac{P(1), P(n) \longrightarrow P(n+1)}{P(n) \text{ for all } n \in N} \qquad \textbf{(IND)}$$

Thus we are only asked to suppose that we have established the validity of $P(1)$, and that we have also verified that the assumption that $P(n)$ holds for some n leads to the conclusion that $P(n + 1)$ holds as well. We are then free to conclude that $P(n)$ is valid for every natural number n.

It is easy to see how this principle follows from our axiomatization of the natural numbers. Thus if we let $X \subseteq N$ denote the set:

$$X = \{n \in N \colon P(n) \text{ is valid}\}$$

we would have $1 \in X$ and $n \in X \Longrightarrow n + 1 \in X$. According to axiom (N5) of Section 3.1, it follows that $X = N$, that is, that $P(n)$ holds for all $n \in N$. Thus we see that (IND) is a direct consequence of our axiomatization of the natural numbers. Or, perhaps it is more accurate to say that the axiomatization has been designed with the Principle of Mathematical Induction in mind. In any case, we are now able to work with the more streamlined version of induction as represented by the (IND) deduction rule. We have only two things to establish for this rule:

---

1. The proposition is true for $n = 1$.
2. The assumption (called the *inductive hypothesis*) that the proposition is true for *some* natural number n implies that it is true for the next largest number $n + 1$.

As a consequence, we are able to conclude that the proposition $P(n)$ is valid for all $n \in N$.

Example 3.9 As an analogy, try to imagine an endless single file of dominos standing on end. We provide Figure 3.1 as an aid to your imagination. Now suppose that we want to be sure that all of these dominos will fall. It is not enough to know that the first thousand or even the first million will fall or have fallen, but we could be sure that they would all fall if we were certain of two things:

1. The first domino is knocked over (in the direction of the others, of course).
2. The dominos are so spaced that whenever any one domino falls, it will automatically knock over its neighbor.

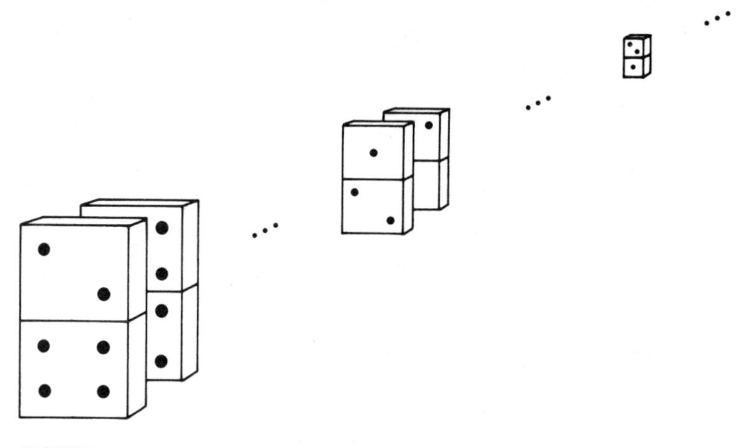

FIGURE 3.1

This is indeed the idea represented in the Principle of Mathematical Induction. (Some politicians have used this domino analogy to explain some possible effects of a weak United States defense posture.) ■

Example 3.10 Perhaps it is best to begin our series of real examples of the use of the principle with a mathematical proposition whose validity has already been established. We refer to the statement

$$P(n): |B^n| = 2^n$$

(from Example 1.54), one that we again seek to verify for all $n \geq 1$.

1. Clearly $P(1)$ is true, that is, we have

$$|B^1| = |B| = 2 = 2^1$$

because $B = \{0, 1\}$.

2. Now suppose that $P(n)$ is true for *some* n [Note that we do not assume that $P(n)$ is true for *all* $n \in N$, because that is what we are trying to prove. Furthermore, we know from part (1) that $P(n)$ is true *for some* n, namely $n = 1$, so that our supposition is indeed a reasonable one]. Can we then show that $P(n + 1)$ is true as a consequence? That is, can we conclude that $|B^{n+1}| = 2^{n+1}$? We note that

$$B^{n+1} = \{b_1, b_2, \ldots, b_n, b_{n+1}): b_i \in B\}$$

and moreover, it is clear that the members of this set are simply the members of B^n with an extra 0 or 1 attached. And we can attach either 0 or 1 to every $(b_1, b_2, \ldots, b_n)$. It follows that

$$|B^{n+1}| = 2|B^n| = 2 \cdot 2^n = 2^{n+1}$$

as required. Note especially how the inductive hypothesis is used in the middle equality of the calculation. ∎

Example 3.11 In an entirely different vein, suppose we try to establish the fact that

$$3 \mid (n^3 - n) \quad \text{for all } n \geq 1$$

that is, that each expression $n^3 - n$ is a multiple of 3. With $n = 1$ it is certain that we have

$$n^3 - n = 1^3 - 1 = 0 = 0 \cdot 3$$

a multiple of 3. Now, if we suppose (the inductive hypothesis) that

$$n^3 - n = 3m$$

for *some* integer $n \geq 1$, then we have

$$\begin{aligned}(n + 1)^3 - (n + 1) &= (n^3 + 3n^2 + 3n + 1) - (n + 1)\\ &= n^3 - n + 3(n^2 + n)\\ &= 3m + 3(n^2 + n)\\ &= 3(n^2 + n + m)\end{aligned}$$

also a multiple of 3. So, indeed, the proposition is proved. ∎

There are many variations that may be taken in the form of the proposition $P(n)$. It needn't be an identity as in Example 3.10. It might be an inequality (as in Example 3.12). Moreover, we are not always required to establish $P(n)$ for all $n \in N$—sometimes we only want to show that $P(n)$ is valid for all natural numbers greater than or equal to k, say. Then, instead of verifying $P(1)$ in (1) of the induction scheme we establish $P(k)$. However, such variations in the general inductive proof are easily handled, once the basic idea is understood.

Example 3.12 As a further illustration of the style of an inductive proof, suppose we verify that for all $n \geq 11$ the inequality

$$3n - 6 < \frac{n(n-1)}{4}$$

holds. With $n = 11$, we have 27 and 27.5 for the left-hand and right-hand sides, respectively, so that $P(11)$ is valid. Then we suppose that the inequality holds for some $n \geq 11$, and we proceed to compute:

$$\begin{aligned} 3(n+1) - 6 = (3n-6) + 3 &< \frac{n(n-1)}{4} + 3 = \frac{n(n-1)+12}{4} \\ &\leq \frac{n(n-1)+(n+1)}{4} < \frac{(n+1)(n-1)+(n+1)}{4} \\ &= \frac{(n+1)[(n-1)+1]}{4} = \frac{(n+1)[(n+1)-1]}{4} \end{aligned}$$

showing that it necessarily holds for $n + 1$ as well. ■

Recursive Definition of Functions

We have already seen examples (in Section 3.1) of a function or an operation defined inductively, using the values of the function at previous arguments to compute its value at the current argument. This is a most important notion throughout mathematics, and its applications are quite widespread, particularly in computer science.

More precisely, a function or sequence $f: N \to N$ is said to be **defined inductively** (or **by recursion**) if there is already a given function $g: N \times N \to N$ such that:

---

1. $f(1) = c$ (some constant element of N)

2. $f(n+1) = g(n, f(n))$ for $n \geq 1$

---

the inductive step being provided by (2). Ordinarily g is "more primitive" than f, that is, it is a function that is already well understood and generally computable by some given mechanical procedure. Of course, if this is the case, then the inductive definition just given will provide a mechanical procedure or *algorithm* for computing each $f(n)$ in turn.

Example 3.13 In Section 3.1 we defined multiplication recursively, assuming that addition was already well understood. Thus we wrote in (M1), (M2):

$$f(1) = m \quad \text{that is,} \quad m \cdot 1 = m$$
$$f(n + 1) = m \cdot n + m \quad \text{that is,} \quad m \cdot (n + 1) = m \cdot n + m$$

providing a recursive definition of the function

$$f = \text{multiplication by } m$$

In terms of the above format, we have:

$$g = \text{addition of } m$$

that is, we obtain:

1. $f(1) = m$ (a constant, when considering the fixed $m \in N$)

2. $f(n + 1) = m \cdot n + m = g(n, f(n))$

provided that $g(r, s)$ is the function:

$$g(r, s) = s + m$$

that is, addition of m to the second argument. ■

Example 3.14 Perhaps the best-known example of a recursively defined function is the *factorial function* of Section 1.5. Thus we may write:

$$f(1) = 1 \quad \text{that is,} \quad 1! = 1$$
$$f(n + 1) = (n + 1) \cdot f(n) \quad \text{that is,} \quad (n + 1)! = (n + 1) \cdot n!$$

to define $n!$ recursively, assuming that multiplication is already well understood. We say this because the function g (in the above format) is essentially multiplication. More precisely, we must have:

$$g(n, m) = (n + 1) \cdot m$$

so that

$$f(n + 1) = (n + 1) \cdot f(n) = g(n, f(n))$$

as required in the above format. ■

It is easy to understand why statements that involve a recursively defined function may quite easily lend themselves to proofs by induction. After all, the whole trick in an inductive proof is to reduce statements about $n + 1$ to statements involving n (so that the inductive hypothesis can be used), and this much is provided immediately in the definition: $f(n + 1) = g(n, f(n))$.

Example 3.15 Suppose that we want to show that $n! \geq n$ for every $n \in N$. We will then be able to write:

1. $1! = 1 \geq 1$

2. $(n + 1)! = (n + 1) \cdot n! \geq (n + 1) \cdot n \geq n + 1$

assuming that $n \geq 1$ in the latter. (See Exercise 15(b) of Section 3.1.) ■

The Summation Symbol

Throughout mathematics, the *summation symbol* $\sum$ has a precise and uniform meaning. If an infinite sequence of integers x_j is given, then as an abbreviation for the sum of the first n terms of the sequence, we write

$$\sum_{j=1}^{n} x_j = x_1 + x_2 + \cdots + x_n$$

The left-hand side can be read "the sum (of x_j) from ($j =$) 1 to n." Thus we have

$$\sum_{j=1}^{5} j^2 = 1^2 + 2^2 + 3^2 + 4^2 + 5^2 = 55$$

$$\sum_{j=1}^{4} (j^2 - j) = (1^2 - 1) + (2^2 - 2) + (3^2 - 3) + (4^2 - 4) = 20$$

$$\sum_{j=1}^{3} j! = 1! + 2! + 3! = 9$$

etc. It happens that the whole meaning of the summation symbol can be given recursively. We have only to write:

---

1. $\sum_{j=1}^{1} x_j = x_1$

2. $\sum_{j=1}^{n+1} x_j = x_{n+1} + \sum_{j=1}^{n} x_j$

---

following the usual format. As a matter of fact, this recursive definition of the summation symbol provides the key to the following example (and to many others as well).

Example 3.16 Suppose we now apply the Principle of Mathematical Induction to verify that for every natural number n we have

$$\sum_{j=1}^{n} j = \frac{n(n + 1)}{2}$$

Note that the expression on the left represents the sum of the first n natural numbers, that is, $1 + 2 + 3 + \cdots + n$. Here, we are dealing with the proposition:

$$P(n)\colon \sum_{j=1}^{n} j = \frac{n(n+1)}{2}$$

1. Clearly $P(1)$ is true because

$$\sum_{j=1}^{1} j = 1 = \frac{1(2)}{2}$$

using the recursive meaning of the summation symbol, part (1).

2. Now suppose that $P(n)$ is true for *some* n [Note again that we do not assume that $P(n)$ is true for *all* $n \in N$, because that is what we are trying to prove]. Can we then show that $P(n + 1)$ is true as a consequence? Yes, because the statement $P(n + 1)$ is simply

$$P(n+1)\colon \sum_{j=1}^{n+1} j = \frac{(n+1)(n+2)}{2}$$

and we can obtain it using part (2) of the recursive meaning of the summation symbol, together with the inductive hypothesis as used in the second equality of the calculation:

$$\begin{aligned} \sum_{j=1}^{n+1} j &= (n+1) + \sum_{j=1}^{n} j = (n+1) + \frac{n(n+1)}{2} \\ &= \frac{2n + 2 + n^2 + n}{2} = \frac{n^2 + 3n + 2}{2} \\ &= \frac{(n+1)(n+2)}{2} \end{aligned}$$

the rest being simple algebraic manipulation. ■

Lest the reader think that identities such as in Example 3.16 are only mathematical curiosities, we quote from Publication 534, *Tax Information on Depreciation*, of the Internal Revenue Service as follows:

Sum-of-the-years'-digits method: Under this method, as a general rule, you apply a different fraction each year to the cost or other basis of each single asset account reduced by estimated salvage value. The denominator (bottom number) of the fraction, which remains constant, is the total of the digits representing the years of estimated useful life of the property. For example, if the estimated useful life is 5 years, the denominator is 15, that is, the sum of $1 + 2 + 3 + 4 + 5$. *To save time in arriving at the denominator, especially when an asset has a long life, square the life of the asset, add the life, and divide by 2. Thus, the asset with a 5-year life has the denominator* $5 \times 5 + 5 \div 2 = 15$. *The numerator (top number)* . . .

Recursive Solution of Problems

In a more general problem solving environment, the idea of recursion is often the key to obtaining an effective solution. We may be given a problem that involves a number of objects (coins, pebbles, or whatever), and the problem statement asks that a certain configuration be achieved, that a certain condition be fulfilled, or something of this sort. If we are able to see how to handle the problem with $n + 1$ objects, provided we can assume that the problem has been successfully treated for n objects, then a recursive solution is usually at hand.

Example 3.17 We take as our primary example the well-known Towers-of-Hanoi puzzle, as illustrated in Figure 3.2. Our task is to move a certain number of disks from peg A to peg C, one at a time, subject to the additional restriction that as the top disk is removed from a peg and placed on another peg, it is placed either on the bottom or on top of a larger disk. We may question whether this can be done at all. Furthermore, if it can be done, we may want to know the total number of moves required in the solution. As it turns out, we will be able to answer these questions, once a recursive approach is taken.

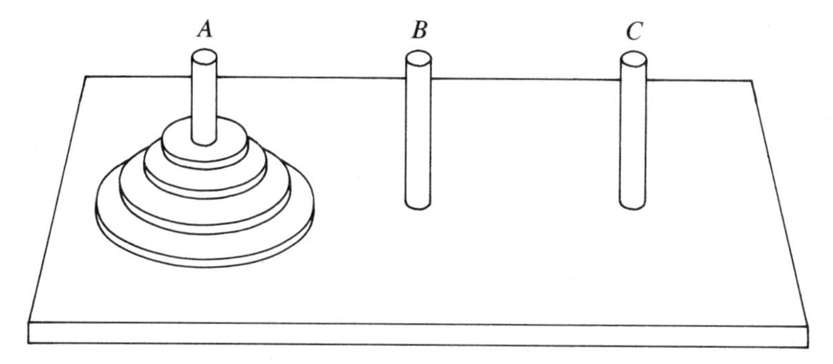

FIGURE 3.2

Suppose we knew how to move n disks in the required manner from one peg to another (say from peg A to peg B). Then *what else* would we have to do to move $n + 1$ disks altogether (from peg A to peg C)? (See Figure 3.3.) Supposing, as we have said, that we have successfully moved the top n disks from peg A to peg B (in a manner that we do not need to describe), then we can move the remaining (n + 1st) disk from peg A to peg C, after which we repeat the process of moving the n disks, only this time, we move them from peg B to peg C. Note how we deal only with the current step, because we assume that the previous steps have already been handled successfully. Yet the solution to the puzzle is then complete.

How many moves are required in this solution? If we let $h(n)$ denote the number of moves needed to transfer n disks from one peg to another, then we are

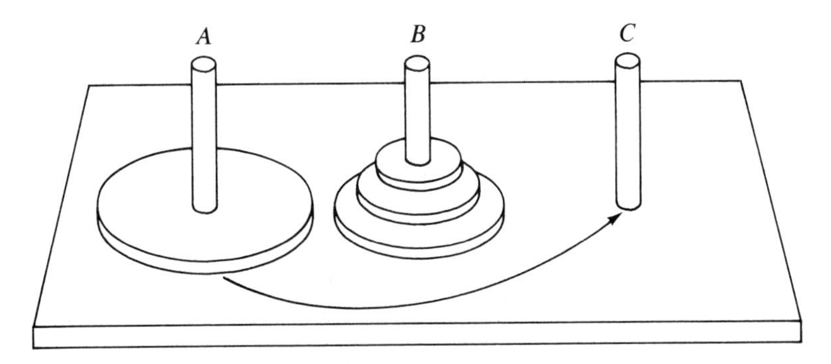

FIGURE 3.3

able to write the following *recursive relation*:

$$h(n + 1) = 2h(n) + 1$$

by reasoning as follows. In moving the $n + 1$ disks altogether, we first moved n of them from one peg to another [in $h(n)$ moves]; then in one move we transferred the remaining disk (from peg A to peg C); finally, we moved the n disks again [in $h(n)$ more moves]. Altogether, this takes $2h(n) + 1$ moves, as just indicated. ■

Example 3.18 Suppose there are four disks. How many moves are involved in the recursive solution? To compute $h(4)$, we will need to use the *termination condition*:

$$h(1) = 1$$

an obvious equality, once we think about it. We may then compute:

$$\begin{aligned} h(4) &= 2h(3) + 1 \\ &= 2(2h(2) + 1) + 1 \\ &= 2(2(2h(1) + 1) + 1) + 1 \\ &= 2(2(2 \cdot 1 + 1) + 1) + 1 = 15 \end{aligned}$$

using the termination condition in the last line. The reader should reconstruct our recursive solution in the case of four disks, just to see that indeed 15 moves are involved. ■

Example 3.19 Let us show in detail how to move three disks (numbered 1 = small, 2 = medium, 3 = large) from peg A to peg C. Recursively, we have:

Disks 1, 2: A $\longrightarrow$ B

Disk 3: A $\longrightarrow$ C

Disks 1, 2: B $\longrightarrow$ C

If we now concentrate on the moving of the two disks, we have the complete solution:

$$\begin{aligned}
&\text{Disk 1: A} \longrightarrow \text{C}\\
&\text{Disk 2: A} \longrightarrow \text{B}\\
&\text{Disk 1: C} \longrightarrow \text{B}\\
&\text{Disk 3: A} \longrightarrow \text{C}\\
&\text{Disk 1: B} \longrightarrow \text{A}\\
&\text{Disk 2: B} \longrightarrow \text{C}\\
&\text{Disk 1: A} \longrightarrow \text{C}
\end{aligned}$$

a total of 7 moves, which agrees with the calculation

$$\begin{aligned}
h(3) &= 2h(2) + 1\\
&= 2(2h(1) + 1) + 1\\
&= 2(2 \cdot 1 + 1) + 1 = 7
\end{aligned}$$

as in the case of $h(4)$. ■

Recursive Definition of Sets

Often we can define an infinite set *in layers*, having conceived of the set as being constructed in just that way, as layer-1, layer-2, layer-3, and so on, ad infinitum. Then it may be the case that each layer may be defined recursively in terms of those that have already been constructed. In this way we obtain an *inductive definition* (or a *recursive definition*) of the set in question.

Example 3.20 In Section 2.7 we gave a recursive definition of the set of well-formed formulas of the propositional calculus, even though the notion of recursion had yet to be introduced. The first layer of this set consisted of the constants (0 or 1), together with propositional variables (p, q, r, etc.) standing alone as formulas. Then it was stated that if a and b were already understood to be wff's, then the expressions

$$a \vee b \qquad a \wedge b \qquad \neg a \qquad a \rightarrow b \qquad a \leftrightarrow b \qquad (a)$$

would also be recognized as wff's. In this way the infinite set of wff's is defined recursively, starting with a first layer, then increasingly more complex formulas are constructed layer by layer. ■

Example 3.21 The set A^* of all words on the alphabet A may be constructed recursively. We may agree that each letter $a \in A$ standing alone constitutes a word of the first layer. (It is convenient to think of the null word λ as being a word—the only word—of layer zero.) Then, by recursion, we agree that if x is a word of A^*, so is each expression xa $(a \in A)$ also a word of A^*. Here, the product xa is just a simple juxtaposition. In any case we obtain the entire infinite set A^* in this way, and we note that the layers correspond to word lengths. In layer-1 are found all of the one-letter words, in layer-2, all of the two-letter words, etc. Evidently these layers are the subsets $A, A^2, \ldots,$ as discussed in Example 1.51 and Section 1.5. ■

EXERCISES 3.2

1 Prove that

$$\sum_{j=1}^{n} j^2 = \frac{n(n+1)(2n+1)}{6}$$

for all $n \geq 1$.

2 Prove that

$$\sum_{j=1}^{n} j^3 = \left[\frac{n(n+1)}{2}\right]^2$$

for all $n \geq 1$.

3 Prove by mathematical induction that

$$|\mathscr{P}(S)| = 2^{|S|}$$

for every finite set S. (*Hint:* In a set with $n + 1$ elements, say

$$S = \{x_1, x_2, \ldots, x_n, x_{n+1}\}$$

there are two kinds of subsets, those that include x_{n+1} and those that do not, and there are just as many subsets of the one kind as there are of the other.)

4 Show that

$$\sum_{j=1}^{n} 2^{j-1} = 2^n - 1$$

for all $n \geq 1$.

5 Give an inductive proof that

$$\sum_{k=0}^{n} \binom{n}{k} = 2^n$$

for all $n \geq 0$. (*Note:* The first case to check is $n = 0$.)

6 Use the recursive definition of the summation symbol to show that:

(a)
$$\sum_{j=1}^{n} x_j = x_1 + x_2 + \cdots + x_n$$

(b)

$$\left(\sum_{j=1}^{n} x_j\right) y = \sum_{j=1}^{n} (x_j y)$$

for all $n \geq 1$.

7 Obtain an explicit representation of the function g used in defining the function:

$$f(n) = \text{exponentiation of } m = m^n$$

as in Exercise 5 of Section 3.1. (*Hint:* Intuitively, g should have something to do with multiplication.)

8 Prove by induction that the function h in the Towers-of-Hanoi puzzle has the closed form

$$h(n) = 2^n - 1$$

for all $n \geq 1$.

9 Establish the *derived* induction rule

$$\frac{P(1), P(j)\ (1 \leq j \leq n) \longrightarrow P(n+1)}{P(n) \text{ for all } n \in N}$$

10 It is claimed that Gauss proved the identity of Example 3.16 by considering the scheme

$$\begin{array}{ccccc} 1 + & 2 & + \cdots + n \\ n + & (n-1) & + \cdots + 1 \end{array}$$

Complete Gauss's proof.

11 The first two terms of the *Fibonacci sequence* are both 1. Thereafter, we obtain the next term of the sequence by adding the two terms immediately preceding. Thus we may write

$$f(1) = f(2) = 1$$
$$f(n+2) = f(n+1) + f(n) \qquad (n \geq 1)$$

showing how the sequence is determined inductively. This sequence was studied by Leonardo Fibonacci (son of Bonaccio) in the thirteenth century. Pinecones, sunflowers, the mating habits of bees, and many other phenomena seem to exhibit properties relating to the Fibonacci sequence. Compute the first few (say the first 10) terms of the Fibonacci sequence.

12 Use induction to show that the Fibonacci sequence satisfies the identity:

$$f(n+1)f(n+2) - f(n)f(n+3) = (-1)^n$$

for all $n \geq 1$. (See Exercise 11.)

13 Let the function $f: N \rightarrow N$ be defined recursively by writing:

$$f(1) = 2$$
$$f(n+1) = 2f(n)$$

Prove by induction that f has the closed form

$$f(n) = 2^n$$

for all $n \geq 1$.

14 Let the function $f: N \to N$ be defined recursively by writing:

$$f(1) = 3$$
$$f(n + 1) = 2f(n) - 3$$

Show that f is a constant function.

15 Consider the set $X \subseteq N$ of natural numbers defined recursively as follows:

(a) $1 \in X$ **(b)** $n, m \in X \Rightarrow n + m \in X$

Show that $X = N$.

16 The set P of well-formed parentheses strings is defined recursively as follows:

i. $(\ ) \in P$

ii. $x, y \in P \Rightarrow (x) \in P$ and $xy \in P$

where in part (ii), the product is simple juxtaposition. List the first few layers of the set P.

17 Give an example of a proposition $P(n)$ for which

$$P(n) \qquad P(n + 1)$$

but nevertheless, all $P(j)$ are false. [*Note:* This will show that to apply induction, it is crucial that $P(1)$ be true.]

18 Prove by induction that the identity

$$1 + 3 + 5 + \cdots + (2n - 1) = n^2$$

holds for all $n \geq 1$, that is, that the sum of the first n odd numbers is n^2.

Sec. 3.3 DIVISIBILITY OF INTEGERS

Only an approximate division is possible in the integers—the computation of a quotient and a remainder.

In the arithmetic of our elementary schooling, we learned how to divide one integer by another to obtain a quotient and a remainder. Obviously, this is not a true division—otherwise there wouldn't be a remainder. In elementary mathematics, it is called *long division, long* because we have to estimate repeatedly the number of times that the divisor goes into the current remainder, until finally we are left with a remainder that is less than the divisor itself, whereupon the process terminates. This is the *division algorithm* that we learned as children. It happens that this same algorithm is the starting point for considering a number of important questions involving integer arithmetic, particularly those that relate to problems of factorization. It is fortunate, therefore, that we all have this same common knowledge dating back to our earliest schooling, as a convenient point of reference. It allows for a clearer presentation than would otherwise be the case.

Goals

After studying this section, you should be able to:

- Illustrate the use of the *Division Algorithm* and show how it applies to the various factorization problems.
- Compute quotient and remainders for given pairs of integers representing divisor and dividend, respectively.
- Define the term "greatest common divisor" and give examples of its application.
- Describe and use the *Euclidean Algorithm* for computing greatest common divisors.
- Define the terms "prime" and "relatively prime."
- Describe the *Factor Algorithm* for finding all of the divisors of a given integer.
- Illustrate the use of the Factor Algorithm to answer various questions of integer divisibility.

Integer Division

Let us examine the problem of integer division anew, as if we had not already seen the results in our study of elementary mathematics. In fact, it happens that there are certain aspects to the problem that are new—new in their relationship to the *integer division* operation that is available in most computer programming languages.

If n and m are natural numbers (or arbitrary integers, for that matter), we say that n **divides** m (evenly), written $n \mid m$, if there is another integer q such that

$$m = q \cdot n$$

that is, if m is an integer multiple of n. We have used this relation of *divisibility* on several occasions already. However, because this even divisibility is rather infrequent, given any n and m, we cannot, strictly speaking, provide the integers with a genuine division operation. (That's why the rational numbers were devised—see Section 3.5.) However, for many computational purposes, a useful substitute operation is available. It goes by the name **integer division** and as we have said, it is found in most computer programming languages, for example, Pascal, FORTRAN, etc. For any quotient m/n, considered as a decimal, we let $\mathrm{tr}(m/n)$ be the integer obtained by *truncation*, that is, obtained by disregarding the fractional part. Thus $\mathrm{tr}(3.14) = 3$, $\mathrm{tr}(-\frac{22}{7}) = -3$, etc. Then, for any pair of integers n, m with $n \neq 0$, we define the **integer quotient** as follows:

$$m \div n = \operatorname{tr}(m/n)$$

Example 3.22 If $m = 22$ and $n = 5$, we have

$$m \div n = \operatorname{tr}\left(\tfrac{22}{5}\right) = \operatorname{tr}(4.4) = 4$$

even though $n \mid m$ is false. That is, there does not exist an integer q such that

$$22 = q \cdot 5$$

but the integer $(q =)$ 4 comes very close. ■

Related to this discussion is the so-called *division algorithm* that we spoke about at the start of this section. Thus if a and b are natural numbers, we may refer to the *quotient* and *remainder* (q and r) when dividing a by b as in long division. If we take $q = a \div b$ and $r = a - bq$ in the statement below, there should be little need for justification of the result. (Nevertheless, see Exercises 4 through 7.) The remainder r is certainly less than b (or otherwise, further division would have been possible).

Division Algorithm For any pair of natural numbers a, b there exist integers q and r such that

$$a = bq + r \quad \text{and} \quad 0 \leq r < b$$

Example 3.23 In Example 3.22 ($a = 22$, $b = 5$), if we let

$$q = a \div b = 4$$
$$r = a - bq = 22 - 20 = 2$$

then certainly we have:

$$22 = 5 \cdot 4 + 2 \quad \text{and} \quad 0 \leq 2 < 5$$

as required. ■

Example 3.24 We should point out that q, r are indeed the quotient and remainder obtained by long (in this case, rather short) division, as in computing:

$$\begin{array}{r} 4 = q \\ 5 \overline{)22} \\ \underline{20} \\ 2 = r \end{array}$$

by the usual means. ■

Greatest Common Divisors

As an application of the division algorithm, we will eventually seek to determine the **greatest common divisor** $\gcd(a, b)$ of a given pair of natural numbers a, b. By this we mean an integer $d = \gcd(a, b)$ that has the following properties:

---

a. $d \mid a$ and $d \mid b$

b. $c \mid a$ and $c \mid b \Rightarrow c \mid d$

Note in particular that the adjective greatest here refers not primarily to d having a greater magnitude than any common divisor c, but instead to d being a multiple of any such c.

Example 3.25 If $a = 24$ and $b = 36$, then $1, 2, 3, 4, 6$, and 12 are all common divisors of a and b. Thus in the case of $c = 3$ we have

$$c \mid 24 \quad \text{and} \quad c \mid 36 \qquad (24 = 8 \cdot 3 \text{ and } 36 = 12 \cdot 3)$$

but $d = 12$ is a multiple of all the common divisors—it is the gcd of 24 and 36. ■

Consider the following situation. In architecture, and in the lumber, automotive, and textile industries, measurements are made and specified in fractions. We would like to know that the answer to a problem is $\frac{17}{32}$, and somehow, the answer 0.53125 is not quite the same.

In such situations, particularly as we attempt to involve computers in the problem solution, there is no apparent means to perform even the simplest grade-school arithmetic (of addition, subtraction, multiplication, and division) of fractions. The usual computer representation of these fractions would be in the form of decimals, as indicated above, and this does not always provide a suitable treatment of the problems involved. There are ways of handling fractions (for example, as pairs of integers) in a computer, but then we would like to calculate, for example:

$$\frac{1}{6} + \frac{2}{15} = \frac{5 + 4}{30} = \frac{9}{30} = \frac{3}{10}$$

always reducing the answer to its *lowest terms*. However, reducing the answer to its lowest terms requires that we know how to compute the greatest common divisor (gcd) of any pair of integers, as in $\gcd(9, 30) = 3$, so that we can divide both numerator and denominator by their gcd in making our reductions. The computational routines are not readily available for performing such arithmetic—at least not directly. The programmer needs to supply these routines. Similarly, to find a *common denominator* for 6 and 15 in the addition above, to avoid an unnecessarily large denominator in the sum, we need to compute the *least common multiple* (lcm):

$$\text{lcm}(6, 15) = \frac{6 \times 15}{\text{gcd}(6, 15)} = \frac{90}{3} = 30$$

again making use of the gcd operation. In general, we have:

$$\text{lcm}(a, b) = \frac{a \cdot b}{\text{gcd}(a, b)}$$

Thus it appears that the gcd operation is fundamental to the development of any fractional arithmetic package, for whatever purpose. It could be that an enterprising young school teacher simply wishes to write a set of programs for the design and testing of examination questions (in the arithmetic of fractions). The gcd function would again be indispensable. Obviously, it is quite a fundamental operation.

Fortunately, a classical algorithm for computing the greatest common divisor of two integers a and b was given centuries ago by none other than Euclid. In a more modern treatment we may describe this in the form of the accompanying algorithm *euclid*. The style of presentation of this algorithm may be new to the student. We do not enter formally into the study of algorithms until Chapter 4. In the meantime, certain stylistic details may need to be overlooked.

algorithm euclid (a, b: *gcd*)

if a is less than b
    interchange their values
compute the remainder on dividing a by b
while we have a nonzero remainder
    replace a by b, b by remainder
    compute remainder on dividing a by b
assign *gcd* as b

Nevertheless, as indicated in this algorithm we first arrange that $a \geq b$ by interchanging their values, should this not be the case. The algorithm then hinges on the observation that every common divisor (say c) of a and b is also a divisor of r (the integer remainder obtained by dividing a by b):

$$a = bq + r$$

$$a = mc \quad \text{and} \quad b = nc \qquad \text{implies that} \qquad r = (m - nq)c$$

and we may show similarly that every common divisor of b and r is also a divisor of a. It follows that the pairs a, b and b, r have the same common divisors, so that having divided a by b, we can begin all over again with a replaced by b and b replaced by r. Because the successive remainders must decrease (to zero), the last value for b is the gcd of the original pair a, b.

Example 3.26 If $a = 180$ and $b = 252$, then we may tabulate the changing values of a, b, r as the algorithm is executed, as in Table 3.1. The reader should trace through the algorithm to see that these are indeed the values assumed. In any case, as a consequence we obtain:

$$\gcd(180, 252) = 36$$

as the last value of b is assigned to *gcd*. ■

TABLE 3.1

a	*b*	*r*	*gcd*
180	252		
252	180	72	
180	72	36	
72	36	0	36

Further Divisibility Questions

Two natural numbers a and b are said to be **relatively prime** if they have no common divisors (other than 1). Evidently, $\gcd(a, b) = 1$ in such cases. As a consequence, there is a simple algorithm to determine whether or not a and b are relatively prime—compute $\gcd(a, b)$ using the Euclidean algorithm:

if $\gcd(a, b) = 1$ then a and b are relatively prime (else not)

—it's as simple as that.

A more fundamental problem (but in fact, one whose solution could be used to solve nearly all of our problems) is the following. Given any natural number n, find a list of all of its divisors (or *factors*, as they are often called). A procedure for solving this problem is learned by students in elementary mathematics as outlined in the accompanying algorithm *factors*. Here, the problem: "check whether m divides n" is dealt with by an appeal to the division algorithm. That is, we simply divide m into n to see if we obtain a remainder of zero. Why do we stop when $m^2 > n$?

```
algorithm factors (n: list)

initialize list as empty
assign m equal to 1
until m² > n
   check whether m divides n
   if so
      adjoin m and n ÷ m to list
   increase m by one
```

Example 3.27 Suppose $n = 30$ and consider Table 3.2, paying particular attention to the developing list. Note that we stop after setting $m = 6$ since $6^2 > 30$. ■

TABLE 3.2

n	m	*list*
30	1	$\{1, 30\}$
	2	$\{1, 2, 15, 30\}$
	3	$\{1, 2, 3, 10, 15, 30\}$
	4	
	5	$\{1, 2, 3, 5, 6, 10, 15, 30\}$
	6	

It is clear that algorithm factors can be used to determine whether or not a number is prime. Recall that an integer $n > 1$ is *prime* if it has no divisors other than itself and one. Moreover, the algorithm can also be used to determine whether two integers are relatively prime or to find the gcd of two given integers. However, in these latter cases, it will be found that the Euclidean algorithm is the better alternative.

Example 3.28 To determine whether $n = 31$ is prime, we may use the algorithm factor as shown in Table 3.3. Because we find only 1 and 31 as divisors, we conclude, indeed, that 31 is a prime number. ■

TABLE 3.3

n	m	*list*
31	1	$\{1, 31\}$
	2	
	3	
	4	
	5	
	6	

Example 3.29 Taking $a = 180$ and $b = 252$, as in Example 3.26, we can use the algorithm factor to compute the following:

$$\text{Divisors of } 180 = \{1, 2, 3, 4, 5, 6, 9, 10, 12, 15, 18, 20, 30, 36, 45, 60, 90, 180\}$$
$$\text{Divisors of } 252 = \{1, 2, 3, 4, 6, 7, 9, 12, 14, 18, 21, 28, 36, 43, 63, 84, 126, 252\}$$

from which the *greatest common* divisor gcd $(180, 252) = 36$ could be determined. However, we note that this computation is considerably more difficult than that of Example 3.26. ■

EXERCISES 3.3

1 Compute the following integer divisions (as in Example 3.22).

(a) $15 \div 4$ **(b)** $15 \div 2$ **(c)** $27 \div 4$ **(d)** $27 \div 3$

(e) $39 \div 5$ **(f)** $39 \div 4$ **(g)** $39 \div 3$ **(h)** $97 \div 7$

2 Obtain quotients and remainders for a, b as given in Exercise 1, using the method of Example 3.24.

3 Obtain quotients and remainders for the following pairs:

(a) $a = 271, b = 14$
(b) $a = 137, b = 12$
(c) $a = 391, b = 22$
(d) $a = 401, b = 17$
(e) $a = 123, b = 12$
(f) $a = 794, b = 40$
(g) $a = 968, b = 22$
(h) $a = 637, b = 17$

using the method of Example 3.24.

4 In a formal proof of the division algorithm, we note that of all the multiples

$$b \cdot 0, b \cdot 1, b \cdot 2, b \cdot 3, \ldots$$

of b, there are only finitely many that are less than or equal to a. Show how this observation depends on the Archimedean Principle of Section 3.1.

5 In a formal proof of the division algorithm (see Exercise 4), taking $x = 0, 1, 2, 3, \ldots$, it is clear that the set

$$\{a - bx: bx \leq a\}$$

is a nonempty set of nonnegative integers and thus contains a least element. Show how this observation depends on the Well-ordering Principle of Section 3.1.

6 Construct a formal proof of the division algorithm. (*Hint:* See Exercises 4 and 5.)

7 Show that the quotient and remainder in the division algorithm are uniquely determined. (*Hint:* Suppose that $a = bq_1 + r_1$ and also $a = bq_2 + r_2$, with each r_i satisfying $0 \leq r_i < b$. Then show that $r_1 = r_2$ and $q_1 = q_2$.)

8 Give a precise definition of the term "least common multiple" and show that the identity

$$\operatorname{lcm}(a, b) = \frac{a \cdot b}{\gcd(a, b)}$$

must hold. (*Hint:* Pattern your definition of lcm after that for gcd.)

9 Use the algorithm euclid to compute greatest common divisors for the following pairs:

(a) $a = 48, b = 42$
(b) $a = 1350, b = 297$
(c) $a = 1144, b = 351$
(d) $a = 154, b = 140$
(e) $a = 297, b = 256$
(f) $a = 8024, b = 412$
(g) $a = 231, b = 770$
(h) $a = 728, b = 829$

10 Show that the gcd function satisfies the following properties:

(a) $a > b \Longrightarrow \gcd(a, b) = \gcd(a - b, b)$
(b) $\gcd(a, b) = \gcd(b, a)$
(c) $\gcd(a, a) = a$

---

for all $a, b \in N$.

11 Show explicitly that if

$$a = bq + r$$

then every common divisor of b and r is also a divisor of a.

12 Use the Euclidean algorithm to determine whether or not the following pairs of natural numbers are relatively prime.

(a) 64 and 77 **(b)** 48 and 51 **(c)** 58 and 91 **(d)** 377 and 493

13 Use the algorithm factors to compute all of the divisors of the following integers:

(a) 56 **(b)** 57 **(c)** 84 **(d)** 96
(e) 124 **(f)** 128 **(g)** 87 **(h)** 68

14 Use the algorithm factors to compute all of the divisors of the following integers:

(a) 3960 **(b)** 427 **(c)** 1281 **(d)** 832
(e) 1591 **(f)** 462 **(g)** 97 **(h)** 9021

15 Use the algorithm factors to determine whether or not the following integers are prime.

(a) 97 **(b)** 143 **(c)** 221
(d) 101 **(e)** 79 **(f)** 323

16 Compute the greatest common divisors of the pairs in Exercise 9, but using the algorithm factors instead.

17 Decide whether or not the pairs in Exercise 12 are relatively prime using the algorithm factors.

18 Compute least common multiples of the pairs in Exercise 9.

Sec. 3.4 BASE CONVERSION

One number system is just as good as another. It just happens that the binary system is more natural for use by computers.

It is important to appreciate the use of the binary number system in all phases of a computer's operation. There are very good reasons for this. A transistor, for example, is either conducting an electrical current or it is not. A magnetic device is magnetized in one direction or in the other. So it is that even at the level of the most simple device, there are only two states. Such considerations make it clear why the binary (referring to two) number system is the most natural one in the case of computers. Because we are accustomed to the use of the decimal system, it is necessary that we learn techniques for conversion from one base to another (for example, from base 10 to base 2, and conversely). Fortunately, these techniques are easily understood and easy to apply, particularly after the study of arithmetic that has preceded.

Goals

After studying this section, you should be able to:

- Discuss the concept of *positional notation* in a number system.
- Describe the use of the division algorithm in reference to the base conversion problem.
- Convert numbers (both integral and fractional parts) from base 2 to base 10.
- Convert numbers from base 10 to base 2 by two different methods.
- Convert numbers from base 8 (octal) and base 16 (hexadecimal) to base 2, and vice versa.
- Explain why the binary number system is more natural for use by computers.

Positional Notation

We have already indicated that it is more natural for computers to use a binary, or base 2, system of arithmetic, inasmuch as its component devices have only two configurations or states. We could refer to these two states as *off* and *on*, but it is easier to denote them by 0 and 1, respectively. They become the *bits* (*b*inary dig*its*) of a binary number system, playing the same role as do the digits of our familiar decimal system. Perhaps the fact that we are endowed with ten fingers accounts for the predominant use of the decimal system in our society. (We say this noting that the term "digit" refers to the finger.) Otherwise, one system is just as good as another. It happens that the aforementioned electronic limitations make the binary system a more natural one in the case of computers.

Fortunately, it is not difficult to become fairly adept in the use of the binary number system, in case you have not used it before. However, first, consider the decimal system. Let's take the number 178; we all recall that the three digits (let us call them $d_2 = 1$, $d_1 = 7$, and $d_0 = 8$) are meant to signify a number of hundreds, tens, and units, respectively, according to a strict **positional notation** system. Thus we may write:

$$\begin{aligned} d_2 d_1 d_0 = 178 &= 1 \times 100 + 7 \times 10 + 8 \times 1 \\ &= 1 \times 10^2 + 7 \times 10^1 + 8 \times 10^0 \\ &= d_2 \times 10^2 + d_1 \times 10^1 + d_0 \times 10^0 \end{aligned}$$

in base 10 [see Figure 3.4(a)]. Using base 2 notation, however, as in the binary system, we could examine a similar expression [see Figure 3.4(b)]:

$$\begin{aligned} 178 &= 1 \times 128 + 0 \times 64 + 1 \times 32 + 1 \times 16 + 0 \times 8 \\ &\quad + 0 \times 4 + 1 \times 2 + 0 \times 1 \\ &= 1 \times 2^7 + 0 \times 2^6 + 1 \times 2^5 + 1 \times 2^4 + 0 \times 2^3 \\ &\quad + 0 \times 2^2 + 1 \times 2^1 + 0 \times 2^0 \\ &= b_7 \times 2^7 + b_6 \times 2^6 + b_5 \times 2^5 + b_4 \times 2^4 + b_3 \times 2^3 \\ &\quad + b_2 \times 2^2 + b_1 \times 2^1 + b_0 \times 2^0 \end{aligned}$$

10^2	10^1	10^0
hundreds	tens	units
1	7	8

(a)

2^7	2^6	2^5	2^4	2^3	2^2	2^1	2^0
hundred twenty-eights	sixty-fours	thirty-twos	sixteens	eights	fours	twos	units
1	0	1	1	0	0	1	0

(b)

FIGURE 3.4

involving powers of 2 rather than powers of 10. Thus it appears that the bits $b_7 = 1$, $b_6 = 0$, $b_5 = 1$, $b_4 = 1$, $b_3 = 0$, $b_2 = 0$, $b_1 = 1$, $b_0 = 0$ provide us with a binary representation: 10110010, for this same number. Indeed, we note that in summing the right-hand side in the above, we obtain:

$$128 + 32 + 16 + 2 = 178$$

which is the original number being considered.

Conversion Between Binary and Decimal

All that we ever have to do in order to convert a binary system number into a standard decimal number is the following: *add up the appropriate powers of two* [remembering the meaning of the various bit positions, as indicated in Figure 3.4(b)].

Example 3.30 If we want to convert 110, 1011, 1101101 from binary to decimal, the procedure is always the same. We simply add up the appropriate powers of two, writing:

$$110 \longrightarrow 4 + 2 = 6$$
$$1011 \longrightarrow 8 + 2 + 1 = 11$$
$$1101101 \longrightarrow 64 + 32 + 8 + 4 + 1 = 109$$

while noting the meaning of the bit positions listed in Figure 3.4(b). (If we need higher-order bits—beyond "hundred-twenty-eights"—we can extend the counting: 256, 512, 1024, etc., as far as needed.) ■

Example 3.31 Using the above techniques, and counting from zero, it is clear that the first few binary numbers may be tabulated as shown in Table 3.4. As practice, the reader should extend the list from 16 to 31. ■

TABLE 3.4

Binary	*Decimal*	*Binary*	*Decimal*
0	0	1000	8
1	1	1001	9
10	2	1010	10
11	3	1011	11
100	4	1100	12
101	5	1101	13
110	6	1110	14
111	7	1111	15

We have just seen that the conversion of a binary number to a decimal number is quite easy. However, how do we perform the reverse transformation? How do we determine that 178 should be written as 10110010 in the binary system? In general, if we wish to convert a number n from decimal to binary, how should we proceed? We will describe two methods. Here is *method one*. We can first find the bit position (k) of the 1 farthest to the left by determining the highest power (2^k) of 2 that is less than or equal to n. In the example we have been considering ($n = 178$), this would be $k = 7$ because

$$2^7 = 128 \leq 178$$
$$2^8 = 256 \not\leq 178$$

From this point, we have only to subtract this power of 2 from n and begin as before, with this difference as a new n, to find the next-to-furthest-to-the-left bit that is 1. In fact, from this point on, we can simply count backwards with k to decide whether to set the remaining bits $b_{k-1} \ldots b_1 b_0$ to 1 or to 0. It all depends on whether or not the current k has a power (2^k) less than or equal to the current n. A little practice (after looking at an example) and you will see that this is quite a straightforward process.

Example 3.32 Using method one on the number $n = 178$, we may tabulate the changing values of n and the decreasing k as shown in Table 3.5. Note that n only

TABLE 3.5

	n	k	2^k	b_k	
	178	7	128	1	$(128 \leq 178)$
$178 - 128 =$	50	6	64	0	$(64 \not\leq 50)$
		5	32	1	$(32 \leq 50)$
$50 - 32 =$	18	4	16	1	$(16 \leq 18)$
$18 - 16 =$	2	3	8	0	$(8 \not\leq 2)$
		2	4	0	$(4 \not\leq 2)$
		1	2	1	$(2 \leq 2)$
$2 - 2 =$	0	0	1	0	$(1 \not\leq 0)$

decreases after we have set a bit to one. As the higher-order bits are determined first (with k decreasing), our resulting conversion may be summarized

$$178 \longrightarrow 10110010$$

(and not the other way around—see method two below). ■

In *method two* we make use of the division algorithm discussed in Section 3.3. In the expression

$$n = b_k 2^k + \cdots + b_1 2^1 + b_0 2^0$$

representing n in the positional notation for the base 2, we note that

$$r = b_0$$
$$q = b_k 2^{k-1} + \cdots + b_1 2^0$$

where q and r are the quotient and remainder upon dividing n by 2. It follows that b_0 is the first remainder, and upon successive divisions (of subsequent quotients) by 2, we will obtain b_1, b_2, etc., the bits of n *in increasing order.*

Example 3.33 If we take the same example ($n = 178$) as before, successive division by 2 yields the following sequence of quotients and remainders:

	q	r
2	178	
2	89	0
2	44	1
2	22	0
2	11	0
2	5	1
2	2	1
2	1	0
	0	1

Remembering that the bits are determined here in increasing order, we summarize by writing:

$$178 \longrightarrow 10110010$$

the same result as before. ■

Conversion of Fractional Parts

To present a complete picture of the base conversion problem, we have to realize that, in general, a number will have both an integral and a *fractional part.* Having only presented routines for converting integers from one base to another, our handling of the problem is not yet complete. In positional notation an arbitrary number N will then be given (say, in the decimal system) as a sequence of digits:

$$N = d_k d_{k-1} \cdots d_2 d_1 d_0 . d_{-1} d_{-2} \cdots d_{-j}$$

separated by a decimal point (as in 178.578125). In effect, we have

$$N = n + p$$

where n is the integral part of N and p is the fractional part. Because we already know how to transform integers from one base to another (particularly between base 2 and base 10), we may concentrate now on the fractional part. Obviously, the results of the separate conversions may be recombined to obtain the complete solution of the integral and fractional parts.

Taking the easier conversion first, suppose we are given a fractional number, for example, $p = 0.100101$, as expressed in base 2. How would we convert p to the decimal system? A little thought reveals that we only need to construct an extension of our earlier table of powers of two (see Figure 3.5) to treat negative powers. Then we have only to add up the appropriate powers of two *as expressed in the decimal system*, consulting Figure 3.5 (extended as necessary).

2^{-1}	2^{-2}	2^{-3}	2^{-4}	2^{-5}	2^{-6}
0.5	0.25	0.125	0.0625	0.03125	0.015625
halves	fourths	eighths	sixteenths	thirty-seconds	sixty-fourths

FIGURE 3.5

Example 3.34 If $p = 0.100101$, then we merely consult Figure 3.5 and add up the appropriate powers of two (expressed in the decimal system):

$$0.100101 \longrightarrow 0.5 + 0.0625 + 0.015625 = 0.578125$$

to compute the decimal equivalent. Similarly, we may convert:

$$0.00101 \longrightarrow 0.125 + 0.03125 = 0.15625$$

$$0.0001011 \longrightarrow 0.0625 + 0.015625 + 0.0078125 = 0.0859375$$

noting that in the previous example, it was necessary to extend the entries of Figure 3.5 by one more power of two. ■

For the reverse conversion, from base 10 to base 2, a simple extension of our method one is suggested. Consider the case of $p = 0.578125$ (where we already know the answer, by virtue of Example 3.34). Evidently we can again find the bit position (j) of the leftmost 1, by determining the first negative power (2^{-j}) of 2 that is less than or equal to p, just as before. In the example being considered, this would be $j = -1$, because we note that:

$$2^{-1} = 0.5 \leq 0.578125$$

Subtracting 0.5 from 0.578125, we continue as before, this time letting $-j$ run through the negative integers $-1, -2, -3$, etc.

Example 3.35 A complete tabulation (again, beginning with $p = 0.578125$) yields the results given in Table 3.6. Indeed this shows that

$$0.578125 \longrightarrow 0.100101$$

as expected. ■

TABLE 3.6

p	$-j$	2^{-j}	b_{-j}	
0.578125	-1	0.5	1	$(0.5 \leq 0.578125)$
$0.578125 - 0.5 = 0.078125$	-2	0.25	0	$(0.25 \nleq 0.078125)$
	-3	0.125	0	$(0.125 \nleq 0.078125)$
	-4	0.0625	1	$(0.0625 \leq 0.078125)$
$0.078125 - 0.0625 = 0.015625$	-5	0.03125	0	$(0.03125 \nleq 0.015625)$
	-6	0.015625	1	$(0.015625 \leq 0.015625)$
$0.015625 - 0.015625 = 0.0$				

The reader may find it interesting to develop a method two for converting fractional parts from base ten to base two. As a hint, we suggest that multiplication by 2 be used in place of the division by 2 that characterized our method two in the case of integral arguments. This would then provide a second alternative for handling the problem. However it is accomplished, one final word of warning should be given. We may not be able to represent the fractional part as a terminating expression in base 2, even though p is a terminating decimal at the start. In such cases, we will have to stop at some point with only an approximate conversion to base 2. This is inevitable, resulting from the fact that $\frac{1}{10}$ is not finitely expressible in base 2.

EXERCISES 3.4

1 Why is it more natural for a computer to use the binary number system?

2 Convert the following numbers from the binary system to the decimal system.

(a) 10000 **(b)** 1111

(c) 100000001 **(d)** 1100011

(e) 10101 **(f)** 1111101

(g) 10111011 (h) 11110111
(i) 1101110110 (j) 11110111011

3 Extend the list in Example 3.31 from 16 to 31.

4 Convert the following numbers from the decimal to the binary system using method one.

(a) 64 (b) 110 (c) 257 (d) 312 (e) 100
(f) 127 (g) 1329 (h) 4096 (i) 4174 (j) 5378

5 Perform the conversions of Exercise 4 using method two.

6 Convert the following fractional parts of numbers from binary to decimal notation.

(a) 0.1101101 (b) 0.111
(c) 0.0011 (d) 0.010101
(e) 0.1101101 (f) 0.00011011
(g) 0.000001 (h) 0.00001001
(i) 0.10000101 (j) 0.11111

7 Convert the following fractional parts of numbers from decimal to binary notation using (the extension of) method one.

(a) 0.606746 (b) 0.65625
(c) 0.046875 (d) 0.500390625
(e) 0.37890625 (f) 0.444678
(g) 0.1953125 (h) 0.3141529
(i) 0.1 (j) 0.333333

8 In developing a method two to convert fractional parts from base 10 to base 2, we note that the expression

$$p = b_{-1}2^{-1} + b_{-2}2^{-2} + \cdots + b_{-j}2^{-j}$$

representing p in the positional notation for base 2, shows that

$$b_{-1} = 1 \Leftrightarrow 2p \geq 1$$

in which case we have

$$2p - 1 = b_{-2}2^{-1} + \cdots + b_{-j}2^{-j+1}$$

and we are ready to multiply by 2 again, successively, to obtain b_{-2}, b_{-3}, etc., the bits of p in the conversion to binary. Note that we disregard any ones to the left of the decimal point in the successive multiplications, using them however to detect a one bit in the binary expansion. Use this method to convert the numbers of Exercise 7 from the decimal system to the binary system.

9 After examining the representations of an arbitrary number n (an integer) in *octal* (base 8) and in binary:

$$\begin{aligned} n &= e_j 8^j + \cdots + e_1 8^1 + e_0 8^0 \\ &= b_k 2^k + \cdots + (b_5 2^5 + b_4 2^4 + b_3 2^3) + (b_2 2^2 + b_1 2^1 + b_0 2^0) \end{aligned}$$

show that it is easy to convert from one base to the other by means of the following formulas:

$$e_0 = 4b_2 + 2b_1 + b_0$$
$$e_1 = 4b_5 + 2b_4 + b_3$$
$$\cdots$$

(*Note:* The octal digits run from 0 to 7. *Hint:* Equate quotients and remainders when dividing by 8.)

10 By analogy (to Exercise 9), derive simple formulas for conversion between the *hexadecimal* system (base 16) and the binary system. (*Note:* The 16 hexadecimal digits may be denoted $0, 1, 2, \ldots, 9, \mathrm{A}, \mathrm{B}, \mathrm{C}, \mathrm{D}, \mathrm{E}, \mathrm{F}$.)

11 Convert the following from binary to octal (see Exercise 9).

(a) 111011 **(b)** 101101001 **(c)** 10111100
(d) 100110101101 **(e)** 1101100101011 **(f)** 100001001101

12 Convert the following from octal to binary (see Exercise 9).

(a) 551 **(b)** 63 **(c)** 777
(d) 7451 **(e)** 6705 **(f)** 4337

13 Convert the following from binary to hexadecimal (see Exercise 10).

(a) 10011100 **(b)** 1111111
(c) 10011101 **(d)** 110110101011
(e) 1011110111111110 **(f)** 1011101111011111111

14 Convert the following from hexadecimal to binary (see Exercise 10).

(a) ABC **(b)** 6E2 **(c)** B7F
(d) 2DD7 **(e)** DEAD **(f)** 73A6

15 Convert the numbers in Exercise 14 from hexadecimal to octal (see Exercises 9 and 10).

16 Convert the numbers in Exercise 12 from octal to hexadecimal (see Exercises 9 and 10).

Sec. 3.5 THE RATIONAL NUMBERS

The rational numbers are the fractions from elementary arithmetic.

In large part, the material of this chapter represents a new look at a familiar subject matter from a more sophisticated point of view. Thus the natural numbers are the positive whole numbers from elementary arithmetic. Similarly, the rational numbers introduced here are simply the fractions we have encountered since our earliest schooling. Here we look at these numbers more carefully, in the hope that we will better understand their algebraic significance. In the case of the natural numbers, an important axiomatization was given, one that illuminated the inductive character of the whole collection. In the present case of the *rationals*, we show how these numbers are constructed out of the integers, with an aim toward solving a fundamental divisibility question.

Goals

After studying this section, you should be able to:

- Show how to construct the rational numbers as pairs of integers, and to define their arithmetic operations.
- Discuss the *fundamental equivalence relation* among rational numbers, and show how to test for equivalence.
- Establish the familiar arithmetic properties (associativity, commutativity, distributivity, etc.) of rational arithmetic.
- Show how the cardinality of the rational numbers is the same as that of the natural numbers.
- Verify that between any two rational numbers there lies another rational number.

Construction of the Rationals

We have seen in Section 3.3 that only an approximate division is available for integers. Thus if a, b are natural numbers, we may determine a quotient q and a remainder r by writing:

$$a = bq + r \qquad (0 \leq r < b)$$

As noted previously, the quotient q is sometimes denoted by $q = a \div b$. However, it should be remembered that this integer $a \div b$ is not, in general, an exact solution to the equation:

$$a = bx$$

(whereas we would think of *a divided by b* as signifying an exact solution). To solve such equations exactly, and thereby properly addressing the divisibility question, we need to extend the number system once again, to permit the treatment of *rational numbers*.

Example 3.36 It has been observed (by the author) that if n people enter a Mexican restaurant and order a plate of nachos, the number of nachos served is never congruent to 0 (modulo n). There's alway one left over. (I call it the Mexican

Remainder Theorem.) If 3 people place an order of nachos, most likely they will bring 7. But 3 doesn't go into 7. Even though $7 \div 3 = 2$, this hardly solves the divisibility question. Who gets the remaining nacho? (Obviously, it too should be divided into thirds. However, this takes us out of the realm of integer arithmetic.) ■

Such questions bring us directly to the study of *rational arithmetic*. In Example 3.36, we would like to treat 7/3 as an exact solution to the problem:

$$7 = 3x$$

and to think of 7/3 as a number in good standing. The trick is to consider pairs of integers to develop an entirely new number system. We look at the set:

$$Z \times N = \{(a, b): a \in Z \text{ and } b \in N\}$$

and think of the pair (a, b) as representing a fraction a/b. Note that a is the *numerator* and b is the *denominator*, and we have restricted the latter to the set N for the obvious reason that negative denominators can always be avoided (as in writing $a/-b = -a/b$). Moreover, we certainly wouldn't want to divide by zero, so N seems to be the best choice here.

Yet, there is still a problem. We don't want to distinguish between fractions that are *reducible*, one to another, as in $14/6 = 7/3$. For this reason, we introduce what turns out to be an equivalence relation on $Z \times N$ by writing:

$$(a, b) \sim (c, d) \Leftrightarrow ad = bc$$

where the test for equivalence (on the right) is performed entirely in the realm of integer arithmetic.

To see that we have introduced a true equivalence relation on $Z \times N$, we verify the necessary properties (of Section 1.4) as follows:

E1. $(a, b) \sim (a, b)$ because $ab = ab$

E2. If $(a, b) \sim (c, d)$ so that $ad = bc$, then

$$cb = bc = ad = da$$

thus showing that $(c, d) \sim (a, b)$

E3. Left as an exercise

The reader should note the use of the arithmetic properties of the integers to derive these results.

In considering the characterization of equivalence relations (as presented in Section 1.4), we note that the set $Z \times N$ is partitioned into equivalence classes by the relation just introduced. Our indistinguishability problem is then settled by defining a **rational number** to be an equivalence class

$$a/b = [(a, b)]$$

with respect to the above relation. Then we give the name Q (for quotients) to the set:

$$Q = \{a/b = [(a, b)]: a \in Z, b \in N\}$$

consisting of all these equivalence classes, regarded as quotients. Formally then, a rational number a/b is an equivalence class of pairs (a, b), with a an integer and b a natural number. This way, we make no distinction between the infinitely many representations that such a number might possess. The set Q of rational numbers is the set of all these equivalence classes. Thus Q is constructed out of the integers themselves.

Arithmetic Operations

It seems that we no sooner solve one problem than another one appears—this time, quite a subtle problem involving the definition of arithmetic operations in the rational number system. It would seem to be quite straightforward. If a/b and c/d are rational numbers, in the sense we have just understood, then it is natural to define *addition* and *multiplication* by writing:

$$a/b + c/d = (ad + bc)/bd$$
$$a/b \times c/d = ac/bd$$

for certainly this agrees with the reader's grade-school experience. On the other hand, there is no assurance that these operations are *well defined* in the sense of Section 1.3. Because we are dealing here with equivalence classes, there could conceivably be a problem because we have defined the results in terms of particular representations of the numbers.

We need to check that if $a/b = a'/b'$ and $c/d = c'/d'$, then the results we have listed (on the right, in the above definitions) are in agreement with those obtained by replacing a by a', b by b', c by c', and d by d', that is, that

$$(ad + bc)/bd = (a'd' + b'c')/b'd'$$

and

$$ac/bd = a'c'/b'd'$$

Otherwise, it could happen that our results will depend on the particular representations we have chosen for a/b and c/d, respectively. Fortunately, this is not the case.

Example 3.37 If $a/b = a'/b'$, then this is only because $(a, b) \sim (a', b')$, that is, $ab' = ba'$. Similarly, if $c/d = c'/d'$, then $cd' = dc'$. It follows that in computing

$$\begin{aligned}(ad + bc)b'd' &= adb'd' + bcb'd' \\ &= ab'dd' + bb'cd' \\ &= ba'dd' + bb'dc' \\ &= bda'd' + bdb'c' \\ &= bd(a'd' + b'c')\end{aligned}$$

we obtain $(ad + bc, bd) \sim (a'd' + b'c', b'd')$, and thus

$$(ad + bc)/bd = (a'd' + b'c')/b'd'$$

as required. We leave the proof that, similarly,

$$ac/bd = a'c'/b'd'$$

as an exercise for the reader. ∎

Arithmetic Laws

There remains the task of verifying that the usual arithmetic laws (I1a,b), (I2a,b), (I3), (I4a,b), and (I5)—from Section 3.1—may be extended to the rational number system. Fortunately, this is not difficult. We find that the fact that these properties hold for the integers will suffice. Of course, this is not surprising when we remember that the rationals were constructed out of the integers directly. However, in reference to (I4a,b) and (I5), it should be kept in mind that we represent an integer n in the rational number system as $n/1$—in particular:

$$0 = 0/1$$

and furthermore, we have $-(a/b) = -a/b$.

Example 3.38 To check the commutativity of addition (law (I1a)), we have only to compute:

$$a/b + c/d = (ad + bc)/bd = (cb + da)/db = c/d + a/b$$

making use of both commutative laws ((I1a) and (I1b)) for the integers. ∎

Example 3.39 To verify that $0 = 0/1$ has the property required in (I4a), we compute:

$$a/b + 0/1 = (a \cdot 1 + b \cdot 0)/b \cdot 1 = (a + 0)/b = a/b$$

using various arithmetic properties of the integers. ∎

Questions of Cardinality

In Section 3.1, after introducing the natural numbers, we proceeded to extend the number system by introducing zero and the negative numbers to obtain:

$$Z = \{\ldots, -3, -2, -1, 0, 1, 2, 3, \ldots\}$$

Now we have extended our number system once again, from Z to Q. In this context, we can ask to what extent the size of our number system has grown, provided of course that such a question makes sense.

How do we measure the size of a set, particularly an infinite set? Very early in our study (in Section 1.5), we stated a fundamental principle—*two sets have the same cardinality* (*size*) *if and only if there is a one-to-one correspondence between them.* This was called the characterization of equicardinality. More than likely, the reader thought of this principle only in reference to finite sets, for example, of sailors and coconuts. In fact, it is customary (and entirely plausible) to extend this principle to any sets whatsoever. What could be more natural than to ask that two sets be paired, member for member, if indeed we are to regard them as size-equivalent.

Consider the set Z, in comparison with N. On the one hand, there would seem to be (roughly) twice as many members of Z as there are members of N. For each natural number n, we find both n and $-n$ in Z. On the other hand, the mapping:

$$\begin{aligned} 1 &\to 0 \\ 2 &\to 1 \\ 3 &\to -1 \\ 4 &\to 2 \\ 5 &\to -2 \\ 6 &\to 3 \\ 7 &\to -3 \\ &\cdots \end{aligned}$$

establishes a one-to-one correspondence of N with Z. And according to the principle we have just agreed to adopt, we have to regard N and Z as being of the same size. This conclusion should not be confusing to the reader, even though it is somewhat counterintuitive.

Now we consider the rational numbers, where our conclusion is even more spectacular. Every integer $n \in Z$ appears in Q in the form $n/1$. Yet you may have the feeling that there are so many more rational numbers (for example, see Theorem 3.4). This feeling is dispelled once and for all by the following statement:

Theorem 3.3 ***Countability of the Rationals*** The set of rational numbers is in one-to-one correspondence with the set of natural numbers.

Proof It is enough to show that this is the case for the positive rational numbers, for then the one-to-one correspondence used in the case of Z can be applied. We imagine that all of the positive rational numbers appear in a two-dimensional array. We list all those rationals with numerator 1 in column 1, those with numerator 2 in column 2, etc., and at the same time, we list all those rationals with denominator 1 in row 1, those with denominator 2 in row 2, etc. In general, the rational number a/b appears in row b and column a, as pictured in Figure 3.6. Now all we need to do is to establish a process for reaching each and every one of the rationals a/b once and only once *in turn*, so that we can associate the *first* one we reach with the natural number 1, the *second* one we reach with the natural number 2, and so on, thus establishing a one-to-one correspondence with N.

$$\begin{array}{ccccccc} 1/1 & 2/1 & 3/1 & 4/1 & 5/1 & 6/1 & \ldots \\ 1/2 & 2/2 & 3/2 & 4/2 & 5/2 & 6/2 & \cdots \\ 1/3 & 2/3 & 3/3 & 4/3 & 5/3 & 6/3 & \cdots \\ 1/4 & 2/4 & 3/4 & 4/4 & 5/4 & 6/4 & \cdots \\ 1/5 & 2/5 & 3/5 & 4/5 & 5/5 & 6/5 & \cdots \\ 1/6 & 2/6 & 3/6 & 4/6 & 5/6 & 6/6 & \cdots \\ \vdots & \vdots & \vdots & \vdots & \vdots & \vdots & \end{array}$$

FIGURE 3.6

We do this by traversing in a snake-like path, starting at the top left corner, moving back and forth through successive diagonals as shown in Figure 3.7. Of

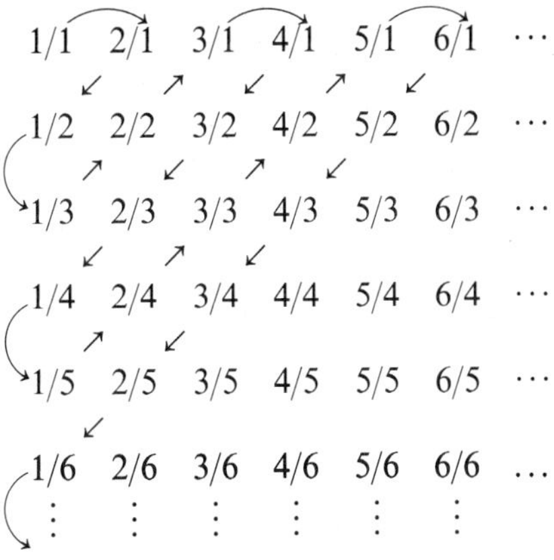

$$\begin{array}{ccccccccccccc} 1/1 & \rightarrow & 2/1 & & 3/1 & \rightarrow & 4/1 & & 5/1 & \rightarrow & 6/1 & & \cdots \\ & \swarrow & & \nearrow & & \swarrow & & \nearrow & & \swarrow & & & \\ 1/2 & & 2/2 & & 3/2 & & 4/2 & & 5/2 & & 6/2 & & \cdots \\ \downarrow & \nearrow & & \swarrow & & \nearrow & & \swarrow & & & & & \\ 1/3 & & 2/3 & & 3/3 & & 4/3 & & 5/3 & & 6/3 & & \cdots \\ & \swarrow & & \nearrow & & \swarrow & & & & & & & \\ 1/4 & & 2/4 & & 3/4 & & 4/4 & & 5/4 & & 6/4 & & \cdots \\ \downarrow & \nearrow & & \swarrow & & & & & & & & & \\ 1/5 & & 2/5 & & 3/5 & & 4/5 & & 5/5 & & 6/5 & & \cdots \\ & \swarrow & & & & & & & & & & & \\ 1/6 & & 2/6 & & 3/6 & & 4/6 & & 5/6 & & 6/6 & & \ldots \\ \downarrow\vdots & & \vdots & & \vdots & & \vdots & & \vdots & & \vdots & & \end{array}$$

FIGURE 3.7

course, the reader can object by saying that we obtain duplications here, as in $2/3 = 4/6$. But in developing our listing:

$$1/1 \quad 2/1 \quad 1/2 \quad 1/3 \quad \not{2}/\not{2} \quad 3/1 \quad 4/1 \quad 3/2 \quad 2/3 \quad 1/4 \quad \ldots$$

we can easily arrange to eliminate such duplications as we go (there are only a finite number of previous fractions to be considered in comparison at any one stage). The result is then a listing of the positive rational numbers without duplication, a listing in one-to-one correspondence with the natural numbers, thus establishing the theorem. □

So, the rational numbers are countable. By this we of course do not mean that we count them in a finite number of steps, as we would a finite set. We mean only that the set Q is commensurate with N, and that in counting with N indefinitely, we encounter (according to the one-to-one correspondence we established) each rational number in turn. That is in fact the whole sense of the proof of Theorem 3.3. In general, this notion of **countability** (having the cardinality of N) serves to circumscribe the range of *discrete* mathematics. If Theorem 3.3 were not true, the rational numbers would not have a place in our study. They would lie outside the realm of discrete mathematics. However, if this notion of countability sets the boundary, we might ask if there is anything outside. Do there exist *uncountable* sets, in this sense? We leave this question for the reader to ponder.

Ordering of the Rationals

Throughout our discussion of the rational numbers, we have been able to generalize the ideas of integer arithmetic without much difficulty. On the whole, we have been guided by the study of fractions from our earliest schooling. The situation is no different now, as we discuss the notion of *order*. In complete agreement with everything the reader has learned, we can say that $a/b < c/d$ if it should happen that

$$ad < bc$$

in the arithmetic of the integers. We do not take the trouble to establish the fact that this definition satisfies the usual properties of order relations—(O1), (O2), (O3) from Section 2.1. Instead, these are left as exercises, and we proceed to consider a more interesting result.

We may contrast Theorem 3.3 (saying that there are not very many rational numbers) with the following theorem.

Theorem 3.4 Between any pair of (distinct) rational numbers, there lies another rational number.

Proof Let a/b and c/d be the rational numbers in question. And without loss of generality, we may assume that $a/b < c/d$ (so that $ad < bc$). Intuitively, we would like to claim that the average of these two numbers meets the requirement of the theorem. Because

$$\frac{a/b + c/d}{2} = \frac{ad + bc}{2bd} = (ad + bc)/2bd$$

is clearly a rational number, it is a genuine candidate. In fact, we have

$$a(2bd) = 2abd = 2bad = bad + bad < bad + bbc = b(ad + bc)$$

which shows that $a/b < (ad + bc)/2bd$. Similarly, we can show that the latter is less than c/d, as required. □

EXERCISES 3.5

1 Show that a/b is an exact solution to the equation

$$a = bx$$

(*Hint:* Treat the integers a, b as rationals $a/1$ and $b/1$.)

2 Verify that property (E3) is satisfied for the relation

$$(a, b) \sim (c, d) \qquad ad = bc$$

on the set $Z \times N$.

3 Complete Example 3.37 by showing that

$$ac/bd = a'c'/b'd'$$

assuming as before that $a/b = a'/b'$ and $c/d = c'/d'$.

4 Verify each of the following laws in the arithmetic of the rational number system:

(a) Law (I1b) **(b)** Law (I2a) **(c)** Law (I2b)

(d) Law (I3) **(e)** Law (I4b) **(f)** Law (I5)

5 Show that the definition for $a/b < c/d$ satisfies the properties (O1), (O2), (O3) listed in Section 2.1.

6 Complete the proof of Theorem 3.4 by showing that

$$(ad + bc)/2bd < c/d$$

as required.

7 Complete the argument (as suggested in the proof of Theorem 3.3) that shows that there is a one-to-one correspondence of N *with all of* the rational numbers. (*Hint:* First show that we can establish a one-to-one correspondence of N with the negative rationals in exactly the same way as we have done for the positive rationals.)

8 Show that the set $N \times N$ is countable. (*Hint:* Establish a one-to-one correspondence of N with $N \times N$, patterning your scheme after the proof of Theorem 3.3.)

9 Show by induction that each set N^k is countable ($k \geq 1$). (*Hint:* See Exercise 8.)

10 Exhibit a one-to-one correspondence of N with:

(a) The positive even integers

(b) The positive odd integers

(c) All even integers

(d) All odd integers

(e) All natural numbers divisible by 7

(f) The set A^* (for the alphabet $A = \{a, b\}$)

Sec. 3.6 COMPUTER ARITHMETIC

A computer makes precise computations within an imprecise number system.

It is important to realize that arithmetic computations are performed quite differently by machine than by humans. Barring malfunction, a computer will perform its calculations with absolute precision. Yet, there are inherent limitations to the resulting accuracy caused by the artificial number system employed in its implementation of ordinary arithmetic. By no means is the machine operating in our rational number system. Only a scaled-down version of this arithmetic is available, in a normalized exponential number system of one form or another. Limited in their precision and only approximate in their arithmetic, we will find that such systems permit miscalculations that can overshadow the precision of the execution. We need to have a keen awareness of these errors to interpret properly the results of a computation.

Goals

After studying this section, you should be able to:

- Describe the parameters (base, precision, range) of a *normalized exponential number system.*
- Explain why it is that normalized exponential number systems are finite.
- Describe the representation of numbers in these systems, both conceptually and internally (within the machine).
- Discuss the normalized exponential arithmetic and provide examples illustrating its inexactness.
- Represent numbers in normalized exponential form (both in base 2 and base 10).
- Describe the notions of *absolute* and *relative error* in an approximation.
- Explain the effects of *subtractive cancellation.*
- Show how computational errors can accumulate.

Normalized Exponential Systems

The type of number system that will be used by a computer is dictated to a large extent by hardware considerations. Generally, it is found to be most convenient to partition a computer's memory into "words" of equal length and then, to let each word be capable of storing a number. Because rational numbers may involve arbitrarily large numerators and/or denominators, it is not possible for the computer to provide a simulation of rational arithmetic. Nevertheless, computers are able to simulate a truncated version of the rational number system, using what are called (*normalized*) *exponential representations of numbers.*

Example 3.40 If we wish to handle very large numbers (for example, 3,000,000,000) and also very small numbers (for example, 0.000000003), and we seek to retain a fixed position for a decimal point, as is necessary in performing arithmetic, then our number representation is evidently quite inefficient in its space utilization. We would have to allow so many positions to the left of the decimal point, say, and these would not be used if we were representing a very small number. A similar comment is appropriate in the opposite direction. The solution here is to borrow an idea from the scientific notation of high-school physics, to represent a number x as a normalized mantissa multiplied by an appropriate power of 10 (in the decimal system). Thus we would write:

$$x = 0.3 \times 10^{10}$$

or

$$x = 0.3 \times 10^{-8}$$

in reference to the two numbers appearing at the beginning of the example. ∎

More generally, in a **normalized exponential number system** (with *base* $b > 1$, *precision* $t > 0$, and *range* $s > 0$), every number x has the form:

$$x = f \times b^e$$

where e is called the **exponent** and f the **mantissa** of the number x. Furthermore, the following is required as determined by the three *parameters* (b, t, s) of the system:

---

1. $b^{-1} \leq |f| < 1$
2. f has at most t significant positions (that is, bits, digits, etc.)
3. $-S \leq e < S$

(S is computed from s in a manner to be discussed later.) Note that (1) is the *normalization condition* that requires that there be a digit immediately to the right of the decimal point (for example, the 3 of Example 3.40 follows the decimal point),

(2) is a consequence of the equal-length word requirement, and (3) establishes a range on the exponents available.

Example 3.41 Table 3.7 illustrates the representation of several numbers in a (base 10) normalized exponential system. Here we have $b = 10$ and $t = 4$. Note that mantissas can be positive or negative, and of course, the same is true of exponents. Note further the truncation in the last illustration, in keeping with requirement (2) listed previously. ■

TABLE 3.7

Number	*Normalized exponential form*	*Mantissa*	*Exponent*
333.3	0.3333×10^3	0.3333	3
0.0033	0.3300×10^{-2}	0.3300	−2
−12.34	-0.1234×10^2	−0.1234	2
0.55	0.5500×10^0	0.5500	0
−7777.7	-0.7777×10^4	−0.7777	4

Example 3.42 We may also consider base 2 normalized exponential systems. (In fact, this is more than likely the representation found in a computer because of the considerations discussed in Section 3.4.) A tabulation of such representations is shown in Table 3.8. Here we have a binary system with $b = 2$ (obviously) and $t = 5$. Moreover, it is more than likely that the exponents in such a system would also be represented in binary. We have left them in decimal here for ease of interpretation. ■

TABLE 3.8

Number	*Normalized exponential form*	*Mantissa*	*Exponent*
1010.1	0.10101×2^4	0.10101	4
0.001111	0.11110×2^{-2}	0.11110	−2
−111	-0.11100×2^3	−0.11100	3
0.1	0.10000×2^0	0.10000	0
−0.0101011	-0.10101×2^{-1}	−0.10101	−1

Internal Representation

Let us consider the binary system and imagine that each memory word is divided into three *fields* or blocks of bits. The first field (one bit) is reserved for the sign of the number—usually 0 = positive and 1 = negative. The second field stores the exponent, and the third field stores the mantissa. Figure 3.8 illustrates this situation in the case where $t = 24$ and $s = 7$.

As we see, the parameter s gives the number of positions (here, bit positions) devoted to the exponent. Of course, the larger s is, the larger the range of exponents

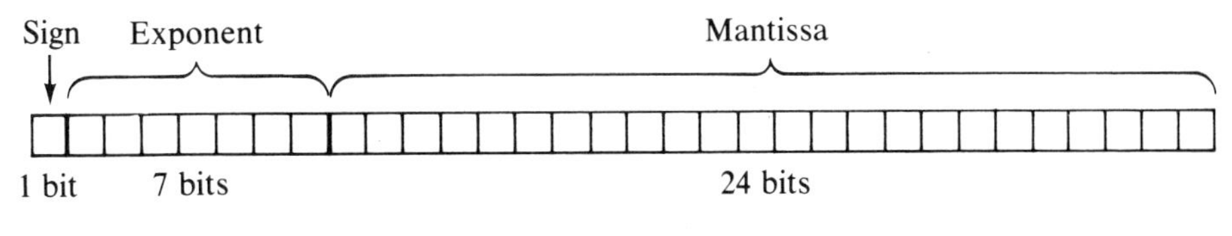

FIGURE 3.8

that can be accommodated. As a matter of fact, the exponent e in the number

$$x = f \times 2^e$$

is usually represented internally by means of its **characteristic**

$$c = e + S$$

where $S = 2^{s-1}$, thus ensuring that only nonnegative integers c need be considered. Moreover, it is clear that exponents e in the range

$$-S \leq e < S$$

may then be treated, as discussed in requirement (3) earlier.

Example 3.43 With $s = 7$, we have the formula:

$$c = e + 64$$

and accordingly, we are able to accommodate exponents in the range:

$$-64 \leq e < 64$$

having noted that $S = 2^{s-1} = 2^6 = 64$. We provide a tabulation in Table 3.9 (again with $s = 7$) that shows the relationship between exponents and their characteristics. ■

TABLE 3.9

e	-64	-63	-62	$\cdots$	-1	0	1	2	$\cdots$	62	63
c	0	1	2	$\cdots$	63	64	65	66	$\cdots$	126	127

Example 3.44 Suppose we are given the number $x = -178.578125$ (in the decimal system), and we ask the question "How is this number stored internally in a binary normalized exponential system, given that $s = 7$ and $t = 24$?" Using the results of Examples 3.32 and 3.35, we know that a conversion of x to the binary system yields:

$$x = -10110010.100101$$

Now the normalized exponential form of x (again, in binary) is

$$x = -0.10110010100101 \times 2^8$$

with an exponent $e = 8$. Accordingly, the characteristic is

$$c = 8 + 64 = 72 \longrightarrow 1001000$$

Note the conversion to base 2. It follows that x has the internal representation shown in Figure 3.9. ■

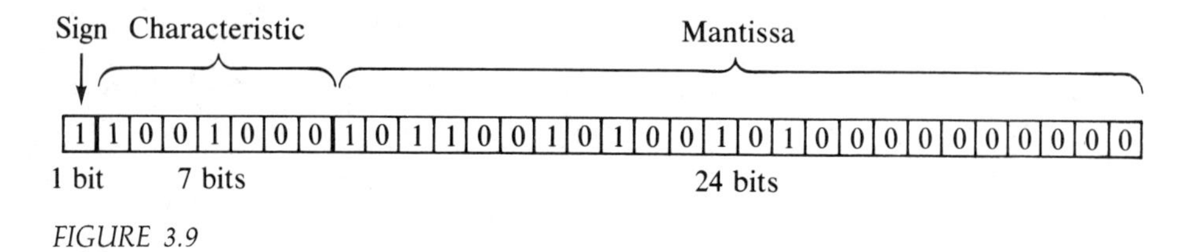

FIGURE 3.9

Arithmetic Operations

We are now ready to discuss the important matter of how computer arithmetic is performed. We assume that all numbers are stored and processed in their exponential form. Sometimes this is called **floating point arithmetic**, because we observe momentarily that the decimal point must float back and forth to line up the arguments appearing in the various operations. Moreover, all mantissas are truncated to t digits, and as we shall see, this creates errors in the computations, even if we began with exact figures at the start.

Suppose, as before, that we have a normalized exponential system with parameters (b, t, s). First let us discuss *computer addition.* If the two numbers to be added happen to have the same exponent, the mantissas are added and the same exponent is used for the sum. This is only natural. On the other hand, if the two numbers have different exponents, then one of the numbers must be adjusted (typically, the one with the smaller exponent) so that the exponents agree before the addition is performed. In any case, the sum is truncated if necessary, so that only t significant positions are retained. Similar remarks apply to the case of *subtraction.*

We assume throughout the next several examples that we are dealing with a decimal system with $t = 4$, even though we know that a binary exponential form is actually used by a computer. This way, the computations are easier to follow.

Example 3.45 For addition, we list several calculations of varying circumstances, hoping that the reader will be able to generalize as necessary. We have:

$$0.1234 \times 10^4 + 0.4017 \times 10^4 = 0.5251 \times 10^4$$

$$0.5542 \times 10^2 + 0.7777 \times 10^2 = 1.3319 \times 10^2 = 0.1331 \times 10^3$$

$$0.1234 \times 10^2 + 0.4017 \times 10^4 = 0.0012 \times 10^4 + 0.4017 \times 10^4 = 0.4029 \times 10^4$$

Note in the second equation that an overflow caused a renormalization of the sum, and note particularly the necessary adjustment of the first argument before the addition was performed in the last equation. Both of these circumstances have led to errors in the sum, as the reader has surely noticed. ■

Example 3.46 As we have said, the situation in the case of subtraction is entirely analogous. Again we cite three illustrations:

$$0.4017 \times 10^{-2} - 0.1234 \times 10^{-2} = 0.2783 \times 10^{-2}$$
$$0.7777 \times 10^{3} - 0.7531 \times 10^{3} = 0.0246 \times 10^{3} = 0.2460 \times 10^{2}$$
$$0.6666 \times 10^{3} - 0.3333 \times 10^{2} = 0.6666 \times 10^{3} - 0.0333 \times 10^{3} = 0.6333 \times 10^{3}$$

hoping that these will be sufficient for understanding the general ideas involved. Note particularly the *subtractive cancellation* in the second instance, which causes the appearance of an insignificant 0 in the result. ■

In establishing rules for computer *multiplication* and *division*, again there are no surprises in store. To multiply two numbers, we multiply their mantissas and add their exponents, whereas in dividing two numbers, we divide their mantissas and subtract their exponents. We have only to remember that results are renormalized as necessary and truncated to t positions, just as before.

Example 3.47 In multiplying the following pairs of numbers, we obtain:

$$0.3355 \times 10^{2} \times 0.4466 \times 10^{3} = 0.1498 \times 10^{5}$$
$$0.4444 \times 10^{-5} \times 0.3579 \times 10^{2} = 0.1590 \times 10^{-3}$$
$$0.1111 \times 10^{2} \times 0.2222 \times 10^{4} = 0.2468 \times 10^{5}$$

noting again the truncation errors in each case. Moreover, in the last instance, it matters whether the machine normalizes before or after truncation. We have assumed it is before truncation. ■

Example 3.48 Similar remarks are applicable to the three following illustrations of division:

$$0.3333 \times 10^{-4} / 0.6543 \times 10^{2} = 0.5093 \times 10^{-6}$$
$$0.4444 \times 10^{7} / 0.1357 \times 10^{4} = 0.3274 \times 10^{4}$$
$$0.8877 \times 10^{2} / 0.2121 \times 10^{4} = 0.4185 \times 10^{-1}$$

To appreciate their significance, the reader should probably carry out the operations on a pocket calculator, just to see the errors involved. ■

Errors of Computation

We have seen that a computer only retains a limited, fixed number of significant digits in its calculations, and it is clear this is a major source of errors in all machine computation. To speak intelligently about this situation, however, some familiarity with the notions of absolute and relative errors is necessary. Again, we borrow from the ideas usually encountered in high-school physics.

The difference between the true value and an approximate value of a number x is called the **absolute error** (in the approximation). Thus if x is approximated by x', the absolute error is

$$e = x - x'$$

though some authors will use the absolute value of this difference instead. In any case, it is usually more meaningful to speak of the **relative error**, the quotient of the (absolute) error by the true value:

$$r = e/x = (x - x')/x$$

and to express the result as a percentage.

Example 3.49 In the second addition of Example 3.45, the true value of the sum is 1.3319×10^2, but because of the truncation in renormalizing, the sum is represented as 0.1331×10^3. The absolute error in this approximation is:

$$e = 0.00009 \times 10^3 = 0.09$$

and the relative error is

$$r = e/x = 0.09/1.3319 \times 10^2 \approx 0.0007$$

or 0.07% (less than $\frac{1}{10}$ of 1%). This may seem small, even insignificant to the reader. On the other hand, small errors occurring over and over again can accumulate to the point where the answer obtained in a long computation can be entirely meaningless. ■

On the one hand, the individual *truncation error* of each individual machine approximation is quite small as we see in Theorem 3.5.

Theorem 3.5 The relative error in truncating to t significant positions is bounded by the figure b^{-t+1}.

Yet, the difficulties inherent in long, repetitive calculation (as is quite common in many applications) is easily seen by considering Theorem 3.6.

Theorem 3.6 The relative error in computing the product of two numbers is the sum of the relative errors in the factors.

Proof If x and y are approximated by x' and y', respectively, we have

$$x' = x(1 - r_x) \quad \text{and} \quad y' = y(1 - r_y)$$

where r_x and r_y are the relative errors in these two approximations. Because

$$\begin{aligned} x'y' &= xy(1 - r_x)(1 - r_y) = xy(1 - (r_x + r_y) + r_x r_y) \\ &\approx xy[1 - (r_x + r_y)] \end{aligned}$$

we have, in neglecting $r_x r_y$ in comparison with the sum:

$$r = r_x + r_y$$

as the relative error in the product. □

Example 3.50 If the approximate sum in Example 3.49 is subsequently multiplied by itself 100 times, the relative error in the result would approach 10%. ■

The Boolean Number System

At the lowest level of computer hardware, in the bit position of our computer arithmetic, a most elementary number system is employed. It is a system of ultimate simplicity, owing to the fact that only two values, 0 and 1, are possible. If we let

$$B = \{0, 1\}$$

denote this set of values, we may define two operations, addition and multiplication, on B as follows:

+	0	1
0	0	1
1	1	0

·	0	1
0	0	0
1	0	1

Thus we have

$$\begin{array}{ll} 0 + 0 = 0 & 0 \cdot 0 = 0 \\ 0 + 1 = 1 & 0 \cdot 1 = 0 \\ 1 + 0 = 1 & 1 \cdot 0 = 0 \\ 1 + 1 = 0 & 1 \cdot 1 = 1 \end{array}$$

in simply reading the tables. The reader may check that the usual arithmetic laws of commutativity, associativity, etc., are satisfied by this rudimentary number system. We call it the **Boolean number system** after George Boole (1815–1864), an English mathematician and logician.

In a chapter devoted to discrete number systems, it is more than appropriate that these Boolean numbers finally make an appearance. What could be more discrete? We will find that this elementary number system plays a crucial role in some of our later developments. Of course, we have tacitly assumed that the student will be able to perform such arithmetic calculations as are required in the mechanics that underlie much of our earlier discussions of this section.

Example 3.51 In adding 5 and 4 (in binary arithmetic) we would have:

$$\begin{array}{r} 1\ 0\ 1 \\ 1\ 0\ 0 \\ \hline 1\ 0\ 0\ 1 \end{array}$$

with a "carry" into the fourth bit position. But note that one plus one equals zero in this system of arithmetic, just as we have specified in the addition table for our Boolean number system. ■

EXERCISES 3.6

1 Discuss the significance of the requirement:

$$b^{-1} \le |f| < 1$$

in a normalized exponential system. In what sense are numbers $x = f \times b^e$ normalized by this requirement?

2 Express the following numbers in normalized exponential form (in the base 10), using $t = 4$ and truncating if necessary.

(a) 3333.33 **(b)** -1000.0 **(c)** 0.73
(d) 0.00084271 **(e)** -3.14 **(f)** 999.9

3 Express the following numbers in normalized exponential form (in the base 2), using $t = 5$ and truncating if necessary.

(a) 11111. **(b)** 101.1101 **(c)** -0.0011
(d) 0.0010101 **(e)** 1101.1 **(f)** -0.0111

4 Obtain a table (as in Example 3.43) to convert the allowable exponents to their characteristics given the following:

(a) $s = 6$ **(b)** $s = 8$ **(c)** $s = 5$

5 Obtain internal representations (in binary) as in Example 3.44 for the following decimal numbers.

(a) 4174.1 **(b)** 312.500390625
(c) -257.444678 **(d)** 1329.606746
(e) -4174.333333 **(f)** 110.65625
(g) 5378.1953125 **(h)** -127.3141529

Assume the parameters $s = 7$ and $t = 24$ as before.

6 Obtain the results of computer addition (as in Example 3.45) for the following pairs of numbers, assuming $t = 4$ as before.

(a) $0.3516 \times 10^2 + 0.2609 \times 10^2$ **(b)** $0.4173 \times 10^{-3} + 0.6983 \times 10^{-3}$

(c) $0.5179 \times 10^{-3} + 0.6333 \times 10^{-6}$ **(d)** $0.8444 \times 10^5 + 0.7999 \times 10^7$

(e) $0.6317 \times 10^6 + 0.9999 \times 10^2$ **(f)** $0.8097 \times 10^{-2} + 0.6666 \times 10^1$

7 Obtain the results of computer subtraction (as in Example 3.46) for the following pairs of numbers, assuming $t = 4$ as before.

(a) $0.3516 \times 10^2 - 0.2609 \times 10^2$ **(b)** $0.6134 \times 10^{-3} - 0.2177 \times 10^{-3}$

(c) $0.7217 \times 10^4 - 0.6666 \times 10^1$ **(d)** $0.8417 \times 10^{-6} - 0.9999 \times 10^{-8}$

(e) $0.7134 \times 10^{-2} - 0.7125 \times 10^{-2}$ **(f)** $0.8316 \times 10^5 - 0.9191 \times 10^0$

8 Obtain the results of computer multiplication (as in Example 3.47) for the following pairs of numbers, assuming $t = 4$ as before.

(a) $0.5432 \times 10^3 \times 0.3333 \times 10^{-5}$ **(b)** $0.2228 \times 10^3 \times 0.1117 \times 10^1$

(c) $0.3333 \times 10^{-2} \times 0.8176 \times 10^5$ **(d)** $0.1111 \times 10^4 \times 0.1111 \times 10^3$

(e) $0.9999 \times 10^{-3} \times 0.8888 \times 10^7$ **(f)** $0.3317 \times 10^7 \times 0.1234 \times 10^{-7}$

9 Obtain the results of computer division (as in Example 3.48) for the following pairs of numbers, assuming $t = 4$ as before.

(a) $0.2233 \times 10^3 / 0.6611 \times 10^{-5}$ **(b)** $0.1119 \times 10^3 / 0.4488 \times 10^1$

(c) $0.8317 \times 10^{-2} / 0.1111 \times 10^5$ **(d)** $0.2121 \times 10^4 / 0.9803 \times 10^3$

(e) $0.1039 \times 10^{-3} / 0.8173 \times 10^7$ **(f)** $0.8888 \times 10^7 / 0.1427 \times 10^{-7}$

10 Suppose that 91,637,146 miles, the mean distance between the earth and sun, is approximated as 92,000,000 miles. Find the absolute error and the relative error in the approximation.

11 Find a bound on the relative truncation error in a binary machine representation with the parameter as follows:

(a) $t = 32$ **(b)** $t = 16$ **(c)** $t = 8$

12 Suppose that we want to find the difference $x - y$ of two nearly equal numbers, say

$$x = 222.88 \quad \text{and} \quad y = 222.11$$

and we are dealing with a (decimal) computer representation where all numbers are truncated to four significant digits ($t = 4$). Find the relative error in the resulting approximation to the difference. (*Note:* Such examples illustrate the effects of subtractive cancellation. The loss of significant digits when two nearly equal numbers are subtracted is the source of some of the most series computational errors.)

13 Find a bound on the relative error when two approximate numbers are added (in terms of the relative errors in the arguments).

Chapter Four

THE NOTION OF AN ALGORITHM

Contents

4.1 Problem-solving Principles
4.2 Toward an Algorithmic Language
4.3 Extensions of the Language
4.4 Proofs of Correctness
4.5 Recursive Algorithms
4.6 Introduction to Computational Complexity

Stephen Kleene (1909–)

Stephen Kleene was born in Hartford, Connecticut in 1909. His father was a professor of economics and his mother an accomplished poet. Kleene received his doctorate at Princeton University in 1934, where he was a student of Alonzo Church. These were exciting times in his chosen field of mathematical logic. Kleene's major lifelong interest—recursive function theory—was created during his stay at Princeton as a synthesis of his own work and that of Church, Gödel, Post, and Turing. This theory seeks to answer the question, What is an algorithm? This question has proved to be most important in the development of computer science. Kleene, through his writings, has been most influential in bringing this theory to its current height of development. He has spent most of his illustrious academic career at the University of Wisconsin, where he holds the rank of Emeritus Professor.

Writings

We know examples in mathematics of general questions, such that any particular instance of the question can be answered by a preassigned uniform method. More precisely, in such an example, there is an infinite class of particular questions, and a procedure in relation to that class, both being described in advance, such that if we thereafter select any particular question of the class, the procedure will surely apply and lead us to a definite answer, either "yes" or "no," to the particular question selected.

A method of this sort, which suffices to answer, either by "yes" or by "no," any particular instance of a general question, we call a decision procedure *or* decision method *or* algorithm *for the question. The problem of finding such a method is called the* decision problem *for the question. Similarly, we may have a* calculation procedure *or* algorithm *(and hence a* calculation problem*) in relation to a general question which requires for an answer, not "yes" or "no," but the exhibiting of some object.*

When we formalize (such) a theory, to do so rigorously, it is practically necessary to reconstruct the theory in a special symbolic language, i.e. to symbolize *it. Instead of carrying out the steps described above on the theory as we find it in some natural word language, such as Greek or English, we build a new symbolic language specially for the purpose of expressing the theory. The natural word languages are too cumbersome, too irregular in construction and too ambiguous to be suitable.*

This new language will be of the general character of the symbolism which we find in mathematics. In algebra, we perform deductions as formal manipulations with equations, which would be exceedingly tedious to perform in ordinary language, as some of them were before the invention by Vieta (1591) and others of the modern algebraic notations. The discovery of simple symbolic notations which lend themselves to manipulation by formal rules has been one of the ways by which modern mathematics has advanced in power. However, the ordinary practice in mathematics illustrates only a partial symbolization and formalization, since part of the statements remain expressed in words, and part of the deductions are performed in terms of the meanings of the words rather than by formal rules.

Introduction to Metamathematics **(Kleene)**

Sec. 4.1 PROBLEM-SOLVING PRINCIPLES

A problem is a class of questions. Each question in the class is an instance of the problem.

We have all experienced the sense of satisfaction for having solved a problem, be it large or small. Those of us who consider ourselves to be particularly inclined toward the sciences or toward mathematics in our educational pursuits have been asked to solve problems of a considerable variety throughout our schooling. But each time we are faced with a new problem, we set out to solve it, wondering whether an idea will finally come to us that will unlock the puzzle and lead the way to a solution. We feel lucky when we are successful, for often the idea seems to come as a shot out of the blue. And yet such inspirations rarely occur by chance. They are more often the result of our having asked the right questions, of having seen the problem in just the right perspective. Such a questioning attitude can—and should—be developed as an art form, thereby creating the basis for a kind of problem-solving methodology, a unified approach to all of one's problem-solving pursuits.

In this section we review such a methodology, hoping it will help you treat problems whose solutions need to be represented by algorithms. The notion of an algorithm is crucial to understanding further topics in computer science; moreover, it should be well understood by all students in the mathematical sciences, because the algorithmic point of view has become of increasing importance in every area of specialization.

Goals

After studying this section, you should be able to:

- Outline the guiding principles of a problem-solving methodology.
- Define the notion of an *algorithm*.
- Devise plans (develop an outline) for solving elementary problems whose solutions are to be represented by algorithms.
- Identify inputs and outputs from a given problem statement.
- Describe the strategy "keep the best" as it relates to optimization problems.
- Discuss the notion of the *state* of a computation.
- Define the concept of an *elementary* process.
- Describe the way in which elementary computational processes cause a transformation of state.

Three Sample Problems

It is easier to answer a question than to solve a problem. Someone might ask, What is the sum of 63 and 141? and we have no difficulty providing an answer. But if we are asked to describe a general method for adding two given numbers we then may have a problem doing so. We might even have to think back to our early school education in order to outline a general procedure that will provide the correct answer in every *instance* of the problem. As was highlighted in the opening paragraph for this section, a **problem** is a class of questions, and each question in the class is said to be an **instance** of the problem.

These three sample problems will serve to focus our attention in this discussions that follow.

Example 4.1 Find the length of the hypotenuse of a right triangle, given the lengths of its two legs. Suppose we denote the hypotenuse as c and the legs as a, b as is usually done. Then, as discussed above, we do not think of $a = 3.0$ and $b = 4.2$, except as representing an instance of the problem. Instead we should imagine that two arbitrarily chosen numbers a, b will be given as *input*, and we are to describe a general method for determining the *output*, c. ■

Example 4.2 Determine the largest and smallest numbers from a given list. Here again we do not assume a specific list of numbers—we have to be prepared to accept any list as input. ■

Example 4.3 Decide whether a given number is **prime**, that is, it is different from one and is evenly divisible only by itself and one. Observe that the output is a simple true or false answer (for this reason, the problem is said to be a **decision problem**). Once again, an instance of the problem will involve a specific number, presumably a positive integer, as input. But our solution must be capable of handling all such instances. ■

In reviewing these examples, one important feature stands out most clearly, and it is a general rule. The inputs and outputs are representatives (instances) chosen from a well-understood class or *type*. In the last example, for instance, the input is understood to be of type positive integer (or natural number, whichever we prefer that these be called). The output, however, is Boolean type: {true, false}. As mentioned above, this last feature distinguishes the so-called *decision problems*.

A second and equally important feature illustrated by our set of examples is that there exists a possible multiplicity of input or output. In imagining all the various types we might consider for input or output of a problem, they needn't be singular. For instance, the input to the problem of Example 4.2 is a *list* of numbers (of some type or other). Furthermore, there is no apparent specified bound on the length of such lists. Accordingly our solution must accommodate the range of possibilities that are acceptable as instances of such an input.

Said in another way, the precise nature of the expected input and the desired output must be well understood in advance if we are to devise a reliable solution to the problem. This much cannot be over-emphasized, for it is perhaps the most common cause of failure among beginners. Therefore: *Know your problem*. This is the first and foremost principle among those that guide us to a successful problem solution. The following section surveys a list of such principles. This will provide you with a useful review of the problem-solving process, particularly as it applies to the discovery of algorithmic solutions.

Guiding Principles

You will be necessarily disappointed if you expect to be presented with an exact and universal problem-solving methodology—one that would be directly applicable to each and every problem ever to be encountered. Indeed, some famous mathematicians and philosophers (notably Descartes and Leibnitz) thought for a time that such a dream could, in some sense, be realized. They even developed many useful techniques toward achieving this goal. But now we know that such hopes are not likely to be fulfilled. What we do have instead, derived from these earlier investigations and formulated in modern terms by the contemporary mathematician George Polya (in his book, *How to Solve It*), is a simple but useful **heuristic**—a system of methods, rules, and suggestions that serve as an aid to our invention and discovery in all problem-solving investigations.

These guiding principles are so broad in scope that to reproduce and illustrate them here would be an almost impossible task, even if that were desirable. Some of the most useful of these principles are paraphrased in the summary presented in Figure 4.1. One notes immediately from this summary that the overall structure of the strategy is simple enough, almost obvious:

- Understand the problem.
- Devise a plan.
- Implement the plan.
- Look back.

Yet many a would-be problem solver goes astray simply because one or another of these admonitions has not been given proper attention. We try to solve a problem that we do not fully understand. Or we start to carry out a plan that has not been fully developed. Or we fail to look back through our final solution, never bothering to notice that there was an easier way. It may be true that we consider ourselves successful at solving problems. But anyone who truly hopes to improve his or her capability would do well to embrace these principles fully and to put them into practice on an everyday basis. The rewards may be surprising!

Consider our first principle: *Understand the problem.* It is foolish and generally a waste of time to attempt to solve a problem that you do not even begin to understand. Certainly you must know the precise nature of the inputs and the desired output before it is sensible to proceed. You should be able to introduce suitable notation and, in most cases, to draw a figure as an illustration of an instance of the problem, before you can be sure that you truly see what is involved. Otherwise it is time to say, I don't understand this problem, and perhaps to seek outside

1. *Understand the problem.*
   Draw a figure.
   Introduce suitable notation.
   Identify inputs and outputs.
   Establish desired relationships.
2. *Devise a plan.*
   Find connections between input and output.
   Can you derive anything of use from the inputs?
   Have you seen it before?
   Do you know a related problem?
   Look at the unknown (the output).
   Restate the problem.
   Perhaps first consider a related but simple problem.
   Make a good guess.
   Perhaps introduce an auxiliary problem.
3. *Implement the plan.*
   Outline your plan.
   Use precise language.
   Check each step.
   Make corrections.
4. *Look back.*
   Test your implementation.
   Could you now devise a better plan?
   Can you use the result in solving another problem?
   Can you verify your results?

FIGURE 4.1 Guiding principles for solving problems.

help before proceeding. For a larger problem the situation may be different. Because the act of trying to solve a problem often contributes to our understanding of it, it may be best to simply jump in and give it a try—no matter how large the "waters." (Of course it may be that a problem is, for some reason, beyond our present capabilities—but that is another matter entirely.)

The most creative and therefore the most difficult phase of the whole problem-solving process is to *devise a plan.* Fortunately, however, we are given a wealth of suggestions in Figure 4.1, and we might simply try one of these after another, hoping we finally will be led to the idea that provides a break-through.

Example 4.4 Trivial though it seems, let us briefly discuss the problem of Example 4.1, to find the length of the hypotenuse of a right triangle, given the lengths of its two legs. Presumably we would not have had difficulty in drawing a figure (as in Figure 4.2) that illustrates the relationships between the inputs a, b and the desired output c. We then would undoubtedly answer "Yes" to the question, "Have you seen it before?" If you remember the Pythagorean theorem from high-school mathematics, our plan is obvious: Square the two legs and compute the square

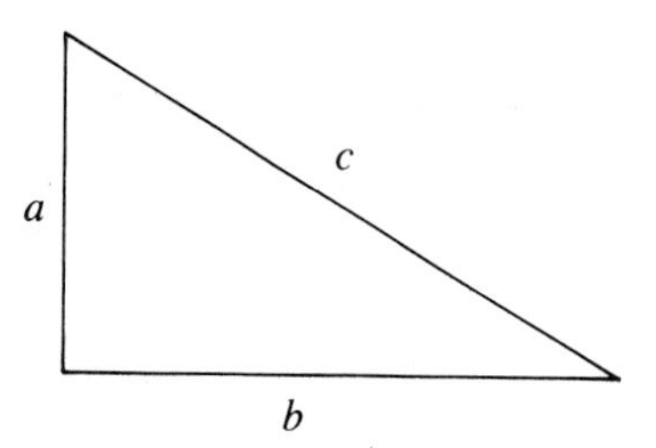

FIGURE 4.2

the sum (of these squares). In symbols:

$$c = \sqrt{a^2 + b^2}$$

root of—a straightforward solution if ever there was one. We therefore seem to be finished. For purposes of preparation for the discussions to follow, however, it is a good idea that we cast this solution in the form of the following algorithm *hypotenuse*, using a style that we develop in some detail in the next section. Certainly in this simple setting there can be no possibility of misunderstanding in reading the algorithm as it is presented here.

algorithm hypotenuse $(a, b{:}\, c)$

compute SUM of a^2 and b^2
set c to squareroot of SUM ■

For the simple problem that we have just discussed, the way to proceed was fairly obvious from the start. For another problem the way might not be so clear, and so we are advised to explore the suggestions of Figure 4.1 in more detail, hoping that one of them will trigger a response that leads to devising a plan.

In general we can say that we have a plan when we have an **outline**, a fairly definite arrangement for the calculations or constructions we will need to perform in order to obtain the desired output. In fact the main achievement in the solution of a problem is often the discovery of an idea on which the outline of a plan may be based.

Example 4.5 Consider the problem of Example 4.2, that of determining the largest and smallest numbers in a given list. In this case the way is not quite so clear. We may be well advised to consider a related but simpler problem, as part of step 2 in Figure 4.1 suggests. Surely it would be simpler if we were asked only to find the largest number alone. But even then we have some thinking to do. We may not have seen such a problem before, even though in fact the problem is typical of many *optimization problems*—find the largest, the best, the most economical, etc.

But imagine that the list is printed on a deck of ordinary cards, one number per card, and as we have said we seek only to determine the largest number so represented. We might think to look at the first card, whose number incidentally is the largest one we have seen so far, not having yet looked at the others. Trivial as it seems, there is the germ of an idea in this observation. We should write this number down and keep it until we encounter a larger one. As we look through the deck, one card after another, we have only to compare each face value with the number we are currently keeping as the "*best*" (i.e., largest) so far. When the face value of a pulled card is larger, we save it as the new best; otherwise we continue looking. At the end of the deck, regardless of its length, the number we are saving is surely the largest one of all. This is a strategy that works. I call it "Keep the best you've seen so far" because it reminds me of the behavior of the unscrupulous Romeo who keeps his current conquest on a string, just waiting until something better comes along. ■

You are encouraged to discover and give names to problem-solving strategies as they are developed, just as we have done here. In so doing you will enlarge upon the list of suggestions in Figure 4.1, thereby enriching our modest reservoir of wisdom and, more importantly, personalizing it as well. To continue in this vein, we might suggest another synthetic proverb: "If something works once, consider using it again." By way of illustration, look once more at the problem we have just solved. It was not the original problem because first we simplified things. But if now we were actually required to find the smallest as well as the largest number in the list, it is clear that our same strategy will work again, with "best" interpreted as "smallest." Experienced problem solvers will have a bundle of these tricks up their sleeves and will know which ones to try in a given circumstance. Beginners, however, often must observe and imitate what others have done until finally their own experiences are sufficient for the task at hand. In the final analysis, realize that one learns to do problems simply by doing them, building on past experience and acquiring new techniques along the way.

Having found an idea that can lead to the broad outline of a plan of attack, we then proceed to *implement the plan.* Often the outline must eventually lead to a precise sequence of statements in a computer programming language. For this reason some degree of precision must be observed in the construction of the outline itself. We will not be using a full-fledged programming language here but, nevertheless, this insistence on precision must be observed, for otherwise we will not even be able to understand one another in the phrasing of our algorithmic solutions.

Example 4.6 Consider again the problem originally encountered in Example 4.2, that of determining the largest and smallest numbers of a list. In view of the continued discussions of Example 4.5, an outline might take the form:

```
read first card and call its value MAX
as long as deck is not finished
   read card VALUE
   if VALUE > MAX then
      take new MAX from VALUE
```

(again supposing, as in Example 4.5, that we are finding only the largest number of the list). Note that it is our intention that the indented portion be executed repeatedly *as long as* we have not yet reached the end of the deck. We think of

VALUE as an *auxiliary variable* (neither input nor output) used to store the face values as they are read. That is, each time we read a card (after the first card) it is assumed that we take on a new VALUE, the number on the face of the card just read. Then, as we see, we compare this VALUE with MAX (the "best so far") to see whether or not we should revise our MAX. The reader can see that, indeed, the phrasing of our outline together with the conventions inherent in the wording capture the intent of our plan as described in Example 4.5. ■

If we have understood our problem, devised a plan for its solution, and then implemented our plan, we should be done. But in Figure 4.1 we are encouraged to *look back*. Once a problem has been solved to our satisfaction we should not be too eager to set it aside. Often we can learn a great deal when we take a retrospective view. For example we may see things that could not be seen at the start. Or perhaps we may see a more natural approach to solving the problem, or even an easier solution. In such cases it is a good idea to back up and solve the problem again from this new point of view. One should strive to become a skilled professional and not a mere technician. For this reason elegance, clarity, style, and all those other attributes usually thought of in connection with the arts are equally important and appropriate to develop here. If the problem was worth solving at all, then its solution is something to be read, understood, and appreciated *by others*. Thus it is important for us not to have merely solved a problem, but to have done so well.

We should aim to find the best solution, if that is meaningful. Furthermore, whenever possible we should find a better solution than that which first came to mind. Yes, look back. Often it is only through a careful review of your first solution that a significant improvement can be seen.

Example 4.7 Consider again the problem of the largest and the smallest. According to the discussion following Example 4.5, the outline given in Example 4.6 can be repeated with only slight changes to allow our finding the smallest number of the list. But we then would be scanning the deck twice. Can we do better? Is there a way that we can find both answers in one pass? Suppose that we save two numbers (MAX and MIN) as we leaf our way through the deck, one the largest we have seen so far and the other the smallest we have seen so far. That's it—a decided improvement, if you imagine yourself performing the task by hand. And such obvious improvements in efficiency are nearly always transformed into a savings in processing time in a computer implementation, should we proceed that far. Since time means money in such situations, we see that looking back can indeed pay dividends. ■

Definition of an Algorithm

With most of the definitions we have given throughout the text, the meaning of the defined term is immediately clear and does not require further explanation. At most an example or two may be necessary in order to clarify the concept. Our definition of the term *algorithm*, however, will require considerable explanation, due to its very nature—despite our extensive use of this term in the preceding chapters. Moreover we will not claim to have provided a final and precise definition of this term, even after our discussion is completed. Computer science students will continue to grapple with this notion throughout their further studies, coming closer and closer to understanding its true meaning with use and practice.

Suppose we have a problem, in the sense just discussed. An **algorithm** for that problem is an organized sequence of instructions for answering any question that is an instance of the problem. It is presumed that the algorithm is written in some precise language, though it is not required that this be a bona fide computer programming language. (*Note:* The term algorithm does not presuppose a connection with machines of any kind, and can be understood quite independently of such notions.)

The exact question to be answered by the algorithm is determined by our setting the values of the *inputs* of the algorithm, prior to its execution. Upon termination of this execution the answer to the question is given by the value of the *output* of the algorithm. It is further supposed that the execution takes place in some step-by-step manner, where each step is an *elementary process*, in a sense that we will explain later. Furthermore, the whole execution should involve no chance occurrences of any kind. It also is required that the execution of the algorithm necessarily will terminate after a finite number of steps, regardless of the values of the inputs. In conclusion, note that all these attributes are necessary for a solution to be considered algorithmic.

Example 4.8 The Euclidean algorithm of Section 3.3 surely possesses all of these attributes. As elementary processes, we can identify the interchange of values of two variables, the computation of remainders in an integer division, etc. These are *deterministic operations*—there is no element of chance involved. Moreover the computations eventually terminate, regardless of the inputs a, b, inasmuch as the remainders continually decrease until a zero remainder results, whereupon the execution comes to an end. ■

Example 4.9 In Example 4.6 we find such processes as the reading of a value from the input list, the assigning of the value of one variable to another, etc. These are typical of those processes we would wish to call *elementary*. Moreover we again are assured that the algorithm will terminate, since the list will be exhausted at some point, regardless of its length, and this is the criterion for ceasing the repeated processes occurring in the algorithm. ■

Elementary Processes

What does the term elementary process really mean? "Elementary"—for whom? Imagine a clerk, or some other identifiable computing agent who might be able to execute our algorithms simply by following the instructions as they are written. Imagine further that this clerk is someone less intelligent (or, at least, less experienced) than ourselves. (In this way we are likely to be more precise in our phrasing of the algorithm.) Pretend for now that this clerk is a high-school student equipped with a pocket calculator, a piece of paper, and a pencil. Although we will provide him with all the supplies he might require, we do not suppose that he will be ingenious in any way. He will simply execute the instructions as we give them.

An **elementary process** is one that can be realized in a constant unit of time, independent of the size of the inputs to the problem. Our clerk can replace (on his sheet of paper) the value of one variable by that of another in a time that is independent of anything else—he simply crosses out the old value and replaces it with the new. Similarly, he will perform an integer division on his pocket calculator in an instant of time, independent of the size of the arguments involved. While there is a degree of idealization in such assumptions, the simplicity of such an accounting scheme will prove to be a great advantage.

Example 4.10 To bring this idea into clearer focus, suppose we are designing an algorithm in which we are dealing with a list of n numbers as input. We are required to find the largest of these numbers as a subprocess, or part of the algorithm taken as a whole. One line of our algorithmic solution might then read:

find the largest number and call it MAX

but it is clear that this process could not be considered elementary. The computation of MAX, or at least the only way we know of computing it, takes a period of time proportional to n, the size of the input list. So we could hardly say that this process can be executed in a unit of time that is independent of the size of the input. Exactly the opposite is true: The bigger the input, the longer the time. ■

State of a Computation

We have seen that the design of an algorithmic problem solution will involve the identification and naming of input and output variables and, in general, the additional naming of auxiliary variables as aids to the computation. Furthermore we have noted that these variables may be of a variety of types. Nevertheless, as the execution of an algorithm is in progress, we can speak of the **state** of the computation, in reference to the totality of values currently held by the class of variables involved. This is not a difficult notion; consider the following example for clarification.

Example 4.11 Consider once more the Euclidean algorithm of Section 3.3, paying particular attention to Example 3.26. Expanding that table of the changing values of the variables one step at a time, we obtain Table 4.1. There are four variables in the algorithm: a and b are input variables, gcd is an output variable, and r, the remainder, is an auxiliary variable. A state is simply a row in this table, showing as required the current values of all the variables as the execution is in progress. Thus the second row shows the values $a = 252$ and $b = 180$, whereas r and gcd

are undefined (since they have not yet been given values). This is the state of the computation just following the interchanging of the values of a and b. The next row shows the *state transformation* that takes place for having computed r as the remainder on dividing a by b (an elementary process). We then have the new state: $a = 252$, $b = 180$, $r = 72$, and *gcd* is undefined. And so it goes. ■

TABLE 4.1

a	b	r	*gcd*
180	252		
252	180		
252	180	72	
180	72	72	
180	72	36	
72	36	36	
72	36	0	
72	36	0	36

In summary we can say that an elementary process causes a **transformation** of the state from one set of values to another. By way of contrast, an algorithm also may involve decisions being made on the basis of an interrogation of some **condition** of the state, that is, some relationship among the variables that may or may not hold at a given point in the execution. But these interrogations do not have any effect on the state. We may think that our clerk simply comes away with a *true* or *false* answer to the interrogation, having inspected the values of the variables for himself.

Example 4.12 Referring to the Euclidean algorithm once more, we note the phrases:

if a is less than b

and

while we have a nonzero remainder

These require that our clerk examine the state to see whether indeed we have $a < b$ or $r \neq 0$, respectively. Our clerk makes no changes in the values of any of the variables; rather, he simply makes the necessary determinations, based on the current state, and comes away with a *true* or *false* answer. This answer then will affect the flow of the execution of the algorithm. For example, the clerk will perform the interchange of a and b if (and only if) the current values of a and b are such that $a < b$ is true. He will continue the repeated execution of the two indented statements if it should happen that $r \neq 0$—otherwise he will proceed to the next statement in sequence, the assigning of the value of b to the output variable *gcd*. But in the decision-making process leading up to these actions, the testing of the condition involved, no change of state takes place. ■

EXERCISES 4.1

1 Describe a method (provide an outline) for:

(a) Sharpening a pencil

(b) Finding someone's telephone number

(c) Translating English words into Pig Latin

(d) Sewing a button on a shirt

(e) Shaving one's face (male) or legs (female)

2 Describe a method for finding the area of a triangle, given the lengths of its sides.

3 Describe a method (provide an outline) for determining whether or not a given pair of integers is relatively prime. (Recall that by this it is meant that they have no common divisors other than the number 1.)

4 Describe a method for determining whether a given positive integer is perfect, that is, whether it is equal to the sum of its divisors (other than itself), as in the case of $6 = 1 + 2 + 3$.

5 Describe a method for barbecuing chicken.

6 Describe a method for determining whether or not an arbitrarily given word (a finite sequence of letters in the ordinary English alphabet) spells the same word backwards as it does forward, as in the case of radar (that is, determine whether the given word is a *palindrome*).

7 Given three numbers, describe a method for determining if they are the lengths of the sides of some triangle.

8 Identify the inputs and outputs in each of the following problems:

(a) Exercise 2 **(b)** Exercise 3 **(c)** Exercise 4

(d) Exercise 5 **(e)** Exercise 6 **(f)** Exercise 7

9 What is wrong with the following proof that the only way two quantities can be equal is when both of them are zero?

$$
\begin{aligned}
x &= y && \text{(by assumption)}\\
x^2 &= xy && \text{(multiply both sides by } x)\\
x^2 - y^2 &= xy - y^2 && \text{(subtract } y^2 \text{ from both sides)}\\
(x - y)(x + y) &= (x - y)y && \text{(factor)}\\
x + y &= y && \text{(cancel } x - y)\\
x &= 0 && \text{(subtract } y \text{ from both sides)}\\
y &= 0 && \text{(from the assumption)}
\end{aligned}
$$

[*Hint: Check each step!*]

10 How can we cut a round one-layer birthday cake into eight equal pieces with three straight cuts of a knife:

(a) Giving unequal portions of frosting?

(b) Giving equal portions of frosting?

[*Hint:* Suppose that the first cut divides the cake into two equal pieces, the second into four equal pieces, and the third into eight equal pieces.]

11 In numbering the pages of a large telephone book, the printer uses 2989 digits. How many pages are there in the telephone book? [*Hint:* First, *make a good guess.* How many digits would the printer use if there were 1000 pages?]

12 Outline a plan for each of the following:

(a) Making popcorn in a skillet

(b) Smoking a pipeful of tobacco

(c) Writing a term paper

(d) Getting up in the morning

13 If an algorithm treats as input a positive integer n, decide whether each of the following would be considered an elementary process:

(a) Determining whether a given integer j divides n evenly.

(b) Finding all the divisors of n.

(c) Deciding whether or not n is prime.

(d) Multiplying n by itself.

14 If an algorithm treats as input a list of n numbers, decide whether each of the following would be considered an elementary process:

(a) Adding up all of the elements in the list

(b) Finding the smallest member of the list

(c) Sorting the list members into ascending order

(d) Looking at the first member, then calculating its square root

Sec. 4.2 TOWARD AN ALGORITHMIC LANGUAGE

A precise pseudolanguage will be used for the phrasing of our algorithms in a uniform and consistent manner.

Now that we have reviewed the problem-solving process and we have some confidence that we can formulate a problem solution in more elementary settings, it is important that we seek to make our algorithmic language more precise. The language we will begin to develop here is really a *pseudolanguage* in that it is not an actual spoken, written, or artificial programming language. Rather it is a blending of English with mathematical symbolism, structured by means of grammatical constructs borrowed from the modern computer programming languages, of which Pascal is perhaps the most representative. You have seen examples of its use already. All that we need to do is to formalize these constructs and agree on their precise meaning so that we all will understand the same algorithm in the same way. By doing so we will establish guidelines that will help you phrase algorithms in a consistent manner. This will ensure that we all can communicate, confident that everyone comprehends the same, intended meaning.

Goals

After studying this section, you should be able to:

- Describe the *compound processes* available in our pseudolanguage—those of *sequence, selection,* and *repetition.*
- Discuss the idea of a *conditional process* as a special case of the selection construct.
- Distinguish the three forms of repetition constructs available in the pseudolanguage and describe situations where one of these forms is preferable over the others.
- Use the sequence, selection, and repetition constructs in developing an algorithmic solution to a problem.
- Understand and use the *top-down design methodology* for obtaining a modularized algorithmic solution, one that is well-organized into component subproblems.
- Describe the *derived constructs*—the *iterated sequence,* the *case construct* (as a generalized selection), and the *for construct* (as a specialized form of repetition)—and illustrate their use in the design of algorithms.
- Discuss the conventions followed when using indentation in the phrasing of a pseudolanguage algorithm, and explain their purpose.

Compound Processes

You already have seen examples of algorithms written in the pseudolanguage to be developed here. You therefore have some idea of what to expect in the following discussions—which, of course, was intentional. The purpose of this section is to help you gain a better appreciation for the overall structure of an algorithm as it should be expressed in our pseudolanguage.

Typically an algorithm will be built up out of elementary processes using three grammatical constructs (those of sequence, selection, and repetition) in order to form *compound processes* of increasing complexity.

Example 4.13 In the Euclidean algorithm of Section 3.3, two (or we could say three) elementary processes are combined in sequence to form the compound process:

replace a by b, b by remainder
compute remainder on dividing a by b

This process in turn is performed repeatedly (or by repetition) in writing:

while we have a nonzero remainder
  replace a by b, b by remainder
  compute remainder on dividing a by b

so as to obtain an even more complex compound process. This process is one of four processes that are combined in an outer-level sequence to make up the whole algorithm. ■

One notes from this example and from others we have discussed previously that we indicate that two or more processes are to be executed sequentially by simply lining up their descriptions one underneath the other. Thus if R and S are processes (possibly themselves compound), the **sequence** $R \circ S$ is to be executed (by our clerk) according to the flow diagram shown in Figure 4.3(a). As we have noted before, the textual representation of such a sequence is to be rendered as shown in Figure 4.3(b), with the description of S lined up immediately below that of R.

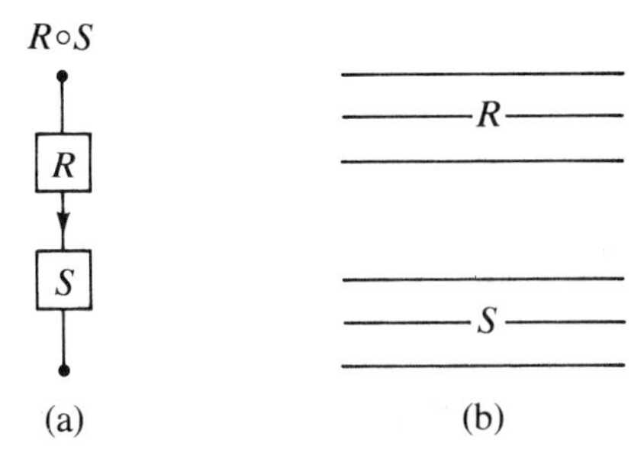

FIGURE 4.3

Example 4.14 As we mentioned in Example 4.13, the Euclidean algorithm is composed of four processes in sequence. Their separate descriptions are written one directly underneath the other, as you saw when we presented the algorithm in Section 3.3. One of these processes is the repetition just discussed. Another is the *conditional process*:

if a is less than b
  interchange their values

where we are to execute the elementary process of interchange on the condition that $a < b$. Such compound processes are but a special case of the selection construct that we now discuss. ■

Suppose that R and S are processes, elementary or compound (it makes no difference). And suppose that C is some condition that may or may not be satisfied by the current state of the computation. Then the **selection** "R if C else S" is to be executed according to the flow diagram shown in Figure 4.4(a). Thus if C is 'true' we do R and if C is 'false' we do S. In the linguistic representation in Figure 4.4(b) we indent the processes R and S for ease of readability, with the *if* and *else* clauses standing out at the margin—that is, at the level where the selection process is intended. Thus it may be that the selection is itself a part of a sequence. The margin for writing the selection is governed by the location of the other processes in the sequence, remembering that these are listed one directly underneath the other.

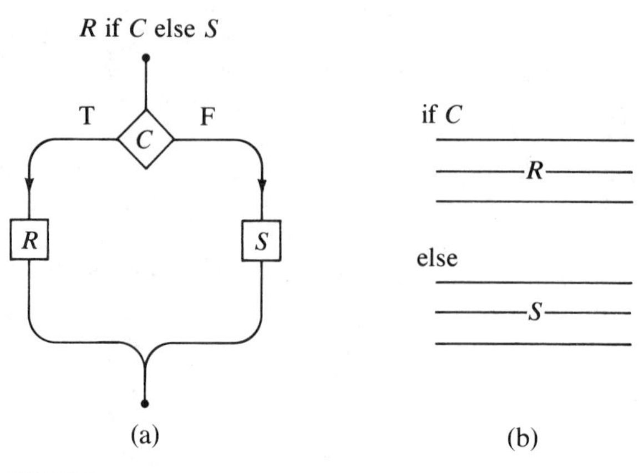

FIGURE 4.4

Often it will be the case that one alternative (which we can always choose to be the 'false' alternative, by negating C if necessary) is vacuous, in the sense that we want to do nothing if C is false. Then in Figure 4.4(a) the right-hand branch would be drawn as a straight line (with the box for S missing) and, similarly, the else clause and the process S would not appear in the textual representation on the right. We then speak of a *conditional* process inasmuch as R is executed (or not) depending on the truth or falsity of the condition C. As we have said it is best to regard the conditional construct as a special case of the selection.

Example 4.15 Consider the decision problem, where we wished to determine whether a given positive integer is prime (introduced in Example 4.3). Recalling that a prime number is a natural number (different from one) that is evenly divisible only by itself and one, we might begin our algorithmic design by noting that $n = 1$ is a rather special case (n being the input variable). Accordingly we might write:

```
if n = 1
   give answer as 'false'
else
   count divisors and decide
```

using the selection construct just discussed. In this phrasing we begin to illustrate a broad and important design technique: *top-down methodology*. This is how we can describe a process in words that will be (*must* be) further refined at a subsequent level of the analysis. Thus we might write: Count divisors and decide. Since this statement obviously is lacking in detail we will have to return to it at a later point to specify what it is that we actually mean. Certainly our clerk is not so intelligent or experienced that he will be able to execute such a statement as it stands. ■

In reference to the capability we intend to provide for executing processes repeatedly, we must first acknowledge the importance of flexibility. We offer repetitions in two flavors here; later in this section another flavor is presented. This way you can choose the form that best fits the requirements of your particular problem.

Suppose that S is a process (possibly a compound process) that we would like to execute repeatedly a number of times as governed by some condition C. Then our **repetition constructs** are of two forms for the present: "S while C" and "S until C," the difference being captured by the flow diagrams in Figure 4.5—in (a) and (b) respectively. As we see, the test for continuing the repetition of S is made *before* the execution of S in the case of S while C, whereas we test the condition C *after* executing S when performing S until C. One form will be more appropriate for a given situation, and it is the algorithm designer's responsibility to make the choice. For instance we note that S is always performed at least once in the case of S until C, whereas this may not be true in using S while C— supposing that C is 'false' at the initial entry to the repetition in (a). We make no particular grammatical distinction, however, between these two forms in our linguistic renderings on the right-hand side of Figure 4.5. It is left for the reader to remember the difference in meaning as outlined earlier in this paragraph, just from seeing the word "while" as opposed to "until."

We note one more difference between these two repetition constructs. We terminate a "while" when C is 'false,' whereas the "until" type of repetition is terminated when C is 'true.' But of course these conventions are consistent with the ordinary meaning of the words while and until. In any case C can always be negated in order to fit the formats given here, so it is not as though we are operating under any serious restriction.

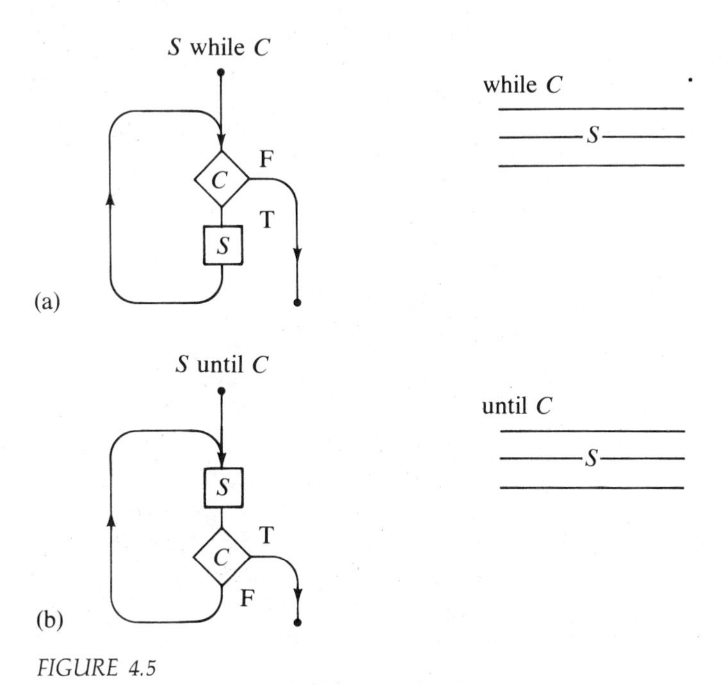

FIGURE 4.5

Example 4.16 In Example 4.6 we have a (compound) process that is to be repeated as long as the deck has not been exhausted. If we are forced to choose—as indeed we should if the pseudolanguage is to be used in its strictest fashion—should we use the while or the until form here? You should see that the S until C construct is not well suited to the task. If the list were only of length 1 then we would have read the first (and only) card in the opening statement of the algorithm. In using the until form of repetition, remember that we execute S before testing C (see Figure 4.5(b))—we then would be asking our clerk to read a card that isn't there! So it is better to use the while form in the phrasing of the accompanying algorithm *largest*. Note that we have dropped the word "then" in the conditional statement of Example 4.6 so as to agree with the linguistic style recommended above. Note further that we give the algorithm a name—largest—and that we identify the input and output variables parenthetically, separating them by a colon in an opening *title line*. This style will be used in all our subsequent examples.

```
algorithm largest (list: MAX)

read first card and call its value MAX
while deck is not finished
   read card VALUE
   if VALUE > MAX
      take new MAX from VALUE
```
■

Example 4.17 In the algorithm "factors" of Section 3.3 we used the until form of the repetition rather than the while. Having first set $m = 1$, and assuming that our input n is a natural number, we want to adjoin 1 and n to our list of factors, whatever else we do. This implies that we will want to execute our compound (repeated) process at least once in any case and, accordingly, the until form is suggested. It does mean that we are testing the first m ($m = 1$) as a divisor unnecessarily, but that's no big deal. ■

Top-down Methodology

An algorithm can be so complex that it is hard to write down its details right out of your head. An intelligent use of our pseudolanguage will enable the designer to set down the overall plan and to provide the overall structure, but to ignore irrelevant details until later. One can describe subprocesses embedded below the outer level of the algorithm in a convenient stylized version of the English language. The designer then can return to these at a later point, providing more and more detail. He or she simply begins again, treating such subprocesses as problems in their own right, as if starting anew. In this way we move from the top level of the design into further levels of detail, down to the level where each process is elementary and each condition is, in the same sense, an elementary condition to be tested. This is the **top-down** algorithmic design methodology. It is, as we will see, a *modular* approach, one that tends to ensure a readable, well-organized solution.

Example 4.18 Suppose we continue the discussion of Example 4.15, the prime number decision problem. In writing:

```
if n = 1
   give answer as 'false'
else
   count divisors and decide
```

we have only provided the overall structure of a solution. We must eventually come to grips with the phrase "count divisors and decide" and provide additional detail. Thus we might think to provide a counting variable, call it COUNT, utilized as follows:

```
set COUNT to 0
test each number and update COUNT accordingly
decide answer on the basis of final COUNT
```

In the middle process, the idea is to test each number (in the range from 2 to $n - 1$) to see if it divides n. If so we would increment the COUNT by one; otherwise, we would not. But we don't need to be specific yet. We have only to see that this sequence of processes, when substituted for the line "count divisors and decide," provides for a somewhat more detailed analysis than before:

```
if n = 1
   give answer as 'false'
else
   set COUNT to 0
   test each number and update COUNT accordingly
   decide answer on the basis of final COUNT
```

We see here that the whole plan is correct, provided we have the proper interpretation of each phrase in mind.

In the case of the last process it is surely our understanding that if the final count is still 0 we will have determined that the answer is 'true' (for then n will have had no divisors except itself and 1), whereas we will know that the answer is 'false' otherwise. As a matter of fact this bit of analysis shows how we may use a selection construct to refine this last process one level further by writing:

```
if COUNT = 0
   give answer as 'true'
else
   give answer as 'false'
```

in place of the phrase "decide answer on the basis of final COUNT." Until the phrase "test each number and update COUNT accordingly" is similarly refined, however, our solution is not complete. Since we have in mind the testing of each number in the range from 2 to $n - 1$, this process is not elementary, and according to the criterion we have established above, it is in need of further refinement. We defer this task for now, however. ■

Derived Constructs

In the case of all three of our methods for composing statements—the sequence, the selection, and the repetition—we may derive further constructs that are an aid to effective algorithmic design. In none of these cases are the derived forms essential; we could get along without them. Nevertheless they definitely enhance our expressive capability.

With the sequence we really offer nothing new; rather, we only confirm what we have been doing all along—namely, allowing any number of processes $S_1, S_2, \ldots, S_n$ to be joined in an *iterated sequence*. Originally we only indicated that this was permissible in the case of two processes. Obviously such pairings can be iterated in defining:

$$S_1 \circ S_2 \circ \cdots \circ S_n = (\cdots (S_1 \circ S_2) \circ \cdots) \circ S_n$$

and the meaning is clear. In executing $S_1 \circ S_2 \circ \cdots \circ S_n$ we of course start with S_1, then S_2, and so on, until finally S_n is executed. In the linguistic rendering we simply write the descriptions of the n processes one underneath the other (as we have been doing all along).

For the selection, the situation is somewhat analogous. Again we have only the immediate capability of choosing between two alternatives (with the R if C else S construct). There are many situations where it is convenient to be able to select from any number of cases $C_1, C_2, \ldots C_m$ of some general condition C. It may be that if C has the configuration C_1 we will want to perform one process S_1; if it has configuration C_2 then another, say S_2, and so on. In Figure 4.6 we give this derived construct a name, we indicate the intended flow diagram (Figure 4.6(a)), and we provide the recommended pseudolanguage phrasing (Figure 4.6(b)). As the flow diagram in Figure 4.6(a) suggests, only one of the processes $S_1, S_2, \ldots, S_m$ is actually executed, and that depending on the case—whether $C_1, C_2, \ldots$ or C_m obtains.

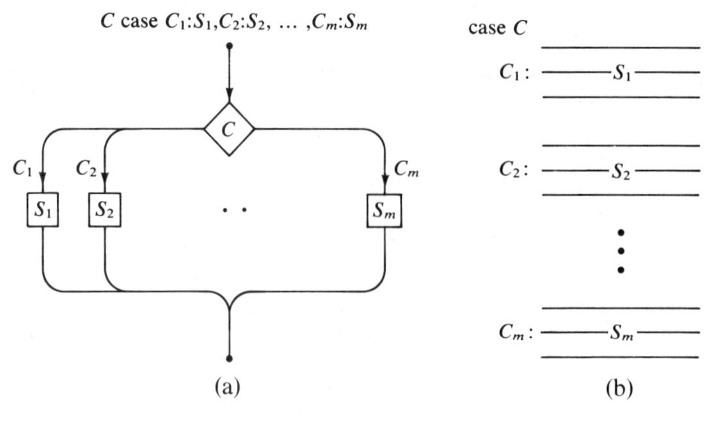

FIGURE 4.6

Example 4.19 Consider an algorithm with three numbers c, b, a as inputs that represent the lengths of a triangle. Suppose we want to classify the input triangle—to determine whether it is equilateral, isosceles, etc. We will assume that the lengths are to be presented in such a form that $c \geq b \geq a$, and that an error class is indicated otherwise. Note that we also must have $c < b + a$, or again there is an error. (Do you know why?) Furthermore we find it most convenient to make use of our newfound case construct in formulating the accompanying algorithm *triangles*, as shown. Note that the five cases are mutually exclusive and exhaustive.

algorithm triangles (c, b, a: class)

if $c \geq b \geq a$ and $c < b + a$
   case comparison
      $c = b = a$: designate class as "equilateral"
      $c = b > a$ or $c > b = a$: designate class as "isosceles"
      $c > b > a$ and $c^2 = a^2 + b^2$: designate class as "right scalene"
      $c > b > a$ and $c^2 < a^2 + b^2$: designate class as "acute scalene"
      $c > b > a$ and $c^2 > a^2 + b^2$: designate class as "obtuse scalene"
else
   designate class as "error" ■

This last property, that of exhausting all the possibilities, is most desirable. Otherwise we will assume that nothing is to be done for the cases that have not been described. For example suppose we are trying to implement the child's nursery rhyme, "Thirty days hath September, April, June, and November. All the rest...," for computing the number of days in a month. Assuming that the month is simulated as an integer in the range from 1 to 12 (representing January through December), we write:

```
case month
   4, 6, 9, 11: assign days as 30
   1, 3, 5, 7, 8, 10, 12: assign days as 31
   2: compute days
```

(Note our convention for listing together those cases that are to be treated identically.) If for some reason the meaningless month value 13 is encountered we will assume that no computation is performed. Alternatively an error message could be transmitted by enclosing the above case construct within a selection.

For those cases that do appear, any process, simple or compound, can be prescribed. Thus in the case of February above we may refine the phrase "compute days" with the selection construct:

```
if year divisible by 4
   assign days as 29
else
   assign days as 28
```

so as to properly account for leap years.

Finally we consider repetitions once more. It would seem that the while and until constructs would be enough. (As a matter of fact, they are sufficient for all purposes.) Nevertheless a derived third form of repetition is most useful in those situations where we want to repeat a process a certain fixed number of times under the control of some auxiliary variable. A *control variable* thus will be introduced in a control clause C, specifying a definite range of values for this variable. The given process S then will be executed repeatedly, for these and only these values. In the pseudolanguage we agree to use the phrasing shown in Figure 4.7(b). The meaning of this new construct, "S for C," is given by the flow diagram in Figure 4.7(a). Here we may think of the (elementary) processes I and J as causing an initialization of the control variable, and a jump in its value (in preparation for the next iteration) as being specified by the control clause C. For those who have not seen this before in their high-school programming classes, it is quite likely that this broad and general description of the for construct is more than a little bit confusing. The following example should help to clarify matters.

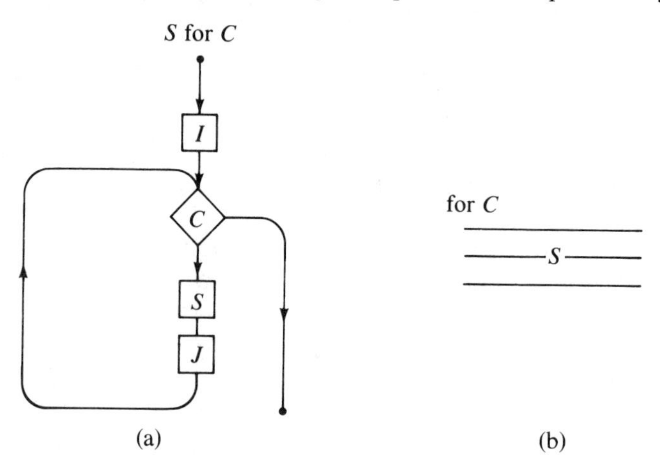

FIGURE 4.7

Example 4.20 It so happens that the for construct is just what we need to complete the prime number decision algorithm, as discussed in Examples 4.3, 4.15, and 4.18. We recall that we needed to refine the phrase, "test each number and update COUNT accordingly." More specifically our idea was to test each number in the range from 2 to $n - 1$ to see if it divides n evenly, then to increase our COUNT by 1 if that were indeed the case. Using the for construct we then capture this idea by writing:

for j running from 2 to $n - 1$
  if $j \mid n$
    increase COUNT by one

Thus j is introduced as a control variable, controlling the number of times that the indented (compound) statement is to be repeated. As a variable j simply takes on the values $2, 3, \ldots, n - 1$, and with each such value we will be asking, Does 2 divide n?, Does 3 divide n?,..., Does $n - 1$ divide n?, in succession. When the answer is "Yes" we will be incrementing the COUNT; otherwise we will not. Surely it is then the case that COUNT will store the number of divisors of n (in the range from 2 to $n - 1$) as required.

Putting all the pieces together from the earlier treatments of this problem we obtain the solution given in the accompanying algorithm *prime*. We would hope that our extended discussion of this problem, particularly the stepwise approach we have taken, might go a long way toward illustrating the top-down design methodology, besides serving as a vehicle for introducing the for construct.

```
algorithm prime (n: answer)

if n = 1
      give answer as 'false'
else
      set COUNT to 0
      for j running from 2 to n − 1
            if j | n
                  increase COUNT by one
      if COUNT = 0
            give answer as 'true'
      else
            give answer as 'false'
```

■

Pseudolanguage Guidelines

Now that we have introduced the overall style of our pseudolanguage and have illustrated its use, it is a good idea to summarize the ideas covered so far. The guidelines provided in Figure 4.8 might help to standardize our form. They also will serve to streamline our phrasing, cutting away unnecessary detail. Understand that we are not asking you to be so rigid that your style is cramped. Our sole aim is to impose sufficient uniformity to guarantee that every user understands the meaning of a pseudolanguage algorithm in the same way.

Guideline 1 seeks to get rid of unnecessary clutter. Guidelines 2, 3, and 4 only remind us of the style we have already introduced for these constructs. Guidelines 5 and 6 state our strong plea for the intelligent use of the English language. This, after all, is the main distinguishing feature of our pseudolanguage, in comparison to the computer programming languages. We thus are encouraged to describe our conditions, our processes, etc., as clearly as possible, although not necessarily by using precise mathematical or programming language formalisms. In fact, English is preferable. This way we are more likely to write what we mean, not whatever it is that fits into a particular framework. Note in guideline 6 how we insist that the description of processes begin with verbs. This serves to distinguish them from

1. Avoid punctuation, particularly at ends of lines.
2. Use vertical alignment for sequences.
3. Identify selections (and conditional processes) as indicated:

```
if ===                          case ===
  -----------                     ---: -----------
  -----------                     ---: -----------
  -----------                      :          :
else ===                          ---: -----------
  -----------
  -----------
  -----------
```

4. Identify repetitions as indicated:

```
while ===                       for ===
  -----------                     -----------
  -----------                     -----------
  -----------                     -----------
until ===
  -----------
  -----------
  -----------
```

5. Write conditions, for clauses and cases (===) clearly, in English or mathematical language.
6. Write processes (-----) in English or mathematics, beginning with a verb, choosing verbs that are expressive of the action to be taken.
7. Make a most careful use of indentation.

FIGURE 4.8 Guidelines for writing in pseudolanguage

the other lines in the pseudolanguage algorithm. Finally, we should pay close attention to the warning in guideline 7 to be most careful with our use of indentation. This is in fact a matter that deserves further illumination.

The reader could hardly have escaped noticing that we have used *indentation* in all of our examples with a specific purpose in mind. It is used to indicate that we are passing down to another level of the algorithm, to the treatment of a process that is embedded beneath the compound structure being employed. We cannot overemphasize the importance of a strict adherence to this technique. Its use provides a graphic display of the structure of an algorithm. We thereby are able to identify the body of a repetition or the various alternatives of a selection in one quick glance. Without such a system of indentation this would be impossible. Although we recommend an indentation of about two spaces in all instances, this is mainly a matter of taste and good judgment. Whatever the scheme, a uniform system of indentation will clarify the logical flow of an algorithm and enhance its readability. That is the main purpose of our indentation scheme.

Finally, though somewhat off the subject, let us return to a brief discussion of the notion of *type*. In the modern computer programming languages (e.g., Pascal) one is required to identify the type of every variable appearing in a program. There are good reasons for this; however, we do not want to enter into such arguments here. In the case of our pseudolanguage such type declarations would be out of place because they tend to decrease the algorithm's readability. Instead we prefer that the type of a variable be *inferred* from the context of the discussion. In the algorithm prime of Example 4.20, no one is really in doubt as to the type of n (natural number) or answer (of Boolean type). It therefore would seem to be redundant and something of a distraction were we to emphasize these matters in the phrasing of our algorithms. Again: Let the context of the discussion make these things clear.

EXERCISES 4.2

1 As an alternative to the prime decision algorithm in Example 4.20, suppose we introduce a Boolean variable *nodivisors*, with the idea of setting *nodivisors* to 'false' as soon as a divisor of the input n has been found. If then we replace the for construct with a while as in the accompanying revised algorithm prime, what should we write for the two missing (labeled ?) processes? In what sense is this algorithm more efficient than that of Example 4.20?

```
algorithm prime (n: answer)
if n = 1
   give answer as 'false'
else
   initialize j at 2
   set nodivisors to 'true'
   while j ≤ n − 1 ∧ nodivisors
      if j | n
         ?
      else
         ?
   take answer from nodivisors
```

2 Trace the changing state of the computation in the algorithm of Example 4.20 with the input:

(a) $n = 17$ **(b)** $n = 91$ **(c)** $n = 35$ **(d)** $n = 23$

3 Follow the instructions for Exercise 2 using the algorithm of Exercise 1.

4 Write a pseudolanguage algorithm that will convert an input length, expressed in feet and inches, to centimeters. (*Note:* There are 2.54 centimeters to the inch.)

5 Write a pseudolanguage algorithm that will find the smallest divisor (>1) of a given positive integer as input.

6 Follow the instructions for Exercise 2 using the algorithm of Exercise 5.

7 With *deposit*, *years*, (annual) *rate*, and periods-per-year (*ppy*) as inputs, write a pseudolanguage algorithm that will compute the output *balance* by accumulating the compound interest earned.

8 Trace the changing state of the computation in the algorithm of Exercise 7, given the inputs:

deposit = 1000.00

years = 2

rate = 0.09

ppy = 4 (quarterly compounding)

9 Write a pseudolanguage algorithm to act as a right triangle verifier. The inputs a, b, c are interpreted as the lengths of the three sides and it is understood that $c \geq b \geq a$. If not, or if for some other reason the lengths do not represent a triangle at all, one should conclude that an input error has occurred, and this should somehow be noted. (*Note:* Strictly speaking, the three sides form a right triangle, with c as hypotenuse, if $c = \sqrt{a^2 + b^2}$. But instead let us only ask that $c - \sqrt{a^2 + b^2}$ be small, say less than 1% of the hypotenuse.)

10 In basic astronomy we learn of the imaginative work of the German Johannes Kepler (1571–1630) who, as a young mystic, mathematician, and astronomer, hoped to discover a harmony to the universe. His work culminated in the formulation of three basic laws, the third claiming that in a planetary system, for any planet, the ratio of the square of its revolutionary period to the cube of its distance from the sun is a constant. In symbols:

$$\frac{p^2}{d^3} = K$$

Write a pseudolanguage algorithm that uses Kepler's formula to estimate d (in a.u.—see *Note*) for a planet in our solar system, given the value of p (in years). (*Note:* If we take one of the planets to be the Earth, measuring its period in years (= 1), and its distance from the sun in *a*stronomical *u*nits (again, = 1, by convention), we obtain the value $K = 1$.)

11 Suppose that in 1980 the population of the U.S. was 200 million and growing at the rate of 5% per year. Write a pseudolanguage algorithm that will find the year when the population will first exceed 300 million if this growth rate were to continue.

12 Use equivalent flow diagrams to show that:

(a) Any until construct "*S* until *C*" can be replaced by a sequence involving a while construct.

(b) Any while construct "*S* while *C*" can be replaced by a conditional statement and an until construct.

13 Show that the for construct "*S* for *C*" can be replaced by an equivalent sequence involving a while construct. (*Hint:* See Figure 4.7(a).)

14 Show that you can replace the case construct "C case C_1:S_1, C_2:S_2, ..., C_m: S_m" by a nesting of selection constructs. (*Hint:* See Figure 4.6(a).)

15 In a simplified stock market model we are to examine three successive quotations of a stock as input, representing a period of two days, and then we are to decide:

i. To buy if both days show an increase, the second bigger than the first

ii. To sell if both days show a decrease, the second bigger than the first

iii. Otherwise, to do nothing

Write a pseudolanguage algorithm that will implement this strategy.

16 The total number of dots appearing in the diagram of Figure 4.9 when we stop at the end of any number of rows is called a **triangular number**. Thus the first four triangular numbers are 1, 3, 6, and 10. Write a pseudolanguage algorithm for deciding whether a given positive integer n is triangular.

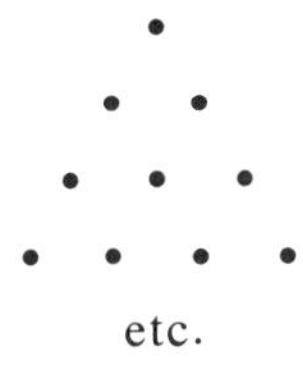

FIGURE 4.9

17 Consider the Fibonacci numbers of Exercise 11 in Section 3.2. Write a pseudolanguage algorithm for determining the first Fibonacci number to exceed 1000.

18 The **golden ratio** $(1 + \sqrt{5})/2 = 1.618033989\ldots$ is the limiting value of the ratio of two successive Fibonacci numbers (see Exercise 11 in Section 3.2). It has been a guide to artistic proportions since ancient times. The Parthenon at Athens has dimensions that fit almost exactly such a rectangular ratio. It is also quite common to find rectangles of this proportion in modern art. Modify the algorithm of Exercise 17 so that you determine as well the final value of the ratio of successive Fibonacci numbers (when the last number first exceeds 1000), for comparison with the limiting value.

19 A positive integer is said to be **square-free** if it has no perfect squares as factors (other than 1). Thus 7 and 15 are square-free but 28 is not. Write a pseudolanguage algorithm for deciding whether a given positive integer n is square-free.

20 Write pseudolanguage algorithms, using for statements, in order to sum the first 100 terms in the following series:

(a) $1 + \frac{1}{2} + \frac{1}{3} + \frac{1}{4} + \frac{1}{5} + \frac{1}{6} + \cdots$

(b) $\frac{1}{1 \cdot 2} + \frac{1}{2 \cdot 3} + \frac{1}{3 \cdot 4} + \frac{1}{4 \cdot 5} + \frac{1}{5 \cdot 6} + \cdots$

(c) $1 - \frac{1}{3} + \frac{1}{5} - \frac{1}{7} + \frac{1}{9} - \frac{1}{11} + \cdots$

21 In a simplified payroll accounting system, assume that we use the hours and minutes worked in order to compute an employee's paycheck according to the following schedule:

$5.50 per hour, for the first 40 hours
$8.25 per hour for overtime
2% deduction for union dues

Write a pseudolanguage algorithm for accomplishing this task, with hours and minutes as input.

22 Write a pseudolanguage algorithm that will find the *range* (the difference between the largest and smallest values) in an input list of numbers.

23 As a portion of a computer dating service, imagine that we are to read a deck of cards giving the heights of various individuals and we are to find the person whose height is closest to 6 feet. Each card will contain an identification number followed by two integers, the first representing feet and the second inches, as in denoting someone's height as 5′10″. Develop a pseudolanguage algorithm that will find this ideal person's identification number. (*Hint:* Keep the best!)

24 A list of cards contains the following information pertaining to the students in your class:

i. Year in college (an integer)
ii. Major field of study (a character code—A, B, C, D, E, F, G, or H)
iii. Previous college mathematics study (a Boolean value)

Write a pseudolanguage algorithm that will determine:

(a) The percentage of students in each year of college
(b) The percentage of students in each major field
(c) The percentage of students with previous college mathematics study

(*Hint:* In (1) and (2), use the "case" construct.)

25 You are to write a pseudolanguage algorithm for solving quadratic equations:

$$ax^2 + bx + c = 0$$

making use of the discriminant $b^2 - 4ac$ to decide whether there are one or two roots. (What about complex roots?) The coefficients a, b, c are to appear as inputs.

26 Write a blackjack dealer's algorithm in pseudolanguage. In blackjack, the dealer must follow a definite set of rules:

i. If the total count in a hand is 16 or less, the dealer must take another card.
ii. If the total count is between 16 and 22 (noninclusive), the dealer must stay.
iii. If the total is over 21, the dealer busts (i.e., he or she loses the game).
iv. A one (an ace) is counted as 11 if it does not put the dealer over 21; otherwise it is counted as 1.

Suppose that the input is a list of integers from 1 to 13, with 11 (jack), 12 (queen), and 13 (king) counted as 10.

27 Write a perfect number decision algorithm in pseudolanguage. (*Note:* See Exercise 4 in Exercises 4.1.)

28 Write a pseudolanguage algorithm for processing a single transaction at a bank. For the individual's account, assume that the current balance and the number of transactions already processed appear as inputs, together with a transaction amount (for credit or debit), used in updating the above balance and number. A service charge of 0.25 is to be assessed if the current balance is below 200.00 or the number of transactions has exceeded 50. A Boolean variable *credit* is used as an additional input to indicate whether we are processing a deposit (credit = 'true') or the cashing of a check (credit = 'false').

Sec. 4.3 EXTENSIONS OF THE LANGUAGE

A most important consideration is that our language be flexible and extendable.

From what has been said already it is clear that our algorithmic pseudolanguage should be extendable regarding the incorporation of mathematical symbolism relating to new algebraic structures as encountered in our studies. This will be a continuing process, one that you can digest bit by bit as our discussions proceed. In this section, however, we refer to those language extensions that are motivated by an attempt to enlarge the expressive capability along lines that parallel those developed for the modern programming languages. First we must introduce *subscripted variables*, or *arrays* as they often are called. This will enlarge substantially the range of problems that we are able to discuss. Furthermore we need to provide a mechanism for treating algorithms as independent modules that can be called into action by other algorithms. It is only through this capability that we are able to fully exploit the modular approach that leads to an effective algorithmic design. Moreover, it is only in this context that the important notion of *recursion*—having an algorithm "call" itself—can be brought into clearer focus. (This concept comes later—see Section 4.5.) To those of you who already have a fairly extensive programming background from high school or self-study, these ideas will be quite familiar. To the rest of you, we can only say that these extended capabilities are entirely natural and thus easily used, following as they do in an obvious way from all that has preceded.

Goals

After studying this section, you should be able to:

- Discuss the notion of a *subscripted variable* and provide examples of its use.
- Characterize those problem situations where the use of a subscripted variable is advantageous.
- Describe the general *search problem* and discuss several ways in which the problem might be solved, commenting on the advantages of one solution over another.

- Describe a method for *sorting* the members of a list.
- Explain the advantages in being able to treat algorithms as independent modules that can be called into action by other algorithms.
- Describe the *calling mechanism* to be used in our pseudolanguage, distinguishing the case where the called module is treated as a function.
- Use algorithms as modules or building blocks in the design of complex algorithmic solutions to larger problems.

Subscripted Variables

We have mentioned the need to extend our language in order to provide the capability for effectively treating new mathematical subject matter as it is encountered in our study. And of course, we should surely allow for the treatment of those mathematical topics already discussed—sets, functions, relations, etc.—in some meaningful way. Thus it should be permissible to speak of variables that are of "set" type, and to allow for such elementary processes as would seem natural in these settings. Certainly it would be appropriate to see the processes:

adjoin x to S

or

designate S as A union B

if A, B, and S were set variables in some algorithm pertaining to set computation. Development and use of such mathematical extensions of our language are definitely encouraged.

But we have in mind an immediate objective that is more fundamental than anything of this sort. We want to ensure that our algorithmic language has the built-in capability for treating *subscripted variables*, or *arrays* as they are called in the modern programming languages.

You might reasonably wonder how it is that the large computer memories are ever fully utilized if we can store only one item of information under a given name. Regardless of the problem, it seems that we have always managed to get by with a half dozen or so separate variable names. But, in truth, this is only because we have greatly restricted the nature of the problems being considered. Whenever we were faced with a large quantity of data, as in the input list of Example 4.2 and the subsequent treatments of this problem, we always have managed to be able to examine and process one item, then examine and process another, and so on, never finding it necessary that all the items be accessible at the same time. If we ever find a situation where this rule does not prevail, then the chances are good that we will want to introduce a subscripted variable. Moreover whenever we speak of lists, tables, matrices, etc. it is again likely that we will want to regard the data as being manipulated in the form of a subscripted variable.

Consider for a moment the problem of sorting 1000 numbers into ascending order, supposing that we have a scrambled arrangement of these numbers on a deck of cards or on some other form of input medium. Whether you realize it or not, this is a typical computing application, though perhaps one that generally does not occur in so elementary a context as suggested here. In fact someone once asked the question, If we were to stop all the computers in the world, and if we could ask them what they were doing at the time, what would be their response? In spite of their being called computers it has been suggested that most of them would not be computing at all; rather, they would have been sorting and rearranging information in one way or another. This may be an overstatement, but it indicates the importance of sorting in every phase of the applications.

Taking this much for granted, let us consider this elementary sorting problem in greater detail. At a later point we can begin to identify some of the advantages for having sorted data. It would seem that if we were going to do anything at all with our list of numbers, we would have to examine them as a collection. We say this because we certainly cannot do as we have done in our previous processing of lists; that is, we cannot simply examine and process, examine and process, one at a time, because we do not know the proper relative position of any one number without considering its relationship to all the others.

So it seems that we might need 1000 variable names (think of it!) just to store the list for eventual processing. We could try to use names A, B, C, D, E, etc. Or, perhaps, the names $A1, A2, A3, \ldots, A1000$ would be better. This latter alternative is getting close to the idea we have in mind. But we think of just a single variable (named A, say), with 1000 indices or *subscripts*. The **subscripted variable** as a whole has a name, in this case, A. But if we want to refer to an individual component its name will consist of two parts—the variable itself and an index or subscript (here, an integer or an integer variable or expression with a value in the range from 1 to 1000), written below the line as in conventional mathematical usage.

Example 4.21 We may refer to A_3 or A_{713} if A is understood to be a subscripted variable with 1000 components. In actual fact A is 1000 variables; but the convenience in being able to refer to each component by the same name cannot be overstated. Carrying these ideas one step further, we can speak of A_i or A_{i+7} where i is an integer variable. What will this mean? It is quite clear. One has only to look at the current value of i. Thus if i currently has the value 3, then A_i refers to A_3 and A_{i+7} refers to A_{10}. ∎

Example 4.22 If we want to add one to each component of the subscripted variable A, as interpreted above, we have only to use the for construct in writing:

for i varying from 1 to 1000
 increase A_i by 1

It is quite common to use a for statement whose control variable runs over the indices of a subscripted variable. Note that in considering 1000 different variables, even if they had been named conveniently, say $A1, A2, \ldots, A1000$, we would have had to write 1000 statements:

increase $A1$ by 1
increase $A2$ by 1
...
increase $A1000$ by 1

in order to accomplish our task! But in using the for construct we automatically repeat the statement:

increase A_i by 1

once with the value $i = 1$, then with $i = 2$, and so on, so that the net effect is as desired. The saving of time and amount in writing is remarkable, to say nothing of the final product's enhanced readability. ■

Example 4.23 The use of subscripted variables, particularly in connection with the for construct, are many and quite varied. For example if A is the same array as before, consisting of 1000 numbers, we can compute their *sum*,

```
set SUM to 0
for i varying from 1 to 1000
   increase SUM by A_i
```

their *average*,

```
set SUM to 0
for i varying from 1 to 1000
   increase SUM by A_i
compute AVE as SUM/1000
```

their *maximum value*,

```
set max to value of A_1
for i varying from 2 to 1000
   if A_i > max
      take max from A_i
```

or whatever else is desired. (Note the use of the keep-the-best strategy in this last illustration.) ■

The subscripted variable A that we have used in all of our examples would be considered a list of some kind. We would say that such instances are *one-dimensional*, inasmuch as we can imagine its component values strung out in a row. More generally one might want to consider a *two-dimensional* subscripted variable, thought of more as a table of values having rows and columns.

Example 4.24 A variable *board* with components $board_{ij}$ might be used in a chess-playing algorithm, the idea being that i and j are to range from 1 to 8, symbolizing the row and column positions on a chessboard. Accordingly it would be appropriate to have the 64 components take on values from the set:

$$\{\text{'P'}, \text{'N'}, \text{'B'}, \text{'R'}, \text{'Q'}, \text{'K'}, \text{'}\square\text{'}\}$$

in representing the chesspieces pawn, knight (N), bishop, rook, queen, king, and the artificial chesspiece, blank, used in indicating that a position is vacant. ■

In a totally different vein we might begin to establish a part of the rationale for our often preferring that large (one-dimensional) quantities of data be sorted. In so doing we will gain some useful experience in the handling of subscripted variables. Suppose that we are again dealing with our list of 1000 numbers, and we are asked to determine whether a given number x is present in our list (and, if convenient, to determine its location within the list). Here we have the *search problem,* one that is quite commonly encountered in computer science applications.

Example 4.25 There is at least one obvious way this problem might be handled. We can simply look at all the numbers, one at a time, using a Boolean variable *found* in writing:

```
set found as 'false'
for i varying from 1 to 1000
  if x = A_i
    set found to 'true'
```

The perceptive reader, however, will suggest an improvement in this approach: Use the while construct instead of the for to write:

```
set found as 'false'
initialize i to 1
while (i ≤ 1000) and ¬ found
  if x = A_i
    set found to 'true'
  else
    increase i by 1
```

In this way we then look only until x is found (if indeed, it is there). Of course in the worst case—where x is the last element of the list or not there at all—there is no improvement. But perhaps equally important here is the fact that we have located x (when it is there) with the final value of i in the execution of the while construct. That was not possible with our first solution. ■

One might ask, Can we still do better? Suppose our array is sorted. Then we can suggest an entirely different approach, one that offers advantages of both convenience and efficiency. You may have heard of an old popular radio show called *20 Questions.* A panel of experts would be asked to identify an unknown subject (a building, a book, a person, or whatever), by asking up to 20 questions, each having a Yes or No answer. With a judicious choice of questions they would learn more and more about the mystery subject with each answer given by the moderator. If the panel had determined that the subject was somewhere in the United States, the next question would invariably be, Is it east of the Mississippi? This is an illustration of a most useful problem-solving technique known variously as *bisection, divide and conquer*, or some other similar name. The point is that although we don't learn the answer in one step, we do rule out half of the domain in any case.

Example 4.26 To see how this technique might be applied to our search problem, we first must suppose that our list A is sorted into numerical order, for example in ascending order from the smallest entry to the largest. In the accompanying algorithm *bisection* we are given A and x as inputs and we use a Boolean variable *found* as our answer (together with an integer *location*). We make use of three indices (again, integer variables)—*top, bottom*, and *mid*—as shown. Thus having set *top* and *bottom* to the first and last index of the array, respectively, we repeatedly compute the middle index (in analogy to Mississippi), the one lying halfway between the two. Comparing the value found in the array at this midpoint with the value of x, we can decide whether x would be in the first or second half of the current and continually shrinking domain. We then change *bottom* or *top* accordingly, unless of course our values happen to coincide, in which case x has been found.

algorithm bisection (A, x: found, location)

set top and bottom to first and last respectively
initialize found as 'false'
while top $\leq$ bottom and $\neg$ found
  compute mid
  case comparison
    $x < A_{\text{mid}}$: change bottom to mid -1
    $x > A_{\text{mid}}$: change top to mid $+ 1$
    $x = A_{\text{mid}}$: set found to 'true'
assign location as mid

■

In our two previous attempts at a solution to the search problem (Example 4.25), considering a list of 1000 components, we might have had to traverse our for or while loop 1000 times in order to find x or to know that it wasn't there. In only 10 traversals in the bisection algorithm we will have a decision. We say 10 traversals because 2^k is the size of the domain that we are able to bisectively search in k passes around the loop, simply because we split the domain in half each time (see Exercise 10 in Exercises 4.1). And this indeed represents a savings.

But remember: We were able to improve our efficiency so dramatically only because our list was sorted. The question then arises: If sorted lists are such a good thing, how do we accomplish the sorting?

A Sorting Algorithm

How to accomplish the sorting is a fundamental question, one that does not have an easy answer that suits all possible circumstances. Among all the sorting methods (there are a great many), however, perhaps none is easier to understand than that of the *selection sort*. For this reason it is a good first illustration for beginning students.

Suppose we have a list A (a subscripted variable) with n components. We first select the smallest of all the values $A_1, A_2, \ldots, A_n$, and we interchange it with A_1. (Assuming that we are going to arrange the elements in ascending order, we thus can be assured that our first entry is right.) Then we find the smallest value from among $A_2, \ldots, A_n$, and we interchange it with A_2. Surely the picture is becoming clear. When we have finally done this for A_{n-1} and A_n, we will have managed to sort the list in order, as required.

If we were to express this repetitive idea—find the smallest and interchange—in our pseudolanguage, we would be led to the top-level phrasing:

```
for each i from 1 to n − 1
   find index of smallest entry from i to n
   interchange it with A_i
```

noting that we really need to find the *index* of this smallest value so that we can properly designate the list element to be involved in the interchange. We have used the for construct again, with a control variable i to be used in counting the number of passes through the repetition ($n - 1$ of them, according to the analysis above). Being assured that this top-level treatment is logically correct, we now have to give further consideration to the two embedded processes, treating each of them as if they were new problems in their own right.

A little reflection shows that the interchange process is already an elementary process. The most we will want to do is to clarify the term "it" at a later point. For the other embedded process, we are fortunate in being able to say that we have seen it before—or, at the very least, we have seen something like it before. Indeed we find that our familiar keep the best strategy will work once again, this time keeping a subscript (*small*) in writing:

```
set small at i
for each j from i + 1 to n
   if A_j < A_small
      take small from j
```

Thus we keep the best (i.e., the index of the smallest) by continually comparing the A_j's with the value at the subscript that is best so far.

Putting it all together, making the appropriate substitutions in our original top-level design we obtain the accompanying pseudolanguage algorithm *selectionsort*. Note that we have replaced "it" with A_{small}, the smallest entry in the subscript range from i to n, as required.

```
algorithm selectionsort (A)
for each i from 1 to n − 1
  set small at i
  for each j from i + 1 to n
    if A_j < A_small
      take small from j
  interchange A_small with A_i
```

Algorithms as Modules

Throughout all our discussions you undoubtedly have noticed a continued emphasis on the use of modularity in our algorithmic designs. The idea is to decompose a problem into meaningful and manageable subproblems. But this technique only reaches its full potential when we are able to consider these subproblems as independent *algorithmic modules* to be called into action by other algorithms. When these are used properly one is able to conceptualize and develop a larger algorithm through the design of the smaller component modules (again—divide and conquer!). All that is left to consider are their interrelationships. This is a decided design advantage, as we shall see.

But sometimes it works the other way. We may find that an algorithm that has already been designed for one purpose can be reused as a modular building block in another algorithmic design.

Example 4.27 Suppose we are to design an algorithm that accepts three numbers $z \geq y \geq x$ as inputs (presumably representing the lengths of the sides of a triangle), and we are required to compute the area A. Of course we always can resort to the formula we learned in high school:

$$A = \sqrt{s(s - x)(s - y)(s - z)}$$

where s is the semiperimeter: $s = \frac{1}{2}(x + y + z)$. But suppose we also are required to classify the triangle according to whether it is equilateral, isosceles, etc. "Ah," we say, "this has been done already." Recognizing this we can call on the triangles algorithm (see Example 4.19) in writing the new algorithm *area*, shown below. Note that we are able to use simpler area formulas in the cases where we deal with an equilateral, isosceles, or right triangle. But, most importantly, the reader should notice how we leave the determination of the triangular class to our earlier triangles algorithm. This has the additional effect of enhancing the readability of the current algorithm. Of course one could have simply modified the triangles algorithm so that it would compute areas at the same time that it determines the class. But then we would have to tamper with an algorithm that is already known to work. The convenience of being able to use something that has already been properly designed is an important consideration to make when deciding what is needed for a real-life application.

algorithm area (z, y, x: A, kind)

call triangles (z, y, x: kind)
case kind

'equilateral': set A to $\frac{\sqrt{3}}{4} x^2$

'isosceles': if $z = y$

set A to $\frac{x}{4}\sqrt{4y^2 - x^2}$

else

set A to $\frac{z}{4}\sqrt{4y^2 - z^2}$

'right scalene': set A to $xy/2$
'acute scalene', 'obtuse scalene': compute s as $\frac{1}{2}(x + y + z)$
set A to $\sqrt{s(s - x)(s - y)(s - z)}$ ■

In the modern computer programming languages, elaborate mechanisms are required in order to permit one algorithm to *call on another*, as we have just done. Because we are dealing with a pseudolanguage, the details are a good deal easier to handle, and we can rely on the reader's common sense in helping to put the necessary ideas across. You surely have noticed in Example 4.27 how it is that we call the algorithm triangles with *arguments* z, y, x for input and *kind* for output, whereas the input and output variables for that algorithm had different names. We don't require that there be an agreement in names. This enables us to call the algorithm over and over again, once as triangles (z, y, x: kind) and as triangles (w, v, u: type) on another occasion, all within the same calling algorithm if we so choose. Our intent is clear and doesn't require any further explanation.

Example 4.28 In developing a payroll routine one might be dealing with several lists of data, for example, a list of employee identification numbers, a list of payroll accounts, etc. It is entirely conceivable that one would need to sort these data on various occasions throughout their processing. Accordingly we might see several calls to a sorting algorithm (perhaps our selectionsort), as indicated below:

algorithm payroll (---: ---)
⋮
call selectionsort (ID)
⋮
call selectionsort (ACC)
⋮
call selectionsort (ID)
⋮

The advantages to being able to use arbitrary names as arguments to the called module are immediately apparent. Whether we call selectionsort with the argument *ID* (the name of the employee identification list) or with *ACC* (the name of the payroll account file), the sorting still will be accomplished. It is as if *A*, the name of the subscripted input variable for selectionsort, is only a place holder, capable of assuming any number of substituted names. The whole subalgorithm—in this case the selectionsort—is then executed with the substituted name in place of *A*. But again it is just good common sense to think of things in this way. ■

Example 4.29 In Example 4.26 we designed a bisection algorithm that assumed that the input list already had been sorted. Alternatively we could call on a sorting routine as shown in the revised algorithm *bisection* below. The new bisection algorithm is self-contained: it does not have to depend on a sorted list as input.

```
algorithm bisection (A, x: found, location)

call selectionsort (A)
set top and bottom to first and last respectively
initialize found as 'false'
while top ≤ bottom and ¬ found
   compute mid
   case comparison
      x < A_mid: change bottom to mid − 1
      x > A_mid: change top to mid + 1
      x = A_mid: set found to 'true'
assign location as mid
```

■

An important variation of this technique, the calling of one algorithm by another, is that in which we view the called module as a **function**. The main difference regards the calling mechanism. In a *function call* the name of the function (originally itself just an algorithm) and a list of its input arguments will appear within the context of some process being described in the calling algorithm:

... name (list of arguments) ...

in much the same manner that we might refer to a mathematical function, for example, *abs*(x), $n!$, or whatever. This implies that the function name itself will obtain a value in its original algorithmic description.

Example 4.30 In order to consider an elementary example, let us suppose that we want to turn the Euclidean algorithm of Section 3.3 into a function called gcd. As a matter of fact, the algorithm is perhaps best thought of in this way: two integers determining a third, the latter as a function of the first two. We then write the accompanying function gcd, where the reader has but two things to note. In the opening title line we refer to a function rather than an algorithm and, at the same time, the former output variable becomes the name of the function (algorithm). Next, the name obtains a value, as indicated by the *return statement* of the last line. Having done this we are free to call on this function in our writing of other algorithms, in the manner illustrated previously.

```
function gcd (a, b)

if a is less than b
   interchange their values
compute the remainder on dividing a by b
while we have a nonzero remainder
   replace a by b, b by remainder
   compute remainder on dividing a by b
return gcd as b
```
■

In order to be more explicit, suppose we look back to the discussion in Section 3.3 (in reference to the applications of rational arithmetic in architectural, textile, automotive, and various other situations). One can think of rational numbers (pairs of integers) as being represented by short lists of integers (i.e., lists of length 2). A rational number a then would have two components, a_1 and a_2, thought of as its numerator and denominator, respectively. We then are in a position to see how our *gcd* function can be used as an aid in performing the necessary rational arithmetic operations.

Example 4.31 In adding a and b and calling the result c we first would need to compute a common denominator c_2. The product of a_2 and b_2 would work perfectly well, but to avoid an unnecessarily large denominator we might do better in factoring out their gcd in the process, as shown in the algorithm *ratadd* below. Then c_1, the numerator of the sum, can be determined in the usual fashion by considering the equation:

$$\frac{a_1}{a_2} + \frac{b_1}{b_2} = \frac{c_1}{c_2}$$

Finally, in the last three lines of the algorithm, we can arrange to reduce c to its lowest terms, making use of the gcd function once again. Altogether the algorithm provides a concrete illustration of the calling mechanism for functions while also serving to describe some of the details involved in programming a rational arithmetic package. Beyond this the reader should imagine a whole arsenal of algorithms, that is, ratadd, ratsub, ratmpy, and ratdiv, all being called to perform in some appropriate area of application.

```
algorithm ratadd (a, b: c)

set c2 = a2b2/gcd (a2, b2)
set c1 = (c2/a2)a1 + (c2/b2)b1
compute d = gcd (c1, c2)
divide c1 by d
divide c2 by d
```
■

Example 4.32 Decision algorithms often are made over into Boolean functions wherein the name takes on the values 'true' or 'false'. For instance the algorithm prime of Example 4.20 could be reinterpreted as a function:

function prime (n)

in which we assign our 'true' and 'false' values to the name *prime*, using return statements and dispensing entirely with output variable *answer*. Then it would be possible to call this function in the midst of the condition clause of some other algorithm, writing for example:

```
if prime (p)
   process R
else
   process S
```

The effect is obvious. In this way we are able to make use of Boolean conditions that are not elementary but, instead, are described by complex decision algorithms of their own. ■

EXERCISES 4.3

1 A person's name is stored and left-justified in a one-dimensional subscripted variable *person* as a string of characters in the form:

firstname initial. lastname

Blank characters are used as separators, with the middle initial and its period (perhaps) missing. Write a pseudolanguage algorithm that will transfer this information into a similar subscripted variable in the alternative form:

lastname, firstname initial.

Assume that both variables have 25 character components.

2 Given a 10×10 two-dimensional subscripted variable A of numbers A_{ij}, write pseudolanguage algorithms for computing:

(a) The largest element in row i

(b) The smallest element in column j

(c) The sum of the elements in row i

(d) The sum of the elements in column j

(e) The sum of all the elements

(*Note:* In a two-dimensional array it is customary to think of the first index as signifying the row and the second the column. *Hint:* In (a)–(d), i or j, respectively, should be treated as an input variable.)

3 Given two sorted lists of numbers already stored in the subscripted variables A and B, write a pseudolanguage algorithm for merging the two lists into one sorted list, C. (*Note:* Assume that the size of C is at least as large as $n + m$, n and m being the sizes of A and B, respectively. *Hint:* Repeatedly compare the top elements of the two lists, moving the top down as an element is selected.)

4 Use the two-dimensional subscripted variable *board* of Example 4.24 and a pair of integer inputs, i and j, interpreted as representing the position of the piece:

bishop ('B')   queen ('Q')
knight ('N')   rook ('R')   king ('K')

on the chessboard. (The piece is treated as an additional input variable.) Write a pseudolanguage algorithm that will indicate the position of the piece on the board (using the appropriate symbol, as above) and also will describe its sphere of influence (using a + symbol), leaving all other squares blank. (*Hint:* First blank out all the squares, then take care of the sphere of influence and then, finally, mark the position of the piece. For the last two of these three processes you will want to use a case construct.)

5 In statistics the **mean** $\bar{x}$ of a list of n numbers $x_1, x_2, \ldots, x_n$ is defined by writing:

$$\bar{x} = \frac{1}{n} \sum_{i=1}^{n} x_i$$

and their "**standard deviation**" σ is given by the formula:

$$\sigma = \sqrt{\frac{\sum_{i=1}^{n} (x_i - \bar{x})}{n - 1}}$$

Write a pseudolanguage algorithm for computing $\bar{x}$ and σ, with n and the subscripted variable x as inputs. (*Note:* Some algebraic manipulation of the formula for σ can lead to a more efficient computation.)

6 Using a two-dimensional subscripted variable A as input, assume it stores numbers in n rows and n columns. Write a pseudolanguage algorithm that will compute the sum of all the numbers in A that are below the diagonal (i.e., those elements A_{ij} with $i > j$). (*Note:* See the Note to Exercise 2. *Hint:* As i varies from 2 to n in a for statement, j should run from 1 to $i - 1$ in an embedded for statement.)

7 In the first pass through an *exchange sort* we compare A_j with A_{j+1} for $1 \leq j \leq n - 1$, in succession (assuming n elements to be sorted), exchanging their values whenever they are out of order. In this way the largest entry of A will have been moved to the bottom in only one pass, and may be ignored subsequently. Similarly the next largest entry will be moved to the next-to-last position in the second pass. After $n - 1$ passes the entries will have been sorted in ascending order. Develop a pseudolanguage algorithm for this sorting routine.

8 Show how to modify and improve the efficiency of the exchange sort (of Exercise 7), using a Boolean variable to detect the fact that the array is prematurely sorted. [*Hint:* It may be that there were no exchanges performed on a given pass.]

9 In the *bubble sort* of a list A of n elements, we compare A_j with A_{j+1} for $1 \le j \le n-1$. But whenever the pair is out of order, the smaller or lighter one is allowed to move upward (i.e., to bubble up) until it is in the proper position relative to the others. Develop a pseudolanguage algorithm for this sorting routine. (*Hint:* To bubble up A_{j+1} put its value into a temporary variable location and compare this with the values above, moving them down as you go, until the proper point of insertion is found.)

10 Suppose that we have two one-dimensional subscripted variables X, Y of length n, each storing a list of numbers. Corresponding elements then may be thought to represent experimentally measured data points (x_i, y_i), in no particular order. For each of the following sorting methods, show how to modify the pseudolanguage algorithm so that it accepts X, Y and n as inputs, sorting *on* X but keeping the pairs (x_i, y_i) together. (*Note:* After sorting, the x's will be in order (but not necessarily the y's), as if we were reading the data points from left to right in the x, y plane.)

(a) selection sort (of the text)

(b) exchange sort (of Exercise 7)

(c) modified exchange sort (of Exercise 8)

(d) bubble sort (of Exercise 9)

11 Trace the state of the computation in the following sorting routines on the input list $A = (5.4, 4.7, 8.2, 3.6, 5.6, 6.4, 5.5, 9.3)$.

(a) selection sort (of the text)

(b) exchange sort (of Exercise 7)

(c) modified exchange sort (of Exercise 8)

(d) bubble sort (of Exercise 9)

12 Devise a pseudolanguage algorithm to serve as an automatic cashier. Use an integer list as a one-dimensional subscripted variable *change*, representing the number of pennies, nickels, dimes, ..., fifty-dollar bills, to be returned for a given purchase amount and a given payment. (*Note:* Assume that the purchase and payment are at most $100.00, and arrange to minimize the total number of *tokens* returned (i.e., don't give more than four pennies, more than one nickel, etc.).)

13 You expect to be writing a number of banking algorithms, and it seems desirable to have a routine for rounding monetary amounts to the nearest penny. Write a pseudolanguage algorithm for accomplishing this task.

14 Write a pseudolanguage algorithm for:

(a) adding **(b)** subtracting **(c)** multiplying **(d)** dividing

two complex numbers. Treat the complex numbers as pairs of ordinary numbers, stored in two-element subscripted variables, one element representing the *real* part, the other the *imaginary* part. (*Note:* In each case you then will have two input variables and one output variable.)

15 Write a pseudolanguage function for computing the interest due on a certain loan principal held for a given time at a given rate (and a certain frequency at which the loan is compounded). (*Note:* See Exercise 7 of Exercises 4.2.)

16 Rewrite the automatic cashier algorithm of Exercise 12 as a function—change (purchase, payment). (*Note:* The function name *change* is subscripted. How does this influence your thinking, particularly in reference to the fact that the name obtains a value?)

17 Write pseudolanguage functions for computing the mean and standard deviation of a list of numbers. (*Note:* See Exercise 5.)

18 Write a pseudolanguage function for converting a Roman numeral to an ordinary integer. Treat the Roman numeral as a one-dimensional, 10-component character array, being prepared to handle the symbols:

$$I = 1 \quad V = 5 \quad X = 10 \quad L = 50 \quad C = 100 \quad D = 500 \quad M = 1000$$

(*Hint:* If the decimal equivalent of any one character exceeds that of the preceding character, reading from left to right, then a subtraction from some total is suggested (e.g., IV = 5 − 4).)

19 Design a pseudolanguage function for performing a straight-line interpolation of y_0 (at a given x_0), using two neighboring points (x_1, y_1) and (x_2, y_2), thought to be on either side. (*Note:* Be sure to do something sensible if it should happen that x_0 is not between x_1 and x_2. *Hint:* In any case, the value of y_0 should be computed in such a way that all three points lie on the same straight line.)

20 Write a pseudolanguage function lcm for computing the least common multiple of two given integers (see Section 3.3).

21 Write pseudolanguage algorithms:

(a) ratsub **(b)** ratmpy **(c)** ratdiv **(d)** reduce

for performing subtraction, multiplication, and division of pairs of rational numbers, and for reducing a single rational number to its lowest terms.

22 Write a pseudolanguage function power (x, n) for computing x to the power n, for any integer n, positive, negative, or zero.

23 Write a pseudolanguage function for computing n factorial. (*Hint:* Use the for construct.)

24 Rewrite the algorithm prime of Example 4.20 as a Boolean function, as suggested in Example 4.32.

Sec. 4.4 PROOFS OF CORRECTNESS

Testing may show the presence of errors, never their absence.

If we could look 25 or even 10 years into the future, we probably would realize that presently computer science is a field in its infancy. Nevertheless, considering its short span of development, it already possesses many of the attributes that one would associate with a mature science. We wish to examine several of these aspects in the remaining sections of this chapter, but only at a level that is suitable for the presentation at hand. For example, some people believe that the day is fast approaching when the computer scientist will be able to (and be expected to) justify his or her assertions to the same degree as that expected of a mathematician. He or she will be able to demonstrate that an algorithm—whatever its scope—indeed performs as it was intended, and that there is no uncertainty on this account, much in the manner of our providing a formal proof of a theorem in mathematics.

Moreover it may be true (and it may be necessary) that such proofs of correctness will be demonstrated and even generated with the aid of another algorithm, a kind of tool for establishing the necessary documentation. In order for you to begin to appreciate the way in which these developments might come about, it is necessary that we become somewhat familiar with certain formal mechanisms that are reminiscent of those appearing in mathematical logic. It happens that the most promising proof techniques are those that can be given a deductive foundation, not unlike those that we already have studied quite extensively in Chapter 2. This was one of the main reasons for our concentrating so heavily on the deductive logic of algorithms at an early stage of our studies—this way the current presentation (and others of its kind) are more readily understood.

Goals

After studying this section, you should be able to:

- Describe the idea of a *formal design specification.*
- State the rule that provides for the analysis of elementary processes and illustrate its use by examples.
- State the deduction rules for the sequence, selection, and repetition constructs and illustrate their use.
- Clarify and illustrate the fact that "testing may show the presence of errors, never their absence."
- Carry out proofs of correctness, at least for algorithms of moderate complexity.

Testing vs. Verification

In providing a few formal techniques for verifying the correctness of algorithms, we certainly do not mean to propose that these techniques be applied each and every time that an algorithm is developed. That would be akin to asking a mathematician to supply an inductive proof for each and every instance in which the principle of mathematical induction is involved! On most occasions the mathematician *knows* that the induction will work without actually proving it. In the same way the designer of an algorithm generally *knows* that a computational loop will terminate with a certain condition being satisfied by the variables; it is not always necessary to provide a formal verification. Nevertheless, as computer scientists (and we emphasize "scientist"), applied mathematicians, etc., we cannot altogether avoid the obligation to use our own unique understanding of an algorithm in order to formulate precise assertions about its computational behavior, and to communicate these assertions to others.

We frequently have traced the state of a computation in order to test the behavior of an algorithm with specific input values. It is clear that this testing technique can illuminate many of the important features of an algorithm, and even can be used to detect certain logical errors in its formulation. But it rarely will suffice for proving that no such errors exist. It may happen occasionally that only finitely many input values are intended and that these are so sufficiently few in number that an exhaustive testing is actually feasible. But this is clearly the exception and not the rule.

Example 4.33 Consider the Euclidean algorithm of Section 3.3 once more, particularly in reference to Example 3.26. There we have run a test of the algorithm for specific input values ($a = 180$, $b = 252$), tracing the changing state of the computation, and, indeed, the algorithm seems to be performing correctly. On the basis of this test, or perhaps 100 similar tests on randomly selected input data, the student might be tempted to conclude that the algorithm in fact does what it was intended to do; but this would not be a valid conclusion.

One might argue that we could go further, testing the algorithm for all pairs of integers that we ever expect to encounter, say those up to the size 2^{48}, just to pick a number. But even if the entire algorithm could be executed (by a computer) in a microsecond, we would need

$$2^{48}2^{48}10^{-6} \text{ seconds} \approx 2.5 \times 10^{15} \text{ years}$$

to complete the testing! ■

What we need instead is some precise *verification* technique, some way to definitely establish that an algorithm does what was intended. In principle such a technique is in fact available. One begins with the idea that what was intended is considered a known, and therefore can be used as a kind of **design specification**. Typically it will be given in the form:

$$\{P\}\ A\ \{Q\}$$

where P and Q are *assertions* involving the input and output variables of the problem, and A is the algorithm that pretends to solve it. We read this specification as saying: "If P holds before the algorithm is executed, and A is then executed, then Q will hold afterwards."

Example 4.34 Continuing once more with a discussion of the Euclidean algorithm as A, we find that we can provide the specification:

$$\{0 < b \wedge 0 < a\} \text{ euclid } \{gcd = \text{gcd}\,(a, b)\}$$

That is, we are only given that a and b are to be positive integers as inputs, and after the algorithm has been executed the output variable *gcd* should have the value of the greatest common divisor (gcd) of a and b. That is what our algorithm should accomplish. Note the different meanings for gcd here. We refer to the variable named *gcd* and the technical meaning (gcd) in one and the same assertion. We hope this has not caused any confusion. ■

Example 4.35 If we consider the prime decision algorithm of Example 4.20, we might have given the specification:

$$\{n > 0\} \text{ prime } \left\{ \text{answer} = \begin{cases} \text{'true' if } n \text{ is prime} \\ \text{'false' otherwise} \end{cases} \right\}$$

This is just what the algorithm is supposed to accomplish—to decide, with the Boolean answer, whether or not the input n is prime. ■

Of course the problem with all of this is that the algorithm A stands in the way between the assertions P and Q. That is, we have to analyze the whole algorithm in order to demonstrate that indeed $\{P\}\ A\ \{Q\}$ is valid. Our task is a bit easier, however, once we remember that our pseudolanguage algorithms have a definite structure. They are built up from elementary processes, using only the three basic constructs—sequence, selection, and repetition. And so, our task is twofold:

1. To find a way to analyze the elementary processes of an algorithm; and then,
2. To discover appropriate deduction rules that will build assertions inductively over those for the processes embedded within the sequence, selection, and repetition constructs.

If this can be done we have only to ascertain whether the outer-level specification $\{P\}\ A\ \{Q\}$ is consistent with the induced behavior built up from the analysis of A. Of course, these are mighty big words, and we have a long way to go before we see how all of the pieces fit together. But we have made a start.

Analysis of Elementary Processes

At the end of Section 4.1 we introduced the idea of the state of a computation, call it x, in reference to the totality of values stored by all the variables in an algorithm. Moreover we illustrated the fact that an elementary process S will cause a transformation of the state:

$$x \longrightarrow S(x)$$

effecting a change in the values of certain variables.

Example 4.36 If x is the state:

$$x\colon a = 3.4,\ b = 5.1,\ c = 7.7$$

and we take as an elementary process S the interchanging of a and c, we would obtain the transformed state:

$$S(x)\colon a = 7.7,\ b = 5.1,\ c = 3.4$$

showing the effect of the interchange. ■

Now suppose that $P(x)$ is any assertion we would like to make about the state x of a computation. A little reflection will show that the rule:

$$\{P(S(x))\}\ S\ \{P(x)\}$$

must hold in relation to any process S (elementary or not). At first glance one might think that the rule has been stated incorrectly, backwards from what we expect. But say it in words: "If P holds for the transformed state $S(x)$ before the process S has been executed, then P holds for x afterwards [because x *has become* $S(x)$]."

Example 4.37 In Example 4.36, with the process

S: interchange a and c

we might want to say that $c < b$ after S has been executed. But that is only so because $a < b$ before the execution. Thus, if $P(x)$ is the assertion $c < b$ then $P(S(x))$ should make the claim that $a < b$; that is, whatever we want to say about x after S should be true of $S(x)$ before S has been executed. Since S indeed interchanges (the values of) a and c, the analysis:

$\{a < b\}$ interchange a and $c\{c < b\}$

is valid. ■

Example 4.38 If S is the elementary process

S: increase n by 1

and we would like to say that $n > 1$ after S has been executed, then it must have been true that $n + 1 > 1$ ($n > 0$) before. Thus we may write:

$\{n > 0\}$ increase n by 1 $\{n > 1\}$

But note that it is really the assertion $n + 1 > 1$ that we obtain on the left, that is, $P(S(x))$ with $P(x)$ being the statement that $n > 1$. ■

Deductive Rules

Having discussed the way in which the elementary processes should be analyzed, we now must describe the manner in which we are able to deduce assertion-process-assertion *triples*:

{assertion} S {assertion}

over the compound processes, those where S is built up from simpler processes using sequence, selection, or repetition. In this way one is able to build broader and broader such triples, working from the inner levels of an algorithm toward the outer level, until finally one can check whether the desired specification:

$$\{P\}\ A\ \{Q\}$$

holds for the algorithm A, taken as a whole.

First consider the simplest compound construct, the forming of a sequence $R \circ S$, for processes R and S. And suppose we have already derived triples:

$$\{O\}\ R\ \{P\} \quad \text{and} \quad \{P\}\ S\ \{Q\}$$

(Note the agreement in P with the concluding assertion for R and the opening assertion for S.) Then, as no surprise, we claim that the triple:

$$\{O\}\ R \circ S\ \{Q\}$$

is valid for the sequence as a whole. In symbols we write (see Figure 4.10):

$$\frac{\{O\}\ R\ \{P\},\ \{P\}\ S\ \{Q\}}{\{O\}\ R \circ S\{Q\}}$$

as a deduction rule, much in the spirit of those we studied in Chapter 2.

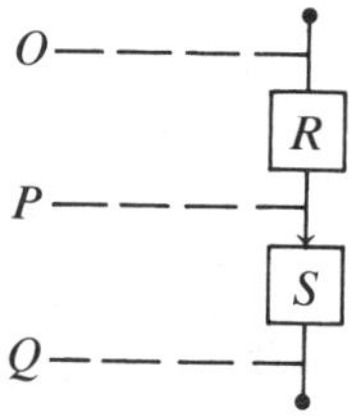

$$\frac{\{O\}\ R\ \{P\}\ ,\ \{P\}\ S\ \{Q\}}{\{O\}\ R \circ S\ \{Q\}}$$

FIGURE 4.10

Example 4.39 If we already have derived the triples

$$\{a > 0\}\ \text{interchange } a \text{ and } c\ \{c > 0\}$$

and

$$\{c > 0\}\ \text{increase } c \text{ by } 1\ \{c > 1\}$$

then we are able to deduce:

$\{a > 0\}$
interchange a and c
increase c by 1
$\{c > 1\}$

Note that we begin to write our triples vertically here, anticipating the fact that they will look this way when our assertions are simply interspersed with the pseudo-language coding. ■

The above deduction rule for the sequence construct is perhaps best understood by asking how it is that we might conclude $\{O\}\ R \circ S\ \{Q\}$, for the two given premises are then immediately suggested. The treatment of the selection construct R if C else S is more complicated, owing to the fact that we must account for the truth or falsity of the condition C. But suppose again that we would like to conclude the validity of some triple:

$$\{P\}\ R \text{ if } C \text{ else } S\ \{Q\}$$

Then the required premises are easily suggested once more. Consulting the flow diagram below, we see that we should have:

$$\{P \wedge C\}\ R\ \{Q\} \quad \text{and} \quad \{P \wedge \neg C\}\ S\ \{Q\}$$

and the desired conclusion would surely follow. Thus we are led to the deduction rule (see Figure 4.11):

$$\frac{\{P \wedge C\}\ R\ \{Q\},\ \{P \wedge \neg C\}\ S\ \{Q\}}{\{P\}\ R \text{ if } C \text{ else } S\ \{Q\}}$$

as a way to analyze the selection construct.

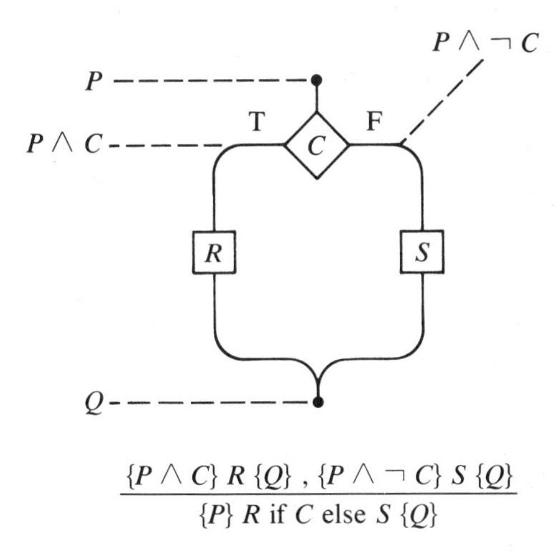

FIGURE 4.11

Example 4.40 We first refer the reader to Example 4.15, where we had outlined a strategy:

```
if n = 1
   give answer as 'false'
else
   count divisors and decide
```

for designing the prime decision algorithm. (Recall that the completed design is given in Example 4.20, with an improvement suggested in Exercise 1 of Exercises 4.2.) In the discussion of Example 4.15 we had not yet settled on the details to be involved in implementing the phrase "count divisors and decide." Nevertheless it would seem that of the two alternatives above we would be able to say:

$$\{n = 1\} \text{ give answer as 'false' } \{n = 1 \wedge \text{answer} = \text{'false'}\}$$

and

$$\{n > 1\} \text{ count divisors and decide } \{Q\}$$

respectively, where Q is the assertion

$$\text{answer} = \begin{cases} \text{'true' if } n \text{ is prime} \\ \text{'false' otherwise} \end{cases}$$

as appearing in Example 4.35.

Note that the concluding assertions in the if and else clauses must agree if we are to apply the R if C else S deduction rule. We observe, however, that:

$$\{n = 1 \wedge \text{answer} = \text{'false'}\} \longrightarrow \{Q\}$$

since 1 is not prime. We thus are able to write:

$$\{n = 1\} \text{ give answer as 'false' } \{Q\}$$

because Q is a weaker assertion than the one originally given. Since

$$n > 0 \wedge n = 1 \longleftrightarrow n = 1$$
$$n > 0 \wedge n \neq 1 \longleftrightarrow n > 1$$

the R if C else S deduction rule is applicable, and we obtain:

$\{n > 0\}$
if $n = 1$
  give answer as 'false'
else
  count divisors and decide
$\{Q\}$

as a valid triple, *provided only that* we decide properly in our refining of the statement "count divisors and decide." Comparing this result with the specification of Example 4.35, one would have to say that we thus have obtained a proof of the *relative correctness* of our prime decision algorithm. ■

Example 4.41 The R if C else S deduction rule also can be used to handle conditional statements, simply by interpreting S as a do-nothing process. For example, consider the conditional statement:

if $a < b$
  interchange their values

in the Euclidean algorithm of Section 3.3. This can be interpreted as saying:

if $a < b$
  interchange a and b
else
  do nothing

We would like to be able to verify the following triple:

$\{0 < b) \wedge (0 < a)\}$
if $a < b$
  interchange a and b
else
  do nothing
$\{0 < b \leq a\}$

for this is what the given conditional statement is supposed to accomplish. We note that if the condition $a < b$ is denoted by C, we have:

$$\{0 < b \wedge 0 < a\} \wedge C \longleftrightarrow 0 < a < b$$
$$\{0 < b \wedge 0 < a\} \wedge \neg C \longleftrightarrow 0 < b \leq a$$

whereas we can surely write:

$$\{0 < a < b\} \text{ interchange } a \text{ and } b\ \{0 < b < a\} \longrightarrow \{0 < b \leq a\}$$

and

$$\{0 < b \leq a\} \text{ do nothing } \{0 < b \leq a\}$$

It follows that the R if C else S deduction rule applies, yielding the triple given above. ■

We recall that the repetition construct came in two flavors (not counting the for construct), the S while C and the S until C. For our purposes it is sufficient to motivate the deduction rule for the while, leaving that for the until as an exercise for the student. In both cases one is asked to suppose that there is an *invariant assertion* I, something that holds before and after the body of the loop (the process we are calling S) has been executed. Using symbols, we thus assume the validity of a triple:

$$\{I \wedge C\}\ S\ \{I\}$$

noting that the condition C causes the loop to be traversed. Since we then exit the while loop in the event of $\neg C$, we formulate the deduction rule (see Figure 4.12):

$$\frac{\{I \wedge C\}\ S\ \{I\}}{\{I\}\ S \text{ while } C\ \{I \wedge \neg C\}}$$

stating that if I holds before we enter the while construct, then $I \wedge \neg C$ should hold afterward. This should be consistent with your understanding of the flow diagram in Figure 4.12.

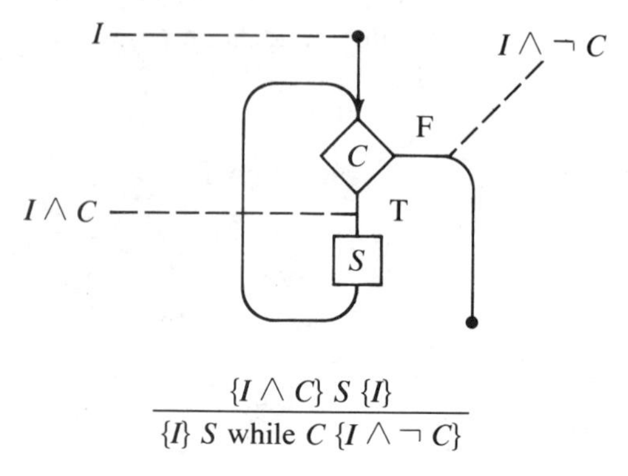

$$\frac{\{I \wedge C\}\ S\ \{I\}}{\{I\}\ S \text{ while } C\ \{I \wedge \neg C\}}$$

FIGURE 4.12

We note that the trick in using this deduction rule is in discovering the invariant assertion I. Nevertheless the explicit indication of the invariant of each repetition in an algorithm can represent the single most valuable aspect of its documentation. We can only hope that you will develop your intuition in thinking about loop invariants through the study of a few well-chosen examples. We provide an informal illustration here, prior to the study of a more substantive example in our closing subsection.

Example 4.42 Consider the algorithm largest from Example 4.16. Recall that we were using the strategy "Keep the best" (i.e., keep the best you've seen *so far*). With this idea in mind, and recognizing that we haven't necessarily seen *all* the entries of the list as the repetition is in progress, we might be led to write the (invariant) assertion:

$$I\text{: MAX is bestsofar} \wedge \text{sofar} \leq \text{all}$$

writing in an informal but suggestive fashion. It is clear that the body of the loop, call it S, behaves in such a way that we have:

$$\{I \wedge C\}\ S\ \{I\}$$

since C ("not finished") has the effect of modifying I to the extent that "sofar < all," whereas S then reads another value (which may be the last), so we are back to I again. As a result the while deduction rule yields:

$$\{I\}\ S \text{ while } C\ \{I \wedge \neg C\}$$

where the latter is the desired conclusion (again, speaking informally):

$$I \wedge \neg C\text{: MAX is bestsofar} \wedge \text{sofar} = \text{all}$$

that is, MAX is "bestofall." ■

Correctness of the Euclidean Algorithm

In the early discussions of this section, particularly in Example 4.41, we used the Euclidean algorithm as an illustration to show that exhaustive testing (of algorithms, in general) is not feasible. We made the point that it is a formal verification of correctness that often is required if we are to establish that our design is sound. The Euclidean algorithm is again an excellent case in point. And the tools for our establishing a proof of correctness are now in hand. But we will find that we will need to use everything we know about this algorithm in order to obtain a detailed verification.

In Example 4.41 we obtained the triple:

$\{0 < b \wedge 0 < a\}$
if $a < b$
  interchange their values
$\{0 < b \leq a\}$

for describing the behavior of the opening conditional statement of the algorithm. With the latter assertion as the opening statement in analyzing the elementary process that follows:

$\{0 < b \leq a\}$
compute the remainder on dividing a by b
$\{I\}$

we ask: What statement I can be made following this division? (We call it I because it should function as the invariant assertion of the subsequent while loop.) We first observe that the remainder—call it r, for the sake of simplicity—obtains a value on performing the division, whereas previous to the execution of this process it was undefined. If we review the discussion preceding Example 3.26 we see that the critical question is whether we then have $r = 0$ or not. Thus if $r = 0$ it will have been the case that b divides a evenly and, hence, $\gcd(a, b) = b$. If $r \neq 0$, however, that earlier discussion shows that $\gcd(a, b) = \gcd(b, r)$, the fact on which the success of the Euclidean algorithm hinges. All of this suggests that we try the assertion:

$$\{0 < b \leq a \wedge [(r = 0 \wedge \gcd(a, b) = b) \vee (0 < r < b \wedge \gcd(a, b) = \gcd(b, r))]\}$$

in the role of I.

We observe—and this was again a part of our thinking in arriving at I as above—that if this is to be our invariant assertion, then we are at least fortunate in that we achieve a convenient dichotomy with respect to C:

$$I \wedge C: 0 < b \leq a \wedge (0 < r < b \wedge \gcd(a, b) = \gcd(b, r))$$
$$I \wedge \neg C: 0 < b \leq a \wedge (r = 0 \wedge \gcd(a, b) = b)$$

as required in an effective use of the while deduction rule. As a matter of fact, the analysis:

$\{I \wedge C\}$
replace a by b, b by remainder
$\{0 < b < a\} \longrightarrow \{0 < b \leq a\}$
compute the remainder on dividing a by b
$\{I\}$

shows that I is indeed invariant as we pass through the body of the loop, thanks to the discussion above. It follows that the while deduction rule is applicable, yielding:

$\{I\}$
while nonzero remainder
  replace a by b
  compute the remainder on dividing a by b
$\{I \wedge \neg C\}$

where the latter assertion has been listed earlier.

Finally, in reference to the last process of the algorithm, we observe that the triple:

$\{I \wedge \neg C\} = \{0 < b \leq a \wedge (r = 0 \wedge \gcd(a, b) = b)\}$
assign b to gcd
$\{gcd = \gcd(a, b)\}$

is valid. It follows, by combining all of the preceding analysis, that we are able to state:

$$\{0 < b \wedge 0 < a\} \text{ euclid } \{gcd = \gcd(a, b)\}$$

for the algorithm as a whole. Since this was the design specification as listed in Example 4.34, our task is complete.

EXERCISES 4.4

1 For the S until C construct, motivate the deduction rule:

$$\frac{\{J\}\ S\ \{I\},\ \{I \wedge \neg C\}\ S\ \{I\}}{\{J\}\ S \text{ until } C\ \{I \wedge C\}}$$

and draw an appropriate annotated flow diagram.

2 Complete the proof of correctness of the prime decision algorithm

(a) As described in Example 4.20

(b) As described in Exercise 1 of Exercises 4.2

3 Write formal design specifications for each of the following:

(a) The perfect number decision algorithm as designed in Exercise 27 of Exercises 4.2

(b) The triangular number decision algorithm as designed in Exercise 16 of Exercises 4.2

(c) The square-free number decision algorithm as designed in Exercise 19 of Exercises 4.2

4 Develop a proof of correctness for each of the following:

(a) The perfect number decision algorithm as designed in Exercise 27 of Exercises 4.2

(b) The triangular number decision algorithm as designed in Exercise 16 of Exercises 4.2

(c) The square-free number decision algorithm as designed in Exercise 19 of Exercises 4.2

(*Note:* See Exercise 3.)

5 Complete each of the following to obtain valid triples:

(a) $\{x > 0 \wedge y < 2\}$ interchange x and y

(b) $\{x > 0 \wedge y < 0\}$ replace x by $y - 3$

(c) $\{x > 0 \wedge y < 0\}$ decrease x by 2

(d) $\{x = 2 \wedge y > 0\}$ replace y by $x - 1$

(e) $\{S = B \cup C \cup D\}$ replace A by $B \cup C$

6 Complete each of the following to obtain valid triples:

(a) interchange x and z $\{x > 0 \wedge y = 17 \wedge z < 0\}$

(b) replace x by $x + y + z$ $\{x > 0\}$

(c) decrease y by 2 $\{x < 0 \wedge y > 17 \wedge z = 3\}$

(d) take square root of x $\{x > 3\}$

(e) replace x by $f(x)$ $\{x > u\}$

7 In what sense is the deduction rule of Exercise 1 superfluous? (*Hint*: See Exercise 12(a) of Exercises 4.2.)

8 Why have we not provided deduction rules for the derived constructs of Section 4.2? (*Hint*: See Exercises 13 and 14 of Exercises 4.2.)

9 Decide whether the following are valid triples:

(a) $\{y^2 \geq x + 3\}$ replace x by $y^2 - x$ $\{x \geq 3\}$

(b) $\{n \text{ is even}\}$ replace m by n^2 $\{4 \mid m\}$

(c) $\{A \neq \varnothing \wedge B \neq \varnothing\}$ replace C by $A \cup B$ $\{C \neq \varnothing\}$

10 Compute $P \wedge C$ and $P \wedge \neg C$ if P and C are as follows:

(a) $P: n \mid 6 \quad C: 2 \mid n$

(b) $P: 5 \leq n < 7 \quad C: 2 \mid n$

(c) $P: B = \varnothing \quad C: A \subseteq B$

11 Complete each of the following to obtain valid triples:

(a)
$\{p = n^2 \wedge m = 3p - 2\}$
increase m by 3
add m to p

(b)
replace C by $\sim C \cap B$
replace A by $\sim A \cap B$
$\{A = C\}$

12 The following *different* algorithm, differing considerably from Euclid's, nevertheless computes the greatest common divisor of a and b. Provide a proof of its correctness. [*Hint:* Try $\gcd(n, m) = \gcd(a, b)$ as an invariant assertion. Then use one or more of the following properties of the gcd function to complete the verification.] (See Exercise 10 of Section 3.3.)

---

i. $n > m \Longrightarrow \gcd(n, m) = \gcd(n - m, m)$
ii. $\gcd(n, m) = \gcd(m, n)$
iii. $\gcd(n, n) = n$

```
algorithm different (a, b: gcd)

set n, m to a, b respectively
while n ≠ m
  if n > m
    subtract m from n
  else
    subtract n from m
take gcd as n
```

13 What is accomplished by the following algorithm? Verify the correctness of your assertion. [*Hint:* Try $f \cdot n^m = n^n$ as an invariant assertion.]

```
algorithm what (n: f)

set f to n
set m to n - 1
while m > 0
  multiply f by n
  decrease m by 1
```

14 What is accomplished by the following algorithm? Verify the correctness of your assertion.

```
algorithm what (n: f)

set f to 1
set m to n
while m > 1
  multiply f by m
  decrease m by 1
```

15 What is accomplished by the following algorithm? Verify the correctness of your assertion.

```
algorithm what (n: f)

set f to 0
set m to 2
while m < n
   multiply m by 2
   increase f by 1
```

Sec. 4.5 RECURSIVE ALGORITHMS

You say I am repeating something I have said before. I shall say it again.

The technique of recursion is a powerful problem-solving tool. We have seen it used in Section 3.2; we will see it used here. You will see it used again and again in subsequent studies. The technique is somewhat more formalized here than in our initial encounter (with the Towers of Hanoi of Section 3.2, etc.). We now will be looking at the possibility of an algorithm (or function) calling itself. (It takes a bit of getting used to.) Ordinarily a student will have to see the idea several times before it becomes totally clear. But we have to start somewhere, and this is as good a place to begin as any.

Goals

After studying this section, you should be able to:

- Understand the idea of a recursive algorithm, the fact that an algorithm may call itself.
- Design algorithmic functions with the use of recursion.
- Discuss the notion of divide and conquer as a problem-solving technique and use the technique in the design of recursive algorithms.

Recursive Functions

We say that we are going to look at the possibility of an algorithm (here, a function) calling itself. We first have to ask ourselves what this might mean. Will we be faced with a Which came first, the chicken or the egg? situation, only introduced as a curiosity, or are there intentional and meaningful circumstances of such recursive calls? As we will see the latter is definitely the case. The whole subject of recursion, particularly in an algorithmic setting, is one of the more elegant notions derived from the modular approach to problem solving. When properly used one is able to achieve a considerable economy of presentation, and after overcoming any initial hesitation the algorithm designer is in command of a most effective problem-solving tool.

Example 4.43 We refer you to the initial discussion of the factorial function in Example 3.14. Recalling that

$$1! = 1$$

and

$$n! = n \cdot (n - 1)!$$

for $n > 1$, we note that the computation of $n!$ according to the recursive definition given here depends on our already having computed $(n - 1)!$. This same technique of passing the buck continues on and on until, finally, the result of 1! is requested. But that is easy. Knowing $1! = 1$, we obtain $2! = 2 \cdot 1! = 2 \cdot 1 = 2$, then $3! = 3 \cdot 2! = 3 \cdot 2 = 6$, etc., and finally $n! = n \cdot (n - 1)!$—whenever we get that far.

This kind of analysis is essential to the understanding of a recursive factorial function when written in our algorithmic pseudolanguage. Thus we have the accompanying recursive function *factorial* in direct correspondence with the recursive definition given above. Here we see that the algorithm actually calls on itself in the last line in order to compute factorial $(n - 1)$ when factorial (n) is requested. Suppose we think of the call to factorial (4) as having been initiated in the middle of some other algorithm. In executing this call the computation is suspended at its last line, pending the result of the recursive call to compute 3!, and so on, until the call of 1! finally returns a value (of 1). At this point the returns will cascade backward, whereupon the desired value:

$$\begin{aligned} 4! &= 4 \cdot 3! = 4 \cdot 3 \cdot 2! = 4 \cdot 3 \cdot 2 \cdot 1! \\ &= 4 \cdot 3 \cdot (2 \cdot 1) = 4 \cdot (3 \cdot 2) = 4 \cdot 6 = 24 \end{aligned}$$

is obtained. Note that in our calculation the first line shows the successive calls and the second line illustrates the cascading returns. It is this kind of reasoning that is characteristic of the handling of a recursive algorithm. Bearing this in mind you should study this first example quite carefully.

```
function factorial (n)

if n = 1
   return factorial as 1
else
   return factorial as n · factorial (n − 1)
```

■

Often an algorithm can be written in both a recursive and a nonrecursive fashion. Thus, in the case of the factorial function, we were asked to derive a (presumably nonrecursive) algorithm for computing $n!$ in Exercise 23 of Exercises 4.3.

In such instances one should attempt to make comparisons of the one rendition to the other, comparing clarity of expression, efficiency of execution, and so on, with a view toward selecting the most appropriate representation for the application at hand.

```
function power (x, n)
if x = 0
   return power as 0
else
   if n < 0
      replace x by its reciprocal
      replace n by its negative
   set p to 1
   for i running from 1 to n
      multiply p by x
   return power as p
```

Example 4.44 Suppose we wish to compute the function x^n for all integer powers n and any given number x. (Assume, however, that $x \neq 0$ if $n < 0$.) We note that if $x = 0$ then we should have $x^n = 0$, regardless of the value of n. If n is negative, however, we can use the fact that

$$x^n = (1/x)^{-n}$$

to reduce the computation to only the case of positive exponents. This is the approach taken in the nonrecursive function *power* given above, one that makes use of the for construct in simply multiplying the partial product p by x over and over again.

By way of contrast we may first observe that the function x^n can be described in a self-referencing fashion:

$$x^n = x^{n+1}/x \qquad (n < 0)$$
$$x^n = x^{n-1} \cdot x \qquad (n > 0)$$

This leads to the recursive function *power* given below, using a case construct to treat the trichotomy. Certainly one has to admit that the recursive treatment is the more elegant, and to anyone of a mathematical bent it is probably a more satisfying solution to the problem than that presented above.

```
function power (x, n)
case n
   n = 0: return power as 1
   n < 0: return power as power (x, n + 1)/x
   n > 0: return power as power (x, n − 1)x
```

■

In both of these recursive algorithms, however, here and in Example 4.43, we must emphasize the importance of the *termination mechanism*. We have provided in each case a mechanism for breaking off the recursion at some point. We obviously cannot continue to have an algorithm call itself over and over again, ad infinitum. Thus, in the present example, the negative and positive sequences of integers:

$$n, n+1, n+2, \ldots \qquad (n < 0)$$
$$n, n-1, n-2, \ldots \qquad (n > 0)$$

ultimately will reach zero, and the recursion is then broken off by the return of the value of 1 in the opening case of the recursive power function. ■

Divide and Conquer

Computer scientists have devised a number of effective techniques for solving large classes of problems. But perhaps none is more successful than that of divide and conquer. This approach, first seen in our treatment of the bisective search algorithm of Example 4.26, reaches its ultimate utility in the context of the recursive treatment of an algorithm.

Example 4.45 By way of illustration we present an algorithm (not a function) for the simultaneous computation of the minimum and maximum entries of a list of numbers, A. Again we certainly could design a nonrecursive solution to the problem. If our list has n members we can find the minimum and maximum entries in something like $2n$ comparisons by a straightforward method. (Once again, keep the best you've seen so far.)

But just suppose that we already knew the minimum and maximum entries in the first half of our list, and that we knew the minimum and maximum entries of the second half as well. Then two simple comparisons would give us the minimum and maximum for the list as a whole. So we then could concentrate on finding the solution for the first and second halves. These would be handled in the same recursive fashion. This is again the technique of divide and conquer, so often suggested in a recursive setting. Written in our pseudolanguage, the technique is handled as shown in the accompanying algorithm *minmax*. Note particularly the computation of the midpoint subscript, intended as the integer part of

$$(\text{first} + \text{last})/2$$

Thus if first = 1 and last = 12 we would have assigned the value 6 to the variable *mid*. Note as well the two subsequent recursive calls to the algorithm itself, calls that then will operate on the two halves of the current list. In the last two lines, where minima and maxima will have been returned for half-lists by previous calls to minmax, we assume that min and max are functions described elsewhere, rather trivial functions requiring just a single comparison for returning their result.

```
algorithm minmax (first, last: m, M)

if last-first ≤ 1
   if A_first < A_last
      set m to A_first
      set M to A_last
   else
      set m to A_last
      set M to A_first
else
   compute mid as the midpoint
   call minmax (first, mid: m1, M1)
   call minmax (mid + 1, last: m2, M2)
   set m to min (m1, m2)
   set M to max (M1, M2)
```

■

The recursive treatment is quite elegant and contains the utmost clarity, having been based on a simple bisection technique. Moreover we will find in the next section that it offers advantages of efficiency as well. Whenever we have two or more ways of treating a problem, all these various attributes should be examined, from one solution to another, so that we will choose the algorithm that seems the best for the situation at hand. But in order to decide which is the more efficient of any two solutions we will need to study the notion of *computational complexity*, the subject of our final section (Section 4.6).

EXERCISES 4.5

1 Design pseudolanguage functions min and max to be used by the minmax algorithm of Example 4.45.

2 Write a nonrecursive minmax algorithm in the pseudolanguage. (*Hint:* See Example 4.45.)

3 Illustrate the successive recursive calls and the cascading returns in the factorial function (Example 4.43) when called:

(a) With the argument 3 **(b)** With the argument 5

(c) With the argument 6 **(d)** With the argument 7

4 Compare the following nonrecursive and recursive treatments of a problem from the standpoint of clarity, efficiency, etc. (*Note:* See Section 4.6.)

(a) Exercise 23 of Section 4.3 vs. Example 4.43

(b) Exercise 2 vs. Example 4.45

(c) The two algorithms of Example 4.44

5 Describe the termination mechanism in the minmax algorithm of Example 4.45.

6 What is accomplished by the following recursive algorithm?

```
algorithm tahw (n)

calculate and list remainder dividing n by 10
set m to integer part of n/10
if m not zero
    tahw (m)
```

7 The Fibonacci sequence of Exercise 11 in Exercises 3.2 can be viewed as the range of values of a function $F(n)$. In pseudolanguage write

(a) A nonrecursive function **(b)** A recursive function

for computing $F(n)$.

8 Let C, D represent two lists of characters, lists of length n. Write a recursive pseudolanguage algorithm for transferring the characters of C into D in reverse order.

9 Write a recursive gcd function in pseudolanguage.

10 Illustrate the successive calls and the subsequent return of values to the recursive power function of Example 4.44 when called with the arguments:

(a) $x = 2, \quad n = 4$ **(b)** $x = 3, \quad n = 4$ **(c)** $x = 2, \quad n = 7$

(d) $x = 2, \quad n = -5$ **(e)** $x = 3, \quad n = -4$ **(f)** $x = 4, \quad n = 4$

11 Illustrate the successive calls and the subsequent return of values to the recursive minmax algorithm of Example 4.45

(a) When A is the list $12, 30, 14, 27, 55, 26, 6, 18, 3, 24$ and the first, last arguments are $1, 10$

(b) When A is the list $14, 3, 12, 18, 49, 1, 75, 6$ and the first, last arguments are $1, 8$

(c) When A is the list $12, 78, 97, 5, 22, 8, 88, 3, 22$ and the first, last arguments are $1, 9$

(d) When A is the list $33, 37, 22, 2, 14, 98, 97, 33, 3, 8, 44$ and the first, last arguments are $1, 11$

Sec. 4.6 INTRODUCTION TO COMPUTATIONAL COMPLEXITY

The running time of an algorithm depends on at least two factors: the size of the input and the complexity of the algorithm itself.

When we are solving a problem with the idea of obtaining an algorithmic solution, it may very well be that there are a number of possible solutions and we often have to choose from among them. On what basis should such a choice be made? There is no easy answer here. On the one hand, an algorithm should be clearly phrased, easy to understand and to implement. On the other hand, we would like an algorithm that is efficient, one that makes good use of computer resources, should we carry the design that far. Often these goals are in conflict.

Clarity and ease of understanding are critical if we are designing an algorithm for instructional purposes, or if we are implementing an algorithm that will be run only a very few times. When we are thinking of an algorithm that may be used over and over again, however, the *running time* becomes critical, often outweighing the other considerations entirely. We focus our attention on this latter question in this closing section. Like the discussion of proofs of correctness in Section 4.4, we thus enter into one of the frontiers of the young computer science discipline, an area where enormous strides have been made. We can only provide the barest outline here, but the reader should rest assured that the discussion is of a most fundamental nature, sure to increase in importance in the years to come.

Goals

After studying this section, you should be able to:

- Discuss the notion of the *growth rate* of a function.
- Define the *order of growth* of a function and list some of the more common distinct orders.
- Determine the order of growth rate for simple functions.
- Define the log n function (to the base 2) and list its first few integral arguments and values.
- State and verify the lemma on polynomial growth rates.
- Discuss the definition of an elementary process in relation to the notion of the order of growth.
- Compute the time complexity of an algorithm.

Running Time and Growth Rate

The running time of an algorithm depends on at least two factors: the size of the input and the complexity of the algorithm itself. As a matter of fact, we will approach our study in such a way that the complexity of an algorithm A is viewed as a **time complexity** $T(A) = T(n)$, expressed as a function of n, the input size. Note that the exact nature of the inputs is not always so important—it is more often only the size that matters. If we are sorting n numbers into ascending order, the exact quantities involved are of little concern; only their number is of interest. That is mainly what affects the running time of the algorithm.

The units of the function $T(n)$ are left unspecified, but one can think of the function as measuring the number of elementary operations performed in executing the algorithm on some idealized machine. Now in spite of what we have said above, the running time occasionally may depend upon the particular input values chosen. To make such matters easier to analyze, we then understand $T(n)$ to refer to the *worst case* running time, that is, the maximum of the running times when taken over all inputs of size n. Still it is not possible to give a number that would measure the actual time, say, in seconds. We then would have to nail down an actual machine on which such measurements were to be made. About all we can say is that the time is *proportional to n*, say, or to n^2, or whatever, but we would not be able to give the constant of proportionality any more than we could specify the time units.

When we are left to describe and compare *growth rates* of functions in such a state of ignorance, it is only the order of the growth that really matters. Thus we will say that the function $T(n)$ is **of order** $f(n)$ if there is a positive number c so that

$$T(n) \leq cf(n)$$

for all sufficiently large n, and we then write $T(n) \sim f(n)$ (read "$T(n)$ is of order $f(n)$"), since the constant c is of little concern.

Example 4.46 A function having a very slow growth rate is that which we denote:

$$f(n) = \log_2 n$$

or simply $\log n$ (with the base 2 understood). It is the same function that you have encountered in high-school mathematics (though usually in the base 10). Thus as n takes on the values $1, 2, 4, 8, 16$, etc.—the powers of 2—we may tabulate the values of $\log n$ as shown in Table 4.2, noting the fact that $\log n$ is simply the inverse of the exponential function (in powers of 2). ■

TABLE 4.2

n	$\log n$
1	0
2	1
4	2
8	3
16	4
⋮	⋮

n	2^n
0	1
1	2
2	4
3	8
4	16
⋮	⋮

Example 4.47 Suppose we consider the function

$$T(n) = 3$$

for all n. Then we have (with $c = 1$):

$$T(n) \leq c \log n$$

for all sufficiently large n; in fact, for all $n \geq 8$. So one might think that we would say "$T(n)$ is of order $\log n$." But in fact we ordinarily intend the notation $T(n) \sim f(n)$

to mean that one has made the *strongest statement* that can be made. Here we could say $T(n) \sim 1$, since it is also true that

$$T(n) \le c \cdot 1$$

for all n, by taking $c = 3$. Since the constant function $f(n) = 1$ is an even slower growing function than the log function (in fact, it doesn't grow at all), it would be more meaningful to have said "$T(n)$ is of order 1." ■

Example 4.48 If $T(n) = \frac{1}{2}(n+1)^2$ then we may write

$$T(n) \sim n^2$$

because

$$T(n) = \tfrac{1}{2}(n+1)^2 = \tfrac{1}{2}n^2 + n + \tfrac{1}{2} \le 2n^2$$

as long as $n \ge 1$ (i.e., for sufficiently large n). ■

Note that such statements as this often can be verified by mathematical induction. But many of them are only instances of the general result that follows:

Lemma

Polynomial Growth Rates A polynomial $T(n)$ of degree k is of order n^k. That is, regardless of the coefficients:

$$a_k n^k + a_{k-1} n^{k-1} + \cdots + a_1 n + a_0 \sim n^k$$

Proof As in the example above, we have:

$$a_k n^k + a_{k-1} n^{k-1} + \cdots + a_1 n + a_0 \le (|a_k| + |a_{k-1}| + \cdots + |a_0|) n^k$$

as long as $n \ge 1$. Consequently in the definition of *order* we may take c as the sum of the absolute values of the coefficients. □

As these results indicate, when we speak of the order of a function we ordinarily strip away the extraneous detail. Thus the orders: $1, \log n, n, n^2, n^3, \ldots, 2^n$, etc., are considered distinct. We do not distinguish between $n^2, n^2 + n, \frac{1}{2}n^2$, etc., however, for they're all of order n^2. The implication here is that 1 is better than $\log n$, $\log n$ is better than n, n is better than n^2, and so on. Of course it could be argued that we have stripped away too much. To be sure, n^2 is smaller than $1000n$ (for $n = 1, 2, 3$, and a few more values). But sooner or later (in fact, when $n = 1000$), n^2 will catch up. Thus our hierarchy of orders has a real meaning as larger and larger problems are considered. The tabulation in Table 4.3 makes this point most clearly in the broader context of the various multiples of distinct orders listed above.

TABLE 4.3

n	$\log n$	$10n$	$T(n)$ $2n^2$	$n^3/4$	2^n
1	0	10	2	.25	2
2	1	20	8	2	4
3		30	18	6.75	8
4	2	40	32	16	16
⋮	⋮	⋮	⋮	⋮	⋮
8	3	80	128	128	256
9		90	162	182.25	512
10		100	200	250	1024
⋮	⋮	⋮	⋮	⋮	⋮
64	5	640	8192	65536	2^{64}

Thus, to some extent, it all depends on the size of input that we intend to treat. An algorithm with growth rate 2^n is as good as one with the order $n^3/4$ if $n = 4$. But from $n = 5$ onward we are better off with the $n^3/4$ algorithm, and there is always some point at which any given cubic order algorithm is better.

All this can be viewed from the opposite perspective. As computers become faster and faster with each generation, there is the natural inclination to try and solve larger and larger problems, those that would have been too costly to treat before. The order of the growth rate of an algorithm will determine the size of problem that we can handle in a reasonable time. Suppose we are allowed 100 seconds of processing time and that our functions $T(n)$ measure time in seconds. Then we are limited in our *maximum problem size* as shown in the first column of Table 4.4. Now except for the $\log n$ algorithm, there is little difference in the maximum problem size from one order to another. But suppose the day comes when we are able to buy a new machine that runs 10 times faster at no additional cost. It is as if we were allowed 1000 seconds of processing time (on the old machine) as opposed to the 100 seconds we were given before. Then the maximum problem sizes are as computed in the second column of our table. The differences are then quite dramatic, again showing that it is the *order of the growth rate* that really matters.

TABLE 4.4

$T(n)$	*Maximum Problem Size* 100 sec	 1000 sec
$\log n$	2^{100}	2^{1000}
$10n$	10	100
$2n^2$	7	22
$n^3/4$	7	17
2^n	6	10

Computation of Time Complexity

Having established the importance of the notion of time complexity, we now describe a routine method for computing

$$T(A) = T(n)$$

for a given algorithm A as a function of n, the input size. Once again we are aided by the fact that the algorithms of our pseudolanguage are built up from the elementary processes, using only a few basic compounding structures—the sequence, the selection, and the repetition. It is here that the idea of an elementary process is most clearly understood.

We recall that a process S is considered elementary if it can be executed in a unit of time that is independent of the input size. Now we see the main purpose in our having made this distinction. We are able to write:

$$T(S) \sim 1$$

for every elementary process S! In fact this is exactly what the definition of elementary states—that there is *some constant c* so that

$$T(S) = T(n) \leq c \cdot 1$$

for all n. It doesn't matter which c it is since we aren't measuring actual time units anyway. It only matters that the complexity function $T(n)$ is a constant—independent of n. One might say that we have rigged the definition of elementary with this result in mind. In fact, that would not be far from the truth.

With this much understood we only have to extend the definition of T inductively over the class of compounding constructs available in our algorithmic language in order to be able to compute $T(A)$ for any algorithm A. In so doing we hope that our definitions are consistent with your intuition. Thus if R and S are processes whose time complexities $T(R)$ and $T(S)$ are already known, we agree to write:

$$T(R \circ S) = T(R) + T(S)$$
$$T(R \text{ if } C \text{ else } S) = T(C) + \max(T(R), T(S))$$

and finally:

$$\left.\begin{array}{l} T(S \text{ while } C) \\ T(S \text{ until } C) \\ T(S \text{ for } C) \end{array}\right\} = |C|T(C) + \sum T_C(S)$$

where $T(C)$ is the complexity of the condition checking of C, and $|C|$ is the number of times that a corresponding while, until, or for loop will be executed, *in the worst case*. Note that we should perhaps add an additional $T(C)$ to the figure used for the S while C construct, since we have to test its condition one more time than the number of times the loop is actually traversed. But only in rare circumstances will this lead to a difference in the order of the resulting complexity function.

In the expression for the complexity of a repetition, we refer to the complexities $T_C(S)$ in order to indicate that one traversal of a loop may be more costly than another. It may be that the complexity of the process S depends on the value of some variable appearing in C, a value that may be changed in the course of executing S. Such situations are troublesome to handle, but it is hoped that our examples will help to clarify the matter.

Moreover it should be noted that $T(C) \sim 1$ in the event that C is an elementary condition, as is quite often the case. Consequently, the *simplified expressions*:

$$T(R \text{ if } C \text{ else } S) = \max(T(R), T(S))$$

$$\left.\begin{array}{l} T(S \text{ while } C) \\ T(S \text{ until } C) \\ T(S \text{ for } C) \end{array}\right\} = \sum T_C(S)$$

ordinarily will give the same order of complexity as the more detailed expressions above.

Example 4.49 Consider the prime algorithm A of Example 4.20. We have

$$T(A) = 1 + \max[1, T(\text{else})]$$

since the whole algorithm is a selection construct at its outer level. For the else alternative we use the sequence and selection formulas in writing:

$$\begin{aligned} T(\text{else}) &= 1 + T(\text{for}) + 1 + \max(1, 1) \\ &= 3 + T(\text{for}) \end{aligned}$$

making use of the fact that the embedded processes are elementary. In handling the S for C construct we note that all $T_C(S)$ are identical, namely $T_C(S) = 1 + 1 = 2$. Because there are $n - 2$ of these, we obtain:

$$T(\text{for}) = (n - 2) \cdot 2 + (n - 2) \cdot 2 = 4n - 8$$

in counting $T(C) = 2$, as seems reasonable. Upon substituting upwards in levels, we obtain:

$$T(\text{else}) = 3 + 4n - 8 = 4n - 5$$

and, finally, this gives

$$\begin{aligned} T(A) &= 1 + \max(1, 4n - 5) \\ &= 1 + 4n - 5 = 4n - 4 \sim n \end{aligned}$$

as our order of complexity. In summary, we are able to say that our prime algorithm is linear, that is, proportional to n in complexity. ■

Example 4.50 The reader should check to see that the simplified expressions for T would have led to the same order of complexity in Example 4.49. An exception to this rule can be given in case we are dealing with the selection:

```
if prime (n)
   process R
else
   process S
```

as discussed in Example 4.32. For then (as we see in Example 4.49) our condition C is not elementary. Supposing that R and S are elementary processes, however, and assuming that prime is a function based on the algorithm of Example 4.20, we would have to write:

$$\begin{aligned} T(R \text{ if } C \text{ else } S) &= T(C) + \max(T(R), T(S)) \\ &= 4n - 4 + \max(1, 1) \\ &= 4n - 4 + 1 = 4n - 3 \sim n \end{aligned}$$

since the complexity of C dominates that of R and S. ■

Example 4.51 Consider the selectionsort algorithm of Section 4.3. Here the outer level construct is a repetition. Noting that the inner for loop is such that its complexity T(for each j) is independent of j, we obtain (using the simplified expressions):

$$T(A) = \sum T_C(S) = \sum_{i=1}^{n-1} T_i(S)$$

where

$$\begin{aligned} T_{i=1}(S) &= 1 + (n-1)(1) + 1 \\ T_{i=2}(S) &= 1 + (n-2)(1) + 1 \\ &\vdots \\ T_{i=n-1}(S) &= 1 + (1)(1) + 1 \end{aligned}$$

Summing and substituting we have:

$$\begin{aligned} T(A) &= 2(n-1) + \sum_{i=1}^{n-1} (n-i) \\ &= 2(n-1) + n(n-1)/2 \sim n^2 \end{aligned}$$

using Example 3.16 and our lemma on polynomial growth rates. ■

Example 4.52 In the bisective search algorithm of Example 4.26, we first treat the outer level sequence in writing:

$$\begin{aligned} T(A) &= 1 + 1 + T(\text{while}) + 1 \\ &= 3 + T(\text{while}) \end{aligned}$$

(*Note:* $4 + T(\text{while})$ would be more correct, since the first statement is really two elementary processes.) Now, because we require $\log n$ traversals of the loop in order to process n elements (according to the discussion following Example 4.26), we obtain:

$$\begin{aligned} T(\text{while}) &= (\log n)(1) + \log n(1 + 2 + \max(1, 1, 1)) \\ &= \log n(1 + 4) = 5 \log n \end{aligned}$$

in counting the case comparison of complexity 2, as seems reasonable. Substituting in the above, we have:

$$T(A) = 3 + 5 \log n \sim \log n$$

Note that this compares favorably with a linear search technique (say, Example 4.25), one whose complexity is simply proportional to n. ■

Example 4.53 As our final example it is instructive to look at a comparison of recursive and nonrecursive algorithms for accomplishing the same task. For this purpose we examine the minmax algorithm of Example 4.45. In order to simplify the calculations we make the observation that the complexity function $T(A)$ should be proportional to the number of comparisons that are made. Accordingly we will write $T(n)$ to mean only this number of comparisons (as a function of n), rather than a total count of all the elementary operations involved. In so doing it is best to make the further simplifying assumption that n is a power of 2. It might mean that we are off by 1 somewhere in our counting, but it can't have much of an effect on our answer.

With all these matters understood, we are able to express $T(n)$ in the form:

$$T(n) = 2T\left(\frac{n}{2}\right) + 2$$

in reading the last four lines of minmax and remembering that the min and max functions will involve one comparison each. We note as well that

$$T(2) = 1$$

in considering the first alternative of the algorithm. Now one might wonder how all of this can help—we seem to have a recursive equation for $T(n)$, not an explicit solution. But suppose we assume that the solution is of the form

$$T(n) = an + b$$

realizing that a nonrecursive solution gives a count of $2n$ (approximately). Our recursive equation then yields:

$$an + b = 2\left(a \cdot \frac{n}{2} + b\right) + 2$$
$$= an + 2b + 2$$

from which we obtain

$$b = -2$$

The condition $T(2) = 1$ then allows us to solve for a, that is,

$$2a - 2 = 1$$
$$a = \frac{3}{2}$$

so that altogether we obtain:

$$T(n) = \frac{3}{2}n - 2$$

In comparing this result with the count of $2n$ in the usual nonrecursive method, the ratio

$$\frac{\frac{3}{2}n - 2}{2n} \sim \frac{\frac{3}{2}}{2} = \frac{3}{4}$$

shows a 25% gain in efficiency in using the recursive algorithm. ■

Such apparent improvements, however, must be taken with a grain of salt. Thus ir a real computing environment they must be weighed against a certain overhead charge that invariably accompanies the use of recursion. It happens that each recursive call will invoke a new set of memory allocations, and before we are done the cost of this additional storage might outweigh any apparent savings in computing time. We point this out only to provide a more balanced presentation of the real-life situation.

EXERCISES 4.6

1 Show the calculations that lead to the figures of the Maximum Problem Size table (Table 4.4)

(a) in the first column. **(b)** in the second column.

2 Calculate a third column for the Maximum Problem Size table (Table 4.4), assuming the equivalent of 10,000 sec. of processing time.

3 Determine the order of growth of the following functions $T(n)$:

(a) $2n^3 - n$ **(b)** 18

(c) $\log_{10} n$ **(d)** $n(\frac{1}{2}n^2 + 4)$

(e) $\sum_{j=1}^{n} j$ (*Hint:* See Example 3.16.)

(f) $10 \sum_{k=0}^{n} \binom{n}{k}$ (*Hint:* See Exercise 5 of Section 3.2.)

4 Calculate the order of $T(R \circ S)$ if

(a) $T(R) \sim n, \quad T(S) \sim n^2$ **(b)** $T(R) \sim 2^n, \quad T(S) \sim n^3$

(c) $T(R) \sim n, \quad T(S) \sim \log n$ **(d)** $T(R) \sim n^2, \quad T(S) \sim n^2$

5 Calculate the order of $T(R$ if C else $S)$ for the complexities $T(R)$ and $T(S)$ listed in Exercise 4. (*Note:* Assume that C is an elementary condition.)

6 Calculate the order of $T(S$ while $C)$ if

(a) Each $T_C(S)$ is of order n, and the loop is executed n times.

(b) Each $T_C(S)$ is of order $\log n$, and the loop is executed at most $\frac{1}{2}n$ times.

(c) $T_C(S) = j \ (1 \le j \le n = |C|)$, and the loop is executed n times. (*Hint:* See Example 3.16.)

(d) $T_C(S) = j^2 \ (1 \le j \le n = |C|)$, and the loop is executed n times. (*Hint:* See Exercise 1 of Exercises 3.2.)

(*Note:* Assume throughout that C is an elementary condition.)

7 For the following algorithms, obtain complexity functions $T(n)$ and determine their order of growth rate.

(a) Algorithm largest in Example 4.16

(b) Algorithm triangles in Example 4.19

(c) Algorithm prime in Exercise 1 of Exercises 4.2

8 For the following algorithms, obtain complexity functions $T(n)$ and determine their order of growth rate.

(a) The algorithm of Exercise 19, Exercises 4.2

(b) The algorithm of Exercise 27, Exercises 4.2

(c) The bisection algorithm of Example 4.29

(*Note:* See Example 4.51.)

9 Express the complexity function for algorithm ratadd (Example 4.31) in terms of that for function gcd (Example 4.30).

10 For the following algorithms, obtain complexity functions $T(n)$ and determine their order of growth rate.

(a) The algorithm of Exercise 3, Exercises 4.3

(b) The algorithm of Exercise 6, Exercises 4.3

(c) The algorithm of Exercise 23, Exercises 4.3

11 For the following algorithms, obtain complexity functions $T(n)$ and determine their order of growth rate.

(a) The algorithm of Exercise 7, Exercises 4.3

(b) The algorithm of Exercise 8, Exercises 4.3

(c) The algorithm of Exercise 9, Exercises 4.3

12 Obtain the complexity function $T(n)$ for the recursive factorial function of Example 4.43 and determine its order of growth rate.

Chapter Five

POLYNOMIAL ALGEBRA

Contents

5.1 Polynomial Calculus
5.2 Graphs of Polynomials
5.3 Interpolation Theory
5.4 Factorization of Polynomials
5.5 Rational Functions
5.6 Difference and Summation Calculus
5.7 Independence and Rank

Joseph-Louis Lagrange (1736–1813)

Joseph-Louis Lagrange was born in Turin, Italy and died in Paris. Lagrange's power over symbols has perhaps never been equalled. He became a professor of mathematics at Turin at the age of 16. He organized the more able of his students into a research society that later became the Turin Academy of Science. His work in celestial mechanics and the theory of equations formed the foundation of these subjects for years to come. He often was challenged by fundamental problems of "arithmetic," discrete problems as they grew to be called. He quickly saw that there were no obvious uniform methods of attack, no general weapon such as those that calculus provided for continuous problems. Nevertheless his investigations were of fundamental importance in the development of modern algebra. Lagrange often received awards for his work from Napoleon himself. But even these would pale in comparison with the esteem with which he is held by contemporary mathematical historians.

Writings

Algebra is a science almost entirely due to the moderns. I say almost entirely, for we have one treatise from the Greeks, that of Diophantus, who flourished in the third century of the Christian era. This work is the only one which we owe to the ancients in this branch of mathematics. When I speak of the ancients I speak of the Greeks only, for the Romans left nothing in the sciences, and to all appearances did nothing.

Diophantus may be regarded as the inventor of algebra. His work contains the first elements of this science. He employed to express the unknown quantity a Greek letter which corresponds in the translations to our n *(but is more commonly written as* x *in current writings). To express the known quantities he employed numbers solely, for algebra was long destined to be restricted entirely to the solution of numerical problems. We find, however, that in setting up his equations consonantly with the conditions of the problem he uses the known and the unknown quantities alike. And herein consists virtually the essence of algebra, which is to employ unknown quantities, to calculate with them as we do with known quantities, and to form from them one of several equations from which the value of the unknown quantities can be determined.*

But in the books which have come down to us (for the entire work of Diophantus has not been preserved) this author does not proceed beyond equations of the second degree, and we do not know if he or any of his successors (for no other work on this subject has been handed down from antiquity) ever pursued researches beyond this point.

More modern efforts of Vieta in Italy, Descartes in France, and others [notably LaGrange himself], toward the solution of equations of higher order (though not entirely complete) are far from having been in vain. They have given rise to the many beautiful theorems which we possess on the formation of equations, on the character and sign of the roots, on the transformation of a given equation into others of which the roots may be formed at pleasure from the roots of the given equation, and finally, to the beautiful considerations concerning the metaphysics of the resolution of equations from which the most direct method of arriving at their solution, when possible, has resulted. All this has been presented to you in (my) lectures.

Lectures on Elementary Mathematics **(LaGrange)**

Sec. 5.1 POLYNOMIAL CALCULUS

Polynomials represent a special class of functions, but in mathematics, they are ubiquitous.

You have undoubtedly encountered polynomials in high-school mathematics courses. They are found throughout mathematics, from the most elementary to the most advanced levels. Our intent here is to broaden and deepen your understanding of polynomials so that you will be able to approach the subject from a more sophisticated point of view than has previously been the case. Most importantly, we wish to introduce an algorithmic aspect to the treatment, one that is perhaps new to the student. In this respect our discussion will build quite naturally on the material of the preceding chapter, at least by way of example, thereby reinforcing that material as well.

Goals

After studying this section, you should be able to:

- Explain the basic terminology used in discussing polynomials: *coefficients, terms, degree,* etc.
- Give the definition of the *zero polynomial* and explain why it is that its degree is not defined.
- Recite the rule for deciding whether two polynomials are equal.
- Compute the sum and the product of two given polynomials, as well as the product of a given polynomial by a constant.
- State the *collecting formula* for obtaining the general term in the product of two given polynomials.
- Evaluate a given polynomial at a specific argument, by the usual method and by Horner's method.
- Compare the two methods of evaluating a polynomial, from the standpoint of computational efficiency.

Basic Terminology

Suppose we begin our discussion with the naive question, What is a polynomial? The answer (the one we hope you expected) is that a **polynomial** (or, more precisely, a *polynomial in x*) is an expression of the form:

$$a_0 + a_1x + a_2x^2 + \cdots + a_nx^n$$

where x is viewed as an indeterminate—but more on this later. Ordinarily the *coefficients* a_i are chosen from one of the familiar discrete number systems of Chapter 3, for example, the integers, the rationals, or whatever; the exponents on x are nonnegative integers, of course. It turns out that the properties of polynomials are to a large extent independent of the underlying number system, but if it helps to think of the coefficients as being rational numbers, say, then so much the better.

Example 5.1 Each of the following:

$$x - 3x^2 + 5x^4$$

$$\frac{1}{4} - x + 2x^4 + \frac{3}{8}x^5$$

is a polynomial (in x). The first has integer coefficients and the second has rational coefficients. (It is also correct to say that the first has rational coefficients.) ■

The expressions $a_i x^i$ are called the *terms* of the polynomial, as given in the definition above. Though we sometimes may substitute numbers for x from the particular number system under consideration, thus treating the polynomial as representing a function, for most purposes it is best to think of x as something that we manipulate in a purely formal manner, without any determination or questioning of its value. It is in this sense that we speak of x as an *indeterminate*. Notationally we may list the terms of a polynomial in either an ascending or a descending order. In elementary algebra it is customary to use the descending order; but for the purposes of certain discussions that follow, we prefer the ascending order.

Example 5.2 Thus

$$2x^3 + 3x^2 - x + 4 \quad \text{and} \quad 4 - x + 3x^2 + 2x^3$$

denote the same polynomial. It just happens that we prefer the order at the right, at least in the present section. ■

Example 5.3 In Example 5.2 it is possible to regard the polynomial as representing a function $a(x)$, say. Then we would have:

$$\begin{aligned} a(-1) &= 4 - (-1) + 3(-1)^2 + 2(-1)^3 = 6 \\ a(0) &= 4 - 0 + 3(0)^2 + 2(0)^3 = 4 \\ a(1) &= 4 - 1 + 3(1)^2 + 2(1)^3 = 8 \\ a(2) &= 4 - 2 + 3(2)^2 + 2(2)^3 = 30 \end{aligned}$$

etc.

As we have said, for many purposes, however, we simply treat a polynomial $a(x)$ as an algebraic expression in x, to be manipulated as suits our immediate purpose. ■

For any polynomial

$$a(x) = a_0 + a_1 x + a_2 x^2 + \cdots + a_n x^n$$

the exponent on the term of highest order (among those terms having a nonzero coefficient) is called the *degree*, written $\deg(a(x))$.

Example 5.4 The specific polynomial just considered in Examples 5.2 and 5.3 has degree 3. The two polynomials introduced in Example 5.1, however, have degrees 4 and 5, respectively. ■

We admit the *zero polynomial*, 0 (i.e., $0 + 0x + 0x^2 + \cdots$) as a polynomial in good standing, but its degree is not defined since it has no nonzero coefficients. For some purposes it is convenient to agree that the zero polynomial has degree $-\infty$, but this is not really very important in the sequel. If a nonzero polynomial has degree 0, then its constant term a_0 (i.e., a_0x^0) is its only nonzero term, and we speak of a *constant polynomial* in such instances. Note the distinction, however, between a polynomial of degree 0 and the zero polynomial itself.

If the nonzero polynomial

$$a(x) = a_0 + a_1x + a_2x^2 + \cdots + a_nx^n$$

is given of degree n, then a_n is said to be its *leading coefficient* (as if the expression had been written in descending order!). Similarly, a_nx^n is called the *leading term*. With some ambiguity (since we have to think of $a_0 = a_0x^0$), we refer to a_0 as the *constant term* in the given polynomial $a(x)$. Note our reference to the whole polynomial expression as $a(x)$, as if to denote a function in the sense of Example 5.3.

Example 5.5 The leading term for the first polynomial considered in Example 5.1 is $5x^4$ and its leading coefficient is 5. Its constant term must be regarded as 0, since we can only write

$$x - 3x^2 + 5x^4$$

in the form:

$$a_0 + a_1x + a_2x^2 + \cdots + a_nx^n$$

(with $n = 4$) by regarding a_3 and a_0 as 0. ■

Two polynomials $a(x)$ and $b(x)$ are said to be *equal* (and we then write $a(x) = b(x)$) if they have the same degree and their corresponding coefficients are equal, term by term. That is, if

$$a(x) = a_0 + a_1x + a_2x^2 + \cdots + a_nx^n$$
$$b(x) = b_0 + b_1x + b_2x^2 + \cdots + b_mx^m$$

are polynomials of degree n and m, respectively, we write $a(x) = b(x)$ if $n = m$ and $a_0 = b_0, a_1 = b_1, a_2 = b_2, \ldots, a_n = b_n$. By convention there is only one zero polynomial, so that to say $a(x) = 0$ is to say that $0 = a_0 = a_1 = a_2 = \cdots$ in the above.

Addition and Multiplication

The operation of addition of polynomials should be familiar to you from your high-school algebra background. Nevertheless we review the operations here, as follows.

Suppose

$$a(x) = a_0 + a_1x + a_2x^2 + \cdots + a_nx^n$$
$$b(x) = b_0 + b_1x + b_2x^2 + \cdots + b_nx^n$$

are given polynomials, *not necessarily of the same degree*, but in which zero terms have been added to the polynomial of lower degree as necessary to effect the appearance of equal degree. We then define the *sum of the two polynomials* by writing

$$a(x) + b(x) = (a_0 + b_0) + (a_1 + b_1)x + (a_2 + b_2)x^2 + \cdots + (a_n + b_n)x^n$$

That is, we simply add the corresponding coefficients to create terms $(a_i + b_i)x^i$ of every order. Note that

$$\deg(a(x) + b(x)) \leq \max(\deg(a(x)), \deg(b(x)))$$

since we artificially have made each polynomial appear to have the same degree ($n = \max$), but then it is possible that cancellation will occur in computing $a_n + b_n$ (i.e., if $b_n = -a_n$).

Example 5.6 Suppose we regard the two polynomials of Example 5.1

$$x - 3x^2 + 5x^4$$

$$\frac{1}{4} - x + 2x^4 + \frac{3}{8}x^5$$

as polynomials with rational coefficients (which, in fact, they are). We then may compute their sum and the result is the polynomial

$$\frac{1}{4} - 3x^2 + 7x^4 + \frac{3}{8}x^5$$

noting the cancellation of the x terms. In fact, we may write:

$$(x - 3x^2 + 5x^4) + \left(\frac{1}{4} - x + 2x^4 + \frac{3}{8}x^5\right) = \frac{1}{4} - 3x^2 + 7x^4 + \frac{3}{8}x^5$$

in strict accordance with the meaning of polynomial equality as defined above. Thus we have:

$$(x - 3x^2 + 5x^4) + \left(\frac{1}{4} - x + 2x^4 + \frac{3}{8}x^5\right) = \left(0 + \frac{1}{4}\right) + [1 + (-1)]x + (-3 + 0)x^2$$
$$+ (0 + 0)x^3 + (5 + 2)x^4 + \left(0 + \frac{3}{8}\right)x^5$$

by the definition of polynomial addition. This is necessarily equal to the indicated result since we have:

$$0 + \frac{1}{4} = \frac{1}{4}$$

$$1 + (-1) = 1 - 1 = 0$$

$$-3 + 0 = -3$$

$$0 + 0 = 0$$

$$5 + 2 = 7$$

$$0 + \frac{3}{8} = \frac{3}{8}$$

in accordance with the requirement for two polynomials to be regarded as equal. ■

Example 5.7 We note that the degree of the sum in Example 5.6 is the maximum of the degrees of the two polynomials in question (in this case, 5). If, however, we were to perform the sum of the two polynomials:

$$x - 3x^2 + 5x^4$$

$$\frac{1}{4} - x - 5x^4$$

the result would be $\frac{1}{4} - 3x^2$, a polynomial of degree 2, owing to the cancellation of the leading coefficients. Thus we see, indeed, that the degree of the sum can be less than the degree of either of the polynomials being added. More dramatically still, if we are adding the polynomials:

$$x - 3x^2 + 5x^4$$
$$-x + 3x^2 - 5x^4$$

then the result, if it is to be a polynomial at all, must be the zero polynomial! When we then write:

$$(x - 3x^2 + 5x^4) + (-x + 3x^2 - 5x^4) = 0$$

this 0 should not be regarded as a number, but as the special polynomial we call the zero polynomial. One then can readily appreciate the technical need for this special polynomial, if only as a fiction. Without it we would not be able to say (in all cases) that *the sum of two polynomials is a polynomial.* ■

The *multiplication of a polynomial $a(x)$ by a constant* is still easier to describe. If c is any number in the underlying system and

$$a_0 + a_1x + a_2x^2 + \cdots + a_nx^n$$

is any given polynomial, then we define:

$$ca(x) = ca_0 + ca_1 x + ca_2 x^2 + \cdots + ca_n x^n$$

noting that each coefficient in $a(x)$ is thereby multiplied by c.

Example 5.8 If we wish to multiply the polynomial

$$x - 3x^2 + 5x^4$$

by -4, say, we obtain the result:

$$-4x + 12x^2 - 20x^4$$

of the same degree as the original polynomial. ■

When dealing with a number system in which the *cancellation law*:

$$cd = 0 \Longrightarrow c = 0 \quad \text{or} \quad d = 0$$

holds, we cannot annihilate the leading term of $a(x)$ unless we multiply by the constant 0. It follows that if $c \neq 0$, we must have

$$\deg ca(x) = \deg a(x)$$

in such systems.

In particular, suppose $c = -1$. Then $ca(x)$ is just the polynomial $a(x)$ with all the signs of its coefficients reversed from positive to negative and vice versa. In fact we may use this device to provide a convenient description of polynomial subtraction. Thus, if

$$a(x) = a_0 + a_1 x + a_2 x^2 + \cdots + a_n x^n$$
$$b(x) = b_0 + b_1 x + b_2 x^2 + \cdots + b_n x^n$$

where, again, the degrees are made equal by adding zero terms to one of the polynomials if necessary, we may interpret

$$a(x) - b(x) = a(x) + (-1)b(x)$$

There then is nothing new to be learned. We simply reverse the signs of $b(x)$ and add the result to $a(x)$ in the usual way. Of course the result is the same as if we had subtracted coefficients of b from those of a, term by term. Presumably this technique is sufficiently familiar to you that it is not necessary for us to provide an example here.

Finally, we must review the *multiplication of polynomials*, one by another. Suppose we are given the two polynomials:

$$a(x) = a_0 + a_1 x + a_2 x^2 + \cdots + a_n x^n$$
$$b(x) = b_0 + b_1 x + b_2 x^2 + \cdots + b_m x^m$$

of degrees n and m, respectively. We note that terms in the product of a given order k can arise in many ways. Thus we may obtain terms in x^4 by multiplying a_0 with b_4x^4, a_1x with b_3x^3, a_2x^2 with b_2x^2, a_3x^3 with b_1x, or a_4x^4 with b_0. For this reason the product of $a(x)$ and $b(x)$ is a polynomial

$$c(x) = c_0 + c_1x + c_2x^2 + \cdots + c_{n+m}x^{n+m}$$

where

$$c_k = \sum_{i+j=k} a_ib_j$$

adding over all indices i and j whose sum is k. That is,

$$\begin{aligned}
c_0 &= a_0b_0 \\
c_1 &= a_0b_1 + a_1b_0 \\
c_2 &= a_0b_2 + a_1b_1 + a_2b_0 \\
c_3 &= a_0b_3 + a_1b_2 + a_2b_1 + a_3b_0 \\
c_4 &= a_0b_4 + a_1b_3 + a_2b_2 + a_3b_1 + a_4b_0
\end{aligned}$$

etc., the latter as explained in detail above. The reader should note that the result agrees with that which would have been obtained by the longhand calculation taught in high school, that is, multiplying each term in $a(x)$ by each term in $b(x)$, then collecting terms of like order. In fact this collection is precisely what is accomplished in the above formula for the general coefficient c_k in the product.

***Example* 5.9** Suppose we have the two polynomials:

$$\begin{aligned}
a(x) &= 3 - 2x + 3x^2 \\
b(x) &= 6 + 3x - 2x^2
\end{aligned}$$

and we wish to obtain their product. We could, of course, do as we have learned before, writing our hand calculations in the schematic form:

$$\begin{array}{rrrrr}
3 - & 2x + & 3x^2 & & \\
6 + & 3x - & 2x^2 & & \\
\hline
18 - & 12x + & 18x^2 & & \\
 & 9x - & 6x^2 + & 9x^3 & \\
 & - & 6x^2 + & 4x^3 - & 6x^4 \\
\hline
18 - & 3x + & 6x^2 + & 13x^3 - & 6x^4
\end{array}$$

where we collect terms in the final addition of the columns. ■

As a matter of fact this method is probably as safe as any in hand calculations. In writing a computer program for performing the polynomial multiplication, however, it is almost surely a better idea to make use of the *collecting formula*:

$$c_k = \sum_{i+j=k} a_i b_j$$

within *a* for loop, used in running over the desired product coefficients c_k, from 0 to $n + m$, as shown in the accompanying algorithm *polyproduct*. Note that in adding the products to c_k, the subscripts on a_i and $b_j = b_{k-i}$ ensure that $i + j = k$ as required.

algorithm polyproduct $(a, b: c)$

for k running from 0 to $n + m$
  set c_k to 0
  for i running from 0 to k
    increase c_k by $a_i b_{k-i}$

Example 5.10 Suppose we imagine that we are interrupting this algorithm in its computation of c_2 when multiplying the two polynomials of Example 5.9. Thus we examine the execution of the outer for loop with the value $k = 2$. A trace of this computation would appear as shown in Table 5.1. We note that the result is the same as that obtained in the middle column of the hand calculation of Example 5.9, as indeed it should be. ■

TABLE 5.1

k	i	$k - i$	c_k	$(= c_2)$
2			0	
	0	2	-6	(adding $a_0 b_2 = -6$)
	1	1	-12	(adding $a_1 b_1 = -6$)
	2	0	6	(adding $a_2 b_0 = 18$)

Polynomial Evaluation

As mentioned previously, there are any number of occasions in which we wish to treat a polynomial as a function. In such instances the question of *polynomial evaluation* becomes of great importance. That is to say, we are given a polynomial

$$a(x) = a_0 + a_1 x + a_2 x^2 + \cdots + a_n x^n$$

and we wish to determine the value of the function $a(x)$ for one or more arguments $x = x_0$. In principle, this presents no particular difficulty. As in Example 5.3 we may simply substitute the given argument x_0 in the expression for $a(x)$ to obtain

$$a(x_0) = a_0 + a_1 x_0 + a_2 x_0^2 + \cdots + a_n x_0^n$$

and in this way we arrive at a corresponding number $a(x_0)$ in the underlying number system.

Example 5.11 For the polynomial

$$a(x) = 4 - x + 3x^2 + 2x^3$$

appearing in Example 5.3, if we choose the argument $x = 3$, we may substitute to obtain:

$$\begin{aligned} a(3) &= 4 - 3 + 3(3)^2 + 2(3)^3 \\ &= 4 - 3 + 3(9) + 2(27) \\ &= 4 - 3 + 27 + 54 \\ &= 82 \end{aligned}$$

according to the ordinary laws of arithmetic. ■

Suppose we examine the details of this kind of calculation more carefully. With each power x^k, the evaluation requires $k - 1$ multiplications, or k multiplications in all, considering the product of x^k with a_k. In considering a polynomial of degree n, we will then require a total of

$$0 + 1 + \cdots + k + \cdots + n = \sum_{k=0}^{n} k$$

multiplications for evaluating all of the terms. But this is a formula we have seen before (Example 3.16) and the sum, as we know, is of order n^2. The question is, can we do better?

We will soon see that the answer is a definite yes. The trick is to rewrite the polynomial

$$a(x) = a_0 + a_1 x + a_2 x^2 + \cdots + a_{n-1} x^{n-1} + a_n x^n$$

in the nested parenthesized form:

$$a_0 + x(a_1 + x(a_2 + \cdots + x(a_{n-1} + x a_n) \cdots))$$

an expression that is clearly equivalent to the original. For we see that each successive coefficient from a_0 to a_n is multiplied by an additional x. That is, a_0 is not multiplied by x at all, a_1 is multiplied once by x, a_2 is multiplied twice by x, and so on, as required. And yet there are only a total of n multiplications (and n additions) involved in the entire calculation! This technique, known as **Horner's method**, represents a distinct improvement over the usual form of evaluation. It is most clearly presented in the form of the accompanying algorithm *polyeval*, where the n multiplications and n additions are clearly in evidence within the "for" loop. Note that we use the output variable "*aofx*" to represent the functional value $a(x)$, as pronounced a of x.

```
algorithm polyeval (a, x: aofx)

set aofx to a_n
for k running from n − 1 downto 0
   multiply aofx by x
   add a_k to aofx
```

Example 5.12 Suppose we consider the same polynomial (as in Example 5.11)

$$a(x) = 4 - x + 3x^2 + 2x^3$$

once more, with the evaluation at $x = 3$. The tabulated result of the calculation with Horner's algorithm then appears as shown in Table 5.2, in complete agreement with the result obtained in Example 5.11. The reader should note that this tabulation is a bit easier to follow than is the direct substitution in the nested expression in parentheses:

$$a(3) = 4 + 3((-1) + 3(3 + 3(2)))$$

the one on which the tabulation of the algorithm is based. ■

TABLE 5.2

x	k	*aofx*
3		2
	2	6
		9
	1	27
		26
	0	78
		82

We have seen that the evaluation of an nth degree polynomial by Horner's method requires n multiplications and n additions. Since we have n additions to perform even in the usual evaluation procedure, the saving is primarily in the number of multiplications involved, a reduction from order n^2 to order n, using the terminology of Section 4.6. Of course, for larger and larger n, this saving becomes more and more pronounced.

EXERCISES 5.1

1 Express $a(x) + b(x)$ as a polynomial.

(a) $a(x) = -2 + 3x, \quad b(x) = 3 - x$

(b) $a(x) = -2 + 3x + x^2, \quad b(x) = -2 + x + 2x^2$

(c) $a(x) = 3 - 2x + x^2, \quad b(x) = -1 - 2x + 2x^2$

(d) $a(x) = -1 + x - 3x^2 + 2x^3, \quad b(x) = -2 + 3x + x^3$

(e) $a(x) = 1 - 2x + 4x^2 + x^3, \quad b(x) = 3 - 2x + x^2$

(f) $a(x) = 4 - 3x + 2x^2, \quad b(x) = -1 - x + x^2 + 2x^3$

(g) $a(x) = 1 - 2x^2 + 3x^4, \quad b(x) = 4x - 3x^3 + x^5$

(h) $a(x) = 3x - x^3 + 4x^5, \quad b(x) = -5 + x^2 + 2x^4$

2 Express $a(x) - b(x)$ as a polynomial, for $a(x)$ and $b(x)$ as given in Exercise 1.

3 For the polynomials $a(x) = 2 - 3x + 2x^2$, $b(x) = 3 - 2x + x^2$, and $c(x) = -5 + 3x - 2x^2$, write each given expression as a single polynomial.

(a) $a(x) + b(x)$

(b) $a(x) - (b(x) + c(x))$

(c) $(b(x) - a(x)) - c(x)$

(d) $a(x) + (b(x) - c(x))$

(e) $a(x) - (b(x) - c(x))$

(f) $c(x) - (a(x) - b(x))$

4 If $a(x)$ is of degree n and $b(x)$ is of degree $n - 2$, what is the degree of $a(x) + b(x)$? Of $a(x) - b(x)$? Of $-a(x)$?

5 If $a(x)$ and $b(x)$ are polynomials, with $a(x_0) = c$ and $b(x_0) = d$, what can you say about the value of $a(x) + b(x)$ when $x = x_0$?

6 If $a(x)$ and $b(x)$ are polynomials, with $a(x_0) = c$ and $b(x_0) = d$, what can you say about the value of $a(x) - b(x)$ when $x = x_0$?

7 If

$$a(x) = 1 - 2x + 3x^2$$
$$b(x) = 1 + (3 + A)x + (B - 2)x^2$$

find A and B so that $a(x) = b(x)$.

8 If

$$a(x) = -1 - 2x^2 + x^3$$
$$b(x) = -A - 2 + (B + 3)x - 2x^2 + (C + 4)x^3$$

find A, B, and C so that $a(x) = b(x)$.

9 Write a formula for $\deg(a(x)b(x))$ in terms of the degrees of $a(x)$ and $b(x)$. (*Note:* Assume that the underlying number system is such that the cancellation law holds.)

10 Show that if

$$a(x)b(x) = a(x)c(x)$$

and $a(x)$ is not the zero polynomial, then $b(x) = c(x)$. (*Note:* Assume that the underlying number system is such that the cancellation law holds.)

11 Compute the product $a(x)b(x)$ for the polynomials given in Exercise 1, using the ordinary hand computation method.

12 Compute the product $a(x)b(x)$ for the polynomials given in Exercise 1, using the collecting formula for the coefficients of the product.

13 Use the collecting formula for the coefficients of the product to compute $a(x)b(x)$.

(a) $a(x) = \frac{1}{4} - 3x + 2x^2, \quad b(x) = 2 + \frac{1}{2}x - \frac{3}{8}x^2$

(b) $a(x) = 2 + x - \frac{1}{2}x^2 + 3x^3, \quad b(x) = \frac{1}{8} - 3x + 9x^2$

(c) $a(x) = -3 + 5x - 4x^2 + 7x^3, \quad b(x) = 2x + x^2 - 6x^3$

(d) $a(x) = 2 + 2x + 5x^3 - x^4, \quad b(x) = -7 + 2x - x^2 - 4x^3$

14 Use the ordinary hand computation method to compute the products $a(x)b(x)$ for the polynomials of Exercise 13.

15 Evaluate $a(x)$ for $x = -2$ using the ordinary substitution method, taking the polynomials $a(x)$ from Exercise 13.

16 Evaluate $a(x)$ for $x = -2$ using Horner's method, taking the polynomials $a(x)$ from Exercise 13.

17 For the polynomial

$$a(x) = -2 + 7x - 5x^2 + 2x^3$$

use the ordinary substitution method to evaluate:

(a) $a(-1)$ **(b)** $a(-2)$ **(c)** $a(2)$ **(d)** $a(3)$

18 Repeat the evaluations of Exercise 17 using Horner's method.

19 How does the cancellation law read when expressed in the contrapositive? (*Note:* See Section 2.1 and especially the derived rule [→C] of Section 2.5.)

20 Evaluate $a(x)$ for $x = 3$ using the ordinary substitution method, taking the polynomials $a(x)$ from Exercise 1.

21 Evaluate $a(x)$ for $x = 3$ using Horner's method, taking the polynomials $a(x)$ from Exercise 1.

22 Show that the cancellation law holds in the following number systems:

(a) The integers **(b)** The rationals

(*Note:* See Sections 3.1 and 3.5, respectively.)

Sec. 5.2 *GRAPHS OF POLYNOMIALS*

A picture is worth a thousand words.

In the applications of polynomials to science and engineering, a graphical picture of the functional behavior can be of great advantage to the analysis. You undoubtedly have learned something about the graphing of mathematical functions in high school. Obtaining the graph of a polynomial, however, presents a number of special problems, not all of them so easily handled. As a matter of fact, some of the techniques necessary to treat these problems must be deferred until the student has had some exposure to differential calculus. Our aim, at this point, is to discuss those aspects of the subject that can be handled by elementary means, without the benefit of the more advanced mathematical machinery and, particularly, without calculus.

Goals

After studying this section, you should be able to:

- Obtain the limiting behavior of a polynomial as its argument becomes large, in either the negative or the positive direction.
- Determine the *vertical intercept* or *y-intercept* of a polynomial.
- Perform the *synthetic division* process, relating the result to the Horner polynomial evaluation technique.
- Obtain bounds on the roots of polynomials.
- Use the *rule of signs* to estimate the number of positive and negative roots.
- Determine the set of rational roots of a polynomial by examining the factors of its leading and constant coefficients.
- Use an algorithmic technique to obtain estimates for irrational roots.
- Use all available information to obtain an accurate sketch of the graph of a polynomial.
- Define the notions of an *algebraic number* and a *computable number*.
- Describe the hierarchy of discrete number systems, from the natural numbers through the computable numbers.

Graph of a Function

We suppose that the general idea of a graph of function is already well known to the student. With a rectangular coordinate system as that shown in Figure 5.1, we are able to provide a geometric representation or picture of a given function

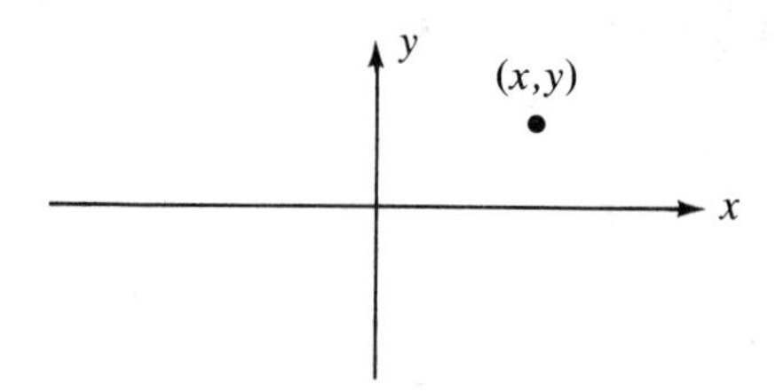

FIGURE 5.1

$y = a(x)$. Each ordered pair of numbers (x, y) with $y = a(x)$ determines a definite point in this coordinate system. The *graph of the function* $y = a(x)$ is the totality of such points.

Here we are mainly concerned with functions $a(x)$ that are polynomials. Since we have already discussed techniques for evaluating polynomials at any given argument x, the means are at hand for our obtaining such a graphical representation. A tabulation of evaluations (Table 5.3) for sufficiently many arguments x will provide any number of points (x, y) that we may plot on the coordinate system. We only have to connect these points with a smooth curve to obtain the desired graphical representation. We say a smooth curve because it is well known that a polynomial cannot jump abruptly from one value to another—it moves continuously through every intermediate value. As a consequence, we may state the following important principle:

Change of Sign Principle If a polynomial $a(x)$ takes on values of an opposite sign at points $x_1 < x_2$, then there is an intermediate point x_0 where $a(x_0) = 0$.

TABLE 5.3

x	$a(x)$

Example 5.13 Suppose we consider the polynomial

$$a(x) = x^2 - 6x + 8$$

as a first example. A tabulation of values of the argument x and the expression $a(x)$ is shown in Table 5.4. After plotting these points on a rectangular coordinate system we are led to the graph of Figure 5.2. Of course we also might investigate the behavior for negative x, but a little experimentation soon will have us convinced that there is little more to learn and that the given graph is sufficient for most purposes.

TABLE 5.4

x	$a(x)$
0	8
1	3
2	0
3	-1
4	0
5	3

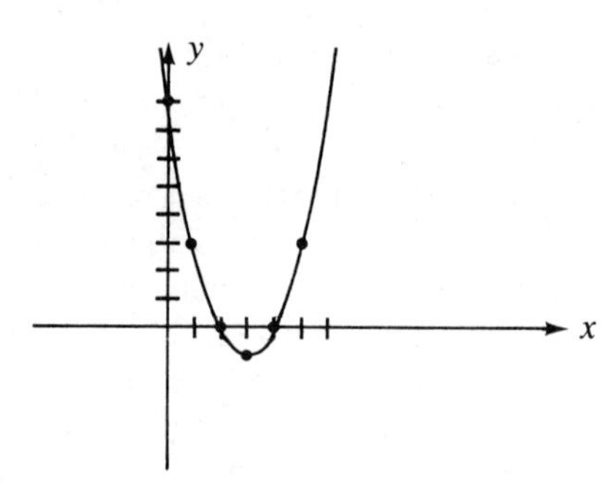

FIGURE 5.2

As an application of the Change of Sign Principle, we note that if the evaluation at $x = 2$ were not given to us we could at least conclude from the evaluations

$$a(1) = 3 \qquad a(3) = -1$$

that there is *some* intermediate point x_0 with $a(x_0) = 0$. It happens that this point is $x_0 = 2$, but that is incidental. ■

Unfortunately, things do not always go as smoothly as in the above example. In any case the computations involved in the evaluations can be prohibitive, even if we are using an order n technique such as Horner's method. What we need is a series of guidelines or aids to the graphing process, a theory that will provide us with general information that we can use to choose our calculations more judiciously. In this chapter we develop this theory in a sequence of steps, learning more and more at each stage until finally you are in possession of a full range of techniques that can be employed as a guide to the polynomial graphing problem.

Limiting Values and Intercepts

We consider once again the general polynomial function

$$a(x) = a_n x^n + a_{n-1} x^{n-1} + \cdots + a_1 x + a_0$$

this time finding it more convenient to list the terms in the descending order. The lemma on polynomial growth rates in Section 4.6 shows that as the argument x grows without limit, in either the positive or negative direction, we have

$$a(x) \sim a_n x^n$$

That is, the leading term dominates all the others. This observation is most useful in determining the limiting behavior of $a(x)$. For we see, in any event, that $a(x)$ becomes infinitely large (in either the negative or the positive sense) as x becomes large positive or large negative.

Essentially there are four cases to consider, depending in combination on whether n is odd or even and whether the leading coefficient a_n is negative or positive. To include all these possibilities, we provide the simple tabulation of Table 5.5, where the pairs of sign entries have an obvious interpretation. They indicate whether $a(x) \sim a_n x^n$ becomes large *negative* or large *positive* as x becomes large to the left (negative) or to the right (positive), respectively.

TABLE 5.5

		n	
		Odd	*Even*
a_n	<0	$+-$	$--$
	>0	$-+$	$++$

Example 5.14 For the polynomial of Example 5.13 we have n even and $a_n > 0$. Accordingly our table exhibits the pair of sign entries "$++$", indicating that $a(x)$ becomes large positive as x becomes large to the left or to the right. And, of course, this observation agrees with Figure 5.2. If we consider the polynomial

$$a(x) = 2x^3 - 3x^2 - 12x + 6$$

we would have n odd and $a_n > 0$. Our table then exhibits the pair "$-+$", indicating that $a(x)$ becomes large negative as x becomes large to the left (negative), but large positive as x becomes large to the right (positive). ■

Such preliminary observations are useful prior to a detailed analysis. As the analysis proceeds, they can be used as a check that the calculations may be correct. Similar comments can be made for the preliminary determination of the *vertical intercept* or *y-intercept*, as it is often called. Here we refer to the point where the graph crosses the y-axis. Since this happens when $x = 0$, a direct substitution into the general expression for $a(x)$ yields

$$a(0) = a_0$$

(the constant term) as the vertical or y-intercept. Again, this is so easy a determination to make that we are well advised to take it into account in advance of any detailed analysis of the graphing problem.

Example 5.15 For the polynomial of Example 5.13, the y-intercept is

$$a(0) = a_0 = 8$$

and we are well advised to have included this as one of the plotted points—$(0, 8)$.

Similarly, for the polynomial introduced in Example 5.14, the y-intercept is

$$a(0) = a_0 = 6$$

and the corresponding point (0, 6) should be plotted at the start.

Note how much we know already about the graph of this polynomial. It passes through the point (0, 6) and becomes large positive as x becomes large positive, and large negative as x becomes large negative, as indicated in Figure 5.3. And we have yet to get our hands dirty! ■

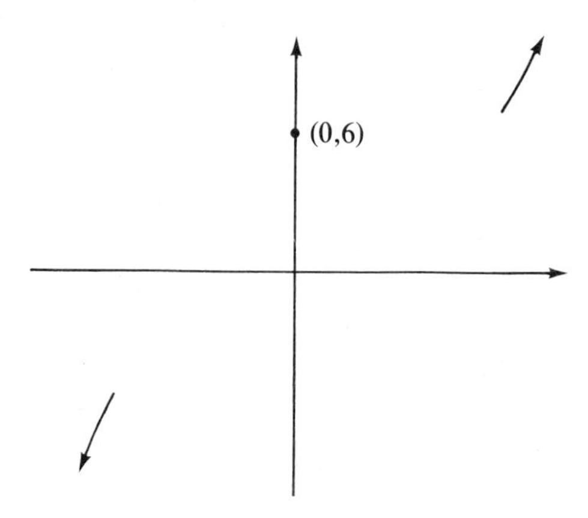

FIGURE 5.3

You must be thinking: If there are vertical intercepts, there also must be *horizontal intercepts*. Indeed there are, in general. These, however, are called the *roots* of the polynomial, and they are considerably more difficult to obtain. In fact it is better that we build up some additional machinery before these matters are discussed. In particular we need to review the *synthetic division* process, one that the student has most likely encountered in high-school mathematics courses. If you find that it is presented differently here, do not worry—the essentials will be the same.

Synthetic Division

Given any linear *divisor* $x - b$ and a general polynomial

$$a(x) = a_n x^n + a_{n-1} x^{n-1} + \cdots + a_1 x + a_0$$

of degree n as *dividend*, one may use the familiar long division process to write:

$$a(x) = (x - b)q(x) + r$$

where the *quotient* $q(x)$ is a polynomial of degree $n - 1$ and the *remainder* r is a constant, that is, simply a number in the underlying system.

Example 5.16 For instance, suppose we divide the polynomial

$$a(x) = 2x^3 + x^2 - 18x - 7$$

by $x - 3$ (choosing $b = 3$). The long division computation then may be arranged in the following familiar pattern:

$$\begin{array}{r|rrrr}
 & 2x^2 & +\ 7x & +\ 3 & \\ \hline
x-3 & 2x^3 + x^2 & -\ 18x & -\ 7 & \\
 & 2x^3 - 6x^2 & & & \\ \cline{2-2}
 & 7x^2 & -\ 18x & & \\
 & 7x^2 & -\ 21x & & \\ \cline{2-3}
 & & 3x & -\ 7 & \\
 & & 3x & -\ 9 & \\ \cline{3-4}
 & & & 2 &
\end{array}$$

indicating that

$$2x^3 + x^2 - 18x - 7 = (x - 3)(2x^2 + 7x + 3) + 2$$

(The reader should check this result by multiplying and adding on the right.) ■

Now we note that this whole division process can be simplified considerably. First of all, the powers of x are only placeholders and need not be written—provided that zero entries are supplied for any missing terms. Moreover, the quotient (the top line) is superfluous since it already is represented by the numbers at the bottom of each column—except the last. Finally, we may as well write b rather than $x - b$, and we then may add rather than subtract in the body of the computation.

Example 5.17 Continuing from the previous example, the simplified computations are summarized below

$$\begin{array}{r|rrrr}
3 & 2 & 1 & -18 & -7 \\
 & & 6 & 21 & 9 \\
 & 2 & 7 & 3 & \boxed{2}
\end{array}$$

where the arrows indicate multiplication by b. Note that the bottom line represents both the quotient and the remainder. We must remember that the terms of the quotient are displaced by one column from their original position; otherwise we are likely to misinterpret the result. ■

Example 5.18 As a further illustration, suppose

$$a(x) = x^4 + 6x^2 - 7x + 8$$

and we wish to divide by $x - 2$. The simplified computations then read as follows (leaving out the arrows):

$$\begin{array}{r|rrrrr} 2 & 1 & 0 & 6 & -7 & 8 \\ & & 2 & 4 & 20 & 26 \\ \hline & 1 & 2 & 10 & 13 & \boxed{34} \end{array}$$

We express the result in the form:

$$x^4 + 6x^2 + -7x + 8 = (x - 2)(x^3 + 2x^2 + 10x + 13) + 34$$

noting that the quotient is always of a degree one less than the dividend—the original polynomial $a(x)$. ■

The simplified computation of these last two examples is known as **synthetic division**. Its great utility is recognized as soon as we realize that the remainder r actually represents the evaluation of the polynomial $a(x)$ at $x = b$. But this is easy to see. We have

$$a(x) = (x - b)q(x) + r$$

and therefore

$$a(b) = (b - b)q(x) + r = 0 + r = r$$

as stated above. In fact, when we recall that the synthetic division process amounts to a series of alternating multiplications and additions, a bell should ring: This is only Horner's method (of polynomial evaluation) in disguise!

Roots of Polynomials

In attempting to determine the graph of a function, whatever its nature, one is aided immeasurably by a knowledge of the roots of the function, that is, those arguments for which the function evaluates to zero. Graphically, such points represent a crossing of the x-axis. More precisely, the number x_0 is called a **root** of the function $a(x)$ provided that $a(x_0) = 0$. In the special case of polynomial functions we will find that the synthetic division process just discussed is both a theoretical and a computational aid. Evidence of this dual advantage is already apparent in the following result, together with its applications.

Theorem 5.1 ***Bound on Roots*** If $a(x)$ has a positive leading coefficient and there are no negative entries in the bottom line of the synthetic division of $a(x)$ by $x - b$, then $b > 0$ is an upper bound for the roots of $a(x)$. Analogously, if we have entries of alternating sign in the bottom line of the synthetic division of $a(x)$ by $x - b$, then $b < 0$ is a lower bound for the roots of $a(x)$.

Proof We write

$$a(x) = (x - b)q(x) + r$$

and recall that the coefficients of $q(x)$ and the number r are represented in sequence as the bottom line of the synthetic division of $a(x)$ by $x - b$. If all these numbers are nonnegative, then $r > 0$, $q(x) > 0$ and

$$a(x) = (x - b)q(x) + r > 0$$

whenever $x > b$. This shows that $a(x)$ cannot have roots $x > b$. A similar argument establishes the analogous statement regarding the lower bound on roots. □

Example 5.19 Consider the polynomial

$$a(x) = 2x^3 + x^2 - 18x - 7$$

of degree 3. A summary of the bottom lines of the synthetic division process for various values of b (notated in the form "b: bottomline") is given below:

1: 2	3	−15	−22		−1: 2	−1	−17	10
2: 2	5	−8	−23		−2: 2	−3	−12	17
3: 2	7	3	2		−3: 2	−5	−3	2
					−4: 2	−7	10	−47

Since the last line at the left is entirely nonnegative, we learn from Theorem 5.1 that $b = 3$ is an upper bound on the roots of $a(x)$. Analogously, because the last line at the right alternates in sign, $b = -4$ is a lower bound on the roots. It follows that any roots x for this particular polynomial must lie within the range $-4 < x < 3$. Actually, in view of the Change of Sign Principle, we can see that we have learned considerably more than this. Recalling that the last entry in the bottom line of the synthetic division process is the value of the function (at $x = b$), and noting that $a(0) = -7$, we are able to say for certain that there are roots between -4 and -3, between -1 and 0, and between 2 and 3. ■

Example 5.20 For the polynomial

$$a(x) = x^4 + 6x^2 - 7x + 8$$

a similar tabulation would begin:

$$1{:}\ 1 \quad 1 \quad 7 \quad 0 \quad 8$$

indicating that $b = 1$ is an upper bound on the positive roots of $a(x)$. In fact, we can say more. Since $-7x + 8 > 0$ for $0 < x < 1$, there can be no positive roots at all. Likewise there can be no negative roots, since it is abundantly clear that $a(x) > 0$ for all $x < 0$. ■

Example 5.21 In considering the polynomial

$$a(x) = x^3 - x^2 + x + 6$$

a tabulation of the bottom lines of the synthetic division yields:

$$\begin{array}{llll} 1{:}\ 1 & 0 & 1 & 7 \end{array} \qquad \begin{array}{rrrr} -1{:}\ 1 & -2 & 3 & 3 \\ -2{:}\ 1 & -3 & 7 & -8 \end{array}$$

According to Theorem 5.1 we may conclude that 1 is an upper bound on the roots and that -2 is a lower bound. Moreover, in view of the Change of Sign Principle, the last column of entries at the right indicates that $a(x)$ has a root between -2 and -1. ■

As our next objective we would like to be able to provide a means for estimating the total number of possible roots, whether positive or negative. But the criterion (Descartes' Rule of Signs) that is most useful in this regard is best preceded by our first providing a statement of the fundamental *factor lemma*, a result that is much used for its own merit.

Lemma

$x - b$ is a factor of $a(x)$ iff b is a root of $a(x)$

Proof To say that $x - b$ is a factor of $a(x)$ is to say that $a(x)$ can be written as a product

$$a(x) = (x - b)q(x)$$

with zero remainder. But then it is clear that

$$a(b) = (b - b)q(b) = 0$$

that is, b is a root of $a(x)$. The proof in the opposite direction is left to you to work out. □

Theorem 5.2

Descartes' Rule of Signs The number of positive roots of a polynomial $a(x)$ is bounded by the number of variations in sign in the coefficients. Analogously, the number of negative roots is bounded by the number of variations in sign in the coefficients of $a(-x)$.

Proof First of all we should understand that the variations in sign are simply the changes—from plus to minus or from minus to plus—in reading the coefficients of the polynomial from left to right. Considering a positive root $b > 0$, we then can argue by induction that it suffices to show that the number of variations in sign in the product

$$a(x) = (x - b)q(x)$$

is at least one greater than the number of variations in sign for the quotient $q(x)$. Note our use of the factor lemma. Suppose that we now represent the above multiplication symbolically:

$$
\begin{array}{cccccccccc}
++ & \cdots & +-- & \cdots & -++ & \cdots & +- & \cdots & +/- & \\
 & & & & & & & & x-b & \\
\hline
++ & \cdots & +-- & \cdots & -++ & \cdots & +- & \cdots & +/- & \\
- & \cdots & --+ & \cdots & ++- & \cdots & -- & \cdots & & -/+ \\
\hline
+ & & - & & + & & - & & & -/+
\end{array}
$$

(where $+/-$ means "plus or minus" and $-/+$ means "minus or plus," *respectively*.) In the last line, representing the signs of $a(x)$, the gaps between the indicated positions where the signs of $a(x)$ are definitely determined may or may not involve additional sign changes—it depends on the magnitudes of the quantities involved. Inasmuch as the constant coefficient of $a(x)$ is opposite in sign to that of $q(x)$, however, we can be assured that at least one additional sign change has taken place, regardless of what happens in these gaps. The proof of the analogous statement regarding negative roots hinges on the fact that the negative roots of $a(x)$ are the positive roots of $a(-x)$. □

Example 5.22 The polynomial

$$a(x) = 2x^3 + x^2 - 18x - 7$$

considered in Example 5.19 exhibits just one change in sign as the coefficients are examined from left to right. According to Theorem 5.2 we then may assert that the polynomial has at most one positive root. As a matter of fact, we know that there is just one, a root between 2 and 3. We had learned that there were two negative roots, one between -4 and -3 and one between -1 and 0. This is confirmed once more by noting that

$$a(-x) = -2x^3 + x^2 + 18x - 7$$

has two changes in sign. ■

Example 5.23 The polynomial

$$a(x) = x^4 + 6x^2 - 7x + 8$$

previously considered in Example 5.20 can have at most two positive roots, according to Theorem 5.2. From our earlier analysis, however, we know that there are none. Moreover, since

$$a(-x) = x^4 + 6x^2 + 7x + 8$$

has no sign changes, Theorem 5.2 would assure us that there are no negative roots. Again, this is consistent with our earlier analysis. ■

Example 5.24 If we consider the polynomial

$$a(x) = x^3 - x^2 + x + 6$$

as introduced in Example 5.21, we observe two sign changes and, accordingly, there can be at most two positive roots. As a matter of fact, there are none. Why? At the same time, since

$$a(-x) = -x^3 - x^2 - x + 6$$

has just one sign change, Theorem 5.2 indicates that there is at most one negative root. In fact there will be a root between -2 and -1, as we already had decided. ■

Finally, we need to begin to zero in on the roots, trying to determine their actual values so that we can plot the corresponding points on the x-axis of our graphs. From this standpoint the following result is most helpful toward identifying all the rational roots. Throughout the discussion we assume that all rational numbers p/q are expressed in lowest terms so that p and q have no factors in common. We also assume that our polynomials have integer coefficients; if they are rational we can multiply through by the least common multiple of the denominators, and in so doing we won't change the roots.

Theorem 5.3

Rational Roots If p/q is a root of the polynomial

$$a(x) = a_n x^n + a_{n-1} x^{n-1} + \cdots + a_1 x + a_0$$

having integer coefficients, then $p \mid a_0$ and $q \mid a_n$.

Proof If p/q is a root of $a(x)$, then we have

$$a_n(p^n/q^n) + \cdots + a_1(p/q) + a_0 = 0$$

and if we multiply by q^n we obtain:

$$a_0 q^n + a_1 p q^{n-1} + \cdots + a_n p^n = 0$$

Factoring p from all but the first term and transposing, we have

$$-a_0 q^n = p(a_1 q^{n-1} + \cdots + a_n p^{n-1})$$

showing that $p \mid a_0 q^n$. Since p and q have no factors in common, it follows that $p \mid a_0$. If instead of factoring p we had factored q, we would have obtained

$$-a_n p^n = q(a_0 q^{n-1} + \cdots + a_{n-1} p^{n-1})$$

showing that $q \mid a_n p^n$. Again, because p and q have no common factors, we would be able to conclude that $q \mid a_n$, thus completing the proof. □

Example 5.25 Consider the polynomial

$$a(x) = 6x^3 + 11x^2 - 3x - 2$$

of degree 3. Candidates p/q as rational roots must meet the conditions:

$$p \mid -2 \qquad q \mid 6$$

according to Theorem 5.3. This narrows the field considerably. We only have to consider

$$\frac{p}{q} = 1, -1, 2, -2, \frac{1}{2}, -\frac{1}{2}, \frac{1}{3}, -\frac{1}{3}, \frac{1}{6}, -\frac{1}{6}, \frac{2}{3}, -\frac{2}{3}$$

There can be no other possibilities. Our earlier theory is a great aid here in further reducing the number of candidates. The bottom line tabulation of Theorem 5.1 begins with

1:6 17 14 16

showing immediately that any positive roots have to be less than 1 (thereby eliminating 1 and 2 from consideration). Furthermore, Theorem 5.2 shows that there can only be one positive root, at most. When we apply synthetic division with the next possibility ($x = \frac{1}{2}$) we obtain:

$$\begin{array}{r|rrrr} 1/2 & 6 & 11 & -3 & -2 \\ & & 3 & 7 & 2 \\ \hline & 6 & 14 & 4 & \boxed{0} \end{array}$$

showing that indeed $\frac{1}{2}$ is a root. Therefore we needn't consider $\frac{1}{3}$, $\frac{1}{6}$, or $\frac{2}{3}$ as roots. Further analysis shows that $-\frac{1}{3}$ and -2 are also roots. It follows that the polynomial may be factored in the form:

$$a(x) = 6x^3 + 11x^2 - 3x - 2 = 6\left(x - \frac{1}{2}\right)\left(x + \frac{1}{3}\right)(x + 2)$$

an observation that leads us in the direction of Section 5.4, if somewhat prematurely. ■

Example 5.26 Consider the polynomial

$$a(x) = 2x^3 + x^2 - 18x - 7$$

as discussed in Examples 5.19 and 5.22. According to Theorem 5.3 a rational root p/q would have to satisfy the conditions:

$$p \mid -7 \qquad q \mid 2$$

so that

$$\frac{p}{q} = 7, -7, \frac{7}{2}, -\frac{7}{2}, 1, -1, \frac{1}{2}, -\frac{1}{2}$$

are the only possibilities. The earlier analysis of Example 5.19 causes us to reject $7, \frac{7}{2}$, and -7 as being out of range. We already had decided that there are three roots, one between -4 and -3, one between -1 and 0, and one between 2 and 3. We therefore can ignore $1, -1$, and $\frac{1}{2}$ as well. This leaves only $-\frac{1}{2}$ and $-\frac{7}{2}$ as possible rational roots. But the corresponding synthetic division calculations:

$-1/2$	2	1	-18	-7
		-1	0	9
	2	0	-18	2

$-7/2$	2	1	-18	-7
		-7	21	$-21/2$
	2	-6	3	$-35/2$

show that neither is the case. We have learned something from all of this, however. Comparing the above evaluations (at $x = -\frac{1}{2}$ and $-\frac{7}{2}$) with those of Example 5.19, we now are able to say that $a(x)$ has three roots—one between $-\frac{7}{2}$ and -3, one between $-\frac{1}{2}$ and 0, and one between 2 and 3 (most likely between $\frac{5}{2}$ and 3). But if the roots are not rational, what are they? ■

Approximate Roots

As we have just seen, it may happen that a polynomial has a root that cannot be expressed as a rational number; it is, as we say, *irrational*—but more on this later. When this is the case, one needs some method for obtaining an approximation to its value, for that is then about the best that one can do. If our interest is only in obtaining an adequate graph of the function, then almost any reasonable method will do. For other purposes, particularly in the applications to science and engineering, it may be necessary to obtain very close approximations, and then the particular method should be chosen with great care. Here we provide only a simple bisection technique, one that is strongly reminiscent of the bisection algorithm of Section 4.3.

We suppose that one already has an interval (low, high) in which a root is known to lie. Ordinarily, for the given polynomial

$$a(x) = a_n x^n + a_{n-1} x^{n-1} + \cdots + a_1 x + a_0$$

the values of a(low) and a(high) then will be of opposite sign. Presumably this information will have come about through analyzing the polynomial according to the techniques that have already been described. We then may suppose that one hopes to estimate the root to a given precision—call it approx—so that it may be considered as an additional input to the estimation procedure. That is, we may want to approximate the root to within $\frac{1}{100}$, say, and this then would be the input value for approx. Our scheme, as we have said, is then really quite simple. We halve the interval at a point called mid, and we evaluate the polynomial there. If the value of a(mid) is of the same sign as that at the left-hand end of the (current) interval, then we make mid the new left end (knowing the root is then between it and the old right end). Otherwise we make mid the new right end—for the analogous reason. The whole procedure then may be described more formally as indicated in the accompanying algorithm *approxroot*. Note that we split the interval in half once more to obtain the final value of the output approximation *root*, after the width of the current interval is found to be less than or equal to approx.

```
algorithm approxroot (a, low, high, approx: root)

set left, right to low, high
while (right − left) > approx
                        left + right
    compute mid  =  ─────────────
                             2
    if a(mid) same sign as a(left)
        replace left by mid
    else
        replace right by mid
                    left + right
set root to    ─────────────
                         2
```

Example 5.27 Consider once more the polynomial

$$a(x) = 2x^3 + x^2 - 18x - 7$$

as discussed in Examples 5.19, 5.22, and 5.26. We already have learned that there are roots between $-\frac{7}{2}$ and -3, between $-\frac{1}{2}$ and 0, and between $\frac{5}{2}$ and 3 and, furthermore, from the analysis of Example 5.26 we know that these roots must be irrational. Suppose we illustrate the use of the algorithm approxroot by estimating the value of the root between $-\frac{7}{2}$ and -3. A trace of the execution of the algorithm for these input values low and high, together with the input approx $= \frac{1}{100}$, is shown in Table 5.6. We conclude that $x = -3.06$ is an approximate root (to within $\frac{1}{100}$). A similar analysis shows that -0.38 and 2.95 are equally good approximations to the other two roots. ■

TABLE 5.6

left	*right*	*mid*	*a(mid)*
−3.5	−3.0	−3.025	−6.59
−3.25		−3.125	−2.02
−3.125		−3.0625	0.06
	−3.0625	−3.09375	−0.96
−3.09375		−3.078125	−0.45
−3.078125		−3.0703125	−0.19
−3.0703125		−3.06640625	

Example 5.28 Once we have found all the roots to a polynomial (and we have first done the preliminary analysis described in this section), we should be in a position to draw a reasonably good graph of the function. For the polynomial of the previous example we use all available information—reviewing Examples 5.19, 5.22, 5.26, and 5.27—to arrive at the graph shown in Figure 5.4. The fact that the points so naturally connect to form a smooth curve leads us to believe that our analysis has been carried out correctly. ■

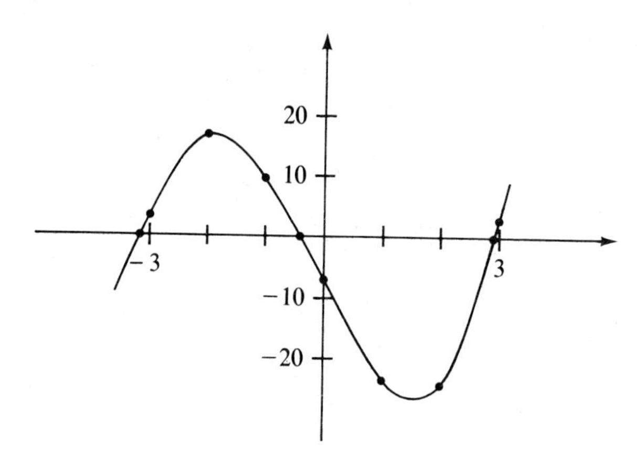

FIGURE 5.4

Algebraic and Computable Numbers

In this brief concluding discussion we would like to give you a slightly different perspective on the ideas we have just encountered. As we do so, you will see that the development is really a continuation of the material presented in Chapter 3, particularly that of Section 3.5. In comparison with the earlier discussions of this section, however, our point of view becomes somewhat more theoretical. And yet we will be speaking only of numbers, so that the level of abstraction will be kept at a minimum.

First we consider the inadequacy of the rational number system. We needn't look very far in order to find a polynomial with an *irrational root*, one that cannot be expressed in the form p/q for integers p and q. Simply consider the polynomial equation

$$x^2 - 2 = 0$$

that is, look for a root of the polynomial $x^2 - 2$. Our experience tells us that the number we want is $\sqrt{2}$ (or $-\sqrt{2}$), since indeed

$$(\sqrt{2})^2 - 2 = 2 - 2 = 0$$

as required. Yet the following result is easily established:

Theorem 5.4 The number $\sqrt{2}$ is irrational.

Proof We give an indirect proof. Suppose instead that $\sqrt{2}$ can be expressed in the form p/q for integers p and q. (We may suppose that p and q have no common factors.) Squaring, we would have

$$2q^2 = p^2$$

which asserts that p^2 is even. But if the square of a number is even, the number itself must be even. Thus p is even and we may write

$$p = 2r$$

for some integer r. Substituting in the above, we obtain $2q^2 = (2r)^2$, or

$$q^2 = 2r^2$$

and thus q^2 is even. By the same argument used above, q must itself be even. Since both p and q are even, they have a common factor—2—contrary to the original assumption. This contradiction completes the proof. □

This simple argument shows that the rational numbers are grossly inadequate for treating the problems of this chapter—even if we limit our attention to polynomials with rational coefficients. At the same time our investigation encourages the following important definition, one that serves to extend the number system adequately for the discussion at hand.

A root of a polynomial having rational coefficients is said to be an **algebraic number**. Of course, any rational number is algebraic, since the linear polynomial equation

$$x - p/q = 0$$

has p/q as its root! Even more importantly the above discussion shows that there are algebraic numbers (e.g., $\sqrt{2}$) that are not rational, so we definitely have achieved an extension of our number system. Yet we obtain the following result, which shows that the algebraic number system is *discrete*—that is, is totally within the realm of our interest.

Theorem 5.5 The algebraic numbers are countable.

Proof A one-to-one correspondence can be given between the natural numbers and the algebraic numbers, using an argument much like that of the proof of Theorem 3.3. We leave determining the details to those of you who are interested in doing so. □

Naturally we would regard the algebraic number system as being somewhat more abstract than the rational system. Yet the new system is totally accessible or constructible in the sense that we may use rational numbers to approximate any one of its members to an arbitrary degree of precision, using an algorithm of the type introduced earlier. Note, however, that if this criterion of approximation is established as a measure of constructibility, we obtain yet a further extension of our number system. That is, suppose we agree that a **computable number** is one that can be approximated to an arbitrary degree of precision by some algorithm. Then our technique for approximating roots shows that every algebraic number is computable. And yet, as the following result shows, there are computable numbers that are not algebraic.

Theorem 5.6 The number π is nonalgebraic.

The proof is beyond the scope of this discussion, as is the argument that shows that π is computable. The latter, however, is encountered by students in their first-year calculus sequence.

In any case we see that we again have achieved a genuine extension of our number system, as we move from one level of abstraction to another. A summary of the various discrete number systems we have encountered along the way is given in the hierarchical diagram of Figure 5.5. In the interest of completeness, however, we need to at least refer to the following somewhat surprising result, stated without proof.

Theorem 5.7 The computable numbers are countable.

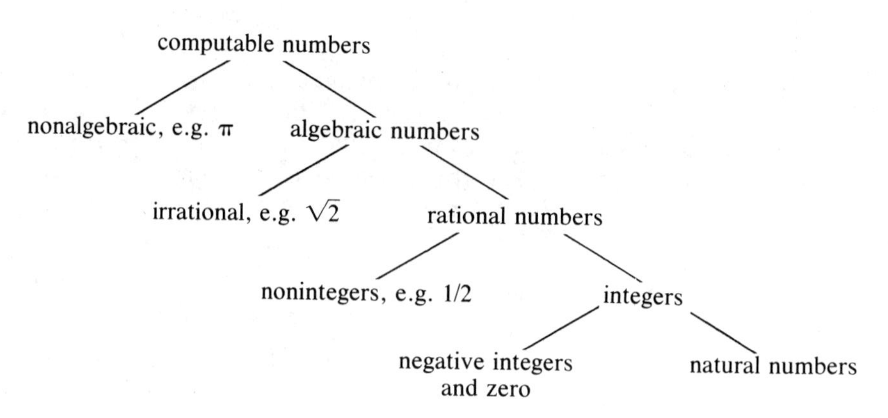

FIGURE 5.5

EXERCISES 5.2

1 Sketch the graphs of the following polynomials by constructing a simple tabulation of values.

(a) $x^2 - 6x + 5$ **(b)** $x^2 - 5x + 4$ **(c)** $x^2 - 7x + 10$

(d) $x^2 - x - 2$ **(e)** $x^3 - 5x^2 + 2x + 8$ **(f)** $x^3 + 6x^2 + 8x$

2 Obtain the limiting behavior for the following polynomials as x becomes large negative and large positive, respectively.

(a) $-x^3 + x^2 - 12x + 5$ **(b)** $x^4 + 10x^2 - 5x + 4$

(c) $-x^4 - x^3 + 2x^2 + 15$ **(d)** $-4x^3 + 3x + 7$

(e) $3x^5 + 12x^4 - 5x^3 - 4x^2 + 7$ **(f)** $-3x^5 + 2x^4 + 4x^3 - 5x^2 + 6$

3 For each of the following polynomials, determine the vertical or y-intercept and the limiting behavior as x becomes large negative and large positive, respectively.

(a) $x^4 - 5x^3 + 4x^2 - 10x + 2$

(b) $-2x^5 + 3x^4 - 5x^3 + 10x^2 - 4$

(c) $-3x^6 + 7x^5 - 4x^4 + 10x^2 - 7x + 6$

(d) $x^6 + 2x^5 - x^4 + 5x^3 - 4x^2 - 7$

(e) $-5x^6 + 2x^5 - 3x^4 + 3x^3 - 2x^2 - 1$

(f) $3x^5 - 3x^4 + 10x^3 - 7x^2 + 9x$

4 Use the long division process to divide each of the given polynomials by the indicated linear divisor.

(a) $x^4 - 4x^3 + x^2 + 7, \quad x - 5$

(b) $-2x^5 + 4x^4 - 3x^3 + 2x + 7, \quad x + 4$

(c) $-x^4 - x^3 + 2x^2 + 7x + 2, \quad x - 2$

(d) $4x^4 + 7x^3 - 5x^2 - 2x + 1, \quad x + 3$

(e) $2x^4 + 9x^3 + 6x^2 - 2x + 2, \quad x - 1$

(f) $-2x^4 + 4x^3 - 5x^2 + 7, \quad x - 3$

5 Use the long division process to divide each of the polynomials of Exercise 3 by the following linear divisors, respectively.

(a) $x - 3$ **(b)** $x - 1$ **(c)** $x + 2$

(d) $x + 2$ **(e)** $x - 4$ **(f)** $x + 3$

6 Use the synthetic division process for the problems of Exercise 4. Then express the results in the form:

$$a(x) = (x - b)q(x) + r$$

7 Use the synthetic division process for the problems of Exercise 5. Then express the results in the form:

$$a(x) = (x - b)q(x) + r$$

8 Use the method described in Theorem 5.1 to obtain bounds on the roots of the following polynomials. Then try to locate the roots between two consecutive integers.

(a) $2x^3 - x^2 - 4x + 2$ **(b)** $3x^3 + x^2 - 9x - 3$
(c) $x^3 - 3x^2 - 4x + 2$ **(d)** $3x^3 + 10x^2 - 2x - 4$
(e) $3x^3 + 3x^2 - 36x - 35$ **(f)** $2x^3 - 14x^2 + 29x - 16$

9 Do the same (as in Exercise 8) for the following polynomials.

(a) $x^4 - x^3 - 24x^2 + 4x + 78$ **(b)** $x^4 - 2x^3 - 9x^2 + 10x + 5$
(c) $x^4 - 8x^3 + 5x^2 + 6x - 3$ **(d)** $x^4 - 8x^3 + 12x^2 + 16x - 16$
(e) $6x^4 + 11x^3 - 25x^2 - 33x + 21$ **(f)** $3x^4 + 2x^3 - 41x^2 - 26x + 26$

10 Use Theorem 5.2 to estimate the number of positive and negative roots for the polynomials of Exercise 8.

11 Do the same (as in Exercise 10) for the polynomials of Exercise 9.

12 Do the same (as in Exercise 10) for the following polynomials.

(a) $2x^3 + 5x^2 - 3x + 4$ **(b)** $3x^3 - 4x^2 - 2x - 1$
(c) $4x^3 + 7x^2 + 6x + 3$ **(d)** $x^4 + x^3 - x^2 - x - 1$
(e) $x^4 - 3x^3 - 2x^2 + 6x - 1$ **(f)** $2x^4 + 3x^2 + 7$

13 Complete the proof of the factor lemma in the opposite direction.

14 Obtain all rational roots of the following polynomials.

(a) $x^3 + 2x^2 - 5x - 6$ **(b)** $x^3 - 7x - 6$
(c) $x^3 + 3x^2 - 6x - 8$ **(d)** $x^3 - 3x^2 - 6x + 8$
(e) $2x^3 - 5x^2 + 4x - 1$ **(f)** $3x^3 - 2x^2 + 19x - 6$

15 Do the same (as in Exercise 14) for the following polynomials.

(a) $x^4 + x^3 - 7x^2 - x + 6$ **(b)** $x^4 - 3x^3 - 2x^2 + 12x - 8$
(c) $x^4 + 2x^3 - 7x^2 - 8x + 12$ **(d)** $x^4 - x^3 - 11x^2 + 9x + 18$
(e) $4x^4 + 16x^3 + 17x^2 - x - 6$ **(f)** $6x^4 - 17x^3 + 2x^2 + 19x - 6$

16 Do the same (as in Exercise 14) for the following polynomials.

(a) $x^4 - x^3 - 3x^2 - 7x - 6$ **(b)** $x^4 + 7x^3 + 9x^2 - 11x - 6$
(c) $x^5 - x^4 - 8x^3 + x^2 + 13x + 6$ **(d)** $x^5 + 3x^4 - 3x^3 - 13x^2 - 4x + 4$
(e) $2x^5 - x^4 - 6x^3 + 15x^2 + 10x - 8$ **(f)** $9x^5 - 39x^4 + 35x^3 + 5x^2 - 4x - 6$

17 Find an approximate value (to two decimal places) for the smallest positive root in each of the following.

(a) $x^3 + 4x^2 + 2x - 1$ **(b)** $x^3 + 4x^2 - 8$
(c) $x^3 - 2x - 1$ **(d)** $x^3 + 6x^2 + 7x - 6$
(e) $x^3 - 3x^2 - 5x - 1$ **(f)** $x^3 + 9x^2 + 18x - 8$

18 Find an approximate value (to two decimal places) for the numerically smallest negative root in each of the following.

(a) $x^3 - 6x^2 + 6x + 2$ **(b)** $7x^3 - 21x^2 + 4$
(c) $x^3 - 6x^2 + 3x + 4$ **(d)** $x^3 - 3x^2 - 6x + 2$
(e) $x^4 + 2x^3 - 5x^2 - 2x + 1$ **(f)** $x^4 + x^3 - 7x^2 + 2x + 4$

19 Find an approximate value (to three decimal places) for the largest root in each of the following.

(a) $x^3 + 9x^2 + 15x - 21$ **(b)** $x^3 - 3x^2 - 12x - 6$

(c) $x^3 + 3x^2 - 6x - 14$ **(d)** $x^3 + 3x^2 - 3x - 7$

(e) $x^4 - x^3 - 14x^2 + x + 1$ **(f)** $x^4 - 13x^2 - 2x + 20$

20 Use all available information to sketch an accurate graph of the polynomials in Exercise 14.

21 Do the same (as in Exercise 20) for the polynomials of Exercise 15.

22 Do the same (as in Exercise 20) for the polynomials of Exercise 16.

23 Provide the one-to-one correspondence required in the proof of Theorem 5.5.

24 Find an algorithm that shows that the number π is computable.

25 Prove the second half of Theorem 5.1.

26 In what sense does the proof of Theorem 5.2 use an inductive argument?

Sec. 5.3 *INTERPOLATION THEORY*

To interpolate: to insert between or among others.

The dictionary definition just given is not quite appropriate to the discussion that follows. It is often true, even in mathematics, that it means precisely this, to insert (usually numbers) between others. In the context of polynomial functions, however, the meaning is slightly different. An *interpolating polynomial* is one that passes through certain points prescribed in advance. Thus, to some degree, the problem we treat herein is an inverse of the problem in the preceding section. Here we are given a partial description of the desired behavior of a function, and we seek a polynomial having the given behavior. Such a technique is more a matter of synthesis, whereas in Section 5.2 the approach was strictly one of analysis. We will find, however, that our focus is narrower here. We really have only one basic objective in mind—and that helps to make the development easier to follow.

Goals

After studying this section, you should be able to:

- Determine a polynomial having certain prescribed roots.
- Gain a new appreciation for the *straight-line equation*, relating it to the general interpolation problem.
- Write down the Lagrange Interpolation Formula for two, three, or any arbitrary number of points.
- Apply the formula to the treatment of specific polynomial interpolation problems.

Prescribed Roots

It is best that we look at the interpolation problem in a number of restricted settings before considering the general situation. First let us try to construct a polynomial that passes through given points on the x-axis. That is, we suppose that we are looking for a polynomial with certain **prescribed roots**—roots that are given in advance. This is really quite a simple problem, once we think about it for a moment. Suppose we would like a polynomial to have specific roots $x_1, x_2, \ldots, x_n$. Since the linear expression

$$x - x_i$$

takes the value 0 at $x = x_i$, the same will be true of any product having this expression as a factor. The idea then should occur to us to multiply all such expressions together, one for each of the desired roots. We then obtain the product:

$$a(x) = (x - x_1)(x - x_2) \cdots (x - x_n)$$

a polynomial that clearly takes the value 0 at each of the x_i. We say polynomial, for that is exactly what we obtain if we multiply these factors together. In fact, the result may be written

$$a(x) = a_n x^n + a_{n-1} x^{n-1} + \cdots + a_1 x + a_0$$

where the coefficients a_i will depend on the particular roots we have chosen, and the degree n is their number. Necessarily we will have leading coefficient $a_n = 1$, and (except possibly for the sign) the constant coefficient a_0 will be the product of the roots.

Example 5.29 If we prescribe the roots:

$$x_1 = -1 \qquad x_2 = 1 \qquad x_3 = 3$$

then our polynomial must necessarily be of degree 3 (or more). According to the method just described, we obtain

$$\begin{aligned} a(x) &= (x + 1)(x - 1)(x - 3) \\ &= x^3 - 3x^2 - x + 3 \end{aligned}$$

noting that $(x - (-1)) = x + 1$. Indeed, our answer is a polynomial. Should we desire its graph, we are in much better position than we were in Section 5.2, for we already know the roots. After determining the y-intercept $a(0) = 3$ and analyzing the limiting behavior, we are able to provide the sketch of Figure 5.6, where we see that the graph becomes negatively infinite as x becomes large negative, and positively infinite as x becomes large positive. Had we preferred the opposite behavior, we would only need to multiply the whole polynomial expression by -1. In effect

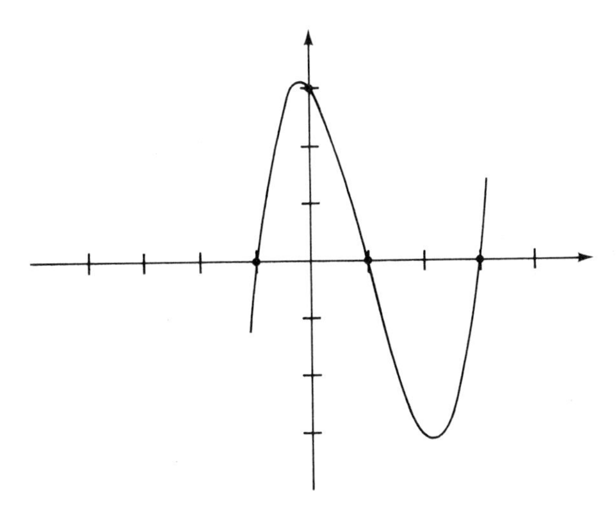

FIGURE 5.6

this will turn the graph upside down, and we will obtain the result:

$$a(x) = -x^3 + 3x^2 + x - 3$$

Furthermore, this does not change the roots. ■

It is clear that things are turned around from what they were in Section 3.2. There we would have been given the polynomial and asked to find its roots. It is in this sense that our present discussion is the inverse of that of the preceding section.

Straight-Line Equation

We now begin to move to the more general setting where the desired interpolated points are arbitrarily situated, not necessarily being restricted to the x-axis. But in order to provide a more graduated development of the theory, we first consider the simple problem of but two points. Recalling that two points determine a line, we thus will be restricting our attention to polynomials of degree 1. After all, the equation of a straight line is a polynomial, albeit a quite simple one. In this simple setting you will be able to make connections to your previous experience while making the necessary preparations for the discussions to come.

The interpolation problem is then a familiar one. We are given two points (x_0, y_0) and (x_1, y_1) and we are asked to find the equation of the straight line passing through them. Considering the graph shown in Figure 5.7, we are led to the expression

$$y = y_0 + \frac{(x - x_0)}{(x_1 - x_0)}(y_1 - y_0)$$

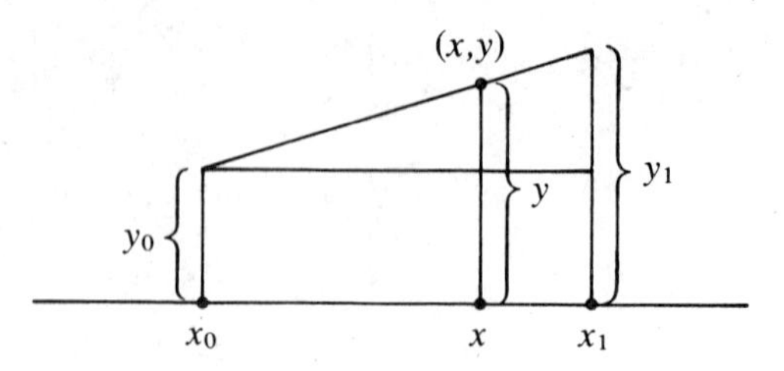

FIGURE 5.7

noting how the proportional fraction comes right out of the geometry of the situation. In recognition of the fact that we have here a function—in particular, a polynomial—we write:

$$a(x) = y_0 + \frac{(x - x_0)}{(x_1 - x_0)}(y_1 - y_0)$$

$$= y_0 \frac{x - x_1}{x_0 - x_1} + y_1 \frac{x - x_0}{x_1 - x_0}$$

where the latter, less familiar representation is included because of its connection to the generalized situation yet to be discussed.

Example 5.30 Suppose our two given points are $(-1, 3)$ and $(2, 5)$. A direct substitution in the above formula (the familiar one) yields

$$a(x) = 3 + \frac{(x + 1)}{3} 2$$

$$= \frac{2}{3} x + \frac{11}{3}$$

You may check that when $x = -1$ we have $a(x) = 3$ and when $x = 2$ we have $a(x) = 5$, as originally specified. ■

Polynomial Interpolation

If two points determine a line, it is entirely reasonable to expect that perhaps three points will determine a parabola, and so on. In fact this is the approach we take in treating the general *polynomial interpolation* problem. We suppose that we are given $n + 1$ points:

$$p_0 = (x_0, y_0)$$
$$p_1 = (x_1, y_1)$$
$$\cdots$$
$$p_n = (x_n, y_n)$$

and we hope to find a polynomial $a(x)$ of degree n that will pass through each of the points. Of course, that is the same as saying that

$$a(x_0) = y_0$$
$$a(x_1) = y_1$$
$$\cdots$$
$$a(x_n) = y_n$$

We need to keep this in mind as we proceed through the balance of this section.

Before we obtain the solution to this general interpolation problem it is best that we first consider yet another special case. Suppose that all but one of the y's is 0, and the remaining one (call it y_j) is 1. That is to say, suppose we are seeking a polynomial $a_j(x)$ that will take the value 1 at x_j and will take the value 0 at n other points $x_0, \ldots, x_{j-1}, x_{j+1}, \ldots, x_n$. Rewriting these requirements in the form:

1. $a_j(x_j) = 1$ **2.** $a_j(x_i) = 0 \qquad (i \neq j)$

we note that condition (2) is strongly reminiscent of the case of prescribed roots, as already discussed. We thus are led to write

$$a_j(x) = c(x - x_0) \cdots (x - x_{j-1})(x - x_{j+1}) \cdots (x - x_n)$$

noting that the inclusion of a multiplying constant c will not affect the fact that condition (2) is satisfied.

In order to determine the appropriate value for c, we then make use of condition (1). Setting $a_j(x_j) = 1$, we obtain

$$1 = c(x_j - x_0) \cdots (x_j - x_{j-1})(x_j - x_{j+1}) \cdots (x_j - x_n)$$

and thus we must have

$$c = \frac{1}{(x_j - x_0) \cdots (x_j - x_{j-1})(x_j - x_{j+1}) \cdots (x_j - x_n)}$$

Putting all the pieces together we have thus obtained

$$a_j(x) = \frac{(x - x_0) \cdots (x - x_{j-1})(x - x_{j+1}) \cdots (x - x_n)}{(x_j - x_0) \cdots (x_j - x_{j-1})(x_j - x_{j+1}) \cdots (x_j - x_n)}$$

as the solution to our restricted interpolation problem. It happens that it is only one small step from here to the solution of the general problem.

Example 5.31 Suppose we consider four points, $x_0 = -1$, $x_1 = 1$, $x_2 = 3$, and $x_3 = 4$. Let x_2 be the point where our polynomial a_j is to take the value 1 (i.e., we take $j = 2$). We then obtain

$$a_2(x) = \frac{(x+1)(x-1)(x-4)}{(3+1)(3-1)(3-4)}$$
$$= -\frac{1}{8}(x+1)(x-1)(x-4)$$
$$= -\frac{1}{8}x^3 + \frac{1}{2}x^2 + \frac{1}{8}x - \frac{1}{2}$$

as our interpolating polynomial. As a check on the derivation, we compute:

$$a(-1) = \frac{1}{8} + \frac{1}{2} - \frac{1}{8} - \frac{1}{2} = 0$$
$$a(1) = -\frac{1}{8} + \frac{1}{2} + \frac{1}{8} - \frac{1}{2} = 0$$
$$a(3) = -\frac{27}{8} + \frac{9}{2} + \frac{3}{8} - \frac{1}{2} = 1$$
$$a(4) = -8 + 8 + \frac{1}{2} - \frac{1}{2} = 0$$

as required. ■

In returning to the general problem, the way is now clear. If we want a polynomial $a(x)$ to pass through $n + 1$ arbitrary points so that

$$a(x_0) = y_0$$
$$a(x_1) = y_1$$
$$\cdots$$
$$a(x_n) = y_n$$

we only have to write

$$a(x) = y_0a_0(x) + y_1a_1(x) + \cdots + y_na_n(x)$$

The resulting expression for $a(x)$ is called *Lagrange's Interpolation Formula*, one of the most important results in the polynomial theory. Though it seems a bit contrived, it is not a trick. It works only because of the special properties of the polynomials a_j. Thus we have

$$a(x_j) = y_0a_0(x_j) + y_1a_1(x_j) + \cdots + y_na_n(x_j)$$
$$= y_ja_j(x_j)$$
$$= y_j \cdot 1$$
$$= y_j$$

for each $j = 0, 1, 2, \ldots, n$.

If we rewrite the Lagrange Formula for $a(x)$ and include the definition of the auxiliary polynomials a_j, we obtain expressions:

$n = 1$

$$y_0 \frac{x - x_1}{x_0 - x_1} + y_1 \frac{x - x_0}{x_1 - x_0}$$

(*Note:* This is precisely the straight-line equation studied earlier.)

$n = 2$

$$y_0 \frac{(x - x_1)(x - x_2)}{(x_0 - x_1)(x_0 - x_2)} + y_1 \frac{(x - x_0)(x - x_2)}{(x_1 - x_0)(x_1 - x_2)} + y_2 \frac{(x - x_0)(x - x_1)}{(x_2 - x_0)(x_2 - x_1)}$$

and the reader can begin to see the general structure of this *detailed form* of the Lagrange Formula. It is just as easy to first derive the auxiliary polynomials $a_j(x)$ and then combine them with the given y_j as in the following.

Example 5.32 Suppose we would like to find a polynomial passing through the points

$$p_0 = (-1, 2)$$
$$p_1 = (1, -1)$$
$$p_2 = (3, 5)$$
$$p_3 = (4, 3)$$

Evidently we will be able to do this using some cubic polynomial. Having already determined the appropriate $a_2(x)$ in Example 5.31, we only have to compute:

$$a_0(x) = \frac{(x - 1)(x - 3)(x - 4)}{(-1 - 1)(-1 - 3)(-1 - 4)} = -\frac{1}{40}x^3 + \frac{1}{5}x^2 - \frac{19}{40}x + \frac{3}{10}$$

$$a_1(x) = \frac{(x + 1)(x - 3)(x - 4)}{(1 + 1)(1 - 3)(1 - 4)} = \frac{1}{12}x^3 - \frac{1}{2}x^2 + \frac{5}{12}x + 1$$

$$a_3(x) = \frac{(x + 1)(x - 1)(x - 3)}{(4 + 1)(4 - 1)(4 - 3)} = \frac{1}{15}x^3 - \frac{1}{5}x^2 - \frac{1}{15}x + \frac{1}{5}$$

Taking the ordinates y_j into account in the Lagrange Formula, we then obtain:

$$\begin{aligned}
a(x) &= y_0a_0(x) + y_1a_1(x) + y_2a_2(x) + y_3a_3(x) \\
&= 2a_0(x) - 1a_1(x) + 5a_2(x) + 3a_3(x) \\
&= -\frac{1}{20}x^3 + \frac{2}{5}x^2 - \frac{19}{20}x + \frac{3}{5} \\
&\quad -\frac{1}{12}x^3 + \frac{1}{2}x^2 - \frac{5}{12}x - 1 \\
&\quad -\frac{5}{8}x^3 + \frac{5}{2}x^2 + \frac{5}{8}x - \frac{5}{2} \\
&\quad +\frac{1}{5}x^3 - \frac{3}{5}x^2 - \frac{1}{5}x + \frac{3}{5} \\
&= -\frac{67}{120}x^3 + \frac{14}{5}x^2 - \frac{113}{120}x - \frac{23}{10}
\end{aligned}$$

As a check on the calculations, we compute the following synthetic divisions (alias Horner evaluations):

$$\begin{array}{r|rrrr}
-1 & -67/120 & 14/5 & -113/120 & -23/10 \\
 & & 67/120 & -403/120 & 129/30 \\
\hline
 & -67/120 & 403/120 & -129/30 & \boxed{2} \\
1 & -67/120 & 14/5 & -113/120 & -23/10 \\
 & & -67/120 & 269/120 & 13/10 \\
\hline
 & -67/120 & 269/120 & 13/10 & \boxed{-1} \\
3 & -67/120 & 14/5 & -113/120 & -23/10 \\
 & & -67/40 & 27/8 & 73/10 \\
\hline
 & -67/120 & 9/8 & 73/30 & \boxed{5} \\
4 & -67/120 & 14/5 & -113/120 & -23/10 \\
 & & -67/30 & 34/15 & 159/30 \\
\hline
 & -67/120 & 17/30 & 159/120 & \boxed{3}
\end{array}$$

showing that indeed

$$\begin{aligned}
a(-1) &= 2 \\
a(1) &= -1 \\
a(3) &= 5 \\
a(4) &= 3
\end{aligned}$$

as originally specified. ■

EXERCISES 5.3

1 Determine polynomials having the following prescribed roots.

(a) $3, -4, 5$ **(b)** $-\frac{1}{2}, 1, 3$

(c) $-1, -2, -3, 5$ **(d)** $3, \frac{7}{2}, 4, \frac{9}{2}$

In each case write the polynomial in the standard form.

2 Determine the equations of straight lines passing through the following pairs of points.

(a) $(-1, -3)$ and $(2, 4)$ **(b)** $(-3, 2)$ and $(3, -6)$

(c) $(-2, 0)$ and $(6, 6)$ **(d)** $(1, 3)$ and $(5, -3)$

(e) $(\frac{1}{2}, -7)$ and $(2, \frac{3}{2})$ **(f)** $(1, 0)$ and $(7, -3)$

In each case write the equation in the standard form.

3 Find a polynomial passing through the following set of points.

(a) $(-1, 3), (2, 4), (5, -1)$ **(b)** $(-5, \frac{1}{2}), (-1, 4), (8, -3)$

(c) $(-6, -4), (-1, 1), (4, -\frac{1}{2})$ **(d)** $(1, 1), (3, -5), (7, 2)$

In each case write the polynomial in the standard form.

4 Find a polynomial passing through the following set of points.

(a) $(-2, 2), (1, 4), (4, 5), (6, -3)$

(b) $(-4, -2), (-1, 5), (3, 2), (6, -2)$

(c) $(1, 2), (3, -4), (6, 1), (10, -3)$

(d) $(-\frac{3}{2}, -5), (-1, \frac{1}{2}), (4, 9), (8, -3)$

In each case write the polynomial in the standard form.

5 Find a polynomial passing through the following set of points.

(a) $(-7, -2), (-2, 3), (1, 1), (4, -8), (8, 2)$

(b) $(-\frac{2}{3}, 7), (1, 6), (3, 1), (5, -\frac{1}{2}), (10, 2)$

(c) $(-8, 8), (-4, 2), (-1, -1), (4, 5), (9, -1)$

(d) $(1, 4), (4, 7), (5, 6), (8, 1), (12, -5)$

In each case write the polynomial in the standard form.

6 Find a polynomial having the indicated roots *and* passing through the indicated points, respectively.

(a) $2, 5$, and $(-3, 7), (9, -1)$. **(b)** -3 and $(2, 6), (8, -4)$

(c) $-1, 4$ and $(-8, 4), (9, -5)$ **(d)** $-2, 3, 5$ and $(7, 7)$

(e) $-2, 5$ and $(-3, 1), (1, 1), (7, -4)$ **(f)** $-3, -1$ and $(2, 2), (4, 5), (8, -1)$

In each case write the polynomial in the standard form.

7 In each case find a polynomial taking the value 1 at the first point and the value 0 at the other points listed.

(a) 4 and $-2, 7, 10$ **(b)** -2 and $-8, 1, 6$

(c) 6 and $-10, 7$ **(d)** -1 and $-7, 8$

(e) 4 and $-3, 3, 5, 9$ **(f)** 8 and $-9, -2, 2, 12$

In each case write the polynomial in the standard form.

8 Sketch the graphs of the polynomials in Exercise 1.

9 Sketch the graphs of the polynomials in Exercise 3.

10 Sketch the graphs of the polynomials in Exercise 4.

11 Sketch the graphs of the polynomials in Exercise 5.

12 Sketch the graphs of the polynomials in Exercise 6.

13 Fill in the algebraic steps showing the equivalence of the two forms of the straight-line equation presented in the text.

14 Write down the detailed form of the Lagrange Interpolation Formula for the case $n = 3$.

15 Write down the detailed form of the Lagrange Interpolation Formula for the case $n = 4$.

Sec. 5.4 FACTORIZATION OF POLYNOMIALS

Divide and conquer.

Each of us has some familiarity with the factorization process from our high-school mathematics program. In factoring quadratic polynomials a definite formula was provided, one that we all have surely memorized. For polynomials of a higher degree, however, the way was never so clear. In fact you probably have the impression that it was mainly a hit-and-miss proposition, with little likelihood of success. We will do our best to correct this impression here. It happens that over the system of rational numbers (as coefficients) there is an algorithm for carrying out the factorization process, whatever the degree of the given polynomial. This algorithm becomes the focus of our attention in this section.

Goals

After studying this section you should be able to:

- Perform the *long division* process, computing quotients and remainders for any pair of polynomials given as dividend and divisor.
- Describe the intent or purpose of the auxiliary algorithms *division*, *interpolate*, *eval*, and *factors*, giving special attention to their role in the factorization process.
- Illustrate the use of the revised Euclidean algorithm, particularly as it applies to polynomials.
- Discuss Kronecker's *factorization algorithm*, showing how it achieves its objective.
- Use this algorithm to obtain factorizations of given polynomials.
- Define the term *irreducible* in relation to polynomials.
- State a fundamental irreducibility criterion and show its application by example.

Quotients and Remainders

It happens that much of our discussion here will parallel that of Section 3.3 on integer division. For this reason it will help if you review that material before proceeding any further. The similarities between integer and polynomial arithmetic are so strong and so compelling that the rewards of your review will be well worth your effort.

As a first instance of this analogy, we cite the *division algorithm* that follows. Here we are given two polynomials $a(x)$ and $b(x)$ as dividend and divisor, respectively. Upon dividing a by b we obtain a unique *quotient* $q(x)$ and a *remainder* $r(x)$, just as in the case of integers. More formally, we have the

---

Polynomial Division Algorithm For any pair of polynomials $a(x), b(x)$ there will exist polynomials $q(x)$ and $r(x)$ such that

$$a(x) = b(x)q(x) + r(x) \quad \text{and} \quad \deg(r) < \deg(b)$$

Note that the condition $(r < b)$ for the integer remainder is replaced here by a statement on degrees. That is, we are assured that the degree of the remainder $r(x)$ is less than that of $b(x)$—otherwise, further division would have been possible.

Example 5.33 In executing the division algorithm we simply apply the familiar long division process—in fact, that *is* the algorithm, as far as we are concerned. Thus if we are asked to divide

$$a(x) = 2x^4 - 4x^3 + x^2 - 3x + 5$$

by

$$b(x) = x^2 - 3x + 2$$

we simply compute according to the well-known technique:

$$\begin{array}{r|l}
 & \phantom{2x^4 - 4x^3 + x^2 - {}} 2x^2 + 2x + 3 \\
x^2 - 3x + 2 & 2x^4 - 4x^3 + x^2 - 3x + 5 \\
 & 2x^4 - 6x^3 + 4x^2 \\
 & \phantom{2x^4 - {}} 2x^3 - 3x^2 \\
 & \phantom{2x^4 - {}} 2x^3 - 6x^2 + 4x \\
 & \phantom{2x^4 - 6x^3 + {}} 3x^2 - 7x \\
 & \phantom{2x^4 - 6x^3 + {}} 3x^2 - 9x + 6 \\
 & \phantom{2x^4 - 6x^3 + 3x^2 - {}} 2x - 1
\end{array}$$

Thus we have

$$q(x) = 2x^2 + 2x + 3 \quad \text{and} \quad r(x) = 2x - 1$$

noting that $\deg(r) = 1 < 2 = \deg(b)$. You should check that indeed

$$\begin{aligned} a(x) &= b(x)q(x) + r(x) \\ &= (x^2 - 3x + 2)(2x^2 + 2x + 3) + (2x - 1) \end{aligned}$$

for the given polynomial $a(x)$. ■

In the special case where $r(x) = 0$, we say that b **divides** a or is a **factor of** a. Already we can begin to see a connection to the factorization problem. In fact the division algorithm will be much used toward our obtaining an algorithmic solution to questions of factorization.

Example 5.34 In reviewing Example 5.33 we can see that if we had been given the slightly different dividend:

$$a(x) = 2x^4 - 4x^3 + x^2 - 5x + 6$$

(with the same divisor) we would have had a zero remainder. It follows that

$$a(x) = (x^2 - 3x + 2)(2x^2 + 2x + 3) = b(x) \cdot q(x)$$

and we would have a factorization of this revised $a(x)$. It is then correct to say that $b(x)$ divides $a(x)$, or that $q(x)$ divides or is a factor of $a(x)$, whichever we wish. ■

Auxiliary Algorithms

It is not all that easy to find a factorization of a given polynomial, assuming such a factorization exists. We will need to have considerable algorithmic machinery at our disposal in order to properly treat the problem. The division algorithm just discussed is a necessary part of this arsenal—yet it is but one of several algorithms we need to introduce.

Fortunately, however, these *auxiliary algorithms* are either ones that we have already discussed, or they are of such a simple nature that we do not need to become too formal in our presentation. Here we list the headings of each of them:

```
algorithm division (a, b: q, r)
algorithm interpolate (p_0, p_1, ..., p_n: a)
function eval (a, x)
function factors (y)
algorithm euclid (a, b: s, t)
```

Already you must have some sense of what to expect.

The first of these is our old friend the division algorithm, where the input variables a, b and the output variables q and r all are considered to be polynomials.

Again, the output of algorithm *interpolate* is a polynomial, but the inputs are a sequence of points $p_i = (x_i, y_i)$, the points through which $a(x)$ is supposed to pass. Thus we simply have summarized the main result of the preceding section, the use of the Lagrange Interpolation Formula, noting that it is surely an algorithmic procedure.

Example 5.35 Recalling Example 5.32, if we submit the points

$$p_0 = (-1, 2)$$
$$p_1 = (1, -1)$$
$$p_2 = (3, 5)$$
$$p_3 = (4, 3)$$

as inputs to algorithm interpolate, then the output will be the polynomial

$$a(x) = -\frac{67}{120}x^3 + \frac{14}{5}x^2 - \frac{113}{120}x - \frac{23}{10}$$

the answer to Example 5.32. ■

Our third auxiliary algorithm, *eval*, is simply an encapsulation of Horner's polynomial evaluation technique, as discussed in Section 5.2. It is convenient to regard the algorithm as a function, however, taking two arguments as input—a polynomial and a number (a and x, respectively). The value of the function eval (a, x) is then also a number, and the intent is that it be $a(x)$, as if we simply had evaluated a at x. As indicated above it would be understood that the evaluation is to be computed using the Horner method.

Example 5.36 Reviewing Example 5.18, if we suppose we are given

$$a(x) = x^4 + 6x^2 - 7x + 8$$

and the additional input $x = 2$, then a call to the function eval(a, x) would give rise to the functional value 34. ■

Next we reintroduce an auxiliary algorithm that is taught in grade-school arithmetic. We say "arithmetic" because unlike the previous algorithms we have discussed, *factors* deals only with integers. Except for a few technicalities we are dealing with the same procedure as that for the algorithm of the same name in Section 3.3. Here, however, it is best to think of factors(y) as a function, taking an integer argument y as input and from it computing the *set* of integer divisors (or factors) of y. As far as can be determined, this is taught by having the student write down all the positive integers from 1 up to y (although the square root of y would be enough). Each such integer is then tested for divisibility into y. Presumably the (ordinary) division algorithm is used for this purpose. Those integers that do not evenly divide y are cancelled out. What remains is the set of (positive) divisors of y. For our purposes such divisors should be notated with a plus-or-minus sign, for we will need them both.

Example 5.37 Consider the integer $y = 30$, known to have a number of divisors. We write down the positive integers from 1 up to 5 (the square root of 30 is between 5 and 6):

1 2 3 ~~4~~ 5

Applying the division algorithm to each we find that all except 4 are divisors of 30. Then we note that positive divisors come in pairs (that is why the square root of y is sufficient as a stopping point). It follows that our set of divisors—the value of factors (y)—is the set:

$$\{\pm 1, \pm 30, \pm 2, \pm 15, \pm 3, \pm 10, \pm 5, \pm 6\}$$

noting the use of plus-or-minus signs. ■

Finally, we discuss a variation of the Euclidean algorithm (see Section 3.3). It so happens that we do not need this algorithm euclid $(a, b\colon s, t)$ in order to discuss the factorization problem. Since we use this algorithm later on in the chapter, however, we should discuss its use here.

Let us first suppose that we are dealing only with integers, just to keep things simple. Recall that the original Euclidean algorithm euclid $(a, b\colon \text{gcd})$ computed the greatest common divisor of a and b. It did this by successive divisions:

$$\begin{aligned} a &= bq_1 + r_1 \\ b &= r_1 q_2 + r_2 \\ &\cdots \\ r_{i-1} &= r_i q_{i+1} + r_{i+1} \\ &\cdots \end{aligned}$$

until finally a remainder of zero was obtained. At that point the previous remainder is the gcd.

What we are interested in seeing here is that each remainder r (and, hence, also the gcd) is a *linear combination* of the original a and b, in the sense that there are integers s and t such that:

$$r = sa + tb$$

Arguing inductively, let us first look at the beginning. We have

$$r_1 = a - bq_1 = (1)a + (-q_1)b$$

showing that $s_1 = 1$ and $t_1 = -q_1$. Again, since

$$r_2 = b - r_1 q_2 = b - (a - bq_1)q_2 = (-q_2)a + (1 + q_1 q_2)b$$

we have $s_2 = -q_2$ and $t_2 = 1 + q_1 q_2$. Now suppose it is true that in general:

$$\begin{aligned} &\cdots \\ r_{i-1} &= s_{i-1}a + t_{i-1}b \\ r_i &= s_i a + t_i b \\ &\cdots \end{aligned}$$

Then according to the formula defining r_{i+1} above we have

$$\begin{aligned} r_{i+1} &= r_{i-1} - r_i q_{i+1} \\ &= (s_{i-1}a + t_{i-1}b) - (s_i a + t_i b)q_{i+1} \\ &= (s_{i-1} - s_i q_{i+1})a + (t_{i-1} - t_i q_{i+1})b \end{aligned}$$

showing that the induction goes through.

It follows that if we begin with the initial assignments:

$$\begin{aligned} s_0 &= 0 \qquad t_0 = 1 \\ s_1 &= 1 \qquad t_1 = -q_1 \end{aligned}$$

we can always compute the next s, t pair according to the identical formulas:

$$s_{i+1} = s_{i-1} - s_i q_{i+1} \qquad t_{i+1} = t_{i-1} - t_i q_{i+1}$$

These computations may be superimposed on the original Euclidean algorithm to finally arrive at integers s, t satisfying

$$\gcd(a, b) = sa + tb$$

Here, of course, we refer to the final values for s and t.

Example 5.38 If we repeat the calculations of Example 3.26 (with $a = 252$ and $b = 180$) but keep track, as well, of the s, t values, we obtain the tabulation of Table 5.7. It follows that we may express the gcd in the form

$$\gcd(a, b) = 36 = -2(252) + 3(180)$$

as a linear combination of the original a and b. ■

TABLE 5.7

a	b	q	r	s	t
				0	1
252	180	1	72	1	−1
180	72	2	36	−2	3
72	36	2	0		

Example 5.39 Suppose we now try the same thing with polynomials. Taking

$$a(x) = x^2 - 2x + 1$$
$$b(x) = x^2 + x + 1$$

we obtain the trace of the computation as shown in Table 5.8. In this case we may again write

$$\gcd(a, b) = 1$$
$$= \left(\frac{1}{3}x + \frac{1}{3}\right)(x^2 - 2x + 1) + \left(-\frac{1}{3}x + \frac{2}{3}\right)(x^2 + x + 1)$$

noting that s, t (as we might have expected) are polynomials, as is the gcd in general. ■

TABLE 5.8

a	b	q	r	s	t
				0	1
$x^2 - 2x + 1$	$x^2 + x + 1$	1	$-3x$	1	-1
$x^2 + x + 1$	$-3x$	$-\frac{1}{3}x - \frac{1}{3}$	1	$\frac{1}{3}x + \frac{1}{3}$	$-\frac{1}{3}x + \frac{2}{3}$

In this last example we say that the polynomials a and b are **relatively prime** since they have gcd = 1 (they have no factors in common). Actually, we use the same terminology, and the effect is the same if the gcd is a constant, whatever its value. But if $\gcd(a, b)$ is the constant c, then we simply divide the resulting s and t by c and use these values. In all such instances, then, the revised Euclidean algorithm

$$\text{euclid}(a, b\colon s, t)$$

computes a pair of polynomials s, t such that

$$sa + tb = 1$$

It is this technique that is important in the following discussions, particularly those of Section 5.5. As a consequence it is just this use of euclid $(a, b; s, t)$ that we should keep in mind.

Polynomial Factorization

The algorithm that we are about to discuss is a classic. It is attributed to the late nineteenth-century mathematician, Leopold Kronecker, whose biography graced the pages of our introduction to Chapter 3. This algorithm seems to have been lost somewhat for a time—there is hardly any mention of it in the mathematical literature. But this algorithm and others of its kind are in the midst of a revival of sorts—and there is good reason for this.

Among all the computer science efforts to create expert systems, perhaps none is more spectacular in its impact than that directed toward automating symbolic computation. With the push of a button, it now is possible to perform all the usual high-school algebraic calculations automatically. This is not an exaggerated claim. In fact, this revolution goes much further. Virtually all the routine computations of the calculus have been implemented by these symbolic manipulation systems—and more are being accomplished with each passing year. The objective of the Kronecker algorithm, to factor a given polynomial, is typical of the problems for which these expert systems are designed. There are more efficient algorithms for this problem, but none is based on more elementary notions. This is one of the reasons why we have chosen to include it in the present discussion.

As we have indicated, we begin with an ordinary polynomial $a(x)$. We ask if we can ascertain whether it has any factors, whether there are polynomials $b(x)$ and $c(x)$ such that

$$a(x) = b(x) \cdot c(x)$$

We suppose that the problem as stated is limited to the case of rational coefficients. Thus, in the above

$$a(x) = a_n x^n + a_{n-1} x^{n-1} + \cdots + a_1 x + a_0$$

and $b(x)$ and $c(x)$ as well, are assumed to have rational coefficients. And that is really our only limitation. We may as well assume that $a(x)$ has integer coefficients, however, for we can always multiply through by a least common multiple of any denominators, and this will not change the problem materially. Moreover, having agreed to this, we can assume (for much the same reason) that the coefficients of $a(x)$ have no integer factors in common. And then we are ready to begin.

Suppose, as above, that $a(x)$ has degree n. We can limit our search for factors to the case of polynomials of degree $k \le [n/2]$, where the square brackets denote the integer part. If $b(x)$ were of a higher degree than this, then $c(x)$ would certainly be of degree $[n/2]$ or less, and we might as well have found c instead of b. For such an integer k, we try to find a set X consisting of $k + 1$ integer points x where $a(x) \neq 0$. This will be easy because it is rare that a polynomial evaluates to zero—after all, that would be a root! For each such point x we also determine the set S_x of integer factors of $a(x)$, using the auxiliary algorithm factors. (The reader should begin to follow the formal description of the algorithm *factor* outlined below.)

Kronecker then argued as follows. If b is to be a factor of a, we must have $b(x) \mid a(x)$ for each of the integer points x just described. Therefore, at each of the $k + 1$ points x, why not try all values of y in S_x as candidates for $b(x)$? Surely these are the only candidates, according to the argument above. With each such choice of points

$$p_i = (x_i, y_i) \quad \text{with } x \text{ in } X \text{ and } y \text{ in } S_x$$

we can interpolate to find the polynomial $b(x)$. Then, either b divides a or it doesn't, and the division algorithm can settle that question. In summary, the whole procedure is as outlined in the accompanying algorithm factor. We note that the output variables are called a_1 and a_2, rather than b and c, for reasons that become clear later. Moreover, the for statements must be read in a special way here. We don't necessarily run through all their options. As soon as we find a factor (when $r = 0$ in division) we quit the algorithm entirely. It then must be supposed that the whole process begins all over again for a_1 and a_2 as inputs, as if we had written a recursive algorithm, thus ensuring that we finally arrive at the complete factorization of a.

algorithm factor $(a: a_1, a_2)$

for k running from $[n/2]$ downto 1
  set x to 0 and X to $\varnothing$
  while $|X| < k + 1$
    if eval $(a, x) \neq 0$
      adjoin x to X
      set S_x = factors (eval (a, x))
    increase x by 1
  for each sequence of $k + 1$ points $p_i = (x_i, y_i)$
      with x in X and y in S_x
    interpolate $(p_0, p_1, \ldots, p_k: b)$
    division $(a, b: q, r)$
    if $r = 0$
      set $a_1 = b$, $a_2 = q$

generates $X = \{x_0, x_1, \ldots, x_k\}$ $a(x_i) \neq 0$ $(0 \leq i \leq k)$

Example 5.40 We initially choose a rather simple example so as not to tax the reader's powers of concentration. Suppose

$$a(x) = x^3 + 2x + 3$$

so that $n = 3$. Then $[n/2] = [3/2] = 1$ so that we begin with $k = 1$, that is, we only have to look for linear factors. The computation of X and the sets S_x are as follows:

$$x_0 = 0 \qquad a(0) = 3 \neq 0 \qquad S_0 = \{\pm 1, \pm 3\}$$
$$x_1 = 1 \qquad a(1) = 6 \neq 0 \qquad S_1 = \{\pm 1, \pm 2, \pm 3, \pm 6\}$$

where the latter represent the possibilities for values $b(x)$.

Within the interior for loop we will at one time select the points:

$$p_0 = (0, 1)$$
$$p_1 = (1, 2)$$

For this selection the interpolation algorithm yields

$$b(x) = x + 1$$

When we then apply the division algorithm we find indeed that the remainder is zero, leading to the factorization:

$$a(x) = (x + 1)(x^2 - x + 3)$$

One then finds that the factors a_1 and a_2 are themselves irreducible, as we say, so that the factorization is then complete as it stands.

Moreover, suppose we have begun with the polynomial

$$\begin{aligned} a(x) &= (x^2 + 1)(x^3 + 2x + 3) \\ &= x^5 + 3x^3 + 3x^2 + 2x + 3 \end{aligned}$$

but we were unaware of this factorization. We then would have had $[n/2] = [\frac{5}{2}] = 2$ and we might have first discovered the factor $x^2 + 1$ by listing:

$$\begin{array}{lll} x_0 = 0 & a(0) = 3 \neq 0 & S_0 = \{\pm 1, \pm 3\} \\ x_1 = 1 & a(1) = 12 \neq 0 & S_1 = \{\pm 1, \pm 2, \pm 3, \pm 4, \pm 6, \pm 12\} \\ x_2 = 2 & a(2) = 85 \neq 0 & S_2 = \{\pm 1, \pm 5, \pm 17, \pm 85\} \end{array}$$

and proceeding as before. Thus we see that the choice:

$$p_0 = (0, 1)$$
$$p_1 = (1, 2)$$
$$p_2 = (2, 5)$$

indeed leads to the polynomial $x^2 + 1$ when we apply the interpolation algorithm. Of course, this particular polynomial will necessarily pass the test of the division algorithm, yielding the output factors:

$$a_1 = x^2 + 1 \qquad a_2 = x^3 + 2x + 3$$

From this point we would repeat the whole algorithm with a_1 and a_2 in place of a. In thus finding that a_1 is irreducible and that a_2 has the factorization originally given, we would obtain

$$x^5 + 3x^3 + 3x^2 + 2x + 3 = (x^2 + 1)(x + 1)(x^2 - x + 3)$$

as our complete factorization. ■

An Irreducibility Criterion

The reader can see that the factorization algorithm, while definitely effective, involves a number of quite extensive computations. For this reason we naturally would welcome any results that might be available for simplifying the task. We present one such result here, an irreducible criterion usually attributed to Eisenstein. Before citing this criterion one must first be aware of a certain terminology. We say that a given polynomial $a(x)$ is **irreducible** if it has no factors (other than itself or a constant). Thus the notion is rather like that of being prime in the case of integers. With this much understood, we then are ready to state the irreducibility criterion of Eisenstein.

Theorem 5.8 ***Eisenstein's Irreducibility Criterion*** Suppose there is a prime p that divides every coefficient of

$$a(x) = a_n x^n + a_{n-1} x^{n-1} + \cdots + a_1 x + a_0$$

except the leading coefficient a_n. Suppose that p^2 does not divide the constant coefficient a_0. Then $a(x)$ is irreducible.

Proof We give a proof by contradiction. Suppose instead that $a(x)$ can be factored as

$$a(x) = b(x) \cdot c(x)$$

for polynomials

$$b(x) = b_r x^r + \cdots + b_0 \qquad c(x) = c_s x^s + \cdots + c_0$$

where we may assume integer coefficients throughout. Since $a_0 = b_0 c_0$ and p divides a_0 but p^2 does not, only one of the pair of integers b_0, c_0 can be divisible by p. Without loss of generality, suppose $p \mid c_0$ but $p \nmid b_0$. Similarly, we have $a_n = b_r c_s$, and since $p \nmid a_n$ we must have $p \nmid b_r$ and $p \nmid c_s$.

We have that p divides the constant coefficient but not the leading coefficient of c. So there must be some first coefficient c_k with $p \nmid c_k$. Looking at the corresponding coefficient in a, we have

$$a_k = b_0 c_k + b_1 c_{k-1} + \cdots + b_k c_0$$

using the collecting formula of Section 5.1. On the right-hand side every term but the first is divisible by p; a_k also is divisible by p. It must follow that $b_0 c_k$ is also divisible by p. But this is impossible, since $p \nmid b_0$ and $p \nmid c_k$, and this contradiction completes the proof. □

Eisenstein's result is used in an obvious way. If we happen to have a polynomial $a(x)$ meeting the given criterion, then it would be a waste of time to attempt to find a factorization—there simply aren't any! Noting the computational time involved in determining this same information by running Kronecker's algorithm, we are indeed grateful to know as much through a more elementary analysis.

Example 5.41 Consider the polynomial

$$a(x) = x^3 + 2x^2 + 4x + 2$$

and note that the prime $p = 2$ divides all the coefficients except the leading coefficient, whereas $p^2 = 4$ does not divide the constant coefficient. It follows from Theorem 5.8 that $a(x)$ is irreducible. We needn't even try to factor it. ■

EXERCISES 5.4

1 Perform the following long divisions.

(a) $x^4 - 4x^3 + 2x^2 - 6x + 7$ by $x^2 - 2x + 6$

(b) $x^5 + 2x^3 + 7x^2 - 9x + 2$ by $x^2 + 4x - 9$

(c) $3x^5 - 5x^4 + 2x^3 - 4x^2 + 2x - 1$ by $x^3 + 2x^2 - x + 1$

(d) $2x^6 + 2x^4 - x^3 + 6x^2 - 5x + 3$ by $x^3 + 2x^2 + x - 3$

(e) $x^5 + 4x^4 - 5x^3 + 9x^2 - 7x - 2$ by $x^3 - 5x + 2$

(f) $5x^7 + 2x^4 - 5x^3 + 10x^2 + 13x + 1$ by $x^3 + 2x - 8$

2 Determine the output of division $(a, b\colon q, r)$ if we are given the following inputs.

(a) $a(x) = x^4 - 4x^3 + 2x^2 - 6x + 7$, $b(x) = x^2 - 2x + 6$

(b) $a(x) = x^5 + 2x^3 + 7x^2 - 9x + 2$, $b(x) = x^2 + 4x - 9$

(c) $a(x) = 3x^5 - 5x^4 + 2x^3 - 4x - 1$, $b(x) = x^3 + 2x + 1$

(d) $a(x) = 2x^6 + 2x^4 - x^3 + 6x - 1$, $b(x) = x^3 + x^2 - 3$

3 Determine the output of interpolate $(p_0, p_1, p_2, p_3\colon a)$ if we are given the following inputs.

(a) $p_0 = (-2, 5)$, $p_1 = (3, 3)$, $p_2 = (5, 9)$, $p_3 = (9, -7)$

(b) $p_0 = (-5, -5)$, $p_1 = (-1, 3)$, $p_2 = (0, -1)$, $p_3 = (8, 8)$

(c) $p_0 = (-8, 9)$, $p_1 = (-2, 2)$, $p_2 = (-1, -1)$, $p_3 = (7, 7)$

(d) $p_0 = (1, 3)$, $p_1 = (3, 5)$, $p_2 = (5, 1)$, $p_3 = (9, -4)$

4 Determine the value of eval (a, x) if we are given the following inputs.

(a) $a(x) = 3x^4 - 5x^3 + 2x^2 - 4x + 7$, $x = -1$

(b) $a(x) = -2x^3 + 4x^2 - 6x + 9$, $x = -2$

(c) $a(x) = x^5 + x^3 - 2x^2 + 4x + 5$, $x = 3$

(d) $a(x) = x^4 + 7x^2 - 4x + 7$, $x = 4$

5 Determine the value of factors (y) if we are given the following inputs.

(a) $y = 36$ **(b)** $y = 60$ **(c)** $y = 47$ **(d)** $y = 100$

6 Determine the (integer) outputs of euclid $(a, b: s, t)$ if we are given the following (integer) inputs.

(a) $a = 252, \quad b = 44$ **(b)** $a = 1844, \quad b = 36$

(c) $a = 460, \quad b = 24$ **(d)** $a = 226, \quad b = 12$

7 Determine the (polynomial) outputs of euclid $(a, b: s, t)$ if we are given the following (polynomial) inputs.

(a) $a(x) = x^3 + 2x^2 - 7x + 3, \quad b(x) = x^2 + 2x - 1$

(b) $a(x) = 2x^3 + 2x^2 - 3x - 8, \quad b(x) = 4x^2 - 7x - 3$

(c) $a(x) = x^4 - 2x^3 + 4x^2 - 2x - 5, \quad b(x) = x^2 - 2x + 5$

(d) $a(x) = x^4 + x^3 - x^2 - 2x - 2, \quad b(x) = x^2 + x + 1$

8 Why would it not be possible to include a root in the set X (the one used in Kronecker's factorization algorithm)?

9 Explain the reason for the particular choice of the points p_i used in Kronecker's factorization algorithm.

10 Why does the outer for loop of Kronecker's factorization algorithm begin at $\lceil n/2 \rceil$?

11 Suppose we never have $r = 0$ in the entire execution of Kronecker's factorization algorithm. What can we then conclude about the input polynomial $a(x)$?

12 How would you change Kronecker's factorization algorithm, factor, so as to exhibit its intended recursive character explicitly?

13 Use Kronecker's algorithm factor to obtain the complete factorization of the following polynomials.

(a) $a(x) = x^5 - 2x^4 - 4x^3 - 14x^2 - 5x - 12$

(b) $a(x) = x^3 - 3x^2 + 4x + 1$

(c) $a(x) = x^3 + 2x + 3$

(d) $a(x) = x^4 + x^3 + 2x^2 + 2x + 1$

(e) $a(x) = x^4 + x^3 - x^2 - 2x - 2$

(f) $a(x) = x^5 - 2x^3 + 3x^2 - 8x - 12$

14 Either decide (using Eisenstein's criterion) that the following polynomials are irreducible, or try to find a factorization (using Kronecker's algorithm).

(a) $x^4 + 5x^3 + 10x^2 + 10x + 35$ **(b)** $x^3 + 6x^2 + 5x + 25$

(c) $x^5 + x^4 + x^3 + x^2 + x + 1$ **(d)** $2x^4 - 3x^3 + 9x^2 - 6x + 15$

(e) $6x^4 - 5x^3 + 10x^2 - 5x + 20$ **(f)** $x^4 - 6x^3 + 2x^2 - 4x + 12$

In any case, decide whether the polynomials are irreducible or not, and state your reasons.

15 Show that in substituting $y = x + c$ for some constant c, the irreducibility of $a(x + c)$ implies the irreducibility of $a(x)$.

16 Find an appropriate substitution (see Exercise 15) to show that the following polynomials are irreducible.

(a) $a(x) = x^4 + 4x + 1$ **(b)** $a(x) = x^4 + 2x^2 - 1$

(c) $a(x) = x^3 - 3x + 1$ **(d)** $a(x) = x^4 - 10x^2 + 1$

(e) $a(x) = x^4 + 1$ **(f)** $a(x) = 5x^4 - 4x^3 + 2x^2 - 20x + 30$

17 Use Kronecker's algorithm to show that each of the polynomials of Exercise 16 is irreducible.

18 Write the formal description of the algorithm division.

19 Write the formal description of the revised Euclidean algorithm euclid.

20 Write the formal algorithmic description of the function factors.

21 Write the formal algorithmic description of the function eval.

22 Write the formal description of the algorithm interpolate.

23 Show that the algorithm division (see Exercise 18) is of complexity order n.

24 Estimate the computational complexity of the algorithm factors (see Exercise 20).

25 Show that the algorithm eval (see Exercise 21) is of complexity order n.

Sec. 5.5 *RATIONAL FUNCTIONS*

Rational functions are to polynomials as rational numbers are to integers.

In these next two sections we examine several interesting topics that depend on a knowledge of polynomial algebra. In this sense they might be considered applications of all that we have learned thus far. Moreover these applications will overlap a number of subject areas of the traditional calculus. Therefore these two sections might be viewed as enrichment modules that help bring some of these ideas from calculus into a sharper focus. It is not necessary that you have already studied calculus in order to appreciate our discussions here—of course, having done so wouldn't hurt either. Such a background certainly would help to make our presentation seem all the more meaningful. But we are quite confident that our treatment of these topics will stand on its own, irrespective of one's point of reference.

Goals

After studying this section you should be able to:

- Obtain the limiting behavior of a rational function as its argument becomes large, in either the negative or the positive direction.
- Determine the y-intercept of a rational function.
- Find the *roots* and *poles* of a rational function, each identified in terms of their multiplicity.
- Describe the differing functional behavior surrounding roots and poles, depending upon the oddness or evenness of their multiplicity.
- Use all available information to obtain an adequate sketch of the graph of a rational function.
- Separate rational functions having relatively prime factors as denominators.
- Obtain the *partial fraction expansion* of a given rational function.

Graphs of Rational Functions

As we indicated in this section's opening headline, a **rational function** $f(x)$ is simply a quotient of two polynomials. Thus we may write

$$f(x) = \frac{a(x)}{b(x)}$$

for polynomials a and b. If necessary for the discussion at hand, we can always assume that $\deg(a) < \deg(b)$. That is because of the division algorithm (i.e., we are free to write)

$$a(x) = b(x)q(x) + r(x)$$

for polynomials q and r, where $\deg(r) < \deg(b)$. Dividing by $b(x)$ we then have

$$f(x) = \frac{a(x)}{b(x)} = \frac{b(x)q(x) + r(x)}{b(x)} = q(x) + \frac{r(x)}{b(x)}$$

and since the methods for the analysis of polynomials are already well understood, we might as well assume that we are to analyze the rational function r/b rather than a/b.

Thus, suppose it is our task to study the means for obtaining the graph of a given rational function $f(x) = a(x)/b(x)$, where

$$a(x) = a_n x^n + \cdots + a_1 x + a_0 \qquad b(x) = b_m x^m + \cdots + b_1 x + b_0$$

that is, $\deg(a) = n$ and $\deg(b) = m$. In analogy to Table 5.5 of Section 5.2 (used in describing the limiting behavior of a polynomial) we then might arrive at the analysis of the limiting behavior of the rational function $f(x)$ as shown in Table 5.9. Again, each of these new tabulations has two entries at each position, showing the behavior as x becomes large negative (the left-hand entry) and large positive (the right-hand entry), respectively.

We make the same interpretations on the left-hand side, when $n > m$, as in our earlier table for polynomials (see Table 5.5), inasmuch as $f(x)$ becomes numerically infinite as x becomes large negative or large positive. On the right-hand side, however, where $m > n$, we find that a new symbolism is necessary. As x becomes large numerically, one sees that $f(x)$ *tends to 0 from above* (↓) *or from below* (↑); this is the meaning of the arrows. Moreover, we have to allow for the special

TABLE 5.9

		$n - m > 0$	
		Odd	*Even*
a_n/b_m	<0	$+\,-$	$-\,-$
	>0	$-\,+$	$+\,+$

		$m - n > 0$	
		Odd	*Even*
a_n/b_m	<0	↓↑	↑↑
	>0	↑↓	↓↓

case $n = m$, where we see that $f(x)$ approaches the value a_n/b_m as x becomes large in either direction. Parenthetically, we note that the corresponding ratio a_0/b_0 of the constant coefficients is the y-intercept of $f(x)$, whatever the values of n and m.

In principle one can then use whichever of these new tables is appropriate in determining the limiting behavior of a given rational function $f(x) = a(x)/b(x)$. But, alternatively, in consideration of our above discussion, we may focus only on the right-hand table if we are willing to replace $a(x)$ by $r(x)$, and then to add in the behavior of the polynomial $q(x)$ as an afterthought.

Example 5.42 Suppose we are asked to obtain the graph of the rational function:

$$f(x) = \frac{x^5 - 3x^4 - 5x^3 + 15x^2 + 5x - 12}{x^2 - 4}$$

Undoubtedly it is easier first to apply the division algorithm, in order to express $f(x)$ in the form:

$$f(x) = x^3 - 3x^2 - x + 3 + \frac{x}{x^2 - 4}$$

$$= (x + 1)(x - 1)(x - 3) + \frac{x}{(x + 2)(x - 2)}$$

as the sum of a polynomial and a new reduced rational function a/b (with the same denominator), where $\deg(a) < \deg(b)$. For this rational function, the lower left-hand entry of Table 5.9 (at the right) shows that the graph must tend to zero from below as x becomes large negative, and from above as x becomes large positive. Moreover, since the denominator has roots at $x = -2$ and $x = 2$, these will be points at which the quotient $a(x)/b(x)$ becomes infinite. The root (at $x = 0$) of the numerator will be a root of the quotient. More detailed analysis leads to Figure 5.8(b).

The graph of our polynomial is obtained according to the methods discussed earlier in the chapter, and the result is shown in Figure 5.8(a). Finally, we have

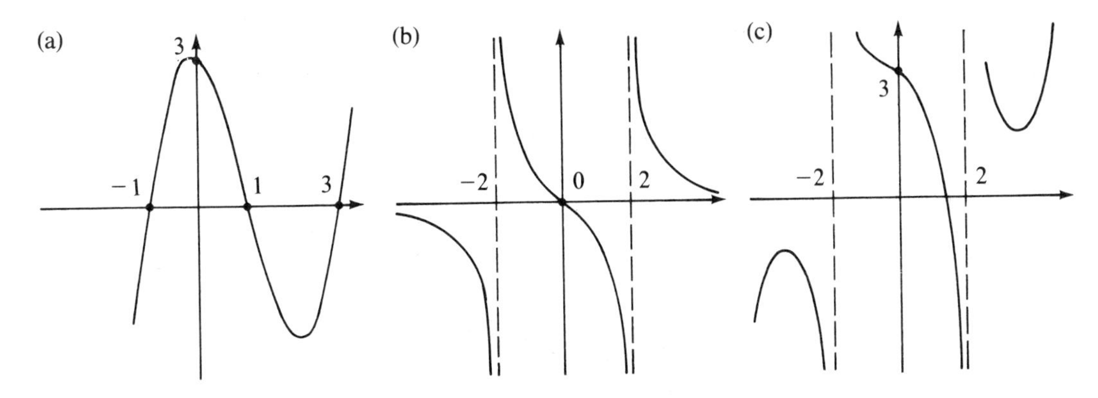

FIGURE 5.8

only to add our two sketches together in order to obtain the graph of the given function $f(x)$ as shown in Figure 5.8(c). ■

We already have encountered two fundamental principles here. Suppose $f(x) = a(x)/b(x)$, and suppose further that a and b have no common roots—otherwise, they could have been cancelled. Then

---

A root of $a(x)$ is a root of $f(x)$.
A root of $b(x)$ is a *pole* of $f(x)$.

---

that is, a point at which $f(x)$ becomes numerically infinite. These observations are of immense help in obtaining the graph of the given rational function $f(x)$, as has been demonstrated already.

Multiplicity of Roots and Poles

Of equal importance is a knowledge of the multiplicity of any roots or poles for the rational function $f(x) = a(x)/b(x)$. Again we suppose that a and b have no common roots. If $(x - x_0)^s$ is a factor of $a(x)$, and provided that this cannot be said of a higher power of $x - x_0$, then x_0 is said to be a *root of multiplicity s* for the function $f(x)$, or for the polynomial $a(x)$. Similarly, if $(x \quad x_0)^s$ is a factor of $b(x)$, but this is true of no higher power of $x - x_0$, then x_0 is said to be a *pole of multiplicity s* for the function $f(x)$. A definite distinction can then be drawn between the behavior surrounding a zero or pole, depending on the oddness or evenness of the multiplicity.

At a root of odd multiplicity the function takes on values of opposite sign to either side, whereas with an even multiplicity these values are of the same sign. Why? Similarly, at a pole of odd multiplicity, the function becomes large positive to one side of the pole but large negative to the other side, whereas with an even multiplicity the values are of the same sign (but still tending to infinity) to either side. Again—why?

Example 5.43 If we modify the reduced rational function of Example 5.42, so that we are looking instead at

$$\frac{x}{x^2 - 4x + 4} = \frac{x}{(x - 2)^2}$$

then we have a pole of multiplicity 2 at $x = 2$. Accordingly the function will tend to an infinity of the same sign on both sides of $x = 2$, and we obtain the graph of Figure 5.9, one that should be compared with Figure 5.8(b). ■

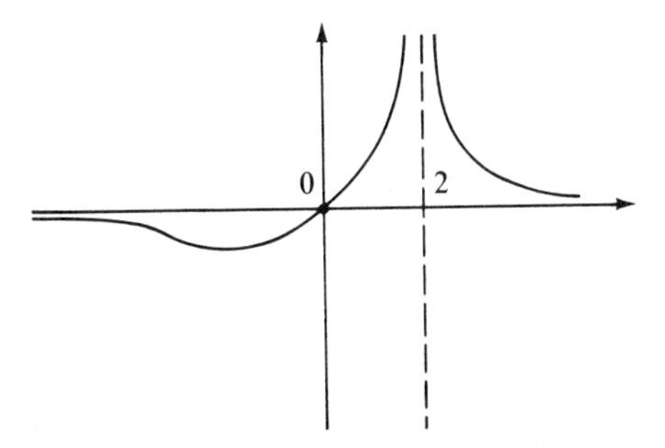

FIGURE 5.9

Example 5.44 Suppose we modify the polynomial $a(x)$ of Example 5.29, writing

$$a(x) = (x + 1)(x - 1)(x - 3)^2$$

so as to create a root of multiplicity 2 at $x = 3$. The graph of $a(x)$ then takes the form shown in Figure 5.10. And, of course, this should be compared with the original graph in Example 5.29 (Fig. 5.8(a)—the same polynomial). Now we find that $a(x)$ takes on values of the same sign (here, positive) on either side of the root at $x = 3$. Moreover, the resulting sketch is then consistent with Table 5.5 of Section 5.2, showing the expected limiting behavior. That is, we now have the limiting behavior "+ +", as predicted for polynomials of even degree with positive leading coefficient. ■

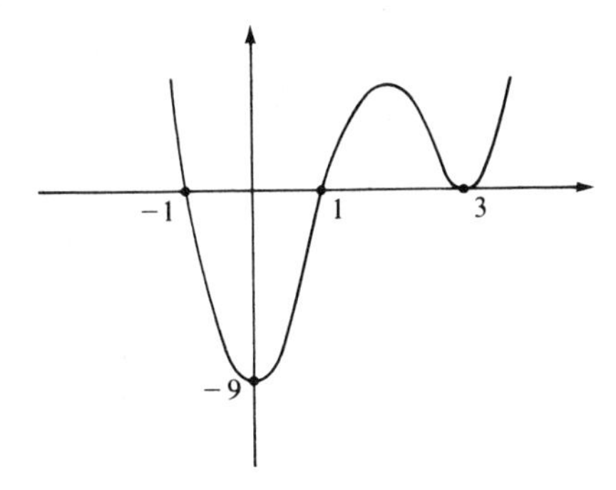

FIGURE 5.10

Powers as Denominators

Rational functions of the type encountered in Example 5.43 appear quite frequently in the applications. That is, we often find that

$$f(x) = \frac{a(x)}{b(x)} = \frac{a(x)}{p(x)^m}$$

where the denominator is a *power* of some polynomial $p(x)$. Again, however, we may assume that $\deg(a) < \deg(b)$. There is then a simple reduction technique for expressing $f(x)$ as a sum of more elementary rational functions, where the denominators are again powers of $p(x)$, but the numerators are of degree less than that of $p(x)$.

We only have to observe that upon dividing $a(x)$ by $p(x)$, we obtain:

$$a(x) = p(x)q(x) + r(x)$$

for polynomials q and r, where $\deg(r) < \deg(p)$. Dividing by $p(x)^m$, we may then write

$$f(x) = \frac{a(x)}{p(x)^m} = \frac{q(x)}{p(x)^{m-1}} + \frac{r(x)}{p(x)^m}$$

and we see that the same technique may then be applied to $q(x)$. Continuing in this fashion, we eventually obtain the *partial fraction expansion*:

$$f(x) = \frac{r_{m-1}(x)}{p(x)} + \cdots + \frac{r_1(x)}{p(x)^{m-1}} + \frac{r_0(x)}{p(x)^m}$$

with each $\deg(r_i) < \deg(p)$. It happens that the function $f(x)$ is often a good deal easier to analyze in this form. The whole procedure is then captured in the formal description of the accompanying algorithm *fraction*. We note that S stands for the sum, wherein the expansion is accumulated. In the last line we add any remaining polynomial $a(x)$ to the expansion. Of course, if we begin with the situation where $\deg(a) < \deg(b)$, then this polynomial will be zero.

```
algorithm fraction (a/p^m: S)

initialize S to 0
for k running from 0 to m − 1
   division (a, p: q, r)
   if r ≠ 0
      adjoin r/p^(m−k) to S
   replace a by q
add remaining a to S
```

***Example* 5.45** Let us consider the rational function

$$f(x) = \frac{a(x)}{b(x)} = \frac{x^7 - 3x^2 + 8}{(x^2 + 1)^3}$$

where, in fact, the condition: $\deg(a) < \deg(b)$ does not hold. A trace of the computation of the fraction algorithm then proceeds as shown in Table 5.10. Thus we see that S accumulates the sum

$$f(x) = x - \frac{3x}{x^2 + 1} + \frac{2x + 8}{(x^2 + 1)^3}$$

This is the desired partial fraction expansion of the original rational function. ■

TABLE 5.10

k	a	p	q	r
0	$x^7 - 3x^3 + 8$	$x^2 + 1$	$x^5 - x^3 - 2x$	$2x + 8$
1	$x^5 - x^3 - 2x$		$x^3 - 2x$	0
2	$x^3 - 2x$		x	$-3x$
	x			

Partial Fraction Expansion

We now intend to show that an arbitrary rational function $f(x) = a(x)/b(x)$ can be separated into a sum of rational functions with powers as their denominators. To each of these component functions, the construction of the fraction algorithm is then available. Altogether then we arrive at a generalized *partial fraction expansion* for the function $f(x)$. The expansion is of great importance in the theory of integration. The resulting representation consists of a polynomial, added to a sum of partial fractions $r(x)/p(x)^m$, where each $p(x)$ is irreducible, and where $\deg(r) < \deg(p)$. Each of the denominators $p(x)^m$ is a factor of the original denominator $b(x)$.

In carrying out the construction, we first will assume that the denominator has been factored, and after grouping common factors together we will have expressed $f(x)$ in the form:

$$f(x) = \frac{a(x)}{p_1(x)^{m_1} p_2(x)^{m_2} \cdots p_k(x)^{m_k}}$$

where each $p_i(x)$ is irreducible. The next step then depends on the following separating lemma:

If $b(x)$ and $c(x)$ are relatively prime polynomials, then

$$\frac{1}{b(x)c(x)} = \frac{s(x)}{c(x)} + \frac{t(x)}{b(x)}$$

for suitable polynomials $s(x)$ and $t(x)$.

Proof Since $\gcd(b, c) = 1$, we have only to choose s and t such that

$$sb + tc = 1$$

recalling our discussion of the main use of the revised Euclidean algorithm (in Section 5.4). It follows that

$$\frac{1}{b(x)c(x)} = \frac{s(x)b(x) + t(x)c(x)}{b(x)c(x)} = \frac{s(x)}{c(x)} + \frac{t(x)}{b(x)}$$

as required. □

Example 5.46 Consider the rational function

$$\frac{1}{(x-3)^2(x+2)}$$

with the relatively prime factors $b(x) = x^2 - 6x + 9$ and $c(x) = x + 2$. We have only to run the revised algorithm euclid (of Section 5.4) as shown in Table 5.11, in order to obtain the pair of functions:

$$s(x) = \frac{1}{25} \qquad t(x) = -\frac{1}{25}(x-8)$$

Note that it was necessary to divide r by 25 (and to do the same to s and t) at the conclusion of the run. In any case, we may then write

$$\frac{1}{(x-3)^2(x+2)} = \frac{\frac{1}{25}}{x+2} + \frac{-\frac{1}{25}x + \frac{8}{25}}{(x-3)^2}$$

This is the desired separation.

algorithm partial $(a/b\colon S)$

initialize S to 0
factor $(b\colon p_1^{m_1}, p_2^{m_2}, \ldots, p_k^{m_k})$
for i running from 1 to $k-1$
  set $c = p_i^{m_i}$ and cancel it from b
  euclid $(b, c\colon s, t)$
  fraction $(sa/c\colon S_i)$
  adjoin S_i to S
fraction $(ta/b\colon S_k)$
adjoin S_k to S

■

TABLE 5.11

b	c	q	r	s	t
				0	1
$x^2 - 6x + 9$	$x + 2$	$x - 8$	25	1	$-x + 8$
			1	$\frac{1}{25}$	$-\frac{1}{25}(x-8)$

This separation technique is the main new ingredient in the algorithm *partial* $(a, b\colon S)$, whose formal description is given above. The separation problem is modified there, however, by simply carrying along a numerator $a(x)$. We are given an arbitrary rational function $f(x) = a(x)/b(x)$ as input. The denominator is factored, and common factors are grouped together as explained earlier. The first power p^{m_1} in the denominator is taken as c and the product of the remaining factors is regarded as b in the modified separation problem. In thus obtaining the representation:

$$\frac{a}{bc} = \frac{sa}{c} + \frac{ta}{b}$$

we may apply the earlier fraction algorithm to the function sa/c (since it has a power as a denominator). We continue this separation technique, using each new ta/b as a starting point, until all the splitting has occurred and until each new sa/c and the final ta/b have been processed by the fraction algorithm. The resulting sum S (the partial fraction expansion of $(f(x))$ is the sum of all of the decompositions S_i generated by the calls to the fraction algorithm. An example will clarify the details of the procedure.

Example 5.47 Consider the rational function

$$f(x) = \frac{x+1}{x^4 - x^3 - x + 1} = \frac{x+1}{(x^2+x+1)(x-1)^2}$$

noting the factorization of the denominator, as would be obtained by an appeal to the factor algorithm. In the separation process we take

$$c(x) = x^2 + x + 1$$

and after cancelling c from the denominator we have

$$b(x) = x^2 - 2x + 1$$

These are the two polynomials required in the call to the algorithm euclid $(b, c: s, t)$. The execution of the algorithm then proceeds as shown in Table 5.12. Since $k = 2$ and

$$sa/c = \frac{\frac{1}{3}(x^2 + 2x + 1)}{x^2 + x + 1} \qquad ta/b = \frac{-\frac{1}{3}(x^2 - x - 2)}{(x-1)^2}$$

TABLE 5.12

b	c	q	r	s	t
				0	1
$x^2 - 2x + 1$	$x^2 + x + 1$	1	$-3x$	1	-1
$x^2 + x + 1$	$-3x$	$-\frac{1}{3}(x+1)$	1	$\frac{1}{3}(x+1)$	$-\frac{1}{3}(x-2)$

these will be our inputs to the fraction algorithm. The reader may check that the output sums are

$$S_1 = 1 + \frac{x}{x^2 + x + 1}$$

and

$$S_2 = -1 - \frac{1}{x-1} + \frac{2}{(x-1)^2}$$

respectively, except for the factor of $\frac{1}{3}$ in each case. It follows that the complete partial fraction expansion is as follows:

$$f(x) = \frac{1}{3}\left\{\frac{x}{x^2 + x + 1} - \frac{1}{x-1} + \frac{2}{(x-1)^2}\right\}$$

after reinserting the common factor $\frac{1}{3}$. ■

EXERCISES 5.5

1 Obtain the limiting behavior and the y-intercept (where appropriate) for each of the following rational functions.

(a) $\dfrac{x}{x^2 - 2x + 1}$ **(b)** $\dfrac{x^4 - 3x^2 + 5}{x^3 - 2x}$

(c) $\dfrac{x^2 - 1}{x^2}$ **(d)** $\dfrac{x^3}{x^2 - 4x - 5}$

2 Do the same (as in Exercise 1) for each of the following rational functions.

(a) $\dfrac{x^4 - 4x^3 + x^2 + 7x + 1}{x^2 - 3x}$ **(b)** $\dfrac{x^5 - 2x^3 + 3x^2 - 4x + 2}{x^2 - x - 2}$

(c) $\dfrac{2x^5 - 4x^3 + 7x^2 - 2x - 1}{x^2 - 6x + 9}$ **(d)** $\dfrac{x^6 - 3x^4 + 7x^3 - 5x^2 + 2x + 3}{x^4 - 6x^3 + 9x^2}$

3 Do the same (as in Exercise 1) for each of the following rational functions.

(a) $\dfrac{x^2 - x - 2}{x^2 - 6x + 9}$ **(b)** $\dfrac{x^3 - 3x^2 + 3x - 1}{x^4 - 4x^3 + x^2 + 6x}$

(c) $\dfrac{x^4 - 3x^3 + 2x^2 - x + 3}{x^2 - 1}$ **(d)** $\dfrac{x^6 - 3x^2 - 7x + 2}{x^4 - 4x^2 + 4}$

4 Describe the poles of the denominator, and roots of the appropriate polynomials, for each of the rational functions of Exercise 1.

5 Do the same (as in Exercise 4) for the rational functions of Exercise 2.

6 Do the same (as in Exercise 4) for the rational functions of Exercise 3.

7 Use all available information to obtain an adequate sketch of the rational functions of Exercise 1.

8 Do the same (as in Exercise 7) for the rational functions of Exercise 2.

9 Do the same (as in Exercise 7) for the rational functions of Exercise 3.

10 Explain why it is that at a root of odd multiplicity a function takes on values of opposite sign to either side, whereas with a root of even multiplicity these values are of the same sign.

11 Explain why it is that at a pole of odd multiplicity a rational function becomes large positive to one side of the pole but large negative to the other side, whereas with an even multiplicity the values are of the same sign (although still tending to infinity) to either side.

12 Obtain the partial fraction expansion of each of the following rational functions, using the algorithm fraction.

(a) $\dfrac{2x-7}{(x-3)^3}$ **(b)** $\dfrac{3x^2-x+2}{(x^2+x+1)^2}$

(c) $\dfrac{2x^2-1}{(x^2-2)^2}$ **(d)** $\dfrac{x-1}{x^3}$

13 Do the same (as in Exercise 12) for each of the following.

(a) $\dfrac{x^3-1}{(x^2+1)^3}$ **(b)** $\dfrac{x}{(x-1)^3}$ **(c)** $\dfrac{x^2+1}{x^4}$ **(d)** $\dfrac{x}{(x+2)^3}$

14 Apply the construction of the separating lemma to each of the following rational functions.

(a) $\dfrac{1}{(x-1)(x+2)}$ **(b)** $\dfrac{1}{(x^2+x+1)(x-3)}$

(c) $\dfrac{1}{(x^2-2)(x^2+2)}$ **(d)** $\dfrac{1}{(x^4+1)(x^2+x+1)}$

15 Obtain the partial fraction expansion of each of the following rational functions, using the algorithm partial.

(a) $\dfrac{x+1}{x^2(x-1)}$ **(b)** $\dfrac{x-1}{(x-2)^3x^2}$

(c) $\dfrac{x^2-3}{(x^2+x+1)^2(x-1)}$ **(d)** $\dfrac{x-2}{x^3(x+1)^2(x-1)^2}$

16 Do the same (as in Exercise 15) for each of the following.

(a) $\dfrac{1}{x^2(x+1)^2}$ **(b)** $\dfrac{x}{(x^2+2)(x^2+1)}$

(c) $\dfrac{x^2+x+1}{(x-1)(x+1)^2}$ **(d)** $\dfrac{x-1}{x^3(x+1)^2(x-1)}$

17 Give a detailed inductive proof showing the derivation of the partial fraction expansion for rational functions having powers as denominators.

18 Give a detailed inductive proof showing the derivation of the general partial fraction expansion for rational functions—the one achieved through the use of the algorithm partial.

19 Develop a method for representing any rational number as a sum of partial fractions of the form a/p^m, where each p is prime and $0 \le a < p$.

Sec. 5.6 DIFFERENCE AND SUMMATION CALCULUS

The difference and summation operations are inverse to each other.

The difference and summation operators play a role in discrete mathematics that is analogous to that of differentiation and integration in the classical calculus. They find applications to many combinatorial investigations. We do not survey all these applications here, however. For our purposes it is sufficient to merely scratch the surface, and the more detailed applications are then left to the student's further studies. Here we devote most of our attention to the difference operator, this being the easier to handle. In so doing we do not limit our attention to polynomial functions; again, we will find that polynomials indeed play a central role in the theory.

Goals

After studying this section, you should be able to:

- Apply the *difference operator* to arbitrary functions.
- State the rules for *differencing* a sum, difference, product, or quotient of functions.
- Define the *elementary factorial polynomials* and explain their significance in the difference calculus.
- Produce the *Newton Expansion Formula* and illustrate its application with examples.
- Illustrate the use of differences in the tabulation of polynomial values at integer points.
- Convert back and forth between factorial polynomials and standard polynomials.
- State the *fundamental theorem* of the difference and summation calculus, and show examples of its use.
- Describe the *summation operator* acting over a finite set of integer indices.
- Use the fact that differencing is the inverse operation to summation in order to compute sums of functions over integer indices.

Difference Operator

In this section we deal with arbitrary functions $f(x)$, but there is a tendency to look at their values only at a number of discrete points, usually points that are equally spaced. There is a definite practical setting where this point of view is appropriate. We may have taken measurements of a number of functional values for $f(x)$, as in some experimental situation. It is then quite often the case that we will have used an equal spacing in x. Indeed these measured values may represent all that we really know about the function.

Whether this is our setting or not, we may define a discrete analogue of the derivative as used in the classical calculus. Given the function $f(x)$ we introduce a **difference operation** (denoted by Δ, the Greek letter 'D') with the definition:

$$\Delta f(x) = f(x+1) - f(x)$$

Then it is clear that Δf is yet another function. It is convenient to write

$$\Delta f = f^+ - f$$

with the understanding that $f^+(x) = f(x+1)$, so that all is consistent.

Example 5.48 Suppose we first begin with a polynomial. Thus we let

$$f(x) = x^3 - 5x + 2$$

as a first example. In computing the difference we obtain

$$\begin{aligned}\Delta f(x) &= (x+1)^3 - 5(x+1) + 2 - (x^3 - 5x + 2)\\ &= 3x^2 + 3x - 4\end{aligned}$$

so that Δf is again a polynomial, but of degree one less than that of f. As a matter of fact we will find that this behavior is a general rule. ■

Example 5.49 Consider the function

$$f(x) = \log(x)$$

as introduced in Section 4.6. Its difference is easily computed; that is, we have

$$\begin{aligned}\Delta f(x) &= \log(x+1) - \log(x)\\ &= \log\frac{x+1}{x} = \log\left(1 + \frac{1}{x}\right)\end{aligned}$$

using a familiar property of the log function:

$$\log\frac{a}{b} = \log a - \log b$$

one that you have seen in high-school mathematics. ■

So far the difference operator is only just that, an operation that yields one function from another. It seems only to be a manipulation. In that sense we are again reminded of the situation where we first learn of the derivative in differential calculus. It takes some time before its significance can be appreciated. And the same is true here. Suppose, however, that $f(x)$ is a constant. Then it is clear that Δf will be identically zero. Conversely if $f(x)$ shows wide changes in its values, then these will be captured in Δf, as if we are obtaining a measure of the local variation of the given function. Again, for those of us who have seen differential calculus, the analogy to the role of the derivative is quite striking.

Before we can pursue the significance of the difference operator any further, however, we need to work on developing our manipulative skills so that these will not be a problem for us later on. In particular we need to learn of the rules for *differencing* sums, products, quotients, etc., and to see how they act as aids to the treatment of more complex functions.

Often a function $f(x)$ will be given by a quite complex expression, built up from simpler functions through the use of the usual algebraic operations of addition, subtraction, multiplication, and division. In such cases it is a definite advantage to be able to appeal to a set of general rules, showing the behavior of the difference operator with respect to the given operations. This behavior is summarized in the following collection of identities:

$$\Delta(f + g) = \Delta f + \Delta g \qquad (cf) = c\,\Delta f$$

$$\Delta(f \cdot g) = f^+\,\Delta g + g\,\Delta f \qquad (f/g) = \frac{g\,\Delta f - f\,\Delta g}{gg^+}$$

where c is a constant. Note that the first two identities provide a rule for handling the subtraction of two functions, that is, we obtain

$$\Delta(f - g) = \Delta f - \Delta g$$

after writing $f - g = f + (-1)g$.

Each of these identities is easily verified. For example, consider the case of multiplication. Using the definition of the difference operator we have

$$\begin{aligned}
\Delta(f \cdot g)(x) &= (f \cdot g)(x + 1) - (f \cdot g)(x) \\
&= f(x + 1)g(x + 1) - f(x)g(x) \\
&= f(x + 1)[g(x + 1) - g(x)] + g(x)[f(x + 1) - f(x)] \\
&= (f^+\,\Delta g)(x) + (g\,\Delta f)(x) \\
&= (f^+\,\Delta g + g\,\Delta f)(x)
\end{aligned}$$

as required. We leave the verification of the other identities as exercises for the reader.

Example 5.50 In Example 5.48 we could have viewed the polynomial as a sum and, similarly, each term could be viewed as a constant multiplying a power of x. With the use of our first two identities, together with the identity for handling subtractions, we then could compute:

$$\begin{aligned}
\Delta(x^3 - 5x + 2) &= \Delta(x^3) - \Delta(5x) + \Delta(2) \\
&= \Delta(x^3) - 5\,\Delta(x) + 0 \\
&= (x + 1)^3 - x^3 - 5[(x + 1) - x] \\
&= 3x^2 + 3x + 1 - 5 \\
&= 3x^2 + 3x - 4
\end{aligned}$$

Of course we obtain the same result as before. ■

Example 5.51 The identity for the division operation is particularly useful in handling rational functions, as we might expect. Thus if we are given the function

$$f(x) = \frac{x+1}{x^2}$$

we can compute the effect of the difference operation as follows:

$$\begin{aligned}\Delta\left(\frac{x+1}{x^2}\right) &= \frac{x^2\,\Delta(x+1) - (x+1)\,\Delta(x^2)}{x^2(x^2)^+} \\ &= \frac{x^2 \cdot 1 - (x+1)(2x+1)}{x^2(x+1)^2} \\ &= -\frac{x^2+3x+1}{x^2(x+1)^2}\end{aligned}$$

It appears that as a general rule differencing increases the complexity of a rational function, whereas we have seen that it has a simplifying effect in the case of polynomials. ■

Factorial Polynomials

In differential calculus, a special and very important role is played by the elementary polynomials x^n. In the difference calculus this role is taken over by another class of functions, the **elementary factorial polynomials**:

$$x^{(n)} = x(x-1)\cdots(x-n+1)$$

At first sight they seem a bit puzzling algebraically. But they are really quite easy to handle. In more detail we have, as the first few examples:

$$\begin{aligned}x^{(0)} &= 1 \qquad \text{(by convention)} \\ x^{(1)} &= x \\ x^{(2)} &= x(x-1) \\ x^{(3)} &= x(x-1)(x-2) \\ x^{(4)} &= x(x-1)(x-2)(x-3)\end{aligned}$$

etc., a sequence of polynomials $x^{(k)}$ of degree k, having roots at the nonnegative integers.

Using these elementary polynomials one can define general **factorial polynomials**:

$$a(x) = a_n x^{(n)} + a_{n-1}x^{(n-1)} + \cdots + a_1 x^{(1)} + a_0$$

in analogy with the standard ones we have studied in this chapter. In fact these polynomials are really only ordinary polynomials in a disguised notation. This is easy to see, once we notice that the elementary factorial polynomials are products that we can interpret as ordinary polynomials, once we carry out the multiplications.

Example 5.52 Consider the factorial polynomial

$$a(x) = x^{(3)} - 5x^{(1)} + 2$$

an expression that should not be confused with that of Example 5.50. Using the definition of the elementary factorial polynomials, we have:

$$\begin{aligned} x^{(3)} &= x(x-1)(x-2) = x^3 - 3x^2 + 2x \\ x^{(1)} &= x \end{aligned}$$

And we thus obtain

$$a(x) = x^3 - 3x^2 - 3x + 2$$

an ordinary polynomial. ■

The importance of the factorial polynomials stems from the simple relationship between the elementary factorial polynomials and their differences. Specifically, we have the following identity:

$$\Delta x^{(n)} = nx^{(n-1)}$$

one that is easily verified, as follows. We use the definition of the difference operator to compute:

$$\begin{aligned} \Delta x^{(n)} &= (x+1)^{(n)} - x^{(n)} \\ &= (x+1)x(x-1)\cdots((x+1)-n+1) \\ &\quad - x(x-1)\cdots(x-n+2)(x-n+1) \\ &= x(x-1)\cdots(x-(n-1)+1)[(x+1)-(x-n+1)] \\ &= nx^{(n-1)} \end{aligned}$$

as claimed above.

Considering the simplicity of this relationship, the following idea comes to mind. Suppose we are asked to apply the difference operation to a (standard) polynomial:

$$a(x) = b_n x^n + b_{n-1}x^{n-1} + \cdots + b_1 x + b_0$$

(reserving our a's as coefficients of factorial polynomials). Surely this can be done, as shown in Example 5.50. But there might be an easier way. If we could convert the standard polynomial to a factorial polynomial—

$$a(x) = a_n x^{(n)} + a_{n-1} x^{(n-1)} + \cdots + a_1 x^{(1)} + a_0$$

then we could use the properties of the difference operator and the above relationship to compute:

$$\begin{aligned} \Delta a(x) &= a_n \, \Delta x^{(n)} + a_{n-1} \, \Delta x^{(n-1)} + \cdots + a_1 \, \Delta x^{(1)} + a_0 \, \Delta 1 \\ &= n a_n x^{(n-1)} + (n-1) a_{n-1} x^{(n-2)} + \cdots + a_1 \cdot 1 + 0 \\ &= n a_n x^{(n-1)} + (n-1) a_{n-1} x^{(n-2)} + \cdots + 2a_2 x^{(1)} + a_1 \end{aligned}$$

If desired, one then could reconvert this expression into the form of a standard polynomial. But again this would require the existence of a simple conversion technique, from the one kind of polynomial to the other. Such conversion techniques do in fact exist, and they can be most helpful in these and other situations. We return to a study of such *polynomial conversion problems* later in this section.

Example 5.53 If we are given the factorial polynomial

$$a(x) = 2x^{(4)} + 9x^{(3)} + 6x^{(2)} - 1$$

then

$$\begin{aligned} \Delta a(x) &= 2 \cdot 4x^{(3)} + 9 \cdot 3x^{(2)} + 6 \cdot 2x^{(1)} \\ &= 8x^{(3)} + 27x^{(2)} + 12x^{(1)} \end{aligned}$$

an easy computation indeed. ■

Example 5.54 Similarly, if we are given

$$a(x) = x^{(4)} - 5x^{(3)} + 3x^{(2)} - 4x^{(1)} + 7$$

then

$$a(x) = 4x^{(3)} - 15x^{(2)} + 6x^{(1)} - 4$$

and again, what could be easier? ■

Newton Expansion

We can expand upon the question we have just considered. Suppose we try to express an arbitrary function $f(x)$ in terms of a linear combination of the elementary factorial functions. Those who have studied the classical calculus will be quick to point out that something like this can be done for the standard (elementary) polynomials: It goes by the name *Taylor expansion*, but it will be recalled that an infinite series is required, in general. Recognizing this, we might try to write

$$f(x) = \sum_{k=0}^{\infty} a_k x^{(k)}$$

using an infinite linear combination of the elementary factorial polynomials as a basis for the representation.

Let us just suppose that such a representation is possible for the arbitrary function $f(x)$. Since every factorial $x^{(k)}$ except $x^{(0)}$ has x as a factor, if we substitute the value $x = 0$ on both sides we find that

$$f(0) = a_0$$

of necessity. If we then take differences of both sides, we obtain:

$$\Delta f(x) = \sum_{k=0}^{\infty} a_k \, \Delta x^{(k)} = \sum_{k=1}^{\infty} k a_k x^{(k-1)}$$

and, again, when we substitute $x = 0$, all but the first term of our summation will disappear, leaving

$$\Delta f(0) = a_1$$

With subsequent applications of the difference operator and with the resulting substitutions at $x = 0$ we find that altogether we have:

$$\begin{aligned} f(0) &= a_0 \\ \Delta f(0) &= a_1 \\ \Delta^2 f(0) &= 2a_2 \\ \Delta^3 f(0) &= 3 \cdot 2a_3 = 6a_3 \\ &\cdots \\ \Delta^k f(0) &= k! a_k \\ &\cdots \end{aligned}$$

where the notation Δ^k simply denotes k applications of the difference operator. Solving for the undetermined coefficients a_k we are thus free to write:

$$f(x) = \sum_{k=0}^{\infty} \frac{\Delta^k f(0)}{k!} x^{(k)}$$

and the resulting representation is called the **Newton expansion** of the given function $f(x)$. We will find that this expansion has a number of interesting applications.

Example 5.55 Let us first consider the Newton formula in the light of the above conversion problem, in particular, that of converting a standard polynomial to a factorial polynomial. Suppose we consider the ordinary polynomial

$$f(x) = x^3 - 3x^2 - 3x + 2$$

one that we quickly recognize as the answer from Example 5.52. This being the case we will have a good check on our computations. In any event we may systematically tabulate differences and the corresponding substitutions at $x = 0$ as follows:

$$\begin{aligned}
\Delta^0 f(x) &= f(x) \\
&= x^3 - 3x^2 - 3x + 2 \qquad & \Delta^0 f(0) = 2 \\
\Delta^1 f(x) &= \Delta f(x) \\
&= 3x^2 - 3x - 5 \qquad & \Delta^1 f(0) = -5 \\
\Delta^2 f(x) &= \Delta(\Delta f(x)) \\
&= 6x \qquad & \Delta^2 f(0) = 0 \\
\Delta^3 f(x) &= \Delta(\Delta^2 f(x)) \\
&= 6 \qquad & \Delta^3 f(0) = 6
\end{aligned}$$

and $\Delta^k f(0) = 0$ for $k > 3$. If we divide by the appropriate factorials as we substitute in the Newton formula, we obtain the finite summation:

$$\begin{aligned}
f(x) &= f(0) + \Delta f(0) x^{(1)} + \frac{\Delta^2 f(0)}{2!} x^{(2)} + \frac{\Delta^3 f(0)}{3!} x^{(3)} \\
&= 2 - 5x^{(1)} + 0 + x^{(3)} \\
&= x^{(3)} - 5x^{(1)} + 2
\end{aligned}$$

Thus we have performed the polynomial conversion:

$$x^3 - 3x^2 - 3x + 2 \to x^{(3)} - 5x^{(1)} + 2$$

and we recognize this as being the correct answer in reviewing Example 5.52. ■

We have just seen that our expansion formula can be of use in addressing the polynomial conversion problem. Another important application of the Newton formula arises in the case of measured data at equally spaced intervals. Often we would like to fit a polynomial through a series of data points, to interpolate as we say. If these correspond to integer points on the x-axis, then the Newton formula is easily adapted to the problem. (If not, then the formula may still be of use, but one may first have to perform a rather complicated normalization—we do not investigate these details here.) An example should serve to clarify this particular application.

Example 5.56 Let us suppose that we have accumulated the measurements shown in Table 5.13. The interpretation of the variables x and y need not concern

TABLE 5.13

x	y
0	41
1	43
2	47
3	53
4	61
5	71
6	83
7	97

us here. But, in any case, had we seen this problem in Section 5.3 our first inclination would be to try to interpolate the data with a seventh degree polynomial, using the Lagrange Formula. In computing a few differences and making use of the Newton Formula instead, we see that this first approach would have been wholly inappropriate. We consider the *difference table* shown in Table 5.14. This shows that a polynomial $y = a(x)$ interpolating the given data must satisfy:

$$a(0) = 41$$
$$\Delta a(0) = 2$$
$$\Delta^2 a(0) = 2$$

whereas we take $\Delta^k a(0) = 0$ for $k > 2$. It follows that the data are interpolated by the second-degree polynomial

$$\begin{aligned} a(x) &= \frac{41}{0!}x^{(0)} + \frac{2}{1!}x^{(1)} + \frac{2}{2!}x^{(2)} \\ &= x^{(2)} + 2x^{(1)} + 41 \\ &= x(x-1) + 2x + 41 \\ &= x^2 + x + 41 \end{aligned}$$

as the reader may easily check. ■

TABLE 5.14

x	y	Δy	$\Delta^2 y$	$\Delta^3 y$
0	41	2	2	0
1	43	4	2	0
2	47	6	2	0
3	53	8	2	0
4	61	10	2	0
5	71	12	2	
6	83	14		
7	97			

Polynomial Conversion

We already have introduced the polynomial conversion problem, and considering the fact that the standard polynomials are useful for some purposes, whereas the factorial polynomials are more appropriate for others, the importance of the problem is easily appreciated. One method for converting from the standard to the factorial representation has already been given in connection with the use of the Newton Formula. This method suffers from the disadvantage that one must compute nth differences in order to perform the conversion of an nth-degree polynomial. One would hope that a more efficient method would be available.

Suppose again that we are given a polynomial, as represented in the standard form:

$$a(x) = b_n x^n + b_{n-1} x^{n-1} + \cdots + b_1 x + b_0$$

and we are trying to find the coefficients a_k of the factorial representation. We are thus led to set up the equality:

$$\begin{aligned} a(x) &= b_n x^n + \cdots + b_1 x + b_0 = a_n x^{(n)} + \cdots + a_1 x^{(1)} + a_0 \\ &= a_n x(x-1)\cdots(x-n+1) + \cdots + a_2 x(x-1) + a_1 x + a_0 \end{aligned}$$

where we make use of the definition of the elementary factorial functions in the last line. In noting that all but the constant coefficient here have x as a factor, we obtain a_0 as a remainder if we divide $a(x)$ by x; that is, we have

$$a(x) = xq_0(x) + a_0$$

in a direct application of the division algorithm. Moreover, each of the terms in $q_0(x)$ has $x - 1$ as a factor, except for the new constant term a_1. The argument continues with each q_{k-1} and $x - k$, regarding the coefficient a_k. Thus, altogether we may write:

$$\begin{aligned} a(x) &= xq_0(x) + a_0 \\ q_0(x) &= (x-1)q_1(x) + a_1 \\ q_1(x) &= (x-2)q_2(x) + a_2 \\ &\cdots \end{aligned}$$

and this suggests that we may obtain each of the desired coefficients a_k by successive synthetic divisions by $x, x - 1, x - 2$, etc. These coefficients then will appear in ascending order as the remainders, one by one. An example will clarify the method.

Example 5.57 Consider the standard polynomial

$$a(x) = 2x^4 - 3x^3 + x^2 - 1$$

and let us set up the synthetic divisions in the following tabular form:

$$
\begin{array}{r|rrrrr}
1 & 2 & -3 & 1 & 0 & \boxed{-1} \\
  &   & 2 & -1 & 0 & \\
2 & 2 & -1 & 0 & \boxed{0} & \\
  &   & 4 & 6 & & \\
3 & 2 & 3 & \boxed{6} & & \\
  &   & 6 & & & \\
4 & 2 & \boxed{9} & & & \\
  & \boxed{2} & & & &
\end{array}
$$

The coefficients of the equivalent factorial polynomial then may be read as the diagonal remainders. And thus we have performed the conversion:

$$2x^4 - 3x^3 + x^2 - 1 \rightarrow 2x^{(4)} + 9x^{(3)} + 6x^{(2)} + 0x^{(1)} - 1$$

using quite an efficient computational technique. ■

The reverse conversion, from factorial to standard polynomials, is quite straightforward. We may simply appeal to the definition of the elementary factorial polynomials, then multiply out all the factors. More efficient methods do exist but a discussion of these would take us too far afield.

The Fundamental Theorem

You no doubt have observed that it is rather difficult to deal with mathematical summations. Part of the reason for this is that we seldom know just what the sum represents. We are unable to obtain a *closed form* representation. And yet this is precisely where the difference (and summation) calculus finds its greatest use.

We suppose that we are given a function $f(x)$ taking values at the nonnegative integers x. Except for our calling the sequence a function, together with a change in starting point and a change to x as the summation index, the situation then is identical to that first encountered in Section 3.2. Again the summation

$$\sum_{x=0}^{n} f(x)$$

means nothing but what it says, the sum of the values of $f(x)$ for each integer x running from 0 to n. Algebraically we thus have

$$\sum_{x=0}^{n} f(x) = f(0) + f(1) + f(2) + \cdots + f(n)$$

showing that summing from 0 to n is *an operator*. By this we mean that in applying the summation to $f(x)$ we obtain another function $g(n)$—a function, however, with a different argument. But again this is sometimes less than we would like to know. We often would like to think that there might be some way of summing the terms and obtaining a cleaner result.

It happens that we can do precisely this, provided that the function being summed can be expressed as a difference. Then a simple closed form representation for the sum can be obtained by an appeal to the following:

Theorem 5.9

Fundamental Theorem of the Difference and Summation Calculus Suppose the function $f(x)$ can be expressed as a difference, that is, $f(x) = \Delta F(x)$ for another function $F(x)$. Then

$$\sum_{x=0}^{n} f(x) = F^{+}(n) - F(0)$$

Proof We simply compute

$$\begin{aligned}\sum_{x=0}^{n} f(x) = \sum_{x=0}^{n} \Delta F(x) &= \sum_{x=0}^{n} [F(x+1) - F(x)] \\ &= F(n+1) - F(n) + F(n) - F(n-1) \\ &\quad + \cdots - F(1) + F(1) - F(0) \\ &= F(n+1) - F(0) \\ &= F^{+}(n) - F(0)\end{aligned}$$

noting the cancellation of all but the two extreme terms. □

It is in the sense of this theorem that the difference and summation operations are inverse to one another (but see also Exercise 23 in Exercises 5.6). Thus, when written entirely in terms of the function $F(x)$, the fundamental theorem may be written in the form:

$$\sum_{x=0}^{n} \Delta F(x) = F^{+}(n) - F(0)$$

as if the summation and difference operations cancel one another out. This is a very suggestive way of looking at things, as the reader will discover in examining Exercise 23.

Example 5.58 We will use the fundamental theorem to derive the following closed form representation for the sum of the elementary factorial functions:

$$\sum_{x=0}^{n} x^{(k)} = \frac{(n+1)^{(k+1)}}{k+1}$$

for each $k \geq 0$. The parameter k causes the sum to depend on both n and k. For each fixed k, however, the summation is merely a function of n, as is ordinarily the case. In obtaining the given result we will make use of the special relationship between the elementary factorial functions and their differences in writing:

$$\sum_{x=0}^{n} x^{(k)} = \sum_{x=0}^{n} \Delta \frac{x^{(k+1)}}{k+1} = \frac{(n+1)^{(k+1)}}{k+1} - \frac{0^{(k+1)}}{k+1}$$

$$= \frac{(n+1)^{(k+1)}}{k+1}$$

We note that this representation is valid for all $k \geq 0$, and its usefulness is clarified in the examples that follow. ■

Example 5.59 Here we already know the answer (Example 3.16), but let us use the result from Example 5.58 to compute the sum of the first n positive integers. With $k = 1$ we have

$$\sum_{x=0}^{n} x = \sum_{x=0}^{n} x^{(1)} = \frac{(n+1)^{(2)}}{2} = \frac{(n+1)n}{2}$$

noting that the $x = 0$ term was included for consistency. ■

Example 5.60 A more revealing application of Example 5.58 can be obtained if we instead compute the sum of the first n squares (see Exercise 1 of Exercises 3.2). For then we will have $k = 2$. We first will need to express x^2 as a sum of factorial polynomials. Having done this we then obtain the following:

$$\sum_{x=0}^{n} x^2 = \sum_{x=0}^{n} (x^{(2)} + x^{(1)}) = \frac{(n+1)^{(3)}}{3} + \frac{(n+1)n}{2}$$

$$= \frac{(n+1)n(n-1)}{3} + \frac{(n+1)n}{2}$$

$$= \frac{(n+1)n(2n+1)}{6}$$

as yet another application of the general formula obtained in Example 5.58. ■

Example 5.61 As a final direct application of the fundamental theorem, we note that $\Delta 2^x = 2^x$ (see Exercise 4 in Exercises 5.6). It follows that we are able to obtain a closed form for the summation of terms 2^x quite easily. We have

$$\sum_{x=0}^{n} 2^x = \sum_{x=0}^{n} \Delta 2^x = 2^{n+1} - 2^0 = 2^{n+1} - 1$$

and we note that this is a result that is often seen (see Exercise 4 of Exercises 3.2) and is often used. ■

EXERCISES 5.6

1 Compute the differences of the following polynomials.

(a) $x^3 + 3x^2 - 2x + 7$ **(b)** $4x^2 - 5x - 7$

(c) $2x^2 - 3x + 9$ **(d)** $x^3 - 4x^2 + x - 1$

2 Compute the differences of the following rational functions.

(a) $\dfrac{x^2+3}{x}$ **(b)** $\dfrac{x^2-x+1}{x-2}$ **(c)** $\dfrac{x}{x^2+2x}$ **(d)** $\dfrac{3x}{x-1}$

3 Compute the differences of the following functions.

(a) $x + \dfrac{1}{x}$ **(b)** 2^x **(c)** $\arctan(x)$ **(d)** $\log(x+1)$

4 Show that $\Delta 2^x = 2^x$.

5 Show that $\Delta(\arctan x) = 1/(1 + x^+x)$.

6 Show that $\Delta \log(x)$ approaches 0 as x becomes large.

7 Verify the following rules.

(a) $\Delta(f + g) = \Delta f + \Delta g$ **(b)** $\Delta(f - g) = \Delta f - \Delta g$

(c) $\Delta(cf) = c\,\Delta f$ **(d)** $\Delta\left(\dfrac{f}{g}\right) = \dfrac{g\,\Delta f - f\Delta g}{gg^+}$

8 Verify the following *linearity property*:

$$\Delta\left(\sum_{i=1}^{n} a_i f_i\right) = \sum_{i=1}^{n} a_i \Delta f_i$$

by induction on n, for constants a_i and functions f_i.

9 Establish the general rule that the difference operator applied to a polynomial of degree n yields a polynomial of degree $n - 1$.

10 Compute the differences of the following factorial polynomials, leaving the results in the factorial form.

(a) $x^{(4)} + 2x^{(3)} - 4x^{(2)} + 8x^{(1)} + 2$

(b) $3x^{(5)} - 5x^{(3)} + 2x^{(2)} - 9x^{(1)} + 1$

(c) $5x^{(5)} + 4x^{(4)} - 9x^{(3)} + 2x^{(2)} - 8$

(d) $3x^{(6)} - 4x^{(5)} + 9x^{(4)} - 5x^{(3)} + 2x^{(1)} - 6$

11 Use the Newton expansion to convert the polynomials of Exercise 1 to factorial form.

12 Do the same (as in Exercise 11) for the following polynomials.

(a) $x^2 + 3x - 8$ **(b)** $4x^3 - 3x^2 + 1$

(c) $5x^2 - 5x + 2$ **(d)** $3x^3 + x^2 - 4x - 3$

TABLE 5.15

(a)

x	y
0	5
1	8
2	14
3	23
4	35
5	50

(b)

x	y
0	−3
1	1
2	9
3	21
4	37

(c)

x	y
0	10
1	5
2	3
3	7
4	20
5	45

(d)

x	y
0	−7
1	34
2	77
3	124
4	177
5	238
6	309

13 Use the Newton expansion to interpolate the data shown in Table 5.15 with polynomials.

14 Convert the following polynomials from standard to factorial form using synthetic division.

(a) $x^4 + 3x^3 - 8x^2 + 2x - 3$

(b) $6x^5 - 3x^4 - 9x^3 + 2x^2 - 3x + 1$

(c) $3x^5 - 4x^4 + 5x^3 - 2x^2 + 7x - 5$

(d) $5x^6 - 3x^5 + 3x^4 - 3x^3 + 3x^2 + 2x - 8$

15 Do the same (as in Exercise 14) for the polynomials of Exercise 1.

16 Do the same (as in Exercise 14) for the polynomials of Exercise 12.

17 Convert the following polynomials from factorial to standard form.

(a) $x^{(3)} + 3x^{(2)} - 4$ **(b)** $x^{(3)} - 3x^{(2)} + 4x - 1$

(c) $8x^{(4)} - 7x^{(2)} + 6$ **(d)** $2x^{(4)} + 2x^{(3)} - x^{(2)} + 5$

18 Do the same (as in Exercise 17) for the polynomials of Exercise 10.

19 Show that

$$\sum_{x=0}^{n} \frac{1}{1 + (x + 1)x} = \arctan(n + 1)$$

(*Hint:* See Exercise 5.)

20 Obtain a closed form expression for the sum of the first n cubes, using the result of Example 5.58.

21 Do the same (as in Exercise 22) for the fourth powers.

22 Do the same (as in Exercise 22) for the fifth powers.

23 Show that

$$f^{+} = \Delta \sum_{x=0}^{n} f(x)$$

24 Using the result of Example 5.50, show that

$$\sum_{x=0}^{n} (3x^2 + 3x - 4) = (n + 1)[(n + 3)(n - 1) - 1]$$

25 Use the fundamental theorem to obtain closed form expressions for the following summations.

(a) $\sum_{x=0}^{n} (x + 1)^2$ **(b)** $\sum_{x=0}^{n} (x - 1)^3$

(c) $\sum_{x=0}^{n} (x^2 + x + 1)$ **(d)** $\sum_{x=0}^{n} (2x - 1)$

(*Hint:* First find a polynomial having the given summand as its difference.)

26 Sketch the graphs of the first few elementary factorial polynomials.

27 Compute differences for the following functions.

(a) $\dfrac{ax + b}{cx + d}$ **(b)** a^x **(c)** x^n

28 Compute kth differences for the following functions.

(a) a^x **(b)** $\dfrac{1}{x}$ **(c)** $\log x$

29 Show that

$$\sum_{x=0}^{n} a^x = \frac{a^{n+1} - 1}{a - 1} \qquad (a \neq 1)$$

Sec. 5.7 INDEPENDENCE AND RANK

The notions of independence and rank provide a gateway to the study of linear algebra.

The subject of linear algebra, long a mainstay in the undergraduate curriculum of mathematics and computer science majors, and others as well, is important for its wide range of applications. And yet it is a study in abstraction. In its general treatment of the notion of *linearity*, only the essence of this important property appears, without regard to the context of the problem or the intended field of application. But here we are fortunate in having discussed one of the specific problem domains at some length, namely, the realm of polynomials and their applications. We are thus in a good position for introducing a few of the salient features of linear algebra in a concrete setting. In our brief discussion here we thus will introduce ideas that are helpful in our final chapter. But more importantly we provide a convenient gateway for the student's subsequent study of linear algebra in its more abstract framework.

Goals

After studying this section, you should be able to:

- Define the notions of *linear combination*, *dependence*, and *independence* for polynomials and for vectors of an arbitrary coordinate space.
- Decide whether a given collection of polynomials or vectors is dependent or independent.
- Use row operations to reduce such a collection to echelon form and thereby to determine the *rank* of the collection.

Linear Combinations of Polynomials

Our opening discussion centers on the possible representation of a given polynomial $p(x)$ as a *linear combination* of other polynomials. In any number of circumstances such a representation can be of considerable advantage in the analysis. If there are constants $c_1, c_2, \ldots, c_k$ such that p can be written

$$p = c_1p_1 + c_2p_2 + \cdots + c_kp_k$$

then p is said to be a **linear combination** of the polynomials $p_1, p_2, \ldots, p_k$. Note that we abbreviate $p(x)$ as p, and similarly with the other polynomials in the expression. Note further that we already have seen such representations in connection with our discussion of the Lagrange Interpolation Formula of Section 5.3.

Example **5.62** Consider the polynomial

$$p(x) = 3x^3 - 2x^2 + 5x + 12$$

as a case in point. Using the methods of Section 5.4 we would find that $p(x)$ is irreducible. It happens, however, that

$$3x^3 - 2x^2 + 5x + 12 = 3(x^3 + x) - 2(x^2 - x - 6)$$

as you may easily verify. Moreover each of the polynomials on the right-hand side may be reduced or factored; that is,

$$\begin{aligned} x^3 + x &= x(x^2 + 1) \\ x^2 - x - 6 &= (x - 3)(x + 2) \end{aligned}$$

so that we are able to obtain the simplification:

$$3x^3 - 2x^2 + 5x + 12 = 3x(x^2 + 1) - 2(x - 3)(x + 2)$$

Thus we are able to express $p(x)$ as a linear combination of reducible polynomials, even though p itself is irreducible. ■

This situation occurs quite frequently. One very simple setting is that of *completing the square.* We are given a quadratic polynomial

$$p(x) = x^2 + bx + c$$

whereas the square of a linear term $x + a$ would have the representation:

$$(x + a)^2 = x^2 + 2ax + a^2$$

In setting $2a = b$ we choose $a = b/2$. We then have only to write

$$p(x) = x^2 + bx + c = (x + a)^2 + (c - a^2)$$

in order to express p as a linear combination (here, a sum) of a squared linear polynomial and a constant polynomial.

Example 5.63 With the quadratic polynomial

$$p(x) = x^2 - 8x + 19$$

we obtain $a = -\frac{8}{2} = -4$. Thus we may write

$$x^2 - 8x + 19 = (x - 4)^2 + 3$$

in noting that $19 - a^2 = 19 - 16 = 3$. ■

Independent Polynomials

When one polynomial is a linear combination of others there is the sense that the whole collection is then redundant in its generating capability. There is a *dependency* among the individual members of the collection. It is this notion that we would like to study at some length.

Example 5.64 Suppose we consider the collection of the three polynomials

$$p_1(x) = 3x^3 - 2x^2 + 5x + 12$$
$$p_2(x) = x^3 + x$$
$$p_3(x) = x^2 - x - 6$$

appearing in Example 5.62. And suppose we are investigating all polynomials of the form:

$$p = c_1p_1 + c_2p_2 + c_3p_3$$

that is, all linear combinations of these three polynomials. Since we know from Example 5.62 that

$$p_1(x) = 3p_2(x) - 2p_3(x)$$

it follows that any such polynomial p may be written in the form

$$\begin{aligned} p &= c_1(3p_2 - 2p_3) + c_2p_2 + c_3p_3 \\ &= (3c_1 + c_2)p_2 + (c_3 - 2c_1)p_3 \end{aligned}$$

as a linear combination of p_2 and p_3 alone. In this sense the polynomial p_1 is redundant. ■

In order to make these ideas more precise, let us agree to say that a collection

$$P = \{p_1, p_2, \ldots, p_k\}$$

of polynomials is **dependent** if there are corresponding constants $c_1, c_2, \ldots, c_k$ (not all zero) such that

$$c_1p_1 + c_2p_2 + \cdots + c_kp_k = 0$$

Note that with

$$c_1 = c_2 = \cdots = c_k = 0$$

the stated relationship is automatically satisfied. But if this is the only way that the relationship holds, that is, if

$$c_1p_1 + c_2p_2 + \cdots + c_kp_k = 0 \Rightarrow c_1 = c_2 = \cdots = c_k = 0$$

then the collection P is said to be **independent**.

Example 5.65 The collection of polynomials

$$P = \{p_1, p_2, p_3\}$$

as discussed in Example 5.64 is dependent. Since

$$p_1 = 3p_2 - 2p_3$$

(as originally shown in Example 5.62), we may simply transpose to write:

$$1p_1 - 3p_2 + 2p_3 = 0$$

We have thus found constants c_1, c_2, c_3 not all zero (in fact, all nonzero) such that

$$c_1p_1 + c_2p_2 + c_3p_3 = 0$$

as required in the definition of dependent. ■

Example 5.66 By way of contrast, for any n the collection

$$Q = \{1, x, x^2, \ldots, x^n\}$$

is independent. For in setting

$$c_0 1 + c_1 x + c_2 x^2 + \cdots + c_n x^n = 0$$

we must surely have

$$c_0 = c_1 = c_2 = \cdots = c_n = 0$$

That is because the zero polynomial (the right-hand side of our expression) is precisely that polynomial with all of its coefficients equal to zero. ■

In general the question of the dependence or independence of an arbitrary collection of polynomials reduces to a problem of the solution of a system of simultaneous linear equations, and leads finally to the study of linear algebra generally. The reader should reexamine the collection of three polynomials from Example 5.62 (not knowing beforehand that one of them is a linear combination of the others) to see that this is indeed the case.

To say that a set of polynomials is dependent suggests that they depend on one another in some way. This is true. If the set

$$P = \{p_1, p_2, \ldots, p_k\}$$

is dependent, so that

$$c_1 p_1 + c_2 p_2 + \cdots + c_k p_k = 0$$

for certain constants $c_1, c_2, \ldots, c_k$ (not all zero), then we may *solve* for one of the polynomials in terms of the others. Thus if $c_i \neq 0$ we may simply transpose and divide to obtain

$$p_i = -\frac{c_1}{c_i} p_1 - \cdots - \frac{c_{i-1}}{c_i} p_{i-1} - \frac{c_{i+1}}{c_i} p_{i+1} - \cdots - \frac{c_k}{c_i} p_k.$$

showing indeed that one of the polynomials is then a linear combination of the others. The converse is also true, as seen in Exercise 5 of Exercises 5.7.

In the discussions that follow we would like to use the principle of abstraction to generalize the treatment thus far, and to free ourselves from the feeling that these concepts apply only to the realm of polynomials. Abstractly, a polynomial is nothing more than a finite sequence or *vector* of scalar quantities:

$$a_1 x^{n-1} + \cdots + a_{n-1} x + a_n \leftrightarrow (a_1, a_2, \ldots, a_n)$$

an idea that has much in common with our treatment of polynomials in Section 5.2, in reference to the synthetic division process. The powers of x are merely placeholders and they add no essential information. Moreover, in this same context, if a

and b are two such vectors

$$\mathbf{a} = (a_1, a_2, \ldots, a_n)$$
$$\mathbf{b} = (b_1, b_2, \ldots, b_n)$$

perhaps representing polynomials, perhaps not, then we may define a *vector addition*

$$\mathbf{a} + \mathbf{b} = (a_1 + b_1, a_2 + b_2, \ldots, a_n + b_n)$$

and a *multiplication of a vector by a scalar*

$$k\mathbf{a} = (ka_1, ka_2, \ldots, ka_n)$$

consistent with the corresponding polynomial operations of Section 5.1. It happens that completely analogous operations appear in a wide variety of contexts, and it is best to generalize and to think abstractly for a change. The advantage of this point of view is brought home most forcefully in the discussions that follow.

Coordinate Vector Spaces

The idea of using pairs or triples of numbers to represent points in the plane or in three-dimensional space has quite a long history, dating at least to the seventeenth century. To be precise we must speak of *ordered pairs* and *ordered triples*, for we are really using the notion of a finite sequence. In reviewing our use of this notation we observe that these sequences, (a_1, a_2, a_3) for example, have two distinct interpretations. Such a triple can refer to a point in space, in which case a_1, a_2, and a_3 are called the *coordinates* of the point, or it can refer to a directed line segment or *vector*, and we then speak of a_1, a_2, and a_3 as its *components*. The coordinates or components are called *scalar quantities*, as distinguished from vectors.

There is no mathematical reason for limiting our attention to the cases of two or three dimensions. We are perfectly free to introduce the idea of an *ordered n-tuple* (generalizing from the terminology "trip*le*," "quadrup*le*," etc.) as being a finite sequence $(a_1, a_2, \ldots, a_n)$, with coordinates or components chosen from a suitable number system S. Again, these n-tuples may be thought to represent points of a **coordinate n-space** or vectors (*of the space* S^n)—it depends upon our point of view. We have only to be somewhat convinced that there is a good reason for such a generalization. And we needn't look very far to find such reasons. In representing polynomials as vectors, we already find the need for n-tuples with arbitrary n.

Having once agreed to such a generalization we then can begin by agreeing that two such vectors

$$\mathbf{a} = (a_1, a_2, \ldots, a_n) \qquad \mathbf{b} = (b_1, b_2, \ldots, b_n)$$

are to be regarded as **equal** (and, as usual, we write $\mathbf{a} = \mathbf{b}$) if

$$a_1 = b_1, a_2 = b_2, \ldots, a_n = b_n$$

In adopting the vector addition and scalar multiplication previously given we will be imitating the polynomial arithmetic (and the elementary arithmetic of vectors in 2-space and 3-space as well). Here the parenthetical observation refers to the vector operations usually learned in a high-school physics course.

Continuing the analogy we may introduce a *zero vector*

$$\mathbf{0} = (0, 0, \ldots, 0)$$

and for each vector $\mathbf{a} = (a_1, a_2, \ldots, a_n)$, its *negative*

$$-\mathbf{a} = (-a_1, -a_2, \ldots, -a_n)$$

satisfying $\mathbf{a} + (-\mathbf{a}) = 0$. As expected, this permits the introduction of a *vector subtraction* along quite natural lines. But we omit the details. What is important is that our generalized vector arithmetic shares a list of algebraic properties satisfied by the original polynomial arithmetic:

Theorem 5.10 If $\mathbf{a}, \mathbf{b}, \mathbf{c}$ are vectors in coordinate n-space, then

V1.	$\mathbf{a} + (\mathbf{b} + \mathbf{c}) = (\mathbf{a} + \mathbf{b}) + \mathbf{c}$	**V5.**	$k(l\mathbf{a}) = (kl)\mathbf{a}$
V2.	$\mathbf{a} + \mathbf{b} = \mathbf{b} + \mathbf{a}$	**V6.**	$k(\mathbf{a} + \mathbf{b}) = k\mathbf{a} + k\mathbf{b}$
V3.	$\mathbf{a} + \mathbf{0} = \mathbf{0} + \mathbf{a} = \mathbf{a}$	**V7.**	$(k + l)\mathbf{a} = k\mathbf{a} + l\mathbf{a}$
V4.	$\mathbf{a} + (-\mathbf{a}) = \mathbf{0}$	**V8.**	$1\mathbf{a} = \mathbf{a}$

where k, l are arbitrary scalars in the underlying number system.

Proof The verification of these laws is straightforward, and we leave the details to the reader. It should be noted, however, that this list of properties becomes the axiomatic framework for introducing the abstract notion of a *vector space*, as it is encountered in elementary linear algebra. □

It should be clear that these coordinate n-spaces, with their corresponding algebras, are completely determined by the character of the underlying number system S. Depending upon the desired application we may choose from among any of several discrete number systems S—

B: Boolean numbers
Z: integers
Q: rational numbers
A: algebraic numbers
C: computable numbers

as outlined in Chapter 3 and in the discussion surrounding the diagram at the conclusion of Section 5.2. It may be that we will need to divide, and then Q is a better choice than Z. Or it may be that we will need square roots, and in such cases we probably will want to use A rather than Q. Again we may be dealing with problems of computer arithmetic where the Boolean numbers are the most appropriate (see also Section 6.2).

Example 5.67 Consider the Boolean number system B, as discussed in the conclusion of Section 3.6. It might be thought that there is no '-1' here, and that, accordingly, B could not serve as a set of scalars for a coordinate vector space. But in noting that

$$1 + 1 = 0$$

we see that 1 serves perfectly well as a *minus one*. And, correspondingly, the property **(V4)** would be interpreted as if it had been written:

$$\textbf{(V4)} \quad \mathbf{a} + \mathbf{a} = \mathbf{0}$$

It would look a bit strange, but all would be consistent. ■

In any event we now have available any of a number of generalized coordinate n-spaces to suit the application at hand. We have a Boolean n-space, an n-space whose vectors have integer components, a rational n-space, an algebraic n-space, a computable n-space, whatever we choose. These will be denoted B^n, Z^n, Q^n, A^n, and C^n, respectively, hereafter. It is hoped that the reader will be able to generalize our earlier discussions of dependence versus independence to each of these settings.

Rank of a Collection

Suppose we now consider a collection of coordinate vectors

$$A = \{\mathbf{a}_1, \mathbf{a}_2, \ldots, \mathbf{a}_m\}$$

from some coordinate space S^n, without regard to their interpretation, whether representing polynomials, points in space, or whatever. Again we may ask the question whether or not the collection of vectors is dependent or independent. In fact, in the case of the former, we will be able to measure the degree of dependence or redundancy in terms of an integer called the *rank* of the collection.

We will find that it is most convenient to work with the **matrix** of the collection, a rectangular arrangement of the individual components of the vectors:

$$A = \begin{bmatrix} a_{11} & a_{12} & \cdots & a_{1n} \\ a_{21} & a_{22} & \cdots & a_{2n} \\ \vdots & \vdots & & \vdots \\ a_{m1} & a_{m2} & & a_{mn} \end{bmatrix}$$

written row by row. Thus

$$\begin{aligned}\mathbf{a}_1 &= (a_{11}, a_{12}, \ldots, a_{1n})\\ \mathbf{a}_2 &= (a_{21}, a_{22}, \ldots, a_{2n})\\ &\cdots\\ \mathbf{a}_m &= (a_{m1}, a_{m2}, \ldots, a_{mn})\end{aligned}$$

and we use the same symbol A, whether in reference to the collection or to the associated matrix. Using a system of *elementary row operations*:

R_i, R_j	(interchange rows $\mathbf{a}_i$ and $\mathbf{a}_j$)
kR_i	(multiply row $\mathbf{a}_i$ by $k \neq 0$)
$kR_i + R_j$	(add k times row $\mathbf{a}_i$ to row $\mathbf{a}_j$)

we will be able to reduce the matrix A to one having a particularly simple form, where the dependence or independence of the original collection is immediately apparent.

Let us first describe this form in some detail, so that we will be able to recognize it when we see it and, moreover, so that we are quite sure just what it is that we are striving for. We will say that a matrix is in **echelon form** provided that the following conditions are met:

1. Any zero rows are grouped together at the bottom; for the others—
2. The first nonzero entry is a '1' (the *leading* 1);
3. Leading 1's progress to the right in reading down the rows.

(*Note:* There then will be zeros below every leading 1—otherwise we would violate (3).)

Example 5.68 Each of the following matrices is in echelon form:

$$\begin{bmatrix}1 & -1 & 2 & 5\\ 0 & 1 & 9 & -3\\ 0 & 0 & 0 & 1\\ 0 & 0 & 0 & 0\end{bmatrix} \begin{bmatrix}1 & 2 & 0\\ 0 & 0 & 1\\ 0 & 0 & 0\\ 0 & 0 & 0\end{bmatrix} \begin{bmatrix}1 & 2 & 3 & 0 & 5\\ 0 & 0 & 1 & 2 & -1\\ 0 & 0 & 0 & 1 & 4\end{bmatrix}$$ ■

We will find that the reduction of an arbitrary matrix to echelon form is quite an orderly computation, albeit one that is somewhat tedious to perform by hand.

Example 5.69 Consider the matrix

$$A = \begin{bmatrix}1 & 1 & 1 & 1\\ 1 & 1 & 2 & 3\\ 3 & 3 & 5 & 4\\ 2 & 2 & 2 & -1\end{bmatrix}$$

The reduction to echelon form proceeds as follows, using appropriate elementary row operations as indicated:

$$\begin{bmatrix} 1 & 1 & 1 & 1 \\ 1 & 1 & 2 & 3 \\ 3 & 3 & 5 & 4 \\ 2 & 2 & 2 & -1 \end{bmatrix} \xrightarrow{-1R_1 + R_2} \begin{bmatrix} 1 & 1 & 1 & 1 \\ 0 & 0 & 1 & 2 \\ 3 & 3 & 5 & 4 \\ 2 & 2 & 2 & -1 \end{bmatrix} \xrightarrow{-3R_1 + R_3} \begin{bmatrix} 1 & 1 & 1 & 1 \\ 0 & 0 & 1 & 2 \\ 0 & 0 & 2 & 1 \\ 2 & 2 & 2 & -1 \end{bmatrix}$$

$$\xrightarrow{-2R_1 + R_4} \begin{bmatrix} 1 & 1 & 1 & 1 \\ 0 & 0 & 1 & 2 \\ 0 & 0 & 2 & 1 \\ 0 & 0 & 0 & -3 \end{bmatrix} \xrightarrow{-2R_2 + R_3} \begin{bmatrix} 1 & 1 & 1 & 1 \\ 0 & 0 & 1 & 2 \\ 0 & 0 & 0 & -3 \\ 0 & 0 & 0 & -3 \end{bmatrix}$$

$$\xrightarrow{-\frac{1}{3}R_3} \begin{bmatrix} 1 & 1 & 1 & 1 \\ 0 & 0 & 1 & 2 \\ 0 & 0 & 0 & 1 \\ 0 & 0 & 0 & -3 \end{bmatrix} \xrightarrow{3R_3 + R_4} \begin{bmatrix} 1 & 1 & 1 & 1 \\ 0 & 0 & 1 & 2 \\ 0 & 0 & 0 & 1 \\ 0 & 0 & 0 & 0 \end{bmatrix}$$

Note that the final matrix is in echelon form. ∎

In using elementary row operations to reduce a matrix to echelon form, it is important to carry out these operations in a systematic fashion. A formal procedure, sure to arrive at the desired result, is the following recursive algorithm bearing the name of Gauss.

```
algorithm gauss (A)
if A ≠ 0
   if first column is zero
      disregard first column yielding Ā
      gauss (Ā)
   else {some a_i1 ≠ 0}
      perform R_1, R_i
      perform 1/a_11 R_1
      for i running from 2 to m
         perform −a_i1 R_1 + R_i
      disregard first row yielding Ā
      gauss (Ā)
```

In examining the accompanying description in detail it will be seen that these are precisely the steps that we have followed in the reduction of Example 5.69.

Ordinarily we begin the computation in the else clause. Our goal is then to first arrange that the upper left-hand entry of the matrix be different from zero (performing the interchange if necessary). The combination of the next two processes:

$$\begin{array}{l} \text{perform } 1/a_{11}R_1 \\ \text{for } i \text{ running from 2 to } m \\ \quad \text{perform } -a_{i1}R_1 + R_i \end{array}$$

when taken together, is known as *pivoting*. We create a leading 1 as the first nonzero entry in the given row. We then use this '1' together with appropriate multiples so as to zero out all the entries below it in the given column. We say "the given row (column)" because we move progressively downward and to the right in establishing our subsequent pivot positions. That is the nature of the recursion. Once we have moved downward or to the right, we disregard all previous rows or columns, respectively, and we begin anew with the corresponding truncated matrix. Finally, the remaining matrix will be zero (or empty—it amounts to the same thing), and the recursion comes to a close.

Suppose we write $A \sim A'$ if the matrix A' is obtained from the matrix A through the application of an elementary row operation. If we then use this same notation in permitting any number of such operations in succession, we obtain an equivalence relation on the class of matrices. The connection of this notion with our earlier ideas of linearity is then revealed by the following result.

Theorem 5.11 If $A \sim A'$ then every vector of A is a linear combination of the (nonzero) vectors in A'.

Proof The most difficult case to consider is that wherein we have employed a row operation of the form $kR_i + R_j$. Row j then has been transformed from $\mathbf{a}_j$ to $\mathbf{a}'_j = k\mathbf{a}_i + \mathbf{a}_j$. Since we then have

$$\mathbf{a}_j = \mathbf{a}'_j - k\mathbf{a}_i = \mathbf{a}'_j - k\mathbf{a}'_i$$

(because $\mathbf{a}_i$ remains unchanged) we see that $\mathbf{a}_j$ is a linear combination of the vectors in A'. And, of course, $\mathbf{a}_i = \mathbf{a}'_i$ for $i \neq j$, so indeed the stated conclusion is valid in this case. The other two types of row operations are easily handled, so we may consider the proof to be complete (see Exercise 10 in Exercises 5.7). □

Suppose we now begin again with a collection

$$A = \{\mathbf{a}_1, \mathbf{a}_2, \ldots, \mathbf{a}_m\}$$

of vectors from some coordinate space S^n. If we have performed the reduction of the associated matrix A to the echelon matrix E, then it follows from Theorem 5.11 that every vector in A is a linear combination of the nonzero vectors (or rows) in E. The number of nonzero rows in E is then called the *rank of the collection A*. This gives a convenient measure of the dependency or redundancy in the original collection. Moreover, in the limiting case we have

$$\operatorname{rank}(A) = m \Leftrightarrow A \text{ is independent}$$

thus providing a finite algorithmic procedure for answering the independence question.

Example 5.70 The reduction of Example 5.69 shows that the collection

$$A = \{\mathbf{a}_1, \mathbf{a}_2, \mathbf{a}_3, \mathbf{a}_4\}$$

of vectors

$$\begin{aligned}
\mathbf{a}_1 &= (1, 1, 1, 1) \\
\mathbf{a}_2 &= (1, 1, 2, 3) \\
\mathbf{a}_3 &= (3, 3, 5, 4) \\
\mathbf{a}_4 &= (2, 2, 2, -1)
\end{aligned}$$

(from the space Q^4, say) has rank 3. Since $m = 4$ we conclude that A is a dependent collection. Indeed we have the dependency:

$$\mathbf{a}_1 - 2\mathbf{a}_2 + \mathbf{a}_3 - \mathbf{a}_4 = 0$$

as you may verify easily. ■

Example 5.71 In the space Q^3 the collection of vectors

$$\begin{aligned}
\mathbf{a}_1 &= (1, 0, 0) \\
\mathbf{a}_2 &= (0, 1, 0) \\
\mathbf{a}_3 &= (0, 0, 1)
\end{aligned}$$

is independent, since its associated matrix is in echelon form as it stands, with rank $3 = m$. ■

EXERCISES 5.7

1 Complete the square for each of the following quadratic polynomials.

(a) $x^2 - 6x + 10$ **(b)** $x^2 + 2x + 5$

(c) $x^2 - 5x + 17$ **(d)** $x^2 - 8x + 1$

(e) $x^2 + x + 3$ **(f)** $4x^2 - 3x + 7$

2 In each case write the first polynomial as a linear combination of the last two polynomials.

(a) $x^2 + 5x + 8$, $x^2 + 2x + 3$ and $2x^2 + x + 1$

(b) $x^2 + 9x - 8$, $x^2 - x + 2$ and $3x^2 + 2x + 1$

(c) $2x^3 - x^2 - 3x - 3$, $x^3 + 2x^2 + x + 2$ and $3x^3 + x^2 - 2x - 1$

(d) $x^3 - 8x^2 - 15x - 2$, $2x^3 - x^2 - 3x + 5$ and $x^3 + 2x^2 + 3x + 4$

3 Decide whether the following collections of polynomials are dependent or independent.

(a) $1, x + 1, x^2 + x, x^3 - x$

(b) $x^2 - 5, x^2 + 2x - 1, x + 2$

(c) $x^2 + 5x + 8, x^2 + 2x + 3, 2x^2 + x + 1$

(d) $1, x^{(1)}, x^{(2)}, x^{(3)}$

(e) $0, x, x^3$

(f) $2, x, x^2, x^2 + 2x + 1$

4 Show that if one of the polynomials in the collection

$$P = \{p_1, p_2, \ldots, p_k\}$$

is the zero polynomial, then P is a dependent collection.

5 Show that if one of the polynomials in the collection

$$P = \{p_1, p_2, \ldots, p_k\}$$

is a linear combination of the others, then P is a dependent collection.

6 Verify each of the following as listed in Theorem 5.10:

(a) Property (V1) **(b)** Property (V2)

(c) Property (V3) **(d)** Property (V4)

(e) Property (V5) **(f)** Property (V6)

(g) Property (V7) **(h)** Property (V8)

7 Which of the following matrices are in echelon form?

(a) $\begin{bmatrix} 1 & 0 & 3 & 2 \\ 0 & 1 & 1 & 1 \\ 0 & 0 & 1 & 3 \end{bmatrix}$ **(b)** $\begin{bmatrix} 0 & 1 & 3 & 4 \\ 0 & 0 & 1 & 0 \\ 1 & 0 & 0 & 0 \end{bmatrix}$ **(c)** $\begin{bmatrix} 1 & 4 & 4 & 5 & 6 \\ 0 & 0 & 1 & 3 & 2 \\ 0 & 1 & 0 & 1 & 3 \end{bmatrix}$

(d) $\begin{bmatrix} 1 & 0 & 2 \\ 0 & 1 & 4 \\ 0 & 0 & 4 \end{bmatrix}$ **(e)** $\begin{bmatrix} 1 & 0 & 1 & 4 \\ 0 & 2 & 1 & 5 \\ 0 & 0 & 0 & 1 \end{bmatrix}$ **(f)** $\begin{bmatrix} 1 & 2 & 3 & 4 & 5 & 6 \\ 0 & 0 & 1 & 3 & 5 & 2 \\ 0 & 0 & 0 & 1 & 1 & 1 \\ 0 & 0 & 0 & 0 & 1 & 0 \end{bmatrix}$

8 Reduce each of the following matrices to echelon form.

(a) $\begin{bmatrix} 5 & -5 & 5 & 15 & -30 \\ -9 & 9 & -15 & -35 & 66 \\ -6 & 6 & -9 & -22 & 42 \end{bmatrix}$ **(b)** $\begin{bmatrix} 1 & 0 & -1 & 2 & -1 \\ 1 & 1 & 1 & -1 & 2 \\ 0 & -1 & -2 & 3 & -3 \\ 5 & 2 & -1 & 4 & 1 \\ -1 & 2 & 5 & -8 & 7 \end{bmatrix}$

(c) $\begin{bmatrix} 3 & 4 & 5 & -2 \\ 2 & -5 & 2 & 0 \\ 7 & 6 & 9 & -2 \end{bmatrix}$ **(d)** $\begin{bmatrix} -2 & -4 & -1 & 3 \\ 5 & 2 & 3 & 5 \\ 1 & -6 & 1 & 8 \end{bmatrix}$

9 Show that in the case of a matrix with n rows and n columns, the Gaussian algorithm is of complexity order n^3.

10 Verify the statement in Theorem 5.11

(a) for the case of a row operation of the form R_i, R_j

(b) for the case of a row operation of the form kR_i

11 Determine whether or not the first given vector is a linear combination of the last two vectors.

(a) $(-1, 11, -1)$, $(2, -3, 4)$, and $(3, 5, 7)$

(b) $(1, -6, -14)$, $(1, 2, 3)$, and $(4, 0, -5)$

(c) $(12, 3, -10)$, $(2, 5, -3)$, and $(4, -9, 5)$

(d) $(0, 44, 58)$, $(4, 8, 10)$, and $(-3, 5, 7)$

(e) $(5, 5, -4)$, $(2, 5, -7)$, and $(4, -3, 6)$

(f) $(3, 9, -13)$, $(1, 1, 1)$, and $(3, 6, -5)$

12 Show that the following sets of vectors are dependent in Q^3.

(a) $(2, 5, -8)$, $(1, 1, 3)$, $(2, -1, 20)$

(b) $(3, 1, 4)$, $(3, 6, -1)$, $(3, -4, 9)$

(c) $(4, -3, -1)$, $(-5, 3, 1)$, $(-1, 0, 0)$

(d) $(1, 2, 3)$, $(5, -7, 3)$, $(-11, 29, 3)$

13 Show that the following sets of vectors are independent in Q^3.

(a) $(-2, 4, 5)$, $(3, 7, -2)$, $(1, 2, 3)$

(b) $(1, 2, 5)$, $(-3, -3, 6)$, $(4, 0, 4)$

(c) $(8, -1, 0)$, $(4, 7, 2)$, $(5, -9, 3)$

(d) $(5, -3, -1)$, $(1, 1, 1)$, $(9, -7, 2)$

14 Determine whether or not the following sets of vectors are independent.

(a) $(2, 4, -7)$, $(3, 6, 0)$, $(-1, -2, 14)$

(b) $(5, 2, 4)$, $(1, 4, 1)$, $(2, -10, 1)$

(c) $(1, 2, 3)$, $(-3, 2, -4)$, $(4, 5, 7)$

(d) $(-2, -6, 1)$, $(0, 2, 5)$, $(8, 34, 21)$

(e) $(3, 6, 1)$, $(2, 2, 7)$, $(3, -6, 1)$

(f) $(1, 1, 1)$, $(6, 2, -3)$, $(2, 6, -3)$

15 Find the rank of each of the following collections:

(a) $(2, 3, -4, 2)$, $(4, 0, 3, 1)$, $(1, 1, 1, 1)$, $(-3, 6, -7, 3)$

(b) $(1, 2, 3, 4)$, $(-3, 5, 7, 1)$, $(0, 11, 16, 13)$, $(3, 6, 9, 12)$

(c) $(-8, 2, 18, 7)$, $(3, -2, -4, 0)$, $(4, 9, -5, 4)$, $(0, 1, 3, -2)$, $(2, 5, -1, 1)$

(d) $(5, -1, -5, 3)$, $(7, 1, -3, 3)$, $(9, 3, -1, 3)$, $(1, 1, 1, 0)$

16 Find the rank of each of the following collections:

(a) $(1, 1, 2, 1)$, $(2, 1, 3, 4)$, $(1, 0, 1, 2)$

(b) $(0, 0, -8)$, $(1, 2, -1)$, $(2, 4, 6)$

(c) $(-3, 6, 12, -9)$, $(1, 0, 1, 3)$, $(1, -2, -4, 3)$

(d) $(2, -9, 7, 0), (0, -4, 2, 1), (2, -1, 3, -2), (2, 3, 1, -3)$

(e) $(5, 9, 8, 1, 4), (1, 0, -2, 2, -1), (5, 7, 4, 3, 2)$

(f) $(1, 0, 0, 0), (0, 1, 0, 0), (0, 0, 1, 0), (0, 0, 0, 1)$

17 Show that the relation $A \sim A'$ (as extended to permit any number of row operations in succession) is an equivalence relation on the class of matrices, each having the same number of rows and columns.

Chapter Six

GRAPHS AND COMBINATORICS

Contents

6.1 Graphs and Subgraphs
6.2 Circuit Rank
6.3 Trees
6.4 Planar Graphs
6.5 Isomorphism and Invariants
6.6 Covering Problems
6.7 Directed Graphs
6.8 Path Problems

Oystein Ore (1899–1968)

Oystein Ore was born in Oslo, Norway, where he graduated from its University in 1922. He continued his mathematical studies at the University of Göttingen, Germany, and the Mittag-Leffler Institute in Sweden. He began establishing his reputation in the United States in 1927 with an appointment as professor of mathematics at Yale University, where he remained for the greater part of his mathematical career. He is known for his hundred or more mathematical research papers that are of the highest quality, and for his more popular writings, particularly the New Mathematics Library book *Graphs and Their Uses*, written for both the high-school and lay audiences. This book is a classic for its success at introducing advanced mathematical ideas to a broader audience than is generally thought to be possible.

Writings

The term "graph" as used here denotes something quite different from the graphs you may be familiar with from analytic geometry or function theory. The kind of graph you probably have dealt with consisted of all points in the plane whose coordinates (x, y), *in some coordinate system, satisfy an equation in* x *and* y. *The graphs we are about to study are simple geometrical figures consisting of points and edges connecting some of these points; they are sometimes called "linear graphs." It is unfortunate that two different concepts bear the same name, but this terminology is now so well established that it would be difficult to change. Similar ambiguities in the names of things appear in other mathematical fields, and unless there is danger of serious confusion, mathematicians are reluctant to alter the terminology.*

The first paper on graph theory was written by the famous Swiss mathematician Euler and appeared in 1736. From a mathematical point of view, the theory of graphs seemed rather insignificant in the beginning since it dealt largely with entertaining puzzles. But recent developments in mathematics and particularly in its applications have given a strong impetus to graph theory. Already in the nineteenth century graphs were used in such fields as electrical circuitry and molecular diagrams. At present there are topics in pure mathematics, for instance, the theory of mathematical relations, where graph theory is a natural tool, but there are also numerous other uses in connection with highly practical questions: matchings, transportation problems, the flow in pipeline networks, and so-called programming in general. Graph theory now makes its appearance in such diverse fields as economics, psychology, biology, and computer science. To a small extent puzzles remain a part of graph theory, particularly if one includes among them the famous four-color map theorem *that intrigues mathematicians today as much as ever.*

In mathematics, graph theory is classified as a branch of topology, but it is also strongly related to algebra and matrix theory.

In the following discussion we have been compelled to treat only some of the simplest problems from graph theory. We have selected these with the intention of giving an impression, on the one hand, of the kind of analyses that can be made by means of graphs and, on the other hand, of some of the problems that can be attacked by such methods. Fortunately no great apparatus of mathematical computations needs to be introduced.

Graphs and Their Uses **(Ore)**

Sec. 6.1 GRAPHS AND SUBGRAPHS

No, these are not the graphs you sketch to describe the behavior of a function!

Graph theory may be considered a part of combinatorial mathematics. It has its origins in a 1736 paper by the celebrated mathematician, Leonard Euler, but this now famous paper remained a neglected and isolated contribution for nearly a century. Interest in graphs began to stir only gradually, as the same mathematical structure arose in seemingly unrelated disciplines, one after another. However, because a number of popular puzzles could be formulated directly in the language of graph theory, it was still some time before the subject came to be taken seriously.

The relevance of graph theory is no longer a subject of debate. The theory has greatly contributed to our understanding of programming, communication theory, electrical networks, switching circuits, biology, chemistry, and psychology. From the standpoint of the applications it is safe to say that graph theory has become the most important part of combinatorial mathematics. It follows that anyone interested in computer applications should be aware of the more important concepts and a few of the most typical algorithms having to do with graphs.

Goals

After studying this section you should be able to:

- Describe a given graph as a *relation*, as a *matrix*, or in a *pictorial representation*.
- Define the notions of *subgraph* and *spanning subgraph*.
- Describe the complete graphs and provide a formula for the number of their edges.
- Define the idea of the degree of a vertex and state a fundamental result on the number of vertices of odd degree in any graph.
- Define the notions of *path* and *circuit* in a graph, and use these to introduce the idea of connectivity and to define what is meant by a *tree*.
- Compute the *connectivity number* of any given graph.

Representation of Graphs

Whenever we encounter a network or diagram whose nodes or vertices are joined by various arcs, it is likely that a graph-theoretic treatment is appropriate to the analysis. This is true when we consider highway maps, electrical networks, flowcharts, transportation networks, and a variety of other situations. If the arcs or edges have an associated orientation or direction, then we enter the realm of the *directed graphs*, to be discussed later in this chapter. For the time being we do not assume such an orientation, and so our edges are merely connecting links from one vertex to another.

Formally, a **graph** $G = (V, E)$ is a mathematical structure consisting of a nonempty set V of *vertices* and a subset

$$E \subseteq V \otimes V = \{\{v, w\} : v \neq w \text{ in } V\}$$

Thus, $V \otimes V$ is the set of all unordered pairs $\{v, w\}$ of distinct members of V, unordered in the sense that we do not distinguish $\{v, w\}$ from $\{w, v\}$. And E, the set of *edges* of the graph G, is some subset of this set. We write $\{v, w\} \in E$ or, equivalently, $v\ E\ w$ to express the fact that there is an edge between the vertices v and w. The latter notation emphasizes the fact that E can be viewed as a binary relation on V, in the sense we discussed in Section 1.4.

In this chapter we consider only finite graphs, that is, graphs $G = (V, E)$ where in each case the set of vertices is a finite set, say $|V| = n$. When the number of vertices is finite, so too is the number of edges. Considering E as a subset of $V \otimes V$, we ordinarily denote its size by $|E| = m$.

Example 6.1 Let us consider the set of vertices

$$V = \{v_1, v_2, v_3, v_4, v_5\}$$

and suppose that we choose the set of edges

$$E = \{\{v_1, v_2\}, \{v_1, v_5\}, \{v_2, v_3\}, \{v_2, v_4\}, \{v_3, v_4\}, \{v_3, v_5\}, \{v_4, v_5\}\}$$

Then $G = (V, E)$ is a graph with

$$|V| = n = 5 \qquad |E| = m = 7$$

that is, a graph with five vertices and seven edges. ■

A *pictorial representation* of a graph $G = (V, E)$ is obtained in the following way. We draw a point for each vertex and we join the points v and w by an arc or line segment just in case $v\ E\ w$ (i.e., $\{v, w\} \in E$). The lines need not be straight, and the points can be arranged however we choose. Consequently, both Figures 6.1(a) and 6.1(b) are valid representations of the graph just introduced.

(a)
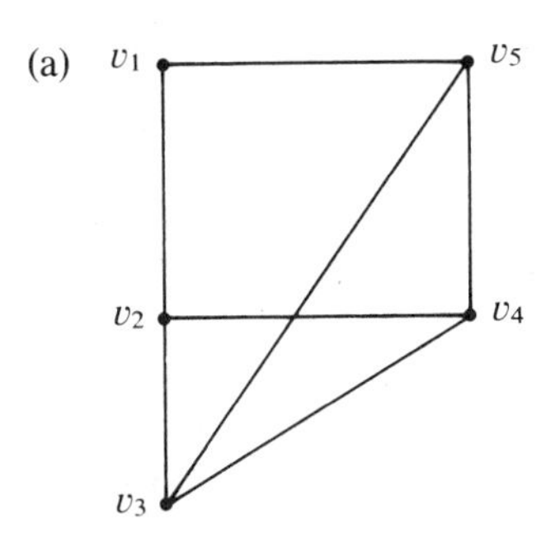

(b)
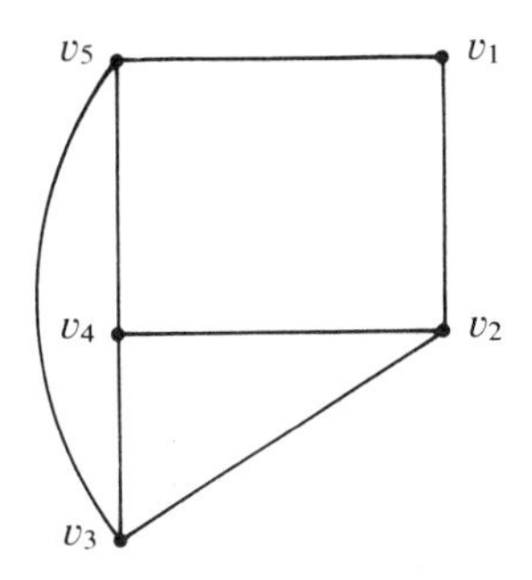

FIGURE 6.1

We must bear in mind, however, that a graph is not a picture but a mathematical structure of the kind first introduced. Therefore the same graph can be pictorially represented in various ways, as we have just illustrated. And erroneous conclusions might perhaps be drawn from a particular picture. Thus it might appear, on the basis of Figure 6.1(a), that our graph has six vertices; but this is not the case. The apparent intersection of the edges $\{v_2, v_4\}$ and $\{v_3, v_5\}$ does not represent a vertex. It is simply a peculiarity of the way in which the graph was represented. We will find it convenient to introduce examples of graphs by means of such pictures. As long as we heed these few words of warning, however, no confusion will arise.

To be absolutely safe, the original set-theoretic or relational representation is preferred. And, in speaking of a relation, an alternate *matrix representation* is possible. If $|V| = n$, we have only to introduce a matrix having n rows and n columns of zeros and ones:

$$E = [e_{ij}]$$

(hoping that our double use of the symbol E will not lead to confusion) in which we assign

$$e_{ij} = \begin{cases} 0 & \text{if } v_i \not{E}\, v_j \\ 1 & \text{if } v_i\, E\, v_j \end{cases}$$

Thus the following left-hand matrix represents the graph $G = (V, E)$ originally introduced as Example 6.1.

$$\begin{array}{c} \\ v_1 \\ v_2 \\ v_3 \\ v_4 \\ v_5 \end{array}\begin{array}{c} \begin{array}{ccccc} v_1 & v_2 & v_3 & v_4 & v_5 \end{array} \\ \begin{bmatrix} 0 & 1 & 0 & 0 & 1 \\ 1 & 0 & 1 & 1 & 0 \\ 0 & 1 & 0 & 1 & 1 \\ 0 & 1 & 1 & 0 & 1 \\ 1 & 0 & 1 & 1 & 0 \end{bmatrix} \end{array} \qquad \begin{array}{c|cccc} v_2 & 1 & & & \\ v_3 & 0 & 1 & & \\ v_4 & 0 & 1 & 1 & \\ v_5 & 1 & 0 & 1 & 1 \\ \hline & v_1 & v_2 & v_3 & v_4 \end{array}$$

Owing to the fact that $\{v, w\} = \{w, v\}$, however, and that our pairs of vertices must be distinct, the lower triangular portion of this matrix retains all of the necessary information, as shown at the right. In our subsequent examples this triangular matrix representation often will be preferred.

Unfortunately, graph theory is one of those areas of mathematics in which the number of definitions seems almost to outweigh the number of significant results. We will try not to overburden you with definitions, although a certain number of them is unavoidable. Here we will present a few definitions relating to the concept of a *subgraph*. One might expect that such an idea would somehow be connected with the notion of a subset, and that is indeed the case. The graph $H = (V_1, E_1)$ is said to be a **subgraph** of the graph $G = (V, E)$ if $V_1 \subseteq V$ and $E_1 \subseteq E$. If $V_1 = V$ we say that H is a **spanning subgraph** of G, and it is then only the case that some of the edges of G have been omitted in deriving H.

Example 6.2 Among the subgraphs of the graph $G = (V, E)$ of Example 6.1, three are pictured in Figure 6.2. Both H_2 and H_3 are spanning subgraphs, but H_1 is not. Of course, each subgraph is a graph in its own right. Thus

$$H_1 = (V_1, E_1)$$

where

$$V_1 = \{v_1, v_2, v_4, v_5\}$$
$$E_1 = \{\{v_1, v_2\}, \{v_1, v_5\}, \{v_2, v_4\}, \{v_4, v_5\}\}$$

and from this representation we see clearly that $V_1 \subseteq V$ and $E_1 \subseteq E$ as required in the definition of a subgraph. ■

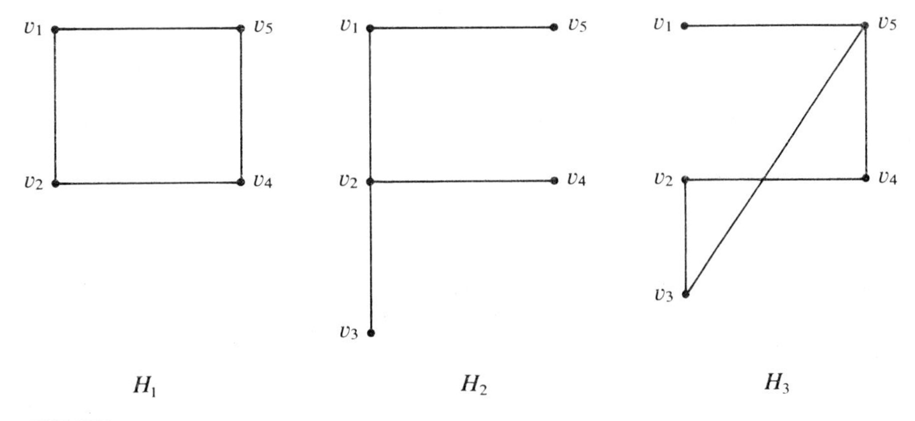

FIGURE 6.2

Complete Graphs

Every graph is a spanning subgraph of some *complete graph*, one in which all possible edges appear. More precisely, $G = (V, E)$ is said to be a **complete** graph on $|V| = n$ vertices if

$$E = V \otimes V = \{\{v, w\} : v \neq w \text{ in } V\}$$

Ordinarily the complete graph on n vertices is denoted K_n (after the German spelling of "complete"). Now if we fix V in considering only the spanning subgraphs $G = (V, E)$ of the complete graph $K = (V, V \otimes V)$, then we may introduce for each G a **complementary graph** $G' = (V, E')$ where

$$\{v, w\} \in E' \quad \text{iff} \quad \{v, w\} \notin E$$

Thus for the same set of vertices, G' has precisely those edges that are not found in G. Its triangular matrix representation will have ones (or zeros, respectively) where the matrix for G has zeros (or ones, respectively).

A graph G is said to be **bipartite** (in two parts) if its vertices can be partitioned into two separate subsets U and W, so that the only edges of G are those of the form $\{u, w\}$ with $u \in U$ and $w \in W$. All such edges need not appear, but if in fact they do then G is called a **complete bipartite** graph. By analogy to the above, the complete bipartite graphs are designated by $K_{m,n}$ when $|U| = m$ and $|W| = n$. It is easily seen that the general bipartite graphs $G = (U \cup W, E)$ are appropriate for the study of matching problems (e.g., applicants to jobs). In more abstract terms, however, they simply provide a mechanism for the representation of relations from one set to another.

Example 6.3 Figure 6.3 shows the complete graphs K_n for $n = 1$ through 6. We recall that the interior intersections in our pictures do not represent vertices for these graphs in the case of K_4, K_5, and K_6. ■

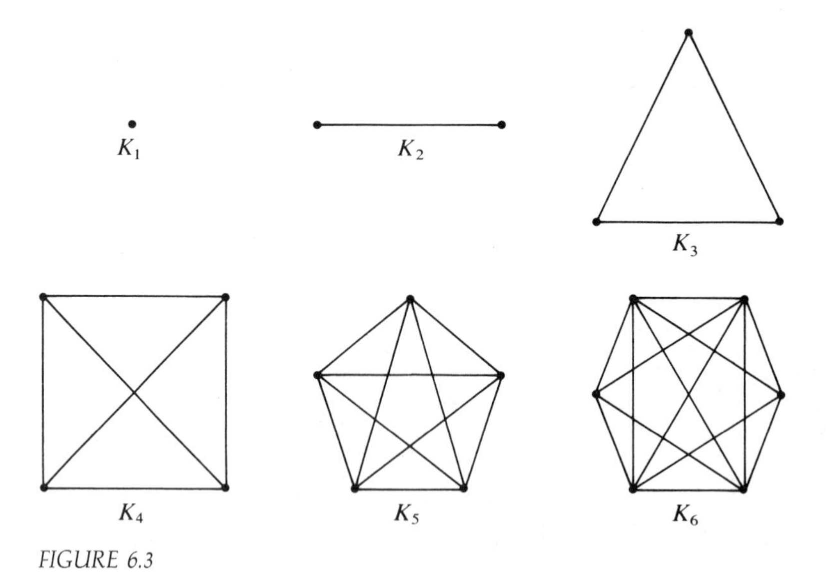

FIGURE 6.3

Example 6.4 A few of the smaller complete bipartite graphs $K_{m,n}$ are shown in Figure 6.4. ■

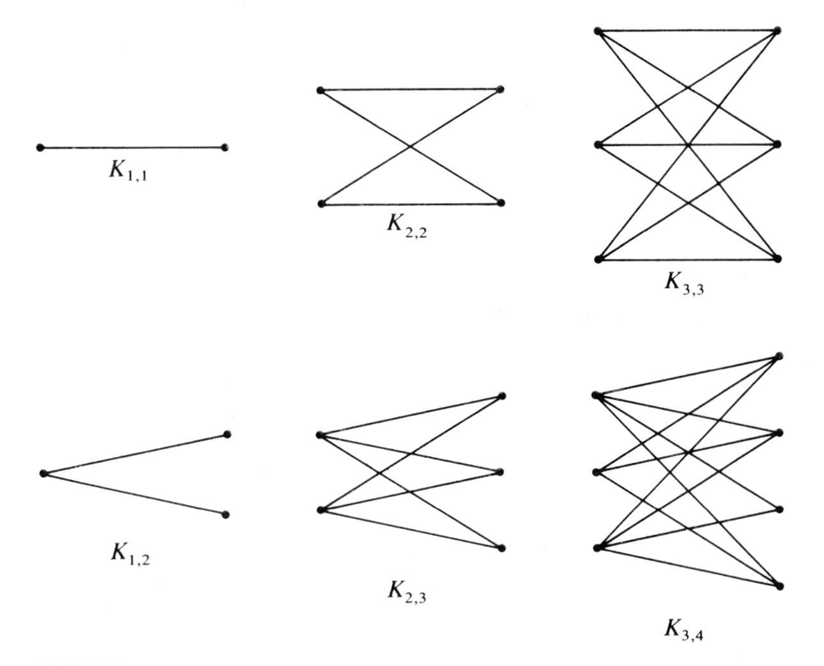

FIGURE 6.4

Example 6.5 Consider once more the graph $G = (V, E)$ of Example 6.1 (as reproduced in Fig. 6.5(a)). The complementary graph G' has the same set of vertices

$$V = \{v_1, v_2, v_3, v_4, v_5\}$$

but the complementary set of edges:

$$E' = \{\{v_1, v_3\}, \{v_1, v_4\}, \{v_2, v_5\}\}$$

A picture of G' is shown in Figure 6.5(b). ■

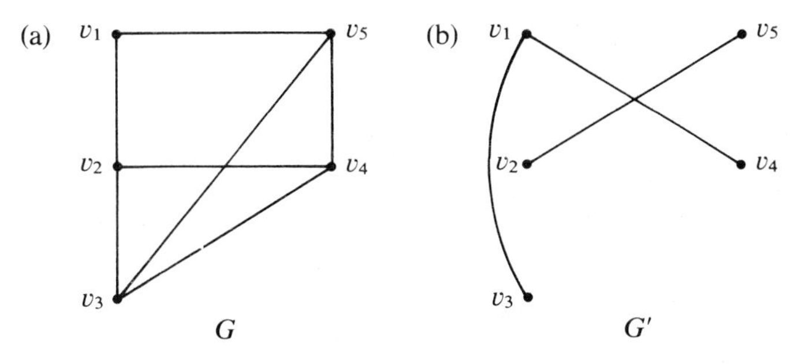

FIGURE 6.5

Example 6.6 The complete graph K_5 has 10 edges, as shown in Figure 6.3, and K_6 has 15 edges. Can we obtain a general formula for the number of edges in K_n? The combinatorial nature of the problem allows for several different determinations. If we recall the manner in which the lower triangular matrix representation was derived (taking half of the matrix after deleting the diagonal), we arrive at the figure:

$$\frac{n^2 - n}{2} = \frac{n(n-1)}{2}$$

because the relation matrix for K_n consists entirely of ones. Alternatively we may add the number of entries in successive rows of the triangular matrix representation. And according to the result of Example 3.16 we obtain

$$1 + 2 + \cdots + (n-1) = \frac{n(n-1)}{2}$$

the same result again. As a third approach, consider the number $(n-1)$ of edges emerging at each of the n vertices. Multiplying $n(n-1)$ would count each edge twice, and we would have to divide this result by two, again leading to the same conclusion. ■

Vertex Degrees

This last calculation makes use of the notion of the *degree* of a vertex. Let $G = (V, E)$ be any graph and consider any vertex $v \in V$. The number of edges $\{v, w\} \in E$ *incident to* or emerging from this fixed vertex v is called the **degree** of v, denoted $d(v)$. As we have just demonstrated, these figures often will enter into a graphical enumeration problem. In order to further illustrate this point, we would like to state two elementary consequences of the definition of the degree, each of which is proved by a simple counting argument.

Lemma In any graph, the sum of the degrees among all of the vertices is given by the tabulation:

$$\sum_{v \in V} d(v) = 2m$$

where m is the number of edges in $G = (V, E)$.

Proof As in the previous example, every edge contributes twice to the summation. □

Theorem 6.1 In any graph, the number of vertices of odd degree is even.

Proof Suppose we let

$$V = V_e \cup V_o$$

denote the decomposition into vertices of even and odd degrees, respectively. Obviously the sum

$$\sum_{v \in V_e} d(v)$$

is even; and so is the sum of the degrees when taken over all the vertices, according to the lemma. It follows that the difference

$$\sum_{v \in V_o} d(v) = \sum_{v \in V} d(v) - \sum_{v \in V_e} d(v)$$

also must be even. Now a sum of odd numbers can be even only if the number of summands is even. It follows that $|V_o|$ is even, as was to be proved. □

Example 6.7 In the graph of Example 6.1, as reproduced in Figure 6.5(a), we have $d(v_1) = 2$, but otherwise

$$d(v_2) = d(v_3) = d(v_4) = d(v_5) = 3$$

which is odd. Therefore we have an even number (4) of vertices of odd degree, as predicted by Theorem 6.1. ■

Paths and Circuits

The real utility of graphs becomes apparent when we begin to discuss routes or itineraries as if the edges were used to represent links of a shipping, distribution, or transportation network. If we look at a road map of the continental United States, we see that it always seems possible to arrange a route between any two given points. The lost traveler is not always so convinced of this fact, however. And if we look at a world road map, then there are certainly points—San Francisco and London, for instance—that cannot be joined by an automobile route. In this regard it thus appears that the connected graphs are of special importance—those graphs for which connecting routes can always be found for any pair of vertices.

Each problem has its own peculiarities when expressed in graph-theoretic terms. Consider the special requirements of the traveling salesperson. He or she would like to visit a certain group of cities without duplication, and then finally to return to the point of origin. In graph-theoretic terms, such a route is called an *elementary circuit*. We will find that all such special requirements usually can be described in one way or another in terms of the various types of paths to be found in the underlying graph.

In any graph $G = (V, E)$ a **path** from v to w is a sequence $\langle v_0, v_1, \ldots, v_k \rangle$ of (not necessarily distinct) vertices $v_i \in V$ with the properties

$$\begin{matrix} v_0 = v \\ v_k = w \end{matrix} \quad \text{and} \quad \{v_{i-1}, v_i\} \in E$$

for each $i = 1, 2, \ldots, k$. If $v_0 = v_k$ then the path is called a **circuit**. If all the vertices v_i are different (except for $v_0 = v_k$ in the case of a circuit), then we speak of an **elementary** path or circuit. A weaker requirement is to ask that none of the edges be duplicated, and we then refer to a **simple** path or circuit.

Example 6.8 In the graph of Example 6.1, as reproduced in Figure 6.5(a),

i. $\langle v_1, v_5, v_3, v_4, v_5 \rangle$ is a simple path from v_1 to v_5

ii. $\langle v_1, v_5, v_3, v_4 \rangle$ is an elementary path from v_1 to v_4

iii. $\langle v_1, v_5, v_3, v_4, v_2, v_1 \rangle$ is an elementary circuit

iv. $\langle v_1, v_5, v_4, v_1 \rangle$ is not a circuit

In the first instance we do not have an elementary path because the vertex v_5 is repeated. In the last instance, $\{v_4, v_1\}$ is not an edge of the graph, so that the given sequence is not a path (hence, not a circuit) of the graph. ∎

Connectivity

A graph is said to be **connected** if every pair of vertices $v \neq w$ is joined by a path. A connected graph without circuits is said to be a **tree**. This latter concept is of immense importance in the applications, particularly to computer science. For this reason we devote a full section of this chapter to its study.

The property of *being joined by a path* is an obvious equivalence relation on the set of vertices of a graph $G = (V, E)$. According to the characterization of equivalence relations in Section 1.4, V is partitioned into nonoverlapping subsets of vertices

$$V = V_1 \cup V_2 \cup \cdots \cup V_p$$

in which every pair of vertices in a particular **component** V_i is joined by a path (i.e., related) in the original graph G. The number (p) of connectivity components is called the **connectivity number** of the given graph G. In a bit of abuse of the notation, we further write $|G| = p$ in this regard.

Example 6.9 The graph G' pictured in Figure 6.5(b) has connectivity number $p = 2$. The edge $\{v_2, v_5\}$ is totally disjoint from the rest of the graph, regardless of how it may appear in the figure. ∎

Example 6.10 A few examples of trees are presented in Figure 6.6. Fortunately we have a case of well-chosen terminology here. They actually look like trees.

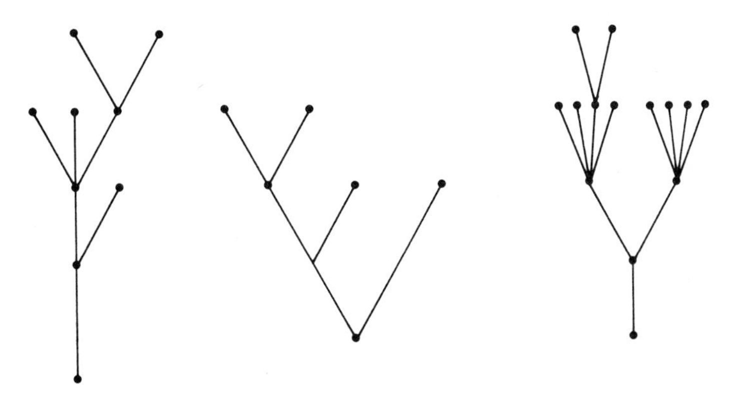

FIGURE 6.6

Note, however, that if the whole of Figure 6.6 is regarded as a graph $G = (V, E)$ then we would have $|G| = p = 3$, that is, the graph would have three separate connectivity components. ■

If we have a picture of a small graph G (as in Examples 6.9 and 6.10), then in most cases it is easy enough for a human observer to determine the connectivity number $|G| = p$. The situation is quite different, however, for larger graphs or with the internal representation of a graph in a computer. In this latter case, the graph is not generally represented pictorially, and the computer is limited in its perceptive capabilities, in any event. What we need is an algorithm for determining this important graphical parameter p. A simple procedure designed for this purpose is as shown in the accompanying algorithm *connectivity*.

```
algorithm connectivity (G: p)

set p to 0
initialize B as ∅
while B ≠ V
   choose an element x from ~B
   set A to {x}
   while there are edges from A to ~A
      for v ∈ A and w ∈ ~A
         if v E w
            adjoin w to A
   union A to B
   increase p by 1
```

In this algorithm, A is the set of vertices of a growing connectivity component. We say "growing" because we continually adjoin vertices w corresponding to edges $\{v, w\}$ with $v \in A$ and $w \notin A$. The set B represents all vertices so far selected in forming all these components. Until $B = V$, we continue to choose an (unselected) element x as a starting point for building a new component. Note that we begin with $p = 0$ and we increment p every time that a new component is exhausted. So we cannot help but arrive at the connectivity number of G at the conclusion of the execution of the algorithm.

Example 6.11 Consider the graph $G = (V, E)$ given by the following triangular matrix representation:

$$\begin{array}{c|ccccccc}
v_2 & 0 & & & & & & \\
v_3 & 0 & 0 & & & & & \\
v_4 & 0 & 1 & 0 & & & & \\
v_5 & 1 & 0 & 0 & 0 & & & \\
v_6 & 1 & 0 & 0 & 0 & 1 & & \\
v_7 & 0 & 1 & 0 & 0 & 0 & 0 & \\
v_8 & 0 & 1 & 0 & 0 & 0 & 0 & 1 \\
\hline
 & v_1 & v_2 & v_3 & v_4 & v_5 & v_6 & v_7
\end{array}$$

Partial intermediate results from a trace of the execution of the algorithm are given as follows:

$x = v_1 \quad A = \{v_1, v_5, v_6\} \quad B = \{v_1, v_5, v_6\} \quad p = 1$

$x = v_7 \quad A = \{v_7, v_2, v_8, v_4\} \quad B = \{v_1, v_2, v_4, v_5, v_6, v_7, v_8\} \quad p = 2$

$x = v_3 \quad A = \{v_3\} \quad B = V \quad p = 3$

Note that the arbitrary selection of the element $x \in B$ will have an effect on the sequence of the computations, but not on the final result $p = 3$. ■

If a graph with n vertices is given, say, by its triangular matrix representation, then the reader can devise a truly elementary algorithm for computing m, the number of edges in the graph. We have just provided an algorithm for determining p, the connectivity number of the graph. The three parameters n, m, and p are important in the detailed analysis of the circuit structure of a graph, as shall be presented in the next section.

EXERCISES 6.1

1 Consider the graphs $G = (V, E)$ represented by the following triangular matrices. In each case, draw a picture of the graph.

(a)

$$\begin{array}{c|cccc}
v_2 & 0 & & & \\
v_3 & 1 & 1 & & \\
v_4 & 1 & 1 & 1 & \\
v_5 & 1 & 1 & 0 & 1 \\
\hline
 & v_1 & v_2 & v_3 & v_4
\end{array}$$

(b)

$$\begin{array}{c|ccccc}
v_2 & 0 & & & & \\
v_3 & 0 & 0 & & & \\
v_4 & 0 & 1 & 0 & & \\
v_5 & 1 & 1 & 0 & 1 & \\
v_6 & 1 & 0 & 0 & 1 & 1 \\
\hline
 & v_1 & v_2 & v_3 & v_4 & v_5
\end{array}$$

(c)
$$\begin{array}{c|ccccc}
v_2 & 1 & & & & \\
v_3 & 1 & 1 & & & \\
v_4 & 1 & 1 & 0 & & \\
v_5 & 1 & 1 & 1 & 1 & \\
v_6 & 0 & 1 & 1 & 1 & 1 \\
\hline
 & v_1 & v_2 & v_3 & v_4 & v_5
\end{array}$$

(d)
$$\begin{array}{c|ccccc}
v_2 & 0 & & & & \\
v_3 & 1 & 0 & & & \\
v_4 & 1 & 0 & 1 & & \\
v_5 & 0 & 1 & 1 & 1 & \\
v_6 & 1 & 0 & 1 & 0 & 1 \\
\hline
 & v_1 & v_2 & v_3 & v_4 & v_5
\end{array}$$

2 Determine the degree of each vertex in the graphs of Exercise 1, and show that Theorem 6.1 holds.

3 Consider the graphs $G = (V, E)$ shown in Figure 6.7(a)–(c). In each case, determine the triangular matrix representation.

(a)

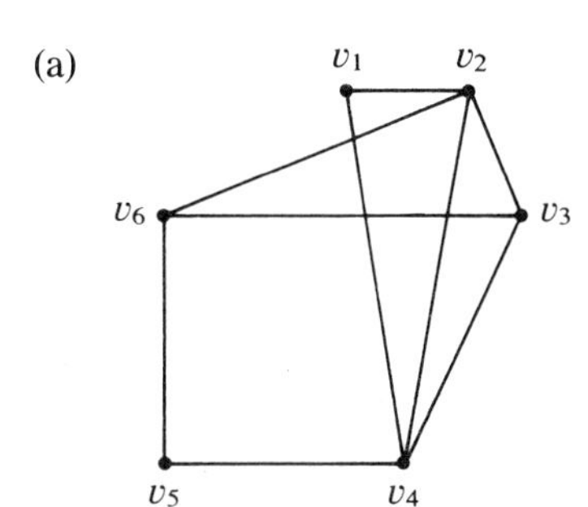

(b)

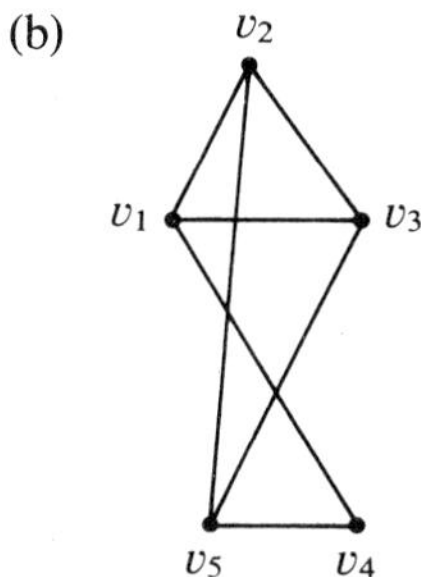

(c)

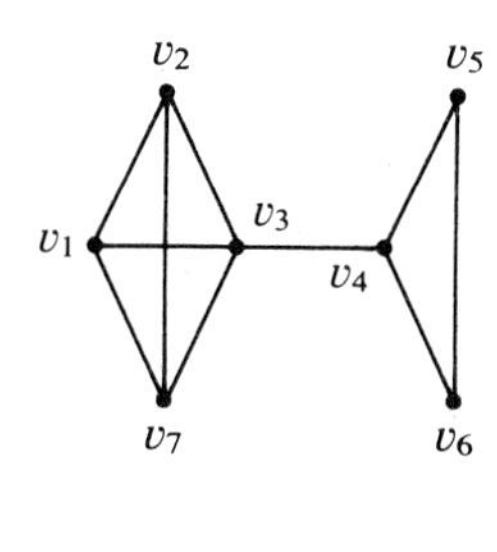

FIGURE 6.7

4 Determine the degree of each vertex in the graphs of Exercise 3, and show that Theorem 6.1 holds.

5 Draw (pictures of) the graphs G' complementary to those of Exercise 3.

6 Do the same (as in Exercise 5) for the graphs of Exercise 1.

7 Obtain a formula for the number of edges in the complete bipartite graphs $K_{m,n}$.

8 For each of the following graphs G, describe two subgraphs H and I.

(a) Exercise 3(a) **(b)** Exercise 3(b) **(c)** Exercise 3(c)

9 Give an example of a simple and a nonsimple path in each of the following graphs.

(a) Exercise 3(a) **(b)** Exercise 3(b) **(c)** Exercise 3(c)

10 Do the same (as in Exercise 9) for circuits.

11 Give an example of a simple but nonelementary path in each of the following graphs.

(a) Exercise 3(a) **(b)** Exercise 3(b) **(c)** Exercise 3(c)

12 Do the same (as in Exercise 11) for circuits.

13 Use the connectivity algorithm to compute the connectivity number p for each of the following graphs.

(a) Exercise 3(b)

(b) The complementary graph to that of Exercise 3(b)

(c)
$$\begin{array}{c|ccccc} v_2 & 0 \\ v_3 & 0 & 0 \\ v_4 & 1 & 0 & 0 \\ v_5 & 0 & 1 & 0 & 0 \\ v_6 & 1 & 0 & 0 & 1 & 0 \\ \hline & v_1 & v_2 & v_3 & v_4 & v_5 \end{array}$$

(d)
$$\begin{array}{c|cccccc} v_2 & 1 \\ v_3 & 0 & 1 \\ v_4 & 1 & 0 & 1 \\ v_5 & 0 & 0 & 0 & 1 \\ v_6 & 0 & 0 & 0 & 0 & 1 \\ v_7 & 0 & 0 & 0 & 1 & 0 & 1 \\ \hline & v_1 & v_2 & v_3 & v_4 & v_5 & v_6 \end{array}$$

14 Design a formal algorithm for determining m, the number of edges in a graph $G = (V, E)$, as given by its triangular matrix representation.

Sec. 6.2 CIRCUIT RANK

We find that we can apply some ideas from elementary linear algebra to the study of graphs.

We already have made some reference to the use of graph theory in the study of electrical networks. In fact, an electrical network has an underlying graph in which every edge lies on a circuit and, consequently, electrical engineers will refer to the subject of circuit analysis as if these were the only circuits. It happens, however, that the techniques that are usually employed in this analysis have broad and far-reaching consequences, for graph theory generally and for each of its domains of application. Nonetheless, the particular application to network analysis is most important, and we will find that it serves as a convenient springboard for launching our more general discussions.

Goals

After studying this section, you should be able to:

- Describe the applications of graph theory to the study of electrical networks.
- Represent the elementary circuits of a graph as vectors in a Boolean coordinate space.
- Define the notion of *independence* as it applies to the *circuits* of a graph.
- Use this notion of independence to define the *circuit rank* of a graph, and to illustrate its significance.
- Compute a basis for the *circuit space* of any given graph.
- Derive an arithmetical formula for the *circuit rank.*

Network Analysis

An electrical network, as usually conceived, has an underlying graph in which every edge lies on a circuit. Each of the edges is labeled with the value of an associated electrical component (battery, resistor, etc.), but the graph itself is of primary importance in any comprehensive network analysis. In particular, the circuit structure of the graph plays a central role in the various methods for determining the electrical currents flowing in the branches or edges of the network.

In one popular method for studying electrical networks (*loop current analysis*), one assumes that certain unknown and fictitious loop currents are flowing simultaneously around the various circuits of the underlying graph. Using *Ohm's Law*:

$$e = iR \qquad \text{(voltage = current} \times \text{resistance)}$$

and the equally familiar *Kirchhoff's Law*:

The algebraic sum of the voltage drops around a circuit is zero

from high-school physics, we are able to write linear algebraic (loop) equations involving these unknown currents. Using techniques from linear algebra (e.g., Section 5.7), we then hope to solve the system and to determine the unknown currents.

Example 6.12 Consider the electrical network of Figure 6.8(a), together with the underlying graph as shown in Fig. 6.8(b). We have assumed names $e_1, e_2, \ldots, e_{12}$ and directions for the unknown voltage drops across each of the branches. Suppose that we imagine loop currents i_1, i_2, i_3, i_4 corresponding to the circuits:

$$c_1: v_6 v_5 v_8 v_9 v_6$$
$$c_2: v_5 v_4 v_7 v_8 v_5$$
$$c_3: v_4 v_1 v_2 v_3 v_6 v_5 v_4$$
$$c_4: v_1 v_4 v_5 v_8 v_9 v_6 v_3 v_2 v_1$$

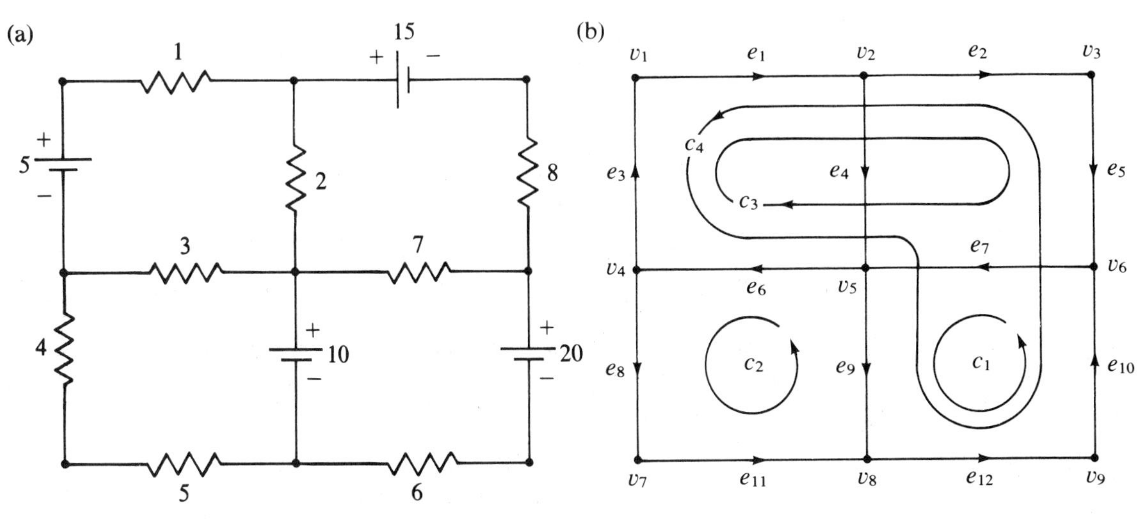

FIGURE 6.8

respectively. Note that the circuits imply a definite direction of traversal, and these directions in turn determine the algebraic sign of the voltage drops to be used in applying Kirchhoff's Law.

The application of Kirchhoff's Law to each of the four loops then will lead to the following system of equations:

$$\begin{aligned}
&L_1\colon e_7 + e_9 + e_{12} + e_{10} = 0\\
&L_2\colon e_6 + e_8 + e_{11} - e_9 = 0\\
&L_3\colon e_3 + e_1 + e_2 + e_5 + e_7 + e_6 = 0\\
&L_4\colon -e_3 - e_6 + e_9 + e_{12} + e_{10} - e_5 - e_2 - e_1 = 0
\end{aligned}$$

If we use Ohm's Law across the resistive branches (and we substitute the values of the batteries), referring to Figure 6.8(a), then the system becomes:

$$\begin{aligned}
&L_1\colon 7(i_1 + i_3) + 10 + 6(i_1 + i_4) - 20 = 0\\
&L_2\colon 3(i_2 + i_3 - i_4) + 4i_2 + 5i_2 - 10 = 0\\
&L_3\colon -5 + 1(i_3 - i_4) + 15 + 8(i_3 - i_4)\\
&\qquad + 7(i_1 + i_3) + 3(i_2 + i_3 - i_4) = 0\\
&L_4\colon 5 + 3(-i_2 - i_3 + i_4) + 10 + 6(i_1 + i_4) - 20\\
&\qquad + 8(-i_3 + i_4) - 15 + 1(-i_3 + i_4) = 0
\end{aligned}$$

With a little more algebra one obtains a system of linear algebraic equations:

$$\begin{aligned}
a_{11}i_1 + a_{12}i_2 + a_{13}i_3 + a_{14}i_4 &= b_1\\
a_{21}i_1 + a_{22}i_2 + a_{23}i_3 + a_{24}i_4 &= b_2\\
a_{31}i_1 + a_{32}i_2 + a_{33}i_3 + a_{34}i_4 &= b_3\\
a_{41}i_1 + a_{42}i_2 + a_{43}i_3 + a_{44}i_4 &= b_4
\end{aligned}$$

That is, $Ai = b$, four equations in four unknowns.

The important thing to observe in all of this is the identity:

$$L_1 = L_3 + L_4$$

that is, one of the equations is a linear combination of two of the others. It follows that the square matrix A will not have rank four. And, accordingly, we will not be able to obtain unique solutions for our four unknown currents. If we had managed to write a system of equations that could be solved uniquely for i_1, i_2, i_3, i_4, then we would have been able to obtain the individual branch currents by superposition of loop currents, that is,

$$\begin{aligned}
&i_1 + i_4\colon \text{the current from } v_8 \text{ to } v_9\\
&i_4 - i_2 - i_3\colon \text{the current from } v_4 \text{ to } v_5
\end{aligned}$$

etc. But the point is we would never get that far with the above system of equations.

In order to obtain a system of equations that admits a unique solution, we need to choose a maximum number of loops or circuits $c_1, c_2, \ldots, c_r$ such that *dependencies* do not occur among any of the corresponding loop equations. In view of the dependency

$$L_1 - L_3 - L_4 = 0$$

in the above, it might be thought that $r < 4$ in the present example. But this is not the case. In fact, if we simply change our fourth circuit only slightly, taking

$$c_4 \colon v_2 v_5 v_8 v_9 v_6 v_3 v_2$$

instead, then we will obtain a new system of four equations without dependencies. Its solution then will be obtainable by using the standard methods of linear algebra. ■

As a byproduct of our analysis of the circuit structure of graphs, we will be able to describe an algorithmic process for the systematic determination of a proper collection of circuits $c_1, c_2, \ldots, c_r$ for use in the electrical network problem. Any ad hoc procedure would become hopelessly difficult as the complexity of the network increases. But graph theory comes to the rescue and finally offers a most orderly and systematic approach.

Independent Circuits

We shall no longer consider electrical networks but, instead, we will return to a discussion of arbitrary graphs $G = (V, E)$. If we designate the edges as a set

$$E = \{e_1, e_2, \ldots, e_m\}$$

then there are at least two ways to look at circuits. Instead of the sequence $v_0, v_1, \ldots, v_k = v_0$ of vertices encountered in traversing the circuit, we may consider the circuit c as the associated collection of edges encountered, writing

$$c = \{e_{i_1}, e_{i_2}, \ldots, e_{i_k}\}$$

as a subset of E, where

$$\begin{aligned} e_{i_1} &= \{v_0, v_1\} \\ e_{i_2} &= \{v_1, v_2\} \\ &\vdots \\ e_{i_k} &= \{v_{k-1}, v_k\} \end{aligned}$$

respectively. Evidently we restrict our attention to simple circuits by this convention, and we lose track of the direction of traversal, but otherwise the circuits are quite adequately described. In fact, we can go one step further in representing c as a vector of the coordinate space $\mathbf{B}^m$ (see the discussion surrounding Theorem 5.10, in particular, Example 5.67). Thus we may write

$$c = (b_1, b_2, \ldots, b_m)$$

with the components determined by the rule:

$$b_j = \begin{cases} 1 & \text{if } e_j \text{ is an edge of the circuit } c \\ 0 & \text{otherwise} \end{cases}$$

Example 6.13 In the graph of Example 6.1, let us name the edges

$$E = \{e_1, e_2, e_3, e_4, e_5, e_6, e_7\}$$

as shown in Figure 6.9. Then we can list the various elementary circuits and their corresponding vectors in this new notation as follows:

$$\begin{array}{ll}
\{e_1, e_2, e_4, e_7\} & c_1 = (1,1,0,1,0,0,1) \\
\{e_3, e_4, e_5\} & c_2 = (0,0,1,1,1,0,0) \\
\{e_5, e_6, e_7\} & c_3 = (0,0,0,0,1,1,1) \\
\{e_1, e_2, e_4, e_5, e_6\} & c_4 = (1,1,0,1,1,1,0) \\
\{e_3, e_4, e_6, e_7\} & c_5 = (0,0,1,1,0,1,1) \\
\{e_1, e_2, e_3, e_5, e_7\} & c_6 = (1,1,1,0,1,0,1) \\
\{e_1, e_2, e_3, e_6\} & c_7 = (1,1,1,0,0,1,0)
\end{array}$$

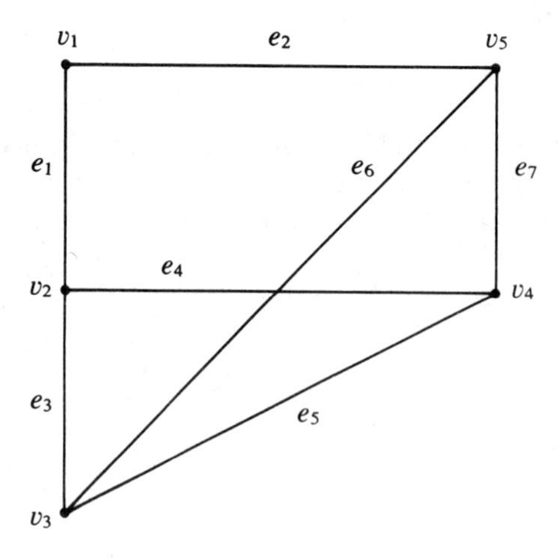

FIGURE 6.9

Alternatively, these circuits might have been represented as subsets of E, for example, $c_1 = \{e_1, e_2, e_4, e_7\}$, $c_2 = \{e_3, e_4, e_5\}$, etc., as shown at the left. ■

Example 6.14 If we label the edges of the network of Example 6.12 as shown in Figure 6.10, then the circuits discussed in that example become identified with the vectors:

$$\begin{aligned} c_1 &= (0,0,0,0,0,0,1,0,1,1,0,1) \\ c_2 &= (0,0,0,0,0,1,0,1,1,0,1,0) \\ c_3 &= (1,1,1,0,1,1,1,0,0,0,0,0) \\ c_4 &= (1,1,1,0,1,1,0,0,1,1,0,1) \end{aligned}$$

respectively. Now observe that we obtain the dependency:

$$c_1 + c_3 + c_4 = 0$$

(the zero vector) when computed in the vector space $\mathbf{B}^{12}$. (Again, see the discussion surrounding Theorem 5.10.) Written otherwise, we have

$$1c_1 + 0c_2 + 1c_3 + 1c_4 = 0$$

showing that the vectors c_1, c_2, c_3, c_4 are indeed dependent in $\mathbf{B}^{12}$. Recall that this is the set of circuits that caused trouble in Example 6.12. ■

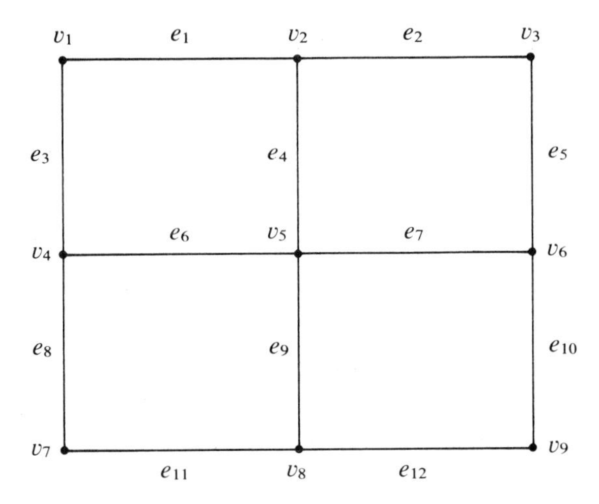

FIGURE 6.10

Motivated by Examples 6.12, 6.13, and 6.14, and the accompanying discussion, we proceed to introduce a vector space framework for the study of the circuit structure of a graph. Let the graph $G = (V, E)$ be given with $|E| = m$. And suppose we let C denote the subset of $\mathbf{B}^m$ consisting of all the elementary circuits of G, considered as vectors. Recalling the discussion of Section 5.7, we also let C denote the matrix having these vectors as rows. The rank of the collection C is called the **circuit rank** of G, written $\rho(G)$.

The importance of the circuit rank stems from the fact that the nonzero rows of the echelon matrix obtained from $\mathbf{C}$ will provide a maximum number of independent circuits in the graph G. And, of course, this number is $\rho(G)$. Clearly we must have

$$\rho(G) = \text{rank}\,(\mathbf{C}) \leq m$$

in any case. But we can in fact compute $\rho(G)$ directly. We have only to reduce the matrix $\mathbf{C}$ to echelon form in the usual way (using the algebra of B). According to Theorem 5.11, the nonzero rows of this echelon matrix will form a *basis for the circuit space* C in that every circuit is a linear combination of these. Their number is the desired circuit rank $\rho(G)$. The whole technique then provides an effective solution to the electrical network analysis problem. Eventually we will see that the circuit rank can be computed by an arithmetical formula, but for the time being the approach just described will have to suffice.

Example 6.15 In the graph of Example 6.13, we already have listed the elementary circuits. If we construct the matrix with these circuits as rows, we may use the techniques of Section 5.7 to obtain an echelon form, as shown below. One should note the extreme simplicity of the calculations, owing to the fact that $1 + 1 = 0$ in B (so that $0 - 1 = 1$ as well).

$$\begin{bmatrix} 1 & 1 & 0 & 1 & 0 & 0 & 1 \\ 0 & 0 & 1 & 1 & 1 & 0 & 0 \\ 0 & 0 & 0 & 0 & 1 & 1 & 1 \\ 1 & 1 & 0 & 1 & 1 & 1 & 0 \\ 0 & 0 & 1 & 1 & 0 & 1 & 1 \\ 1 & 1 & 1 & 0 & 1 & 0 & 1 \\ 1 & 1 & 1 & 0 & 0 & 1 & 0 \end{bmatrix} \to \begin{bmatrix} 1 & 1 & 0 & 1 & 0 & 0 & 1 \\ 0 & 0 & 1 & 1 & 1 & 0 & 0 \\ 0 & 0 & 0 & 0 & 1 & 1 & 1 \\ 0 & 0 & 0 & 0 & 1 & 1 & 1 \\ 0 & 0 & 1 & 1 & 0 & 1 & 1 \\ 0 & 0 & 1 & 1 & 1 & 0 & 0 \\ 0 & 0 & 1 & 1 & 0 & 1 & 1 \end{bmatrix}$$

$$\to \begin{bmatrix} 1 & 1 & 0 & 1 & 0 & 0 & 1 \\ 0 & 0 & 1 & 1 & 1 & 0 & 0 \\ 0 & 0 & 0 & 0 & 1 & 1 & 1 \\ 0 & 0 & 0 & 0 & 1 & 1 & 1 \\ 0 & 0 & 0 & 0 & 1 & 1 & 1 \\ 0 & 0 & 0 & 0 & 0 & 0 & 0 \\ 0 & 0 & 0 & 0 & 1 & 1 & 1 \end{bmatrix} \to \begin{bmatrix} 1 & 1 & 0 & 1 & 0 & 0 & 1 \\ 0 & 0 & 1 & 1 & 1 & 0 & 0 \\ 0 & 0 & 0 & 0 & 1 & 1 & 1 \\ 0 & 0 & 0 & 0 & 0 & 0 & 0 \\ 0 & 0 & 0 & 0 & 0 & 0 & 0 \\ 0 & 0 & 0 & 0 & 0 & 0 & 0 \\ 0 & 0 & 0 & 0 & 0 & 0 & 0 \end{bmatrix}$$

It follows that the circuits c_1, c_2, c_3 are a basis for the circuit space and $\rho(G) = 3$.

■

Circuit Rank Formula

Soon we will see that the circuit rank of a graph $G = (V, E)$ can be computed according to the simple formula

$$\rho(G) = m - n + p$$

whenever

$$|G| = p \qquad |V| = n \qquad |E| = m$$

Let us temporarily denote $r(G) = m - n + p$ and make the following observation. If we augment a graph $G = (V, E)$ by adding an edge to obtain

$$G^+ = (V, E \cup \{e\})$$

where $e = \{v, w\} \notin E$, then we have two cases to consider:

1. v and w are joined by a path in G;
2. they are not.

In the first case $r(G^+) = r(G) + 1$ because m increases by one while p remains unchanged. In the second case, $r(G^+) = r(G)$ since m increases by one while p decreases by one. This simple observation is important in obtaining the following fundamental result.

Theorem 6.2 In any graph $G = (V, E)$ with

$$|G| = p \qquad |V| = n \qquad |E| = m$$

the circuit rank is given by the formula

$$\rho(G) = m - n + p$$

Proof We will show how to find $m - n + p$ independent circuits in G. If $E = \{e_1, e_2, \ldots, e_m\}$, we consider the sequence of subgraphs

$$G_0, G_1, \ldots, G_m = G$$

each having the vertex set V, but in which

$$E_0 = \varnothing \quad \text{and} \quad E_i = E_{i-1} \cup \{e_i\}$$

Thus we simply add one edge at a time, so that each $G_i = G_{i-1}^+$ relative to the addition of $e_i = \{v_i, w_i\}$.

First we compute the initial and final values for the function r, as defined above. Evidently we have

$$r(G_0) = 0 - n - n = 0 \qquad r(G_m) = m - n + p$$

And we notice that v_i and w_i are joined by a path in G_{i-1} if and only if the addition of the edge e_i closes a new circuit c_i. It follows by our preliminary observation that r increases by one with each c_i. Altogether we must obtain $r(G_m) - r(G_0) = m - n + p$ increases and, accordingly, $m - n + p$ circuits c_i in succession. The collection of circuits thus obtained is independent because each c_i has an edge (namely e_i) not found in any of the preceding (see Exercise 7 of Exercises 6.2).

The argument that shows that it is not possible to find more than $m - n + p$ independent circuits will be left as an exercise (see Exercise 9 of Exercises 6.2). □

This result has important theoretical implications. But the proof was constructive and, in fact, it provides an algorithm for the systematic determination of a proper collection of circuits on which to base the loop current method for solving electrical network problems. Other techniques will be suggested as we proceed. Especially in the case of graphs of a particular structure (e.g., the planar graphs of Section 6.4), these other techniques may be more appropriate, more efficient than the one presented here. But the present approach is reasonably effective in the general case. Note that in using the algorithm suggested by our proof, it is necessary that one be able to decide whether or not two given vertices (v_i and w_i here) are joined by a path in a graph (the graph G_{i-1} in this case). We need to know the nature of such a path in order to construct the circuit c_i. Of course, the reader can devise algorithms for answering such questions. It is perhaps best to wait until the algorithms for solving the various path problems are discussed in Section 6.8, however.

Example 6.16 In the graph of Examples 6.1, 6.13, and 6.15, we have

$$p = 1 \qquad n = 5 \qquad m = 7$$

and accordingly

$$\rho(G) = m - n + p = 7 - 5 + 1 = 3$$

in agreement with the calculations of Example 6.15.

In order to illustrate the construction in the proof of Theorem 6.2, let G_i have the edges

$$E_i = \{e_1, e_2, \ldots, e_i\}$$

as shown in Figure 6.11. The calculations

$$r(G_0) = 0 - 5 + 5 = 0$$
$$\vdots$$
$$r(G_3) = 3 - 5 + 2 = 0$$
$$r(G_4) = 4 - 5 + 1 = 0$$
$$r(G_5) = 5 - 5 + 1 = 1$$
$$r(G_6) = 6 - 5 + 1 = 2$$
$$r(G_7) = 7 - 5 + 1 = 3$$

show that r increases by one every time that the augmenting edge closes a new circuit.

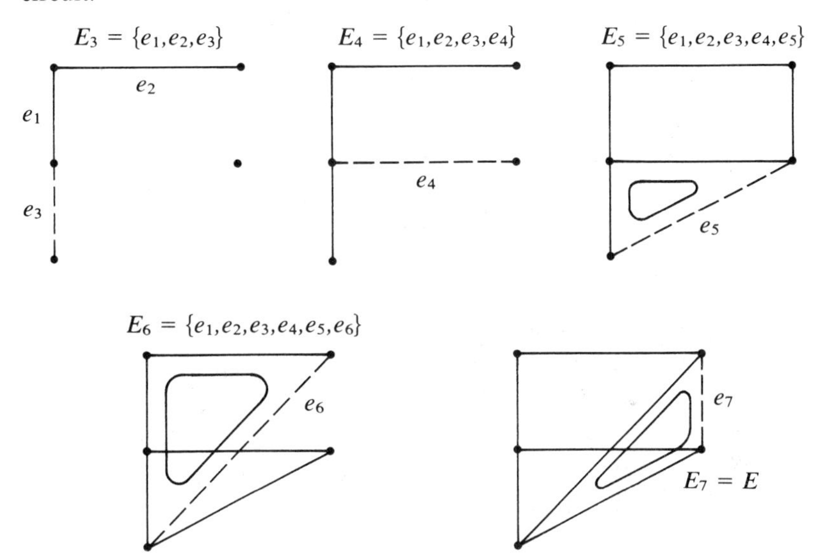

FIGURE 6.11

The circuits obtained by this procedure are not uniquely determined because of the possible multiplicity of paths joining the vertices v_i and w_i in G_{i-1}. They are always an independent set of circuits, however, regardless of the choice of paths, because each circuit has an edge (the newly adjoined one) that is not found in any of the preceding circuits (again, see Exercise 7 in Exercises 6.2). In the present example we have chosen paths that lead to the circuits

$$\{e_3, e_4, e_5\} \qquad \{e_1, e_2, e_3, e_6\} \qquad \{e_5, e_6, e_7\}$$

as shown in Figure 6.11. These are the circuits originally named c_2, c_7, and c_3, respectively, in Example 6.13. Note, however, that this is a different solution from that obtained in Example 6.15. ■

EXERCISES 6.2

1 Solve for the four unknown loop currents of Example 6.12 after making the change in the fourth circuit as suggested at the end of that example. (*Note:* You first will have to revise the equation L_4 accordingly.)

2 Show that the following collections of circuits are dependent in the graph of Example 6.13 (Fig. 6.9).

(a) $\{c_1, c_3, c_4\}$ **(b)** $\{c_1, c_2, c_3, c_4\}$ **(c)** $\{c_4, c_5, c_6, c_7\}$
(d) $\{c_2, c_3, c_4, c_5\}$ **(e)** $\{c_2, c_3, c_5\}$ **(f)** $\{c_1, c_2, c_6, c_7\}$

3 Find a collection of three independent circuits in the graph of Example 6.13, other than the ones obtained in that example or in Example 6.16. Show that your collection is independent.

4 In addition to the circuits listed in Example 6.14 (Fig. 6.10), suppose we name

$$c_5 = (1, 0, 1, 1, 0, 0, 0, 1, 1, 0, 1, 0)$$
$$c_6 = (0, 1, 0, 1, 1, 1, 0, 1, 0, 1, 1, 1)$$
$$c_7 = (0, 0, 0, 0, 0, 1, 1, 1, 0, 1, 1, 1)$$
$$c_8 = (1, 1, 1, 0, 1, 0, 0, 1, 0, 1, 1, 1)$$

Show that the following collections of circuits are dependent.

(a) $\{c_1, c_5, c_6, c_7, c_8\}$ **(b)** $\{c_1, c_2, c_3, c_5, c_8\}$
(c) $\{c_1, c_2, c_3, c_8\}$ **(d)** $\{c_1, c_3, c_5, c_7, c_8\}$

5 Find a collection of four independent circuits in the graph of Example 6.14 (Fig. 6.10). Show that your collection is independent. (*Note:* See Exercise 4 for a more complete listing of circuits.)

6 Use Theorem 6.2 to compute the circuit rank of the following graphs.

(a) Exercise 1(a) of Exercises 6.1
(b) Exercise 1(c) of Exercises 6.1
(c) Exercise 3(a) of Exercises 6.1
(d) Exercise 3(b) of Exercises 6.1
(e) Exercise 3(c) of Exercises 6.1
(f) Exercise 5(c) of Exercises 6.1
(g) Example 6.14 (Fig. 6.10)
(h) The cube with its (four) diagonals

7 Suppose that an ordered collection $\{c_1, c_2, \ldots, c_k\}$ of circuits in a graph G has the property that each c_i has an edge not found in any of the preceding. Prove that the collection is independent.

8 In any graph G, show that there are $n - p$ independent vectors, each perpendicular to all the circuit vectors. (Here, it is assumed that the perpendicularity is characterized by a zero generalized *dot product*.) (*Hint:* First suppose that G is connected ($p = 1$) and find $n - 1$ such vectors by considering a sequence of subsets

$$V_1 \subseteq V_2 \subseteq \cdots \subseteq V_{n-1}$$

of V, each of increasing size $|V_j| = j$. Then associate a vector with each V_j that will represent the set of edges from V_j to $V \sim V_j$. Finally, note that each of these vectors has an even number of edges in common with any circuit vector.)

9 Prove that there cannot be more than $m - n + p$ independent circuits in a graph G. (*Note:* The proof uses linear algebra. *Hint:* Use the result of Exercise 8.)

10 As in Example 6.16, illustrate the construction in the proof of Theorem 6.2 for each of the graphs of Exercise 6.

11 Considering circuits as sets of edges, let c_1 and c_2 be circuits in any graph G, and let $e \in c_1 \cap c_2$. Prove that there exists a circuit c such that $e \notin c \subseteq c_1 \cup c_2$.

Sec. 6.3 TREES

I think that I shall never see a graph as simple as a tree.

The notion of a tree, a connected graph without circuits, has already made its appearance in our earlier discussions. It is somewhat surprising that such a simple idea can have so many applications. And yet we will see that trees abound in our subsequent studies. We study spanning trees, optimal trees, labeled trees, search trees, in fact, trees of every possible description. In computer science particularly, trees are of immense importance. They are used as a data storage model, as a device for sorting in alphanumeric order, as a means for parsing or compiling algebraic expressions, and for a number of other purposes. And so it is well that we learn as much as we can about them. This is most easily accomplished, however, in first concentrating on the uncluttered graph-theoretic setting, where we keep a certain distance from these applications.

Goals

After studying this section, you should be able to:

- List a number of equivalent characterizations of trees as a special class of graphs, and show how it is that each of these is equivalent.
- Describe two methods for obtaining a *spanning tree* for a given graph.
- Discuss the use of spanning trees in solving various optimization problems.
- Use spanning trees to derive a basis for the circuit space of a graph.
- Describe and illustrate the use of *binary labeled trees* in searching and sorting problems.
- Describe and illustrate the use of binary labeled trees in the handling of arithmetic expressions.
- Distinguish the *Polish* (or postfix) *form* of an expression from the ordinary infix form.
- Convert expressions from infix to postfix form, and vice-versa.

Characterization of Trees

Recall that an acyclic (without circuits) connected ($p = 1$) graph is called a **tree**. As graphs, trees are quite simple and quite special, but they have immense importance in the applications. The reader will be easily convinced of this importance as we proceed, but first we wish to establish a number of convenient characterizations of trees. We will do this in the following theorem, where we state six logically equivalent propositions. Particular attention should be paid to (4) and (5). There, when we use the term *edge maximality* (*minimality*) with respect to a certain property, we simply mean that the graph G has the property in question, but if an edge is added (deleted), then the property will no longer hold.

Theorem 6.3 The following statements are equivalent for a graph on n vertices, with $n \geq 2$.

1. G is connected and acyclic (i.e., G is a tree).
2. G is acyclic and has $n - 1$ edges.
3. G is connected and has $n - 1$ edges.
4. G is edge maximal with respect to being acyclic.
5. G is edge minimal with respect to being connected.
6. Every pair of vertices in G is joined by just one path.

Proof In establishing the implications in a cycle, we will find that each of the individual arguments is entirely straightforward, and most of them follow rather easily from the circuit rank formula (Theorem 6.2).

1⇒2 Since $p = 1$ and $\rho(G) = m - n + p = m - n + 1 = 0$ because of the assumptions in (1), we have $m = n - 1$ immediately.

2⇒3 Since (2) implies $\rho(G) = m - n + p = (n - 1) - n + p = 0$, we obtain $p = 1$ by transposition.

3⇒4 With $p = 1$ and $m = n - 1$ we have

$$\rho(G) = m - n + p = (n - 1) - n + 1 = 0$$

showing that G is acyclic. If, however, we were to add an edge, then m would increase by one, but p and n would remain unchanged. As a result $\rho(G)$ would increase to one.

4⇒5 Suppose (4) is true. Then if G were not connected there would exist vertices v and w not joined by a path. Thus, adjoining the edge $\{v, w\}$ would not form a circuit. This contradicts the assumption (4), so we must have $p = 1$. In still assuming (4), we would have

$$\rho(G) = m - n + p = m - n + 1 = 0$$

showing that $m = n - 1$. Now if we were to delete an edge, we would obtain a graph G' with

$$m' = m - 1 = n - 2 = n' - 2$$
$$\rho(G') = m' - n' + p' = 0$$

the latter because we had no circuits to begin with. It follows that

$$p' = n' - m' = n' - (n' - 2) = 2$$

showing that G is edge minimal with respect to being connected.

5 ⇒ 6 Assuming (5), we see that G is connected, so that every pair of vertices v, w is joined by a path. If there were more than one path from v to w, then the deletion of an edge belonging to a second path but not the first would not disconnect G, in contradiction to (5).

6 ⇒ 1 Clearly (6) implies that G is connected. But if, additionally, G had a circuit, then some pair of vertices would be joined by two paths, in contradiction to (6). □

Spanning Trees

The above characterizations of trees can be used in many ways. Here we first show how to construct a *spanning tree* for any given (connected) graph. With the obvious interpretation of our earlier definitions, a **spanning tree** for a graph $G = (V, E)$ is a subgraph of G that is, in fact, a tree on the same set of vertices. We will find that this notion has numerous applications.

Corollary Every connected graph has a spanning tree.

Proof When G is connected, choose an edge that can be deleted (assuming there is one) without disconnecting G. Continue this process until it is no longer possible to delete an edge without disconnecting G. Since according to Theorem 6.3, (5) is equivalent to (1), we then have a (spanning) tree. □

Example 6.17 Let us use this method to find a spanning tree for the graph shown in Figure 6.12(a). We may remove in succession the edges *af*, *ag*, *ef*, *bc*. Then we will have the subgraph shown in Figure 6.12(b). Now we cannot remove *fg* or *cf*, for example, because either way we would disconnect the graph. But we can delete *de*, say, followed by *ad* and *ab*, leaving the subgraph shown in Figure 6.12(c). Now it is impossible to continue. The removal of any more edges would disconnect the graph. The resulting graph is a spanning tree for the original graph. ■

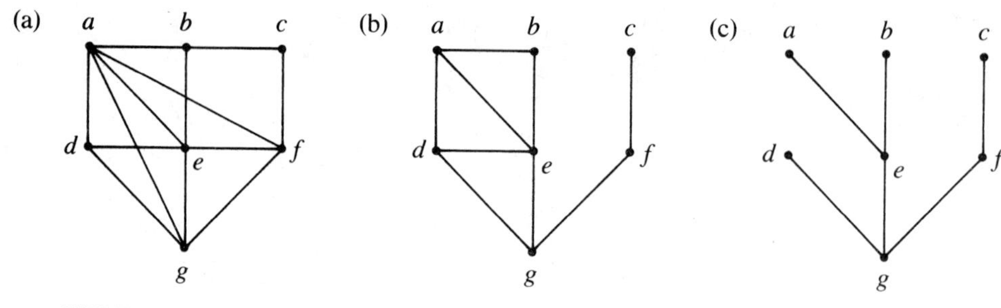

FIGURE 6.12

An algorithmic implementation of this procedure would be quite straightforward. For one thing we can modify our connectivity algorithm (Section 6.1) slightly so as to decide the question of the connectivity of the subgraphs obtained by an edge deletion. We can even determine in advance just how many edges will need to be removed! Note that in Example 6.17 we removed

$$7 = 13 - 7 + 1 = m - n + p = \rho(G)$$

edges altogether. Is this a coincidence? In agricultural terms, is the circuit rank the number of dikes one must remove in order to flood all of the rice fields? (See Fig. 6.12.) In order to answer this question you should observe that the reinsertion of the deleted edges, one at a time, will produce a (unique!) circuit. This is because of the equivalence of (4) and (1) in Theorem 6.3. At the same time the collection of circuits so obtained will be independent because each of these circuits has an edge—the previously deleted one—not found in any of the others. (Once again, see Exercise 7 of Exercises 6.2.)

These last remarks suggest an alternative procedure for finding a basis for the circuit space of any graph G:

---

1. Obtain a *spanning forest*, T, that is, a spanning tree in each of the p connectivity components of G, and let the edges of G that are not in T be called *chords*. The latter will be $\rho(G)$ in number.
2. One by one the reinsertion of each individual chord with T will give rise to a unique circuit in G. The total collection of circuits so obtained will be independent and, again, of size $\rho(G)$.

Once again this procedure gives a solution to the electrical network problem (the one motivating the discussions of Section 6.2). This procedure should be compared with our two previous methods: the one deriving from the proof of Theorem 6.2, and the other as illustrated in Example 6.15.

Example 6.18 In Example 6.17 we have the chords

$$af, ag, ef, bc, de, ad, ab$$

and it follows that the graph of Figure 6.12(a) has the circuit rank $\rho = 7$. When reinserted one at a time these chords give rise to circuits as follows (see Fig. 6.13):

$$af\colon aegfa$$
$$ag\colon agea$$
$$ef\colon egfe$$
$$bc\colon begfcb$$
$$de\colon dged$$
$$ad\colon aegda$$
$$ab\colon aeba$$

You then should check that this collection of circuits is indeed independent. ■

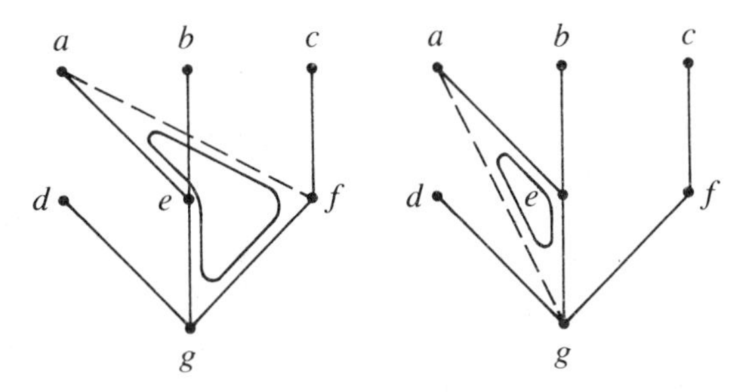

FIGURE 6.13

The proof of the corollary gave an outline of an algorithm for finding a spanning tree in any connected graph $G = (V, E)$, but the whole construction can be approached from the other direction. If $|V| = n$, then we can build a sequence of subgraphs

$$G_0, G_1, \ldots, G_{n-1}$$

each having the vertex set V, but in which $|E_i| = i$. We simply add one edge (from E) after another, being careful that our selections do not produce a circuit. This time it is the equivalence of (4) with (1) in Theorem 6.3 that is applicable. It shows that we again obtain a spanning tree, as soon as the augmentation can no longer be carried out.

Example 6.19 In the graph of Example 6.17 (Fig. 6.12(a)), suppose we already have selected the edges *ab*, *bc*, *ad*, *be*, and *cf*. Then the subgraph G_5 is as shown in Figure 6.14(a). At this stage we cannot select *de*, say, since that would give rise to the circuit *adeba*. We can select *ag*, however, obtaining the subgraph G_6 as shown in Figure 6.14(b). Now no further edges can be added without producing a circuit, and this last subgraph is a spanning tree for the graph of Figure 6.12(a). ■

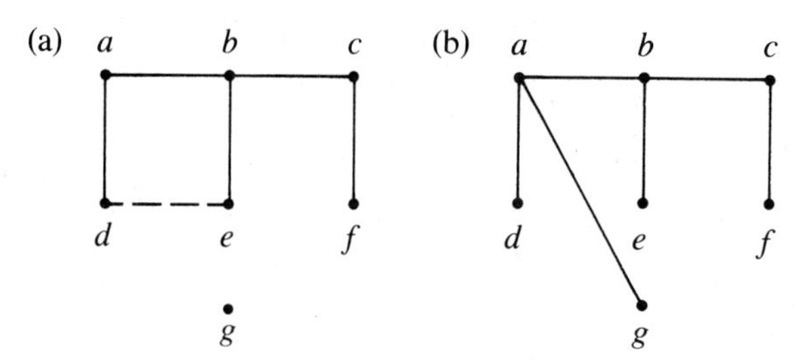

FIGURE 6.14

Optimal Trees

The applications of the concept of spanning trees are many and varied. In order to gain some appreciation for this fact, suppose that the vertices of the set V represent cities, and the edges of E denote pipelines, cables, railways, or any such form of communication or conveyance. In any graph $G = (V, E)$ representing the feasible or the proposed connections, we may be able to associate a mileage or a cost with each edge in representing the length or the expense in joining the various cities. In the *minimal connector problem* we seek to find the most economical network linking all the cities. The solution must surely be a spanning tree (for it would be redundant to have a circuit). With circuits, links could be removed to lower the cost.

To solve the minimal connector problem, we have only to make one slight change in the procedure just introduced. We choose e_1 to be the cheapest edge, and having already determined the subgraph G_{i-1}, we choose e_i to be the most economical link such that

$$G_i = G_{i-1} + \{e_i\}$$

(to introduce a convenient notation) is without circuits. The resulting spanning tree G_{n-1} is called an *economy tree.* We will soon show that these are in fact *minimal* (cost) *spanning trees*, that is, solutions to the minimal connector problem.

Example 6.20 Suppose that we want to design a minimal-length rail network connecting the seven cities shown in the mileage chart of Figure 6.15. In order to build an economy tree we first choose e_1 to be the link between Denver and Cheyenne (100 miles), the smallest distance between any two cities of the table. Then in succession we choose

e_2: Dodge City to Oklahoma City (212 miles)
e_3: Denver to Dodge City (301 miles)
e_4: Dodge City to Omaha (336 miles)
e_5: Denver to Albuquerque (422 miles)
e_6: Cheyenne to Salt Lake City (462 miles)

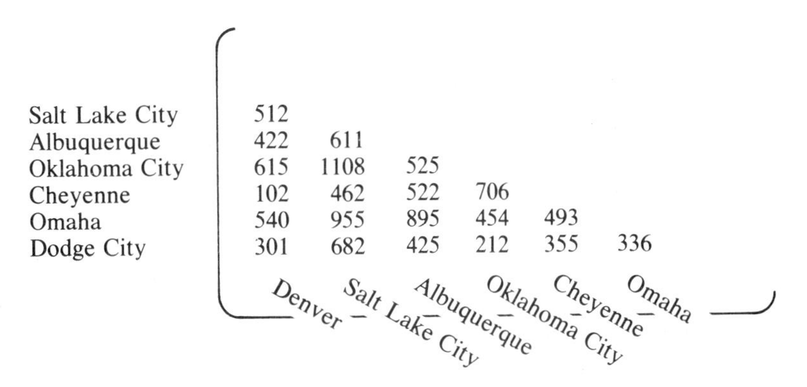

	Denver	Salt Lake City	Albuquerque	Oklahoma City	Cheyenne	Omaha
Salt Lake City	512					
Albuquerque	422	611				
Oklahoma City	615	1108	525			
Cheyenne	102	462	522	706		
Omaha	540	955	895	454	493	
Dodge City	301	682	425	212	355	336

FIGURE 6.15

Finally we obtain the economy tree of Figure 6.16. Note that e_5 was chosen over a cheaper link (Cheyenne to Dodge City) because the latter would have produced a circuit. ■

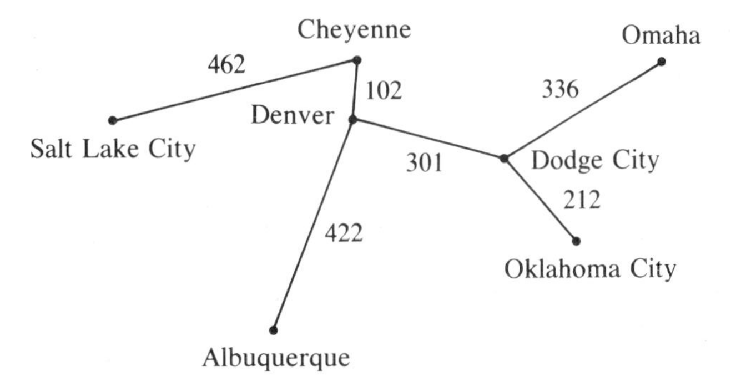

FIGURE 6.16

Theorem 6.4 Every economy tree is a minimal spanning tree.

Proof Suppose we are given an economy tree G_{n-1}. Among all the minimal spanning trees, let T be one that has a maximal number of edges in common with G_{n-1}. We are to compare the *costs*, $c(T)$ versus $c(G_{n-1})$. Each of these trees has the same number $(n-1)$ of edges, so we can let e be an edge (*of minimum cost*) that is in G_{n-1} but not in T. Otherwise $T = G_{n-1}$ and we would be finished. According to Theorem 6.3, the graph $T + \{e\}$ has a unique circuit and there must be an edge f in this circuit that is not an edge of G_{n-1}, simply because G_{n-1} has no circuits.

Suppose we let T' be the spanning tree obtained by exchanging e for f in T. Then the inequality $c(f) > c(e)$ would contradict the fact that T was a minimal spanning tree, and an equality would contradict the fact that among all minimal spanning trees, T had a maximum number of edges in common with G_{n-1} (T' has one more). So we have to conclude that $c(f) < c(e)$.

Just as before, the graph $G_{n-1} + \{f\}$ has a unique circuit, and in this circuit we can find an edge e' that is not an edge of T. An inequality $c(f) < c(e')$ would contradict the whole construction of the economy tree. So we are left to conclude that

$$c(e') \leq c(f) < c(e)$$

but this is in opposition to the choice of e to be of minimal cost among the edges in G_{n-1} but not in T! It follows that there is no alternative to having $T = G_{n-1}$ from the start. □

Labeled Binary Trees

Consider the set B^*, the set of all finite sequences or words on the alphabet $B = \{0, 1\}$ (see Section 1.3). Then the graph $T = (V, E)$ is said to be a **binary tree** provided that the following three conditions hold.

1. V is a finite subset of B^*;
2. If $x0$ or $x1$ is a member of V, then so is the prefix x;
3. The only edges are those of the form $\{x, x0\}$ or $\{x, x1\}$.

Note that conditions (2) and (3) ensure that each vertex is joined by a path to the *root* of the tree, the vertex corresponding to the null word λ. At the other extreme we let the subset

$$L = L(T) = \{x \in V\text{: neither } x0 \text{ nor } x1 \text{ is in } V\} \subseteq V$$

denote the set of all terminal vertices or *leaves* of the tree T. Beginning with our initial use of the term tree in describing a connected acyclic graph, a wealth of botanical terms have found their way into our terminology. We have tried not to be excessive in this regard, but it seems quite appropriate that a few concessions be made in this direction.

Example 6.21 A binary tree is pictured in Figure 6.17. In order to illustrate condition (2) in the definition of binary tree, we note that

$$0101 \in V \Longrightarrow 010 \in V \Longrightarrow 01 \in V \Longrightarrow 0 \in V \Longrightarrow \lambda \in V$$

Thus the condition ensures that the tree contains with any word x (as a vertex), each of its prefixes. This particular tree T has the following set of leaves:

$$L = L(T) = \{10, 110, 0100, 0101\}$$

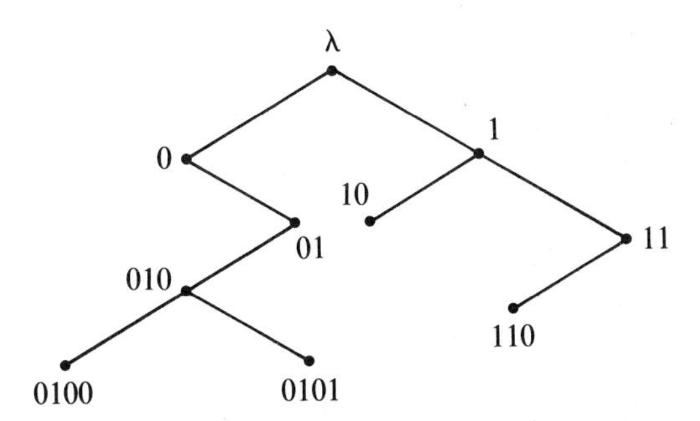

FIGURE 6.17

and, as we see, these are indeed the extremities of the tree. They would look a bit more like leaves, however, if we had drawn our figure the other way round. But we follow the current trend in computer science for drawing trees with the root at the top. ■

In a *labeled* binary tree, the vertices or edges (or both) may carry information that supplements or replaces the usual identifying names. Ordinarily these labels will themselves be members of some other algebraic or arithmetic system, reflecting the particular application being made of the tree model. If we consider the binary tree $T = (V, E)$, it may be helpful to think of this labeling as being accomplished through the application of one or more functions

$$\ell: V \longrightarrow A_V \qquad \ell: E \longrightarrow A_E$$

from the vertices and/or edges of T to an appropriate algebraic or arithmetic system. More often than not, however, we simply indicate the labels on the pictures as if they were to replace the usual identifying names for the vertices and/or edges.

There are any number of applications for binary labeled trees, and we can only afford to spend the time discussing one or two of them. In most cases the binary trees $T = (V, E)$ will be used in connection with certain algorithmic processes, and the actual labeling of the tree will take place during the execution of the algorithm. Most often it is the vertices that become labeled. Since the tree itself is generated during the execution of the algorithm, however, it is hardly correct to say that the labeling is a function, as indicated above. It is more accurate to think of T as an array (in the sense of computer science,—see Chapter 4—not as in linear algebra), with indices x drawn from the set B^*, and with values in the particular algebraic or arithmetic system being employed. When we then write $T[x]$, the meaning will be clear—we will be speaking of the content (the label) of the tree at the vertex corresponding to x.

Since B^* is an infinite set, however, we are certainly not going to label all such vertices in the course of executing the algorithm. Only a finite subset $V \subseteq B^*$ will actually become labeled. And because of the way in which the vertices are labeled, from the root downward, we are assured that this finite subset satisfies condition (2) above, that is, we definitely will obtain a binary labeled tree by the time the process terminates, as it surely must.

As our first application, we consider an alternative to the treatment of the classical searching and sorting problems, as introduced in Section 4.3. We will find that a binary labeled tree representation provides quite an attractive alternative to the use of ordinary linear arrays. Suppose that

$$y = \langle y_1, y_2, \ldots, y_n \rangle$$

is a sequence of items to be sorted. Without loss of generality we may suppose that the individual items are themselves words on some arbitrary alphabet A. All that we need to assume is that there is some dictionary order that prevails among the various words. A *tree sorting* procedure can then be described as in the accompanying algorithm *treesort*.

```
algorithm treesort (y: T)

initialize T as empty
for i running from 1 to n
  set x to λ {the root of T}
  while T[x] ≠ λ
    if y[i] < T[x]
      replace x by x0 {move down to the left}
    else
      replace x by x1 {move down to the right}
  assign T[x] the value y[i]
```

When we say that the tree is initially empty we mean that the array T has the empty word as the label at each and every vertex $x \in B^*$. At each stage of the process, all the words alphabetically preceding the node labeled $T[x]$ are on the left branch from x, and those following $T[x]$ are on the right. With each new word considered we access the tree at $x = \lambda$ and make alphabetical comparisons all the way down the tree, until we come to a node with an empty label, thus having found where to place the new word.

We must be a bit careful in reading this algorithm, however, because there are two alphabets (A and B) being used. The words in B^* indicate positions in the tree, and those in A^* are used as labels, representing the words being read from the text y. An example will serve to clarify these matters.

Example 6.22 Consider the following famous text:

> *Four score and seven years ago, our fathers brought forth upon this continent a new nation,...*

Here we have the sequence $y = \langle y_1, y_2, \ldots \rangle$ with

$$y_1 = \text{four} \qquad y_2 = \text{score} \qquad \ldots \qquad y_{16} = \text{nation} \qquad \ldots$$

TABLE 6.1

i	x	T
1	λ	$T[\lambda]$ = four
2	λ	
	1	$T[1]$ = score
3	λ	
	0	$T[0]$ = and
4	λ	
	1	
	11	$T[11]$ = seven
5	λ	
	1	
	11	
	111	$T[111]$ = years
6	λ	
	0	
	00	$T[00]$ = ago
7	λ	
	1	
	10	$T[10]$ = our

as input to the treesort algorithm. A trace of the execution of the algorithm begins as shown in Table 6.1. For the portion of text indicated above, the algorithm thus produces the binary labeled tree shown in Figure 6.18. ■

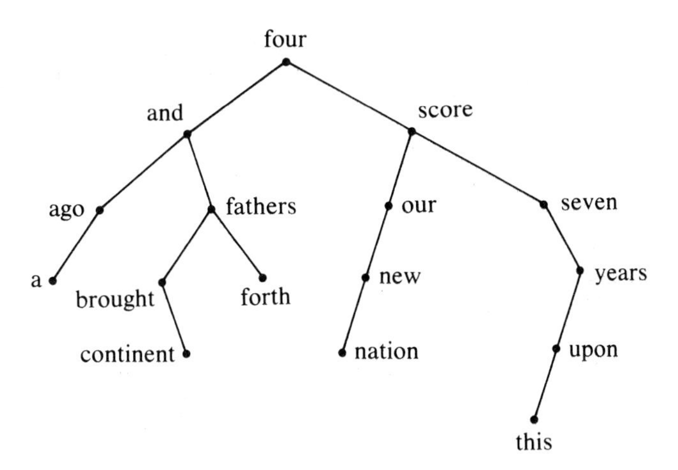

FIGURE 6.18

But how do we interpret the resulting binary labeled tree as representing the sorted sequence? To answer this question it would have been better to have approached the whole subject with the *recursive* definition: a *binary labeled tree* is either empty, or is a triple:

$$\text{tree} = (\text{tree}, \text{label}, \text{tree})$$

where, again, the interior left and right entries of the triple are themselves binary labeled trees in this same sense. More specifically we could write

$$\text{tree at } x = (\text{tree at } x0, x\text{-label}, \text{tree at } x1)$$

for each $x \in B^*$, allowing for the case that any tree at x may be empty. Then in Example 6.22 we would be able to write:

$$\begin{aligned}
\text{tree} = \text{tree at } \lambda &= (\text{tree at } 0, \lambda\text{-label}, \text{tree at } 1) \\
&= (\text{tree at } 0, \text{four}, \text{tree at } 1) \\
&= ((\text{tree at } 00, 0\text{-label}, \text{tree at } 01), \text{four}, (\text{tree at } 10, 1\text{-label}, \text{tree at } 11)) \\
&= ((\text{tree at } 00, \text{and}, \text{tree at } 01), \text{four}, (\text{tree at } 10, \text{score}, \text{tree at } 11))
\end{aligned}$$

etc. Finally, after disregarding the punctuation, we would have obtained the sorted sequence:

a, ago, and, brought, continent, fathers, forth, for,
nation, new, our, score, seven, this, upon, years

or at least this would be the result for the portion of text given above.

The argument just provided (thinly) disguises an actual recursive algorithm for reading out the labels of the tree in the sorted order. (Recall the discussion of recursive algorithms in Section 4.5.) Evidently the procedure is really quite elementary. We must simply arrange to *traverse the tree* (i.e., visit all its vertices) *in-order*:

---

traverse the left subtree
visit the root
traverse the right subtree

---

where, again, the subtree traversals are performed in order, recursively. From the above analysis you will see that if the visitations create the output, the desired sorted sequence of labels is obtained. We leave the details as an exercise (see Exercise 19 in Exercises 6.3).

Finally we note the ease with which new items can be added to the tree structure without upsetting the order of the existing members. (Recall that in adding a new member to a linear array, somewhere in the middle, all the members below the point of insertion must be pushed down.) Furthermore it can be shown that on the average only $n\log n$ comparisons are necessary in performing the tree sort—a figure that certainly compares favorably with the complexity figure of n^2, as obtained for the selection sort (Example 4.51). So we see that there are definite advantages to the use of a tree structure in treating the sorting problem. (The same is true for the related search problems.)

And now, for something completely different, we provide a brief glimpse at the use of binary labeled trees in handling arithmetic expressions. When we use arithmetic or logical expressions in computer programming or in ordinary algebra we find that it is necessary to dictate the order in which the various operations are to be performed. This may be done with parentheses, with a hierarchical convention for the operations (see Table 2.1), or with some combination of these two schemes. In this way we can outline the schedule of operations and clearly indicate the corresponding operands. And yet the real subtleties of a proper interpretation of algebraic expressions are not always fully appreciated. It requires considerable thought in order to design an algorithm that will interpret such expressions correctly. For these and other reasons the ordinary way of writing algebraic expressions is not at all well suited to computer processing. In fact, most systems will immediately translate such expressions into a more convenient form. Let us consider these matters in some detail.

Example 6.23 The expression

$$A \cdot B + C + D/E$$

might be considered ambiguous as it stands. But if we write

$$A \cdot (B + C) + D/E$$

and we also understand that multiplication ($\cdot$) and division ($/$) have a higher priority than addition ($+$) in our hierarchical scheme, then the ambiguity will be removed. ■

The above conventions notwithstanding we will find that the binary labeled trees provide a most convenient means for specifying the order of execution of the various operations in an algebraic expression. Suppose for the moment that we are only considering binary operations, that is, operations with two arguments or operands. Then each execution of an operation may be symbolized as in Figure 6.19(a). If this idea is extended inductively to the expression as a whole, we will obtain a binary labeled tree, one whose leaves identify the distinct occurrences of variables or constants (operands) in the expression, and whose nonterminal nodes are labeled by the various binary operations. In the interpretation of such tree structures, each left and right subtree is to be executed before the operation point at which the two subtrees are grafted (see Fig. 6.19(b)). As a consequence, no parentheses or hierarchical ordering of operations need accompany the tree representation. The structure of the tree tells the whole story.

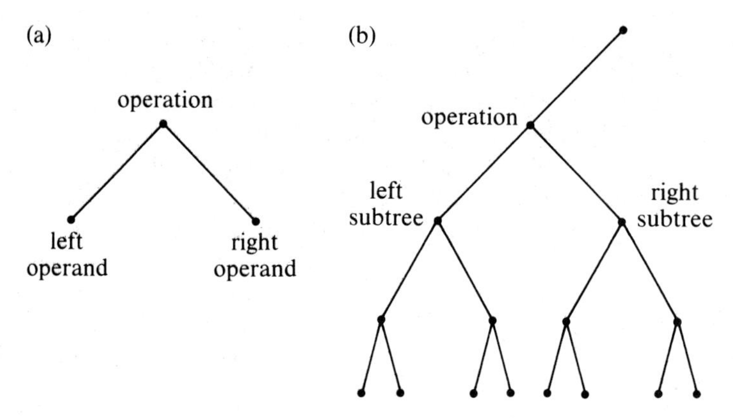

FIGURE 6.19

Example 6.24 Consider once more the expression

$$A \cdot (B + C) + D/E$$

of Example 6.23. With the hierarchical convention already described, the interpretation of this expression is just as clear as that of the *fully parenthesized* version:

$$((A \cdot (B + C)) + (D/E))$$

Let us take the second '+' in this expression as the root of our tree, inasmuch as this is the last scheduled operation. Grafting the subexpressions $A \cdot (B + C)$ and D/E as left and right subtrees at this root, we obtain the diagram of Figure 6.20(a). Continuing inductively, we finally will obtain the binary labeled tree of Figure 6.20(b). The intended schedule of operations and their respective operands are completely specified by the tree structure itself. ■

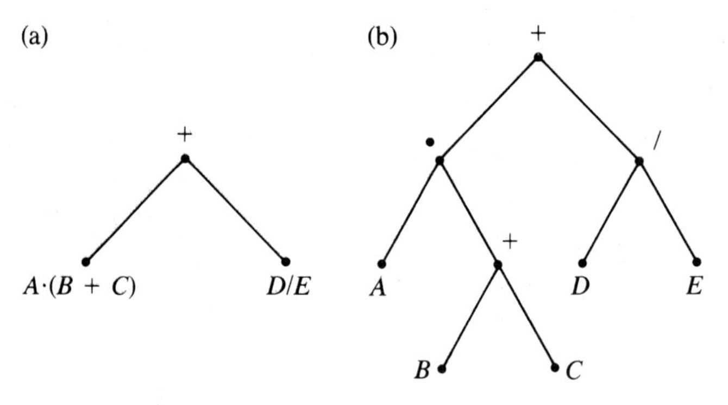

FIGURE 6.20

Example 6.25 The trees of Figure 6.21 represent two more possible interpretations of the original unparenthesized expression $A \cdot B + C + D/E$, forgetting about any possible hierarchical conventions. ■

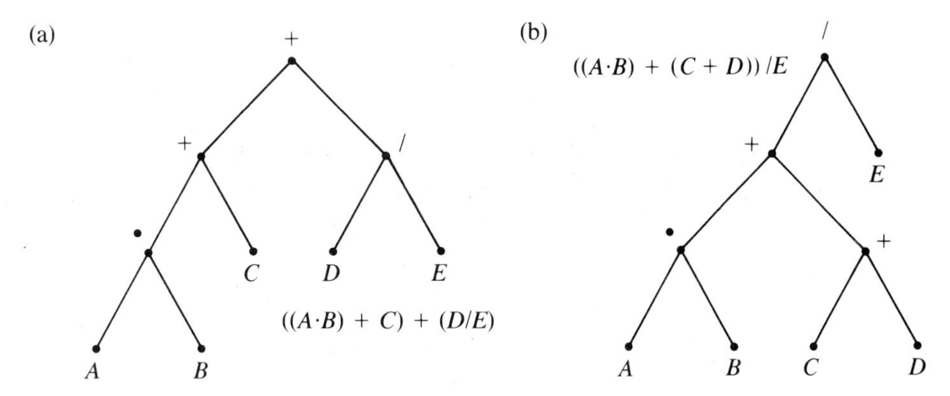

FIGURE 6.21

In certain applications for these ideas we may be less than completely satisfied with these tree representations, helpful though they may be. Certainly the representation may not be as concise as we would like. Fortunately a linear data representation, as developed by the Polish logician Jan Lukasiewicz, will retain all the structural information of the tree. Probably because the word "Polish" is easier to pronounce, this representation has come to be known as the *Polish form* (or the *postfix form*) of the expression. We simply write:

left operand—right operand—operation

instead of the conventional *infix form*:

left operand—operation—right operand

the one we have all grown accustomed to seeing. With the Polish form, there is no need for parentheses. For this reason the Polish expressions are often called *parenthesis-free* expressions.

Example 6.26 Once again we consider the expression

$$((A \cdot (B + C)) + (D/E))$$

of Example 6.24. Transferring the last scheduled operation $(+)$ to the rear we write

$$(A \cdot (B + C))(D/E)+$$

and if the same is done for each subexpression, we will have

$$(A \cdot (B + C)) \longrightarrow A(B + C) \cdot \longrightarrow ABC+\cdot$$

$$(D/E) \longrightarrow DE/$$

Finally we will obtain the Polish expression:

$$A\ B\ C\ +\ \cdot\ D\ E\ /\ +$$

One might say that such parenthesis-free expressions would make sense only if we could reconstruct the tree. But this reconstruction is quite straightforward if we use the following *underlining procedure.* We read the expression from left to right, looking for an operation symbol. When one is found we back up to locate its two operands, and we underline all three symbols. In the case of the present example, we read to the first operator (+) and we back up to find B and C as operands. Underlining these three symbols we have:

$$A\ \underline{B\ C\ +}\ \cdot\ D\ E\ /\ +$$

Now if this group is treated as an operand in itself, then we can start all over again. That is, we draw a line back from the multiplication symbol and under its two operands, A and the group just underlined. Continuing to scan from left to right, looking for new operations, we find the division operator (/) and we back up to find D and E as its operands. Underlining these, we obtain

$$\underline{A\ \underline{B\ C\ +}\ \cdot}\ \underline{D\ E\ /}\ +$$

And finally we have

$$\underline{\underline{A\ \underline{B\ C\ +}\ \cdot}\ \underline{D\ E\ /}\ +}$$

From this representation it is an easy matter to reconstruct the tree of the expression. In the present example we obtain Figure 6.20(b) once more. ■

The inverse of this underlining procedure is equally interesting. Thus we can define a recursive algorithm for deriving the Polish form of an expression from its tree representation. We have only to arrange for a *post-order traversal of a tree*:

---

traverse the left subtree
traverse the right subtree
visit the root

(compare with the in-order traversal introduced earlier). But remember: The subtree traversals must be performed in this same post-order fashion, recursively. Again we leave the details of the algorithm as an exercise for the reader (see Exercise 21).

Example 6.27 Suppose we wish to convert the trees of Example 6.25 (Fig. 6.21) to Polish form. In (a) we obtain, recursively:

$$\begin{array}{lll} \text{left subtree} & \text{right subtree} & + \\ \text{left subtree} & D\ E\ / & + \\ & \cdots & \\ & A\ B \cdot C + D\ E\ / + & \end{array}$$

and, similarly, in (b) we obtain:

$$A\ B \cdot C\ D + + E\ /$$

after several recursive post-order traversals. ■

EXERCISES 6.3

1 Find spanning trees by the method of the proof of the corollary for each of the following graphs.

(a) Exercise 1(a) of Exercises 6.1
(b) Exercise 1(c) of Exercises 6.1
(c) Exercise 3(a) of Exercises 6.1
(d) Exercise 3(b) of Exercises 6.1
(e) Exercise 3(c) of Exercises 6.1
(f) Example 6.13 (Fig. 6.9)
(g) Example 6.14 (Fig. 6.10)
(h) The cube with its (four) diagonals

2 For the same graphs (as in Exercise 1), again obtain spanning trees, this time by approaching the problem from the other direction (as in Example 6.19).

3 Having obtained a spanning tree for each graph, using the method of Exercise 1, use the technique illustrated in Example 6.18 to obtain in each case a basis for the circuit space.

4 Suppose that the entries in the following matrices represent feasible pipeline routes among six cities. If the costs or distances are as shown, design a minimal cost pipeline system connecting the cities. List your edges in their order of selection.

(a)

	a	b	c	d	e
b	130				
c	95	140			
d		170	175		
e	120	100		160	
f	90	125	75	150	110

(b)

	a	b	c	d	e
b	105				
c	100	150			
d	210	210	210		
e	225		230	250	
f		200	210	205	230

5 Show that exactly $\rho(G)$ edges must be removed from a graph G in order to obtain a spanning tree (forest).

6 Characterize all connected graphs having the same number of vertices as edges. That is, what must such a graph look like, and why?

7 Use the treesort algorithm to sort the following texts or partial texts into alphabetical order:

(a) Our father, who art in heaven; hallowed be thy name.

(b) Yesterday, love was such an easy game to play.

(c) O say can you see by the dawn's early light, what so proudly we hailed,...

(d) Warning: The surgeon general has determined that cigarette smoking is dangerous to your health.

(e) When you walk through a storm, hold your head up high; and don't be afraid of the dark.

(f) Wynken, Blynken, and Nod one night sailed off in a wooden shoe,...

8 Use the treesort algorithm to sort the following numbers into numerical order.

(a) $3, 8, 2, 6, 5$ **(b)** $19, 21, 6, 20, 25$

(c) $103, 81, 94, 115, 32, 70$ **(d)** $2, 3, 5, 6, 8$

9 Show a trace of the execution of the treesort algorithm for the sequences given in Exercise 8 as input.

10 Provide an argument showing why $n \log n$ comparisons (on the average) are required in performing a tree sort.

11 Estimate the computational complexity in searching a balanced sorted tree in order to locate a given item. (*Hint:* Suppose the tree is full in the sense that $n = 2^k - 1$ items are stored altogether at levels $0, 1, \ldots, k - 1$.)

12 Obtain binary labeled trees for the following arithmetic expressions:

(a) $(A/B) + C \cdot (D + E)$ **(b)** $(A/B) + ((C \cdot D) + E)$

(c) $A/((B + C) \cdot (D + E))$ **(d)** $(A/(B + C)) \cdot (D + E)$

13 Obtain the Polish form of the expressions in Exercise 12.

14 Use the underlining procedure to transform the Polish expressions obtained in Exercise 13 into their corresponding binary labeled tree representation.

15 Using the underlining procedure, obtain the binary labeled tree corresponding to the following Polish expressions:

(a) $A\ B \cdot C + D\ /$ **(b)** $A\ B \cdot C\ D + / E +$

(c) $A\ B \cdot C\ /\ D\ E + \cdot$ **(d)** $A\ B\ C + D\ E\ / \cdot \cdot$

(e) $A\ B\ C\ D + \cdot E + /$ **(f)** $A\ B\ C \cdot + D\ E\ / +$

16 Use the post-order tree traversal technique to recover the corresponding Polish expressions from the trees obtained in Exercise 15.

17 From the trees obtained in Exercise 15, derive the corresponding parenthesized infix expressions.

18 For each of the binary labeled trees of Figure 6.22, obtain the corresponding infix and Polish forms.

19 Write a formal recursive algorithm for reading out the items of a sorted tree in order.

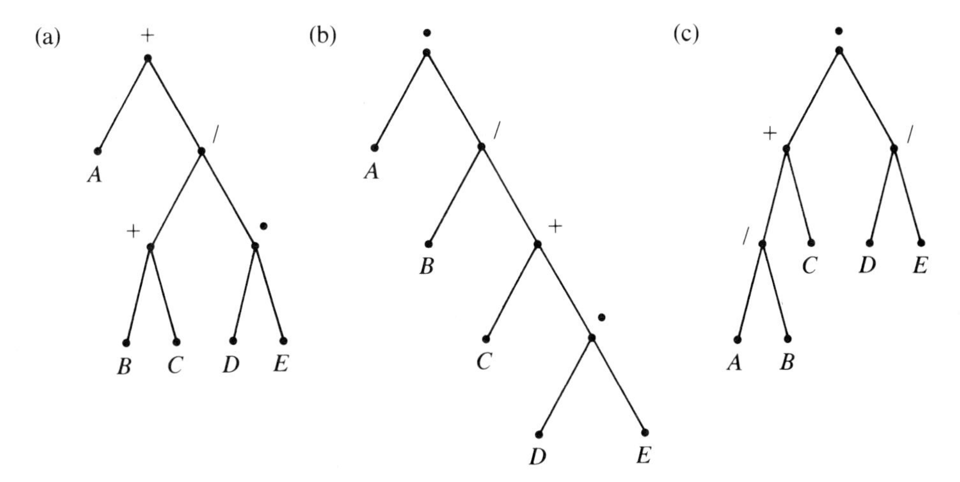

FIGURE 6.22

20 Write a formal algorithm that will implement the underlining procedure, transforming a Polish expression into the corresponding tree structure.

21 Write a formal recursive algorithm for transforming the tree structure of an expression into the Polish form.

22 With a connected graph G and one of its edges e as input, design an algorithm that will decide whether or not "G with e deleted" is connected. (*Note:* Such an algorithm is needed in implementing the technique for constructing spanning trees, as illustrated in Example 6.17.)

23 Design pseudocode algorithms for obtaining a basis for the circuit space of a graph

(a) using the construction implicit in the proof of Theorem 6.2.

(b) using the method illustrated in Example 6.15.

(c) using the method illustrated in Examples 6.17 and 6.18. (*Note:* See Exercise 22.)

24 Compare the algorithms of Exercise 23 from the standpoint of their computational complexity.

25 Verify that the collection of circuits obtained in Example 6.18 is independent.

Sec. 6.4 *PLANAR GRAPHS*

A graph is said to be planar if it can be drawn or pictured in the plane in such a way that its edges intersect only at the vertices of the graph.

It is often the case that a special class of graphs will play a special role in the applications. Already we have seen that this is true for the class of graphs we call *trees*. It is equally true for a broader class of graphs, those we call *planar*. We will see that in studying these rather special graphs we will gain broad insights into the gene .l theory as well.

Goals

After studying this section you should be able to:

- Define the notion of a *planar graph.*
- Describe the two fundamental (small) examples of *nonplanar graphs.*
- State Kuratowski's Theorem as a necessary and sufficient condition that a graph be planar.
- Describe the general Euler formula and provide an interpretation in relation to polyhedra.
- List several tests for planarity and illustrate their use in the case of specific examples.

Planarity

As noted in our opening headline, a graph G is said to be **planar** if it can be given a pictorial representation in which its edges intersect only at the vertices. Intuitively we see that in identifying a planar graph G with a picture $G = (V, E, R)$, we will partition the plane into a collection R of *regions*, one of which is infinite or unbounded—representing the exterior of the graph in the plane. The boundaries or *contours* of the various regions will be formed by the edges of the graph. We may suppose that a definite planar pictorial representation has been chosen, and we let $r = |R|$ denote the number of these regions, including the unbounded one. In fact we will find that r is not dependent on the choice of the planar representation. And, except perhaps in the case of the infinite region, the boundaries or contours will correspond to circuits of the graph.

Example 6.28 The graph of Example 6.1 is planar because it can be drawn as shown in Figure 6.23. Here we have $r = |R| = 4$ because

$$R = \{R_1, R_2, R_3, R_4\}$$

the fourth region being infinite. ■

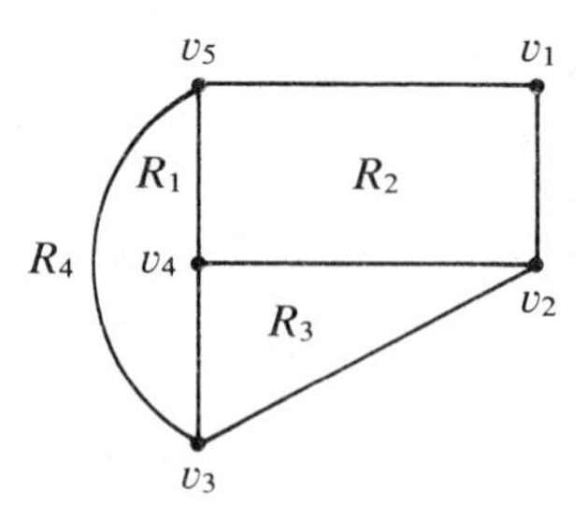

FIGURE 6.23

Example 6.29 The graph of Example 6.17 (Fig. 6.12(a)) is also planar. It can be drawn without intersecting edges as shown in Figure 6.24. Here we have $r = 8$, as the reader will quickly see. ■

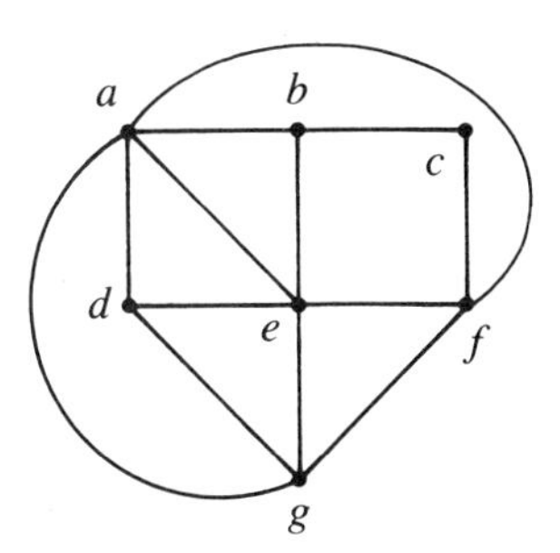

FIGURE 6.24

It is not always so easy to demonstrate that a graph is nonplanar, however compelling the evidence. If it is so that no matter how the graph is drawn, there always will be intersecting edges, this may be quite a difficult fact to prove. Fortunately, certain conditions have been developed for testing planarity; we will touch on a few of these later in this section.

Example 6.30 In the arid southwestern corner of the United States, three houses have been built side by side. Each has its own well at some distance to the rear of the house. Due to lack of adequate rainfall, one or more of the wells frequently runs dry. So it is essential that each family have access to all three wells. As time goes by, the families develop rather strong dislikes for one another. They decide that it would be best for all concerned if they were to each construct paths to all three wells in such a way that they would never have to meet as they walk to and from the wells. But is this possible?

Evidently the proposed network of walkways is the bipartite graph $K_{3,3}$ of Figure 6.4, where there are intersections at every turn. We try to improve the situation with the picture of Figure 6.25, but try as we may we always seem to fall one path short of our objective. Later we will be able to *prove* that the graph $K_{3,3}$ is nonplanar. So it is no wonder that we are having difficulty. ■

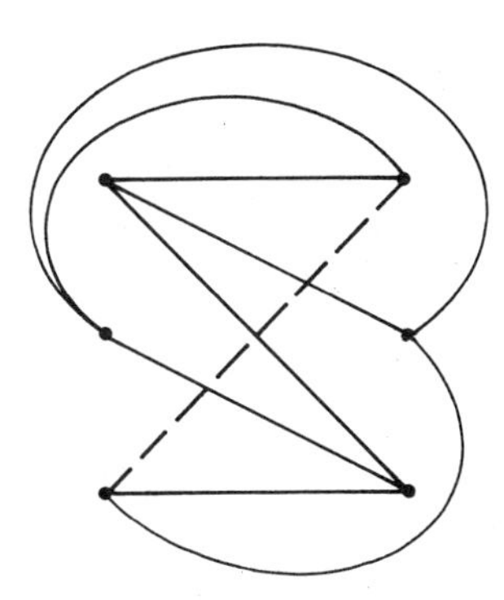

FIGURE 6.25

Questions of planarity arise in certain applications of graph theory to computer science, though we cannot say that they are always of primary concern. Let us mention briefly the printed circuit problem and the flowcharting problem, merely to give an indication of those areas where planarity or near-planarity play a role.

A printed circuit is an electrical network, usually mass fabricated by an electroplating process, the wires being deposited on a nonconducting surface. Such a network can be printed on a single surface if and only if the underlying graph of the network is planar. Actually it is just as common to use both sides of the printed circuit board, in which case the possibilities for network realization depend on the nature of the interconnections permitted from side to side. Suppose that we allow all vertices to appear in duplication to either side, with conducting wires joining the node pairs through holes in the nonconducting board. Even then we will find that there are nonrealizeable networks.

Flowcharts are prepared on a two-dimensional grid so as to allow statements executed sequentially to be placed in a reasonable proximity. In the interest of clarity it is desirable that there be as few overlapping or crossing flowlines as possible. Thus, planarity or near-planarity becomes an objective here as well.

The Euler Formula

As we develop a few of the notions surrounding the study of planar graphs, the reader will gain broader insights into the general theory of graphs. Our first result, Euler's formula, may be known to the student already, particularly when phrased in the language of geometric polyhedra. As with most of our results thus far, the Euler formula follows quite easily from the circuit rank formula (Theorem 6.2), but only after a preliminary result has first been established. We call this result a lemma but, actually, it is quite important in its own right. In particular we should note its obvious utility in connection with the electrical network analysis problem, the one that motivated the discussion of Section 6.2.

Lemma In any planar graph $G = (V, E, R)$, the contours of the finite regions are a basis for the circuit space of G.

Proof We use induction on the number of finite regions. If there is only one finite region (or no finite region), then the statement of the lemma is obviously true. So suppose that the statement holds for all planar graphs with r regions altogether ($r \geq 2$), and let

$$G^+ = (V, E^+, R^+)$$

be planar with $|R^+| = r + 1$. Then we can delete an edge $e = \{v, w\}$ separating a finite region from the infinite region (see Fig. 6.26) to obtain a graph $G = (V, E, R)$ with $E = E^+ \sim \{e\}$ and $|R| = r$. By the inductive hypothesis, the contours of the finite regions of G form a basis for the circuit space, and $\rho(G) = r - 1$. In replacing

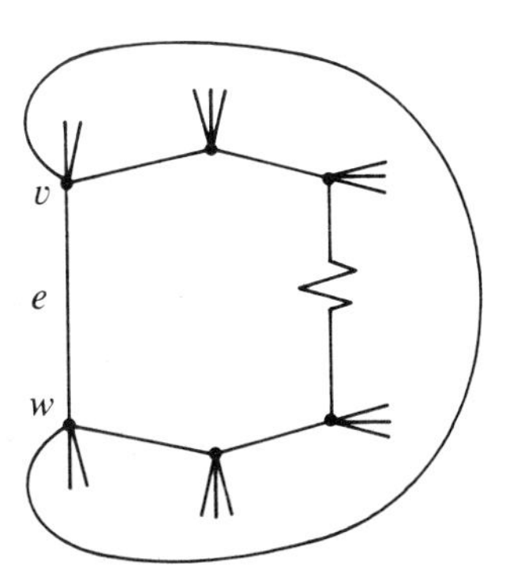

FIGURE 6.26

e to recover the graph G^+, we note that a new circuit is created, and it is independent of the finite contours of G because none of these circuits involves the edge e. Since v and w were already joined by a path in G, we have

$$\rho(G^+) = \rho(G) + 1 = (r - 1) + 1 = r$$

And because this coincides with the number of finite contours of G^+, our proof is complete. □

Theorem 6.5 ***(The Euler Formula)*** In a connected planar graph $G = (V, E, R)$ with

$$|V| = n \qquad |E| = m \qquad |R| = r$$

we have the formula

$$n - m + r = 2$$

Proof According to the lemma, the number of finite regions of G is the same as the circuit rank $\rho(G)$. Using Theorem 6.2 for the case $p = 1$, we obtain

$$r = \rho(G) + 1 = (m - n + 1) + 1 = m - n + 2$$

This is equivalent to the Euler formula. □

Example 6.31 Checking this formula against the data of Examples 6.28 and 6.29, we have

$$n - m + r = 5 - 7 + 4 = 2$$
$$n - m + r = 7 - 13 + 8 = 2$$

respectively. Of course it is more often the case that two of the quantities n, m, r are known, and we then use the formula to determine the third. ■

Example 6.32 In geometry, Theorem 6.5 is known as Euler's polyhedral formula. Consider the stereographic projection of a planar graph onto a sphere, where we imagine the sphere to be resting on the plane at its south pole (see Fig. 6.27(a)). From each vertex v of the graph in the plane (and from points along the edges as well), a ray is drawn to the north pole. The point of intersection with the sphere is taken as the *image* v' of the vertex v. We may speak of stereographic projection in either direction, of the planar graph onto the sphere, or of a polyhedron into a planar graph. In the first case we see that on the sphere there is no difference between the infinite region and any of the others. (The image of the infinite region will contain the north pole.) In fact, the images of the regions are the faces of a somewhat distorted polyhedron. In this context, *Euler's polyhedral formula* is usually written:

$$\text{number of vertices} - \text{number of edges} + \text{number of faces} = 2$$

for any polyhedron. ■

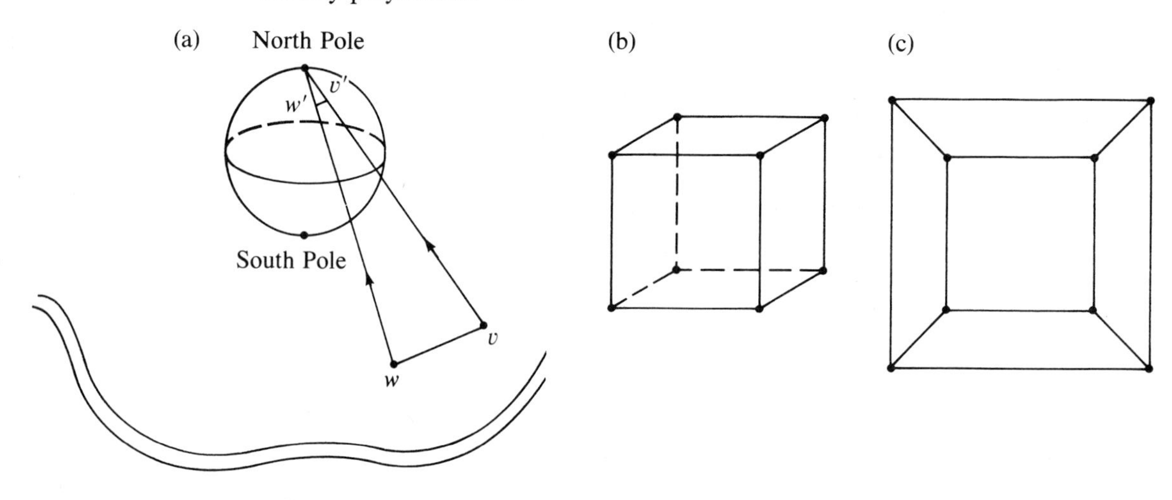

FIGURE 6.27

Example 6.33 Consider the cube, shown as a polyhedron and as its stereographic projection, a planar graph (Fig. 6.27(b) and 6.27(c), respectively). Note in particular that the infinite region of the graph is identified as one of the six faces after projection onto the sphere. In any case we have

$$\text{number of vertices} - \text{number of edges} - \text{number of faces} = 8 - 12 + 6 = 2$$

as predicted by Euler's polyhedral formula. ■

Tests for Planarity

We now attempt to develop a few criteria for determining whether or not a given graph is planar. As we have seen, unless one is able somehow to obtain a pictorial representation for the graph, the question of planarity can be rather difficult to decide.

Theorem 6.6 In a planar graph $G = (V, E, R)$, where

$$|E| = m \qquad |R| = r$$

we must always have the inequality

$$3r \leq 2m$$

If in addition we know that every region has at least four edges in its contour, then we have

$$4r \leq 2m$$

Proof In general, each region has at least three edges in its contour. But when counting the edges that appear in these contours, we have counted each edge twice, and so the given inequality follows. The more stringent inequality is derived in exactly the same way, if it happens that every region indeed has at least four edges in its contour. □

Corollary The so-called *Kuratowski graphs* (Figs. 6.3 and 6.4)

$$K_5 \quad \text{and} \quad K_{3,3}$$

are nonplanar.

Proof If K_5 were planar, then according to Theorem 6.5 we would have

$$r = m - n + 2 = 10 - 5 + 2 = 7$$

But Theorem 6.6 asserts that

$$3r \leq 2m \quad (\text{or } 21 \leq 20)$$

This is an obvious contradiction. Again, if $K_{3,3}$ were planar, the number of regions would be

$$r = m - n + 2 = 9 - 6 + 2 = 5$$

But since no region of $K_{3,3}$ can be triangular, owing to the definition of bipartite, the more stringent inequality of Theorem 6.6 can be used to obtain

$$4r \leq 2m \quad (\text{or } 20 \leq 18)$$

again a contradiction. □

Example 6.34 This last result shows that there can be no solution to the dilemma of the three southwestern neighbors (Example 6.30). ■

Example 6.35 Note that if we insert vertices of degree 2 along an edge of a nonplanar graph (say $K_{3,3}$), as in Figure 6.28, then the resulting graph is just as nonplanar as before. The same is true with the removal of such vertices of degree 2. These constructions have no effect on the planarity or nonplanarity of a graph. This is the sense of the parenthetical insertion in the following remarkable theorem of Kuratowski. ■

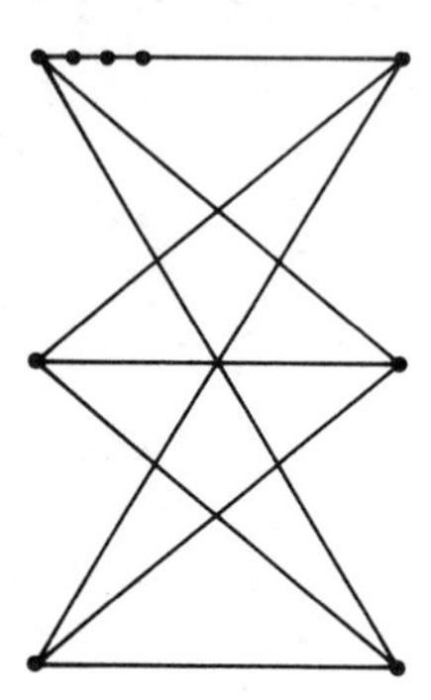

FIGURE 6.28

Theorem 6.7 ***(Kuratowski)*** A graph is planar if and only if it does not contain K_5 or $K_{3,3}$ (perhaps modified by the insertion of vertices of degree 2) as a subgraph.

Proof Our corollary establishes the necessity of the condition. The proof of sufficiency is rather involved and will not be given here. □

Until the appearance of Kuratowski's result in 1930, the problem of characterizing the planar graphs had not been solved. Now there are several other known criteria, but none more striking than this (small) catalog of "forbidden" subgraphs. And yet the conditions of Kuratowski's theorem can be quite difficult to apply in practice. For this reason it is best that we provide a few more necessary conditions for planarity, as follows.

Theorem 6.8 Let $G = (V, E, R)$ be a connected planar graph. Then we must have the inequalities:

1. $m \leq 3n - 6$
2. $d(v) \leq 5$ for some vertex v

Proof In establishing (1) we use Theorems 6.5 and 6.6 as follows:

$$2 = n - m + r \leq n - \frac{3}{2}r + r = n - \frac{1}{2}r \quad (r \leq 2n - 4)$$

$$2 = n - m + r \leq n - m + (2n - 4) \qquad (m \leq 3n - 6)$$

As for (2), we again have $3r \leq 2m$ because of Theorem 6.6. Now if we had $d(v) > 6$ for every vertex v, then an easy counting argument would yield $6n \leq 2m$, and according to the Euler formula we would obtain

$$2 = n - m + r \leq \frac{1}{3}m - m + \frac{2}{3}m = 0$$

an obvious contradiction.

Corollary Let G and G' be complementary graphs on n vertices. If $n < 8$, then at least one of them is planar, whereas if $n > 8$, then at least one of them is nonplanar.

Proof Since $G + G'$ is the complete graph K_n, we may use the result of Example 6.6 to write

$$m + m' = \frac{1}{2}n(n - 1)$$

and it follows that one of the two numbers m, m' must be greater than or equal to $\frac{1}{4}n(n - 1)$. But Example 3.12 shows that

$$\frac{1}{4}n(n - 1) > 3n - 6 \quad (n \geq 11)$$

so we simply use (1) in Theorem 6.8 when $n \geq 11$.

Conversely, one of the two numbers m, m' is less than or equal to $\frac{1}{4}n(n - 1)$. Furthermore you can check that

$$\frac{1}{4}n(n - 1) < 9 \quad (n \leq 6)$$

whereas the Kuratowski graphs K_5 and $K_{3,3}$ have at least nine edges. So we can apply Kuratowski's theorem in case $n \leq 6$.

The remaining cases $n = 7, 9, 10$ have been exhaustively analyzed, and the results are found to be consistent with the statement of the corollary. □

Example 6.36 The graph on nine vertices in Figure 6.29 is planar. We can use the corollary to assert that the complementary graph is nonplanar. The corollary says

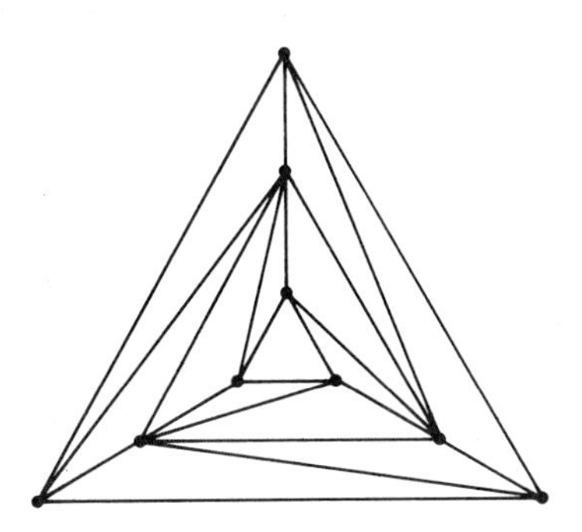

FIGURE 6.29

nothing about the complement of the graph in Figure 6.27(c), however, since it has eight vertices. So far as we know, it may or may not be planar. ■

Example 6.37 Consider the two-sided circuit board problem mentioned earlier. Suppose we have an electrical network whose underlying graph is K_9. Then no matter how we separate the graph into the decomposition $G + G'$, our corollary shows that one of them will be nonplanar. It follows that we cannot realize this circuit within the given physical constraints. ■

EXERCISES 6.4

1 An ancient esoteric practice, Cabalism (Kabbalah) persists even to this day in certain regions of the Near East. It represents one of the world's oldest systems of mystical thought, believed by some to hold the key to all the mysteries of the universe. The essence of the Kabbalah is symbolized in the graph of Figure 6.30 (10 vertices and 22 edges), depicting the emanations of God. Is the graph of the Kabbalah a planar graph? Either prove that it is nonplanar or find a planar representation.

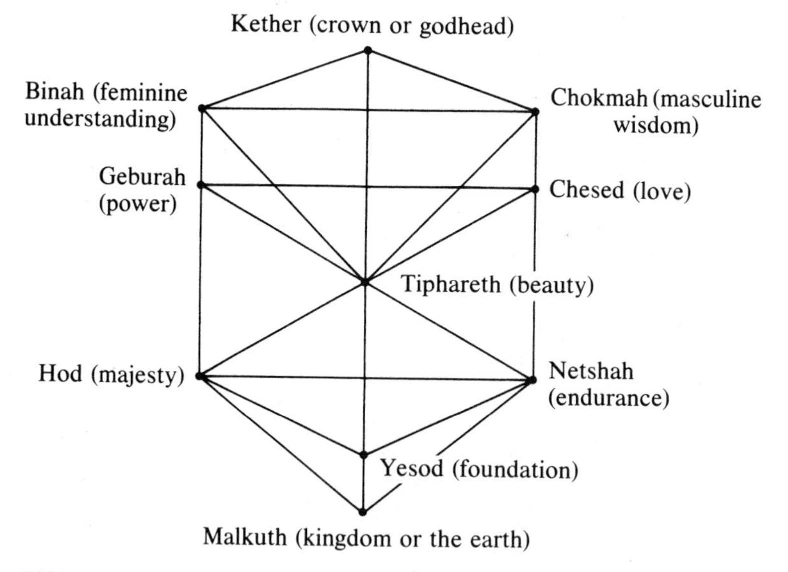

FIGURE 6.30

2 Which of the following gr:.phs are planar? In each case, either prove that the graph is nonplanar or exhibit a planar representation.

(a) Exercise 1(c) of Section 6.1

(b) The cube with its (four) diagonals

(c) The complementary graph to that in (b)

(d) Exercise 3(a) of Section 6.1

(e) $K_{2,4}$

(f) The complementary graph to that in Figure 6.27(c)

3 Without using Kuratowski's theorem, prove that

(a) K_n is nonplanar for all $n > 4$

(b) $K_{m,n}$ is nonplanar for all $m, n > 2$

4 Verify Euler's formula for the graph of the Kabbalah (Exercise 1).

5 Verify Euler's formula for each of the following planar graphs.

(a) the tetrahedron (four triangular faces)

(b) the dodecahedron (twelve pentagonal faces)

(c) the octahedron (eight triangular faces)

(d) a chess board (81 vertices)

(e) a bingo card (36 vertices)

(f) the star of David (12 vertices)

6 Sometimes we are able to say that a graph is nonplanar simply because there are "too many" edges. Explain.

7 Show that

$$\frac{1}{4}n(n-1) < 9$$

for all $n \leq 6$, as required in the proof of the corollary to Theorem 6.8.

8 Given the map shown in Figure 6.31, obtain the associated (planar) graph. (*Note:* The states are considered as vertices and we have an edge between two given vertices if the corresponding states are contiguous.)

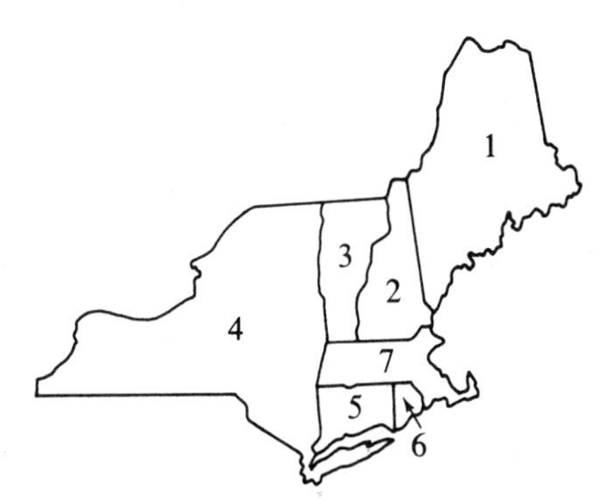

FIGURE 6.31

9 Do the same (as in Exercise 8) for the countries of South America.

Sec. 6.5 ISOMORPHISM AND INVARIANTS

The term isomorphism is a "ten-dollar word" used to express the idea of abstract equivalence.

The notion of *abstract equivalence* is one used fundamentally throughout mathematics. The technical term *isomorphic* (the adjective) is used to express this equivalence between one mathematical object and another. The student first sees the concept in high-school geometry, although the term *congruent* is used there instead. For graphs, the meaning should be clear—one expects that isomorphic graphs can be pictured in exactly the same way, as soon as one can discover a correspondence in the labeling of vertices and edges. The discovery of such a correspondence (or a proof that no such correspondence exists), however, may not spring so quickly to mind. And thus it is useful to have a catalog of *invariants* that can be used to test for isomorphism. Most of these have an independent importance, quite apart from the isomorphism question. For this reason we devote considerable attention to their interpretation and to their effective computation.

Goals

After studying this section you should be able to:

- Define the notion of *isomorphism* for graphs, and discuss its significance.
- Define the notion of an *isomorphism invariant* and describe the usefulness of invariants in deciding the isomorphism question.
- Describe the *numeric graphical invariants* and provide interpretations of their significance in relation to specific problems in the applications.
- State the *four-color theorem* and discuss the historical background of the coloring problem for planar graphs.

The Notion of Isomorphism

How many graphs are there on n vertices? We need to have such figures right at our fingertips in case any sort of exhaustive analysis is required. For example, how many graphs need to be considered in analyzing the cases $n = 7, 9, 10$ in the corollary to Theorem 6.8? Evidently such answers have something to do with the number of entries in the triangular matrix representation of a graph. In fact, since Example 6.6 shows this figure to be $\frac{1}{2}n(n-1)$, and we have two choices for each entry, the corollary to Rule 1 in Section 1.5 leads us to conclude that there are $2^{n(n-1)/2}$ graphs on n vertices. Note that when $n = 7$, this number is already more than two million!

The number can be greatly reduced by considering two graphs to be *essentially the same* when they differ only in the naming or labeling of their vertices.

Thus we say that the graphs $G_1 = (V_1, E_1)$ and $G_2 = (V_2, E_2)$ are **isomorphic** (this is the ten-dollar word) if there exists a one-to-one correspondence

$$f: V_1 \to V_2$$

such that for all pairs of vertices v and w

$$v \; E_1 \; w \Leftrightarrow f(v) \; E_2 \; f(w)$$

and the mapping f is called an **isomorphism** of the graphs G_1 and G_2.

Example 6.38 Let $G_1 = (V_1, E_1)$ be the graph of Example 6.1, that is

$$V_1 = \{v_1, v_2, v_3, v_4, v_5\}$$
$$E_1 = \{\{v_1, v_2\}, \{v_1, v_5\}, \{v_2, v_3\}, \{v_2, v_4\}, \{v_3, v_4\}, \{v_3, v_5\}, \{v_4, v_5\}\}$$

and suppose $G_2 = (V_2, E_2)$ has the vertices and edges

$$V_2 = \{w_1, w_2, w_3, w_4, w_5\}$$
$$E_2 = \{\{w_1, w_2\}, \{w_1, w_4\}, \{w_1, w_5\}, \{w_2, w_3\}, \{w_2, w_5\}, \{w_3, w_4\}, \{w_4, w_5\}\}$$

respectively, as shown in Figure 6.32. Then we can check that the mapping

$$f: V_1 \to V_2$$

given by

$$f(v_1) = w_3$$
$$f(v_2) = w_2$$
$$f(v_3) = w_5$$
$$f(v_4) = w_1$$
$$f(v_5) = w_4$$

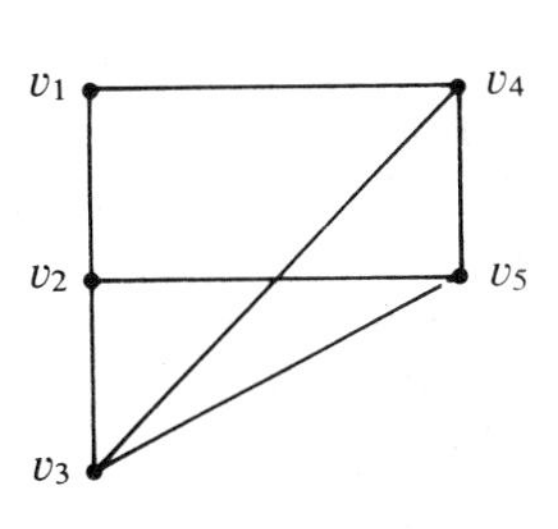

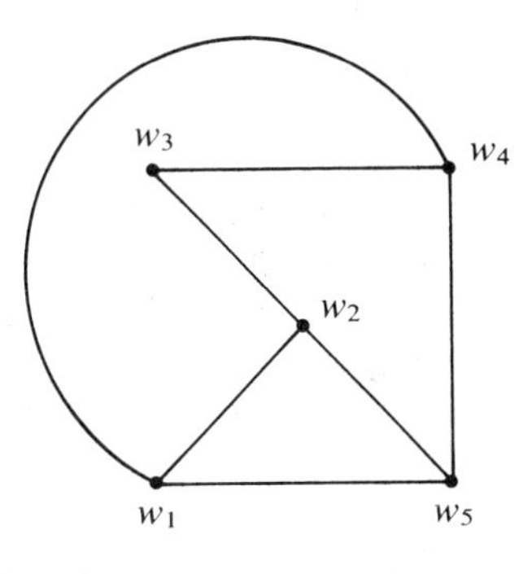

FIGURE 6.32

is an isomorphism of graphs. First of all it is obviously a one-to-one correspondence of the two vertex sets. Moreover we can check that

$$v_i \; E_1 \; v_j \Leftrightarrow f(v_i) \; E_2 \; f(v_j)$$

For instance,

$$\begin{aligned} v_1 \; E_1 \; v_2 \quad &\text{and also} \quad w_3 = f(v_1) \; E_2 \; f(v_2) = w_2 \\ v_1 \; E_1 \; v_5 \quad &\text{and also} \quad w_3 = f(v_1) \; E_2 \; f(v_5) = w_4 \\ v_1 \; E_1 \; v_3 \quad &\text{and also} \quad w_3 = f(v_1) \not{E}_2 \; f(v_3) = w_5 \end{aligned}$$

etc., so that except for the relabeling of the vertices the two graphs are (as we say) *abstractly equivalent*. ■

Example 6.39 The two graphs in Figure 6.33 are not isomorphic, even though they each have five vertices and six edges. Any candidate for an isomorphism f must map the two vertices v, w of degree 3 onto the vertices of degree 3 in the other graph (why?), say $f(v) = v'$ and $f(w) = w'$, as labeled in Figure 6.33. Nevertheless we would have

$$v \; E \; w \quad \text{whereas} \quad v' \not{E}' \; w'$$

Reversing our assignment wouldn't change matters. ■

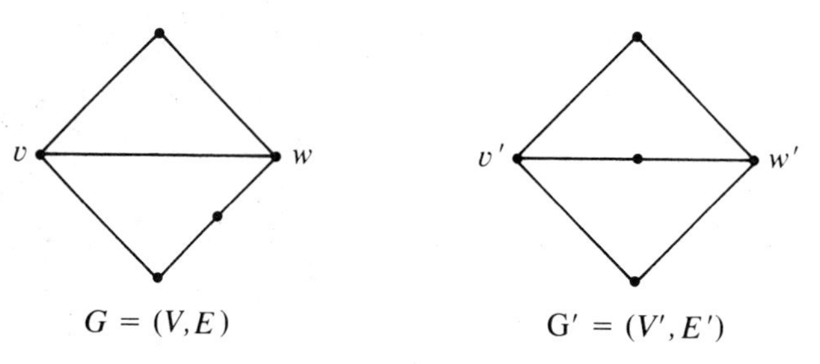

FIGURE 6.33

In reexamining Example 6.38, we might find one facet of the argument particularly bothersome. We could ask, How do we find such a mapping? It would appear that in order to test two graphs for isomorphism, we must examine $n!$ mappings, n being the number of vertices in each of the graphs. We say "each of the graphs" because we cannot have a one-to-one correspondence, let alone an isomorphism, unless the number of vertices agree. And, according to the corollary to Rule 3 of Section 1.5, there are then $n!$ one-to-one correspondences to consider. So again we come up against a rather large number as a potential obstacle to the investigation.

Graphical Invariants

Conceptually, of course, there is no particular difficulty in testing for graphical isomorphism along the lines just suggested, but on purely technical grounds we must reject this straightforward approach of examining all one-to-one correspondences. One is in a much better position to answer the isomorphism question if a few of the graphical invariants are used as tools in the quest. Suppose we write $G_1 \simeq G_2$ if G_1 and G_2 are isomorphic. Then a **graphical invariant** is a function η from graphs to the nonnegative integers (or sequences of integers) for which

$$G_1 \simeq G_2 \Rightarrow \eta(G_1) = \eta(G_2)$$

Note that we already have a few elementary invariants at our disposal:

$n = n(G) =$ the number of vertices in G

$m = m(G) =$ the number of edges in G

$p = p(G) =$ the number of connectivity components in G

In the first instance we can see why we limit our attention to graphs with an equal number of vertices. That is, in reading the contrapositive statement

$$n(G_1) \neq n(G_2) \Rightarrow G_1 \not\simeq G_2$$

we see that graphs with an unequal number of vertices cannot possibly be isomorphic. And, of course, the same is true with the number of edges. Since an isomorphism establishes a one-to-one correspondence of vertices, it also will have set up such a correspondence between the edges as well. It follows that two graphs cannot be isomorphic unless they at least agree in each of these respects.

At the same time, Theorem 6.2 shows that

$$\rho = \rho(G) = \text{circuit rank of } G$$

is another graphical invariant. When we actually employ a graphical invariant η, it is usually in the contrapositive form:

$$\eta(G_1) \neq \eta(G_2) \Rightarrow G_1 \not\simeq G_2$$

as indicated above. Thus, if G_1 and G_2 disagree on a particular invariant, then they cannot possibly be isomorphic and we don't need to examine all of the possible one-to-one correspondences of vertices—only to come to the same conclusion.

If a graphical invariant is to be useful, it should be fairly easy to compute and, eventually, we should provide algorithms for their computation. Certainly these conditions have been met for the invariants introduced thus far. Now consider the row sums (the number of ones in a particular row) of the full relation matrix of a graph. Evidently these figures provide the degrees $d(v)$ of the corresponding vertices, which in turn allow for the computation of the **degree spectrum**

$$\Delta = \Delta(G) = \langle d_0, d_1, \ldots, d_{n-1} \rangle$$

of a graph G. This invariant is a sequence of nonnegative integers, where each

$$d_i = \text{the number of vertices of degree } i$$

is easily computed, once the degrees $d(v)$ are available for each vertex.

In proceeding to define further graphical invariants, our hope would be to discover a collection $\{\eta_1, \eta_2, \ldots, \eta_N\}$ of invariants, such that

$$G_1 \simeq G_2 \Leftrightarrow \eta_1(G_1) = \eta_1(G_2), \ldots, \eta_N(G_1) = \eta_N(G_2)$$

Such a collection is called a *complete set* of invariants because it would allow a complete resolution of the graphical isomorphism question, relying only on the separate computations. Unfortunately a useful complete set of graphical invariants has not yet been found. Nevertheless the utility of the easily computed invariants cannot be denied.

Example 6.40 The row sums of the full relation matrix of Example 6.1 are 2, 3, 3, 3, 3, respectively. Therefore the graph of that example has the degree spectrum $\langle 0, 0, 1, 4, 0 \rangle$. Since the graph G in Example 6.39 (Fig. 6.33) has the spectrum $\langle 0, 0, 3, 2, 0 \rangle$, we know that these two graphs are not isomorphic. Can you see another way to establish this fact more easily? ■

Example 6.41 The two graphs G and G' of Example 6.39 agree on each of the invariants n, m, p; they have the same degree spectrum as well. And yet the two graphs are not isomorphic (as shown in that example). It is necessary to introduce additional invariants in order to distinguish these two graphs. ■

In spite of the fact that there exist graphs with the same degree spectrum that are not isomorphic, the spectrum itself is quite useful in reducing the number of one-to-one correspondences that may have to be considered. Corresponding vertices must have the same degree in any candidate for a graphical isomorphism (why?). Therefore, in the case of graphs of the same degree spectrum

$$\Delta = \langle d_0, d_1, \ldots, d_{n-1} \rangle$$

we need not consider all the $n!$ one-to-one correspondences of vertices in looking for an isomorphism. The vertices of degree i must be mapped onto vertices of degree i and, accordingly (Rule 1 of Section 1.5), only $d_0! d_1! \cdots d_{n-1}!$ mappings will meet this requirement. Unless all the vertices have the same degree, this number will be smaller, often considerably smaller than $n!$.

Example 6.42 Suppose that we return again to the two graphs of Example 6.39. Each has the degree spectrum $\langle 0,0,3,2,0\rangle$, but in looking for an isomorphism only $3!2! = 12$ mappings need to be considered, a figure that is considerably less than the total number ($5! = 120$) of one-to-one correspondences of vertices. ■

In proceeding to introduce further numerical invariants, we now have a new purpose in mind, one beyond the obvious implications for the isomorphism question. Some of the graphical invariants—for example, the *domination, covering, independence*, and *clique* numbers of a graph—are important in the applications because they express an optimality condition of one form or another. Accordingly these definitions involve various maxima or minima on the size of certain collections of vertices in a graph. These rather special collections may be introduced as follows. If $G = (V, E)$ is a graph, we say that the set of vertices

$$W = \{w_1, w_2, \ldots, w_s\} \subseteq V$$

constitutes

(a) a **dominating set** if

$$v \text{ in } V \Longrightarrow v = w \quad \text{or} \quad v\ E\ w \quad (\text{for some } w \text{ in } W)$$

(b) a **covering set** if

$$u\ E\ v \Longrightarrow u = w \quad \text{or} \quad v = w \quad (\text{for some } w \text{ in } W)$$

(c) an **independent set** if

$$w_i \not{E}\ w_j \quad (\text{for all } w_i, w_j \text{ in } W)$$

(d) a **clique** if

$$w_i\ E\ w_j \quad (\text{for all } w_i, w_j \text{ in } W)$$

In a dominating set, every vertex of the graph is, or is adjacent along an edge, to a vertex of the set. A covering set covers all the edges of the graph in that every edge has at least one of its extremities in the set. All the vertices of an independent set are mutually nonadjacent (i.e., not joined by an edge), whereas in a clique just the opposite is true (i.e., every pair of vertices in a clique is joined by an edge of the graph).

In each of these instances we can introduce a corresponding numerical invariant:

$\delta = \delta(G) = \min |W|$: W a dominating set

$\alpha = \alpha(G) = \min |W|$: W a covering set

$\beta = \beta(G) = \max |W|$: W an independent set

$\beta' = \beta'(G) = \max |W|$: W a clique

These minima or maxima are called the **domination number**, **covering number**, **independence number**, and **clique number**, respectively, for the graph in question.

The naming of this last invariant has an interesting sociological connotation. If the relation $v \ E \ w$ is taken to mean that the persons v and w are friends, acquaintances, or whatever, then the vertices of a clique, being mutually adjacent, represent a group of persons who are mutual friends, in the colloquial use of this term. And, of course, the clique number is the size of a largest clique. Analogously, the independence number is the size of a largest independent set. Such a set would be composed of persons who are mutual enemies (nonfriends). This little discussion illustrates the important principle:

$$\beta'(G) = \beta(G')$$

where G' is the graph that is complementary to G. At the same time it explains our notation (β' vs. β), indicating that there is some general relationship between this pair of numerical invariants.

Example 6.43 The graph of Example 6.1 is reproduced in Figure 6.34(a). The collection $\{v_2, v_3, v_4\}$ is a clique for this graph G, since

$$v_2 \ E \ v_3 \qquad v_2 \ E \ v_4 \qquad v_3 \ E \ v_4$$

Evidently the clique number of a graph is the size (n) of the largest complete subgraph (K_n) to be found. Apparently we have the clique number $\beta'(G) = 3$ in this case.

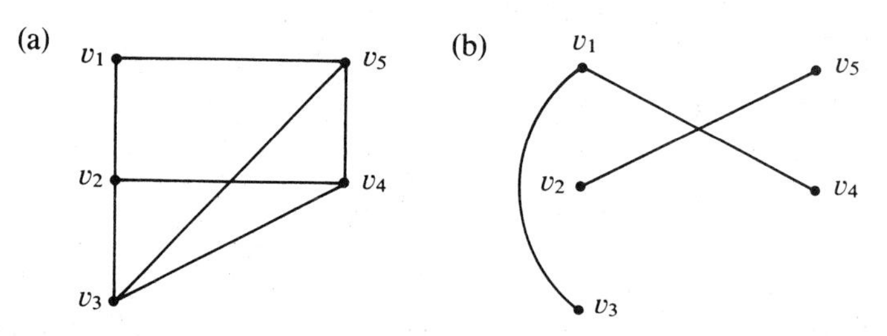

FIGURE 6.34

The complementary graph G' is shown in Figure 6.34(b). Here the same set $\{v_2, v_3, v_4\}$ becomes an independent set of vertices, none of which is joined by an edge. So again we have an illustration of the relationship $\beta'(G) = \beta(G')$. ■

Example 6.44 Consider a communication channel for transmitting symbols from some alphabet A. Due to distortion in the channel, certain symbols are likely to be confused. That is, various pairs of symbols may be received as if they were one and the same. In such a situation it would be advisable not to use the entire alphabet, but to select instead a largest subalphabet having the property that no two letters can be confused. Such a selection would be a largest independent set of vertices in the graph whose edges depict all the pairs of symbols that might be confused. ■

Example 6.45 On a chessboard as shown in Figure 6.35, it is possible to place as many as eight queens in such positions that none is threatened. But if there are nine queens or more, then no matter how they are placed there always will be a threatened queen. It follows that $\beta(G) = 8$ for the graph (on 64 vertices) representing the queen's dominating relation. ■

<table>
<tr><td></td><td></td><td></td><td></td><td></td><td>Q</td><td></td><td></td></tr>
<tr><td>Q</td><td></td><td></td><td></td><td></td><td></td><td></td><td></td></tr>
<tr><td></td><td></td><td></td><td></td><td>Q</td><td></td><td></td><td></td></tr>
<tr><td></td><td>Q</td><td></td><td></td><td></td><td></td><td></td><td></td></tr>
<tr><td></td><td></td><td></td><td></td><td></td><td></td><td></td><td>Q</td></tr>
<tr><td></td><td></td><td>Q</td><td></td><td></td><td></td><td></td><td></td></tr>
<tr><td></td><td></td><td></td><td></td><td></td><td></td><td>Q</td><td></td></tr>
<tr><td></td><td></td><td></td><td>Q</td><td></td><td></td><td></td><td></td></tr>
</table>

FIGURE 6.35

Example 6.46 Suppose that the vertices of a graph $G = (V, E)$ represent various trouble spots in a prison complex, where the edges might correspond to corridors, stairways, etc. The fewest number of guards necessary to control the complex then would be obtained if we could only find a smallest dominating set of vertices. ■

Example 6.47 Suppose, instead, that the edges of $G = (V, E)$ are interpreted as streets (blocks) in a city, and suppose further that it is desired that every street have a fire hydrant at one corner or another (or both), at the ends of the street. Then a minimal number of hydrants needed for this purpose is obtained by determining a smallest covering set of vertices. ■

Example 6.48 Using the graph $G = (V, E)$ of Examples 6.1 and 6.43 (Fig. 6.34(a)) once more, let us see if we can determine the domination and covering numbers. The collection $\{v_2, v_5\}$ is a dominating set, since

$$v_1 \, E \, v_5 \qquad v_4 \, E \, v_5 \qquad v_3 \, E \, v_2$$

Since there is obviously no single vertex that dominates, we must have $\delta(G) = 2$.

In a graph without isolated vertices it is clear that every covering set of vertices is also a dominating set. The converse, however, is not true in general. In this case the dominating set $\{v_2, v_5\}$ is not a covering set because the edge from v_3 to v_4 is not covered. In fact, no pair of vertices covers all the edges of this graph, as you can easily see. Since $\{v_2, v_3, v_5\}$ is a covering set, however, we conclude that $\alpha(G) = 3$. ■

Example 6.49 Consider the graphs G and G' of Figure 6.33. We recall (Examples 6.39 and 6.41) that these graphs are nonisomorphic, and yet they were found to agree on all of our earlier invariants. Since $\alpha(G) = 3$ but $\alpha(G') = 2$, however, we finally have an invariant that distinguishes between the two. ■

In the next section we develop computational techniques for determining all these numerical invariants algorithmically. Surely this is an important matter for us to resolve. For the time being, however, we introduce yet another facet of these investigations. It so happens that there are a number of equalities and inequalities relating the several graphical invariants. We can only afford to present one or two of these results here—but at least these will give an indication of the overall interdependence.

Theorem 6.9 In any graph $G = (V, E)$ we have the relation

$$\beta \geq \delta$$

Proof We will show that an independent set is maximal if and only if it is a dominating set. The inequality then will follow quite easily, since the largest independent set surely will dominate, though it may not be the smallest set with this property.

When we speak of

$$W = \{w_1, w_2, \ldots, w_s\}$$

as a maximal independent set, we mean that we cannot adjoin an additional vertex without losing the independence propery. Such a set must surely dominate because $v \neq w$ and $v \not{E}\, w$ for all w in W would contradict the maximality of W. Conversely, suppose W is a dominating independent set. For any vertex v such that $v \neq w$ for all w in W, we must have $v\, E\, w$ for some w in W, and we thus cannot adjoin v to W while retaining the independence. □

Theorem 6.10 In any graph $G = (V, E)$ with n vertices, we have

$$\alpha + \beta = n$$

Proof This time we show that if $V = A \cup B$ and $A \cap B = \varnothing$, then A is a covering set if and only if B is an independent set. It then will follow immediately that $\alpha + \beta = n$.

With the assumptions that $V = A \cup B$ and $A \cap B = \varnothing$, suppose that A is a covering set. Then consider any two vertices v, w in B. If it were true that $v\, E\, w$, then we would have $v \in A$ or $w \in A$. But either of these would contradict the disjointness of A and B. Consequently, we have $v \not{E}\, w$ for all v, w in B, that is, B is independent. Conversely, if B is an independent set and $v\,E\,w$ then we cannot have both vertices in B. It follows that one of them is in A, and therefore A is a covering set of vertices. □

Example 6.50 Given the graph of Figure 6.34(a), Theorem 6.10 tells us that we should expect to have the independence number

$$\beta = n - \alpha = 5 - 3 = 2$$

recalling the result of Example 6.48. And, sure enough, the largest independent sets to be found are the pairs $\{v_1, v_3\}$, $\{v_1, v_4\}$, and $\{v_2, v_5\}$. ■

Example 6.51 Reviewing Example 6.45 in connection with Theorem 6.9, we note that eight queens can surely dominate a chessboard. Putting one queen on each row will do that. But in fact five queens will suffice, as shown in Figure 6.36. In any case we have $\beta \geq \delta$ as claimed. ■

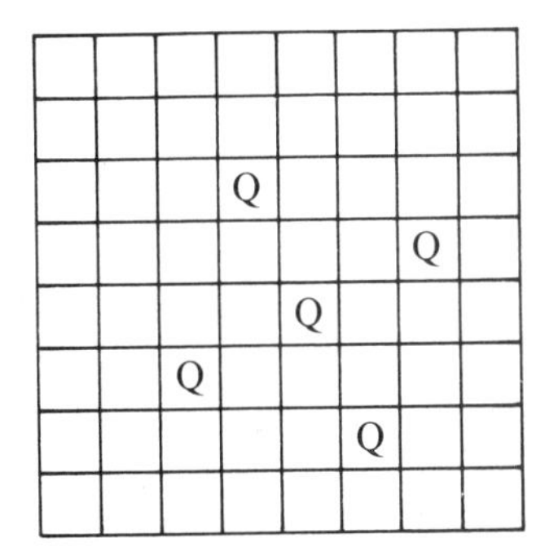

FIGURE 6.36

The Coloring Problem

We have tried to provide vivid illustrations to help demonstrate the meaning of the dominance, covering, independence, and clique numbers of a graph. The meaning of the next invariant to be described here is already suggested by its name—*chromatic number*. And, in fact, the illustration is in technicolor!

Imagine that we have a box of crayons for the purpose of coloring the vertices of a graph $G = (V, E)$. We add the stipulation that vertices that are joined by an edge must be given different colors. What is the fewest number of colors that will suffice? This number is the **chromatic number** $\kappa = \kappa(G)$ for the given graph G. Phrased somewhat differently, let a *proper coloring* of G be a partition

$$V = \bigcup_{i=1}^{s} V_i \qquad (V_i \cap V_j = \varnothing \quad \text{for all } i \neq j)$$

of the vertices, subject to the additional condition:

$$v \text{ and } w \text{ in the same } V_i \Leftrightarrow v \not E w$$

(read the "contrapositive": adjacent vertices should be colored differently). Then $\kappa = \kappa(G)$ is the fewest number of subsets (colors) to be found among such partitions.

Regardless of its applications, we would not have been able to exclude the chromatic number from our sampling of the graphical invariants. That is because this particular number is the subject of one of the most celebrated long-unsolved problems of mathematical history. The problem grew out of the subject of cartography (map making). It is easy to see that if we define an edge relation on a set of countries (as vertices) by agreeing that two vertices are to be joined by an edge just in case the countries are contiguous (sharing a common boundary), then the resulting graph is planar (see Exercise 8 of Exercises 6.4). The question naturally arises, What is the fewest number of colors that need to be used in ensuring that contiguous countries are given a different color? In purely graph-theoretic terms, What is (an upper bound for) the chromatic number of a planar graph? And the answer (in 1976) after two hundred years of study:

Theorem 6.11 ***(Four-color Theorem)*** In any planar graph $G = (V, E)$, we have

$$\kappa(G) \leq 4$$

Proof It long has been known that three colors were not enough to obtain a proper coloring, in general (see Example 6.52) and that five colors would suffice. It was suspected that four colors were sufficient. Finally, in an exhaustive case-by-case analysis, two mathematicians at the University of Illinois were led to a computer-aided proof of this fact. The reader will appreciate being spared of the details. □

Example 6.52 The complete graph K_4 is planar, but obviously requires four colors, that is, $\kappa(K_4) = 4$. This shows that three colors are not sufficient for all the planar graphs. ■

Example 6.53 In scheduling students for classes at a university, consider a graph whose vertices represent the classes, and let there be an edge from v to w if there is a time conflict, that is, because the same student must schedule both classes. Such classes must be arranged to meet at different times. A proper coloring will correspond to a schedule without conflicts. ■

Example 6.54 Consider the graph of Examples 6.1 and 6.43 (Fig. 6.34(a)) once more. Suppose that we color the vertex v_5 red. Then we have to choose a different color, say white, for v_4. Yet a third color, say blue, would have to be selected for vertex v_3, because it is joined by an edge to both v_4 and v_5. Now v_2 cannot be white or blue, but we can color it red. We then would merely have to choose white or blue as a color for v_1 in order to achieve a proper coloring with only three colors. Obviously $\kappa \neq 2$ because of the clique $\{v_3, v_4, v_5\}$. It follows that $\kappa = 3$. ■

EXERCISES 6.5

1 Find all the nonisomorphic graphs having

(a) three vertices **(b)** four vertices **(c)** five vertices

Thus, none of your listed graphs should be isomorphic, and every graph should be isomorphic to one of the graphs on your list.

2 Use a different invariant (than that of Example 6.40) to show that the graph of Fig. 6.1 and the graph G of Fig. 6.33 are not isomorphic.

3 Use various numerical invariants to show that no two of the following graphs are isomorphic:

(a) $K_{2,3}$ **(b)** Figure 6.1

(c) Exercise 1(a) of Exercises 6.1 **(d)** Exercise 1(b) of Exercises 6.1

4 Do the same (as in Exercise 3) for the following graphs:

(a) $K_{3,3}$ **(b)** Exercise 1(b) of Exercises 6.1

(c) Exercise 1(c) of Exercises 6.1 **(d)** Exercise 3(a) of Exercises 6.1

5 Do the same (as in Exercise 3) for the following graphs:

(a) $K_{3,4}$ **(b)** Exercise 3(c) of Exercises 6.1

(c) Figure 6.12(a) **(d)** K_7

6 Determine the degree spectrum of each of the following graphs.

(a) Exercise 1(a) of Exercises 6.1 **(b)** Exercise 3(b) of Exercises 6.1

(c) Exercise 3(c) of Exercises 6.1 **(d)** Exercise 1(b) of Exercises 6.1

(e) Exercise 1(c) of Exercises 6.1 **(f)** Exercise 3(a) of Exercises 6.1

(g) Figure 6.12(a) **(h)** Exercise 1 of Exercise 6.4

(i) the cube with its (four) diagonals

7 Determine the independence number β for each graph of Exercise 6, and exhibit a maximum-sized independent set in each case.

8 Determine the domination number δ for each graph of Exercise 6, and exhibit a minimum-sized dominating set in each case.

9 Determine the covering number α for each graph of Exercise 6, and exhibit a minimum-sized covering set in each case.

10 Determine the chromatic number κ for each graph of Exercise 6, and exhibit a minimum-sized proper coloring in each case.

11 Determine the clique number β' for each graph of Exercise 6, and exhibit a maximum-sized clique in each case.

12 (*For chess players only*) Show directly that $\beta \geq \delta$ for the graph of the knight's dominating relation.

13 Prove that in any graph with n vertices

$$\frac{n}{\beta} \le \kappa \le n - \beta + 1$$

14 In analogy to a set of vertices that covers all the edges and an independent set of vertices, we may consider for connected graphs:

i. a covering set of edges (they cover all the vertices)
ii. an independent set of edges (no two sharing a vertex)

and corresponding minimal and maximal numbers, α_e and β_e, respectively. These are the *edge covering* and *edge independence* numbers. Prove that

(a) $\alpha_e + \beta_e = n$ **(b)** $\frac{1}{2}n \le \alpha_e \le n - 1$

15 Determine the edge covering number α_e (see Exercise 14) for each of the graphs of Exercise 6, and exhibit a minimum-sized covering set of edges in each case.

16 Determine the edge independence number β_e (see Exercise 14) for each of the graphs of Exercise 6, and exhibit a maximum-sized independent set of edges in each case.

17 How many graphs need to be considered in analyzing the cases $n = 7, 9, 10$ in the corollary to Theorem 6.8?

18 Show that in any graph $G = (V, E)$ we have the relation

$$\beta_e \le \alpha$$

(*Note:* See Exercise 14.)

19 Find an example showing that a maximal independent set (i.e., one not contained in a larger independent set) of vertices need not be of maximum size among all independent sets.

20 Characterize those graphs whose independence number is one. That is, what must such a graph look like, and why?

21 In a graph without isolated vertices, show that every covering set (of vertices) is also a dominating set.

Sec. 6.6 COVERING PROBLEMS

We present algorithms for computing the numerical invariants of a given graph.

We have described a number of graphical isomorphism invariants, and we have seen that each of them has an independent significance, quite apart from the isomorphism question. Often one or the other of these graphical parameters will offer a solution to an important optimization problem. Therefore it is crucial that we have some available means for computing these parameters in any given graph. Here we will not be looking for the most efficient algorithms, however, as that is a subject of ongoing research. We will try instead to provide a series of integrated techniques that are intuitively clear, if not representative of the state of the art, from the standpoint of computational complexity theory. The techniques are of somewhat broader scope than is needed for our immediate purpose. In fact, they are used to treat any of a number of covering problems that arise in the applications. So it is worth our while to examine this broader class of problems, and to specialize to the case of our graphical parameters as the occasion warrants. With this point of view we are assured that the ideas will be readily accessible, in consideration of the limited background so far provided.

Goals

After reading this section, you should be able to:

- State the general form of a *covering problem* and describe the covering algebra that is used in treating such problems.
- Interpret the determination of *minimal dominating sets* and *covering sets* as covering problems.
- Show how the *covering algebra* is used to determine the domination and covering numbers of a given graph.
- Illustrate the use of binary labeled trees in computing the independence and clique numbers of a given graph.
- Use the covering algebra in order to compute the chromatic number of a given graph.
- Compute domination, covering, independence, clique, and chromatic numbers of a given graph.

Definition of Covering Problems

Suppose that we are going to order dinner in a restaurant from an à la carte menu. Certainly we should be concerned with the nutritional content of our selections, and so will want to make choices that add up to a well-balanced meal. To simplify the problem, suppose the menu consists of the following items:

A. Spanish omelet	**B.** chicken enchilada
C. Waldorf salad	**D.** steak
E. potato	**F.** liver with onions

Assume that these foods offer the nutritional benefits described in Table 6.2, where the entry '1' or '0' indicates that the nutritional benefit in question is or is not provided, respectively.

TABLE 6.2

	Protein	*Carbohydrates*	*Vitamins*	*Minerals*
A.	1	0	1	1
B.	1	1	0	0
C.	0	0	1	1
D.	1	0	0	0
E.	0	1	1	0
F.	1	0	0	1

We must make a selection of rows from this table, so that among all of the rows selected there is a one in every column. Thus the diner may choose, for example,

A and $B =$ Spanish omelet and chicken enchilada, or
C and D and $E =$ Waldorf salad and steak and potato

either of these constituting a well-balanced meal. We do not permit the diner to over-indulge, however. We do not allow a selection of rows (as A, B, C) for which a row can be deleted while still satisfying all the nutritional requirements. That is to say we are only interested in the *irredundant* coverings. Even so, many of these may exist, and we may wish to know all of them. For one thing we could associate a cost with each entree, and then attempt to choose a feasible selection at minimum cost. These are the kinds of problems that can be treated by our covering theory.

You already are aware that such covering problems arise in a wide variety of situations. We will reinforce this impression as we continue, but before proceeding any further we first should formalize the whole class of problems under consideration. We are given two finite sets, a collection

$$A = \{a_1, a_2, \ldots, a_n\}$$

of *cells*, and another finite collection

$$B = \{b_1, b_2, \ldots, b_m\}$$

whose members are called *points*. Note that we deliberately have chosen a neutral terminology here so as to allow for the various interpretations of the theory. In addition, each cell a has an associated *cost* $\#a$ which we take to be a positive integer, for the sake of convenience. Finally, an *incidence relation* R from A to B is given, and its corresponding *incidence matrix* $R = [r_{ij}]$ consists entirely of zeros and ones. As expected we take

$$r_{ij} = 1 \Leftrightarrow a_i \, R \, b_j$$

As a matter of terminology we say that a_i *covers* b_j (b_j *is covered by* a_i) in either case.

If C is any collection of cells, we take the *cost* $\#C$ *of the collection* to be the sum of the costs of the individual cells in C. Other cost functions also could have been defined, but the one we have chosen is quite common and seems to be the most convenient. Given a *covering problem* (A, R, B), we will say that $C \subseteq A$ is a **minimal covering** if

---

(a) for every point b in B there is a cell a in C with $a\,R\,b$ (and with this condition alone, we say that C is a **covering**);

(b) if C' is also a covering, then $\#C \leq \#C'$.

Moreover, C is said to be an **irredundant covering** if we have the condition:

(c) $C \sim \{a\}$ is not a covering (all a in C)—

---

instead of (b). Just as in the case of the diner, we don't want to consider a covering that would be redundant, in the sense that some cell could be deleted while still maintaining a covering. Our reasons for this are wrapped up in the following fundamental result:

Theorem 6.12 Every minimal covering is irredundant.

Proof If the covering $C \subseteq A$ is redundant, then some cell a in C can be removed, and we still have a covering. But since

$$\#(C \sim \{a\}) = \#C - \#a < \#C$$

C is not a minimal covering. Having established the contrapositive of our statement, the proof is complete. □

Example 6.55 In the diner's problem (A, R, B), we have A as the list of entrees, B the nutritional features, and R the relation that specifies the nutritional components of the various dishes (a zero-one matrix as given in Table 6.2). We may want to attach a numerical cost or price to each entree (that they may not be integers can be taken care of with a scaling factor). In any case we have a covering problem in the sense just outlined. ■

Example 6.56 Consider the discussion of the covering number $\alpha = \alpha(G)$ for a graph $G = (V, E)$, as presented in Section 6.5. To interpret all of this as a covering problem (A, R, B), we have only to take $A = V$, $B = E$, and R the vertex-edge incidence relation:

$$v\,R\,e \Leftrightarrow e = \{v, w\}$$

a relation that is satisfied if v is one of the vertices of the edge e. Here we might think of the costs $\#v = 1$ for every vertex. Then the cost of a covering is simply the number of vertices in the covering. It follows that a minimal covering is one employing the fewest number of vertices. And, in fact, we generally assume this sort of cost situation (the cost of every cell being one) in the absence of other suggested weighting features. ■

Example 6.57 Consider the graph $G = (V, E)$ as given by the pictorial representation of Figure 6.37. The vertex-edge incidence relation is given by the following zero-one matrix:

	e_1	e_2	e_3	e_4	e_5	e_6	e_7	e_8	e_9
a	1	1	0	0	0	0	0	0	0
b	1	0	1	1	1	0	0	0	0
c	0	1	0	0	0	1	0	0	0
d	0	0	1	0	0	0	1	0	0
e	0	0	0	1	0	1	1	1	0
f	0	0	0	0	1	0	0	0	1
g	0	0	0	0	0	0	0	1	1

Note that such matrices always will have the property of having exactly two ones in each column. ■

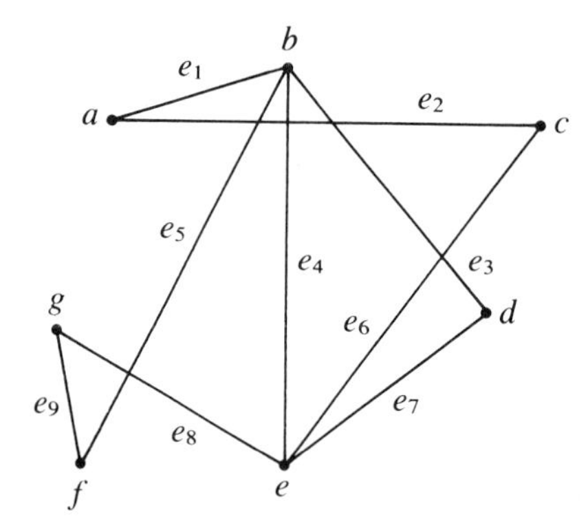

FIGURE 6.37

Example 6.58 Next, consider the discussion of the domination number $\delta = \delta(G)$ of a graph $G = (V, E)$, again as presented in Section 6.5. If we take $A = B = V$ and as R the relation

$$v\ R\ w \Leftrightarrow v = w \quad \text{or} \quad v\ E\ w$$

then the problem of determining a minimal dominating set is reduced once again to a covering problem (A, R, B). ■

Example 6.59 The graph $G = (V, E)$ of Example 6.1 is reproduced in Figure 6.38. For this graph, the relation R that has just been described can be represented in the form of the zero-one matrix:

	v_1	v_2	v_3	v_4	v_5
v_1	1	1	0	0	1
v_2	1	1	1	1	0
v_3	0	1	1	1	1
v_4	0	1	1	1	1
v_5	1	0	1	1	1

Note that in each column, such matrices always will have a number of ones, one in excess of the degree of the corresponding vertex, owing to the inclusion of the diagonal entries. ∎

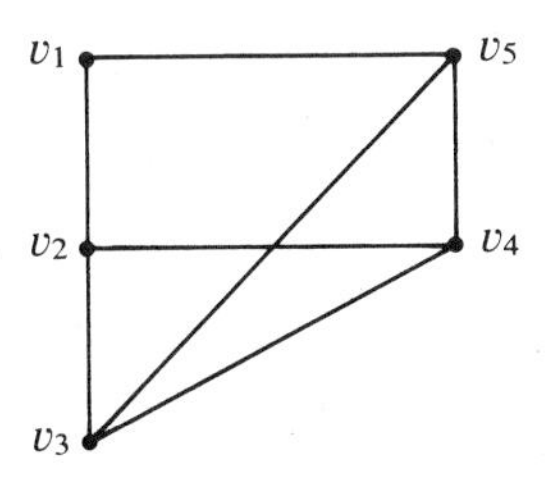

FIGURE 6.38

The Covering Algebra

In the light of Theorem 6.12, the following strategy is suggested regarding the problem of finding a minimal covering for the covering problem (A, R, B): Find *all* the irredundant coverings, and from these select one of minimum cost. In treating this subsidiary problem, that of finding all irredundant coverings, we present an algebraic method that is quite general. We consider simple English-language statements that are either true or false, thus returning for the moment to the framework of Chapter 2. We suppose that we are allowed to form compound statements in the usual way, using the conjunctions *and* and *or* as connectives. We further suppose that such compound statements obey the everyday rules of truth or falsity, as governed by the characteristic truth tables of Section 2.7. For ease of expression, however, we propose that the words *and* and *or* be replaced by the operation symbols '$\cdot$' and '$+$', respectively, in agreement with the notation of Section 2.7. Then an appropriate choice of variables will allow for the formation of compound statements in the usual way.

We intend to use these ideas from logic in writing a polynomial expression that will represent accurately the conditions under which one obtains a covering for the problem (A, R, B). First of all, we assign a variable (a_i seems the most suitable here) to signify the selection of a given cell of A. For each column of the incidence matrix R, we then define a polynomial

$$f_j = f_j(a_1, a_2, \ldots, a_n)$$

given as the sum of those a's that cover the point b_j.

Example 6.60 In the case of the incidence matrix

	b_1	b_2	b_3	b_4
a_1	1	0	0	1
a_2	0	0	1	1
a_3	0	1	1	0
a_4	1	1	0	0

we have

$$f_1 = a_1 + a_4$$
$$f_2 = a_3 + a_4$$
$$f_3 = a_2 + a_3$$
$$f_4 = a_1 + a_2$$

The linguistic interpretation is clear: We can cover b_1 with a_1 or a_4, b_2 with a_3 or a_4, etc. In order to form a compound polynomial for symbolizing the conditions for obtaining a covering, we simply multiply all of these to obtain

$$f = (a_1 + a_4) \cdot (a_3 + a_4) \cdot (a_2 + a_3) \cdot (a_1 + a_2)$$

indicating that we must select a_1 or a_4, and a_3 or a_4, and a_2 or a_3, and a_1 or a_2. ■

In general, we claim that we can obtain all the irredundant coverings for the covering problem (A, R, B) by multiplying out the polynomial:

$$f = f_1 \cdot f_2 \cdot \cdots \cdot f_m$$

as introduced above. In so doing we eventually express f as a sum of products (originally, it is a product of sums) through imposing a *distributive law*:

$$x \cdot (y + z) = x \cdot y + x \cdot z$$

(Note that we already have tacitly assumed the use of *associative* and *commutative laws*:

$$(x + y) + z = x + (y + z) \qquad (x \cdot y) \cdot z = x \cdot (y \cdot z)$$
$$x + y = y + x \qquad x \cdot y = y \cdot x$$

in this connection.) Furthermore we employ the following *idempotent* and *absorption laws*:

$$x + x = x \qquad x \cdot x = x$$
$$x + (x \cdot y) = x \qquad x \cdot (x + y) = x$$

in order to shorten the length of our polynomial expressions, thereby removing redundancy as well. If we think of the economics of the situation, say in relation to the diner's problem, and we suppose that apple pie or (apple pie and berry pie) meet a certain set of nutritional requirements, then an economical solution would not include both ($x + x \cdot y = x$). The other laws have similar interpretations, both here and in the more general context of the covering problem (A, R, B).

Example 6.61 In the case of the diner's problem, we transform the polynomial f from a product of sums to a sum of products as follows (see Table 6.2):

$$\begin{aligned} f &= (A + B + D + F)(B + E)(A + C + E)(A + C + F) \\ &= (B + AE + DE + EF)(A + C + E)(A + C + F) \\ &= (AB + BC + BE + AE + DE + EF)(A + C + F) \\ &= AB + AE + BC + EF + CDE \end{aligned}$$

Suppose we give a detailed description of the first multiplication, just to be sure that the mechanics of the process are clearly understood. Using the distributive law we obtain

$$\begin{aligned} (A + B + D + F)(B + E) &= (A + B + D + F)B + (A + B + D + F)E \\ &= AB + BB + DB + FB + AE + BE + DE + FE \\ &= AB + B + BD + BF + AE + BE + DE + EF \\ &= B + AE + DE + EF \end{aligned}$$

noting particularly the use of the idempotent law ($BB = B$) and the subsequent absorption of the products AB, BD, BF, and BE by B. The last equality for f, that is to be read "A and B, or A and E, or, ...," gives a list of all of the well-balanced meals (irredundant coverings). The diner thus will choose from the meals:

A and B = Spanish omelet and chicken enchilada
A and E = Spanish omelet and potato
B and C = chicken enchilada and Waldorf salad
E and F = potato and liver with onions
C and D and E = Waldorf salad and steak and potato

We should note once more that in each case the selected meals include a '1' from every column—they are indeed coverings. If costs were attached to the individual items of the menu, then we could choose from among these irredundant coverings, a covering of minimum cost. ■

The method just described is quite generally applicable to an arbitrary covering problem (A, R, B). One introduces the product of sums polynomial $f(a_1, a_2, \ldots, a_n)$, and in converting f to a sum of products one then is able to choose a product term representing a minimal covering. This follows as a consequence of Theorem 6.12 and the following result, one which we offer without proof.

Theorem 6.13 Let (A, R, B) be a covering problem with the product of sums polynomial $f(a_1, a_2, \ldots, a_n)$. Then the irredundant sum of products representation of f contains as its summands all of the irredundant coverings of the problem (A, R, B).

Covering and Domination Numbers

We have seen that the problem of determining a minimal covering set or a minimal dominating set of vertices for a graph $G = (V, E)$ is a covering problem in the sense introduced here. Once this interpretation has been made, along the lines of Examples 6.56 and 6.58, respectively, we have only to apply the ideas of Theorems 6.12 and 6.13 to obtain minimal covering or dominating sets, and at the same time to obtain the covering and domination numbers α and δ.

Example 6.62 Consider the graph of Example 6.57 (Fig. 6.37) and the matrix of that example. The conversion of the polynomial $f(a, b, c, d, e, f, g)$ from a product of sums to a sum of products is summarized in the following computation:

$$
\begin{aligned}
f &= (a + b)(a + c)(b + d)(b + e)(b + f)(c + e)(d + e)(e + g)(f + g) \\
&= (a + bc)(b + d) \cdots \\
&= (ab + bc + ad)(b + e) \cdots \\
&= (ab + bc + ade)(b + f) \cdots \\
&= (ab + bc + adef)(c + e) \cdots \\
&= (bc + abe + adef)(d + e) \cdots \\
&= (bcd + adef + bce + abe)(e + g)(f + g) \\
&= (adef + bce + abe + bcdg)(f + g) \\
&= adef + bcef + abef + bceg + abeg + bcdg
\end{aligned}
$$

At this point we can state that the covering number is $\alpha = 4$. Either of these six irredundant coverings is minimal, owing to our cost criterion that simply measures the size of the covering set. ■

Example 6.63 Consider the graph of Example 6.59 (Fig. 6.38) and the matrix of that example. The conversion of the polynomial $f(v_1, v_2, v_3, v_4, v_5)$ from a product of sums to a sum of products is summarized in the following computation:

$$\begin{aligned} f &= (v_1 + v_2 + v_5)(v_1 + v_2 + v_3 + v_4)(v_2 + v_3 + v_4 + v_5)(v_1 + v_3 + v_4 + v_5) \\ &= (v_1 + v_2 + v_3v_5 + v_4v_5)(v_2 + v_3 + v_4 + v_5)(v_1 + v_3 + v_4 + v_5) \\ &= (v_1v_3 + v_1v_4 + v_1v_5 + v_3v_5 + v_4v_5)(v_1 + v_3 + v_4 + v_5) \\ &= v_1v_3 + v_1v_4 + v_1v_5 + v_3v_5 + v_4v_5 \end{aligned}$$

Note our use of the idempotent law in only writing down one of the identical sums representing columns three and four. Either of these five products represents a minimum-sized dominating set. And it follows that the graph has the domination number $\delta = 2$, in agreement with the argument in Example 6.48. ■

Independence and Clique Numbers

The argument at the very beginning of the proof of Theorem 6.10 shows that if A is a covering set of vertices in the graph $G = (V, E)$, then the complementary set $B = V \sim A$ is independent, and conversely. It follows that we can obtain the maximal independent sets by merely complementing the minimal covering sets.

Example 6.64 In the graph of Examples 6.57 and 6.62 (Fig. 6.37), we have obtained the minimal covering sets:

$$\{a,d,e,f\} \qquad \{b,c,e,f\} \qquad \{a,b,e,f\} \qquad (b,c,e,g\} \qquad \{a,b,e,g\} \qquad \{b,c,d,g\}$$

and it follows by the argument just reviewed that the complementary sets:

$$\begin{aligned} A &= \{c,d,g\} & D &= \{a,d,f\} \\ B &= \{c,d,f\} & E &= \{a,e,f\} \\ C &= \{a,d,g\} & F &= \{b,c,g\} \end{aligned}$$

are maximal independent sets (and that the graph has the independence number $\beta = 3$). We have named these sets A, B, C, D, E, F for later reference.

```
algorithm independence (G: T, β)

initialize T as empty
set T[λ] = V and L = {λ}
for j running from 1 to n − 1
  for i running from j + 1 to n
    if E[i, j] = 1
      mark (as M) leaves x containing both v_i, v_j in label
      remove M from L
      for all marked nodes x
        split into branches at x0 and x1
          with labels T[x] ~ {v_i} and T[x] ~ {v_j}
        if these labels not contained in other leaf labels
          adjoin x0, x1 to L
```

■

An alternative means of computing all of the maximal independent sets can be given, using the notion of a binary labeled tree (see Section 6.3). The procedure is as described in the accompanying algorithm *independence*. As it stands, the algorithm determines the independence number β for any graph $G = (V, E)$, and at the same time lists all of the maximal independent sets—those that are not contained in a larger independent set. But a simple modification:

$$\text{if } E[i, j] = 0$$

rather than $E[i, j] = 1$, will produce instead the clique number β' and a list of all the maximal cliques.

As with most formal descriptions of algorithms, the procedure is nowhere near as complicated as it might appear. The binary tree becomes labeled with subsets of V, the set of vertices of the graph $G = (V, E)$. Initially the tree T is unlabeled; or, if you prefer, you may consider all of its nodes to be labeled with the empty subset, as we have indicated. Then the root of the tree is labeled with the entire set V. The subset $L \subseteq B^*$ keeps track of the leaves of the tree as the labeling proceeds. The two for statements simply search through the triangular matrix representation of the graph, looking for pairs of vertices with $v_i \ E \ v_j$ ($E[i, j] = 1$). Then the set M marks all the leaves of the tree whose labels contain this pair as a subset. Subsequently, M is deleted from L because a subset that contains this pair cannot be independent. Suppose that one of these marked nodes of the tree is $x \in B^*$. Then the most interior sequence of computations in our algorithm will label the nodes below and to the left and right from x with the maximal subsets of $T[x]$ *not containing both* of the vertices v_i and v_j (see Fig. 6.39). At the same time we check to see whether the labels of these new leaves are not themselves contained in other labeled leaves of the tree. After all we are interested only in the maximal independent sets. Depending on the outcome of this

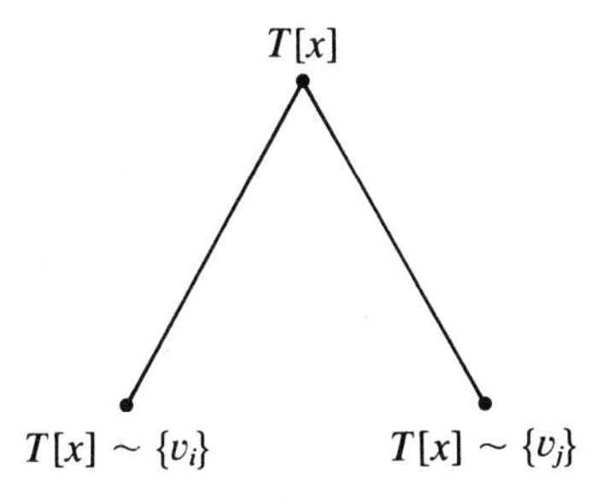

FIGURE 6.39

inquiry, L may be modified accordingly. Then we are ready to examine a new entry $E[i, j] = 1$.

Example 6.65 In order to illustrate the various steps of the independence algorithm, let us consider the graph of Example 6.57 (Fig. 6.37). Initially, only the root of our eventual tree is labeled, as shown in Figure 6.40(a). At this point, $L = \{\lambda\}$. Then we shall begin to examine the entries in the triangular matrix representation:

	a	b	c	d	e	f
b	1					
c	1	0				
d	0	1	0			
e	0	1	1	1		
f	0	1	0	0	0	
g	0	0	0	0	1	1

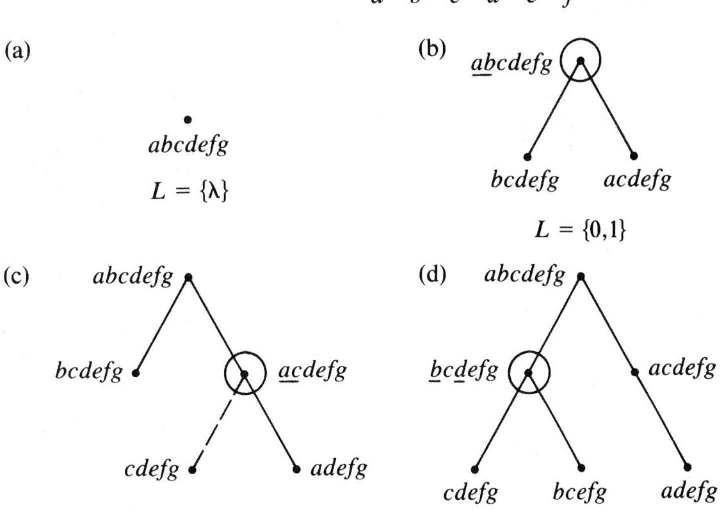

FIGURE 6.40

looking for ones. The arrangement of our for statements is such that we scan the columns from left to right, top to bottom.

The first 1 we encounter is the one that shows that $a\ E\ b$. We look at the leaves (leaf) of the tree to see if any are labeled with subsets containing $\{a, b\}$. In Figure 6.40(b) we perform the required splitting operation as described in Figure 6.39. Now our set of leaves is $L = \{0, 1\}$. The next 1 we encounter is the one corresponding to the relation $a\ E\ c$. There is one leaf (at $x = 1$) for which

$$\{a, c\} \subseteq T[x] = \{a, c, d, e, f, g\}$$

so we set $M = \{1\}$, and this node is deleted from L. The algorithm then sets

$$T[10] = T[x0] \longleftarrow T[x] \sim \{a\} = \{c, d, e, f, g\}$$
$$T[11] = T[x1] \longleftarrow T[x] \sim \{c\} = \{a, d, e, f, g\}$$

In the next line it checks to see whether the new labels would be contained in those of other leaves of the tree. Since

$$T[10] \subseteq T[0] \quad (\text{with } 0 \in L)$$

the node 10 does not become a new leaf of the tree (see Fig. 6.40(c)). Now we have $L = \{0, 11\}$. This result of the splitting due to the next 1 entry ($b\ E\ d$) is shown in Figure 6.40(d), at which point we have $L = \{00, 01, 11\}$.

In hand computation the entire tree can be generated as one figure. In this case, the whole computation would be summarized as in Figure 6.41. The circled nodes in the figure correspond to the marked nodes (elements of M) at the time of recognition of the last 1 entry ($f\ E\ g$) in the matrix. And thus, at the con-

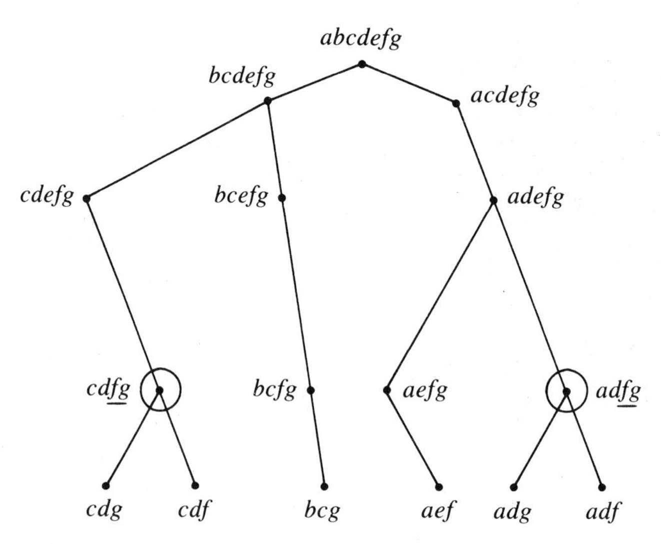

FIGURE 6.41

clusion, we have

$$L = \{0010, 0011, 0111, 1101, 1110, 1111\}$$

and finally (in examining the labels of the leaves), $\beta = 3$. The labels of all the leaves at the conclusion of execution of the algorithm will correspond to the maximal independent sets. In this instance these are the subsets

$$\{c, d, g\} \qquad \{c, d, f\} \qquad \{b, c, g\} \qquad \{a, e, f\} \qquad \{a, d, g\} \qquad \{a, d, f\}$$

respectively, in agreement with our result in Example 6.64. ■

The Chromatic Number

If we want to use the covering theory to compute the chromatic number (and a list of all minimal proper colorings) for a graph $G = (V, E)$, we first must assume that all the maximal independent sets have been found. These may be found by either of the methods we have just discussed. Suppose the maximal independent sets are named $B_1, B_2, \ldots, B_n$. Then we may introduce a covering problem (A, R, B) in which $A = \{B_1, B_2, \ldots, B_n\}$ and $B = V$. We write $B_i \, R \, v$ just in case $v \in B_i$. Because the vertices of an independent set are not joined by edges, one thinks of the independent sets as the colors, and so it is natural to try to cover the vertices with the independent sets, so interpreted. Otherwise the procedure is much the same as before.

Example 6.66 Consider the graph discussed in Example 6.57 (Fig. 6.37) and in the subsequent examples along the way. We have already obtained the maximal independent sets

$$\begin{aligned} A &= \{c, d, g\} & D &= \{a, d, f\} \\ B &= \{c, d, f\} & E &= \{a, e, f\} \\ C &= \{a, d, g\} & F &= \{b, c, g\} \end{aligned}$$

in Example 6.64 (or in Example 6.65, by another method). The appropriate covering matrix is as follows:

	a	b	c	d	e	f	g
A	0	0	1	1	0	0	1
B	0	0	1	1	0	1	0
C	1	0	0	1	0	0	1
D	1	0	0	1	0	1	0
E	1	0	0	0	1	1	0
F	0	1	1	0	0	0	1

In converting the product of sums polynomial f to a sum of products, we may summarize the computation:

$$\begin{aligned} f &= (C + D + E)F(A + B + F)(A + B + C + D)E\,(B + D + E)(A + C + F) \\ &= (CF + DF + EF)(A + B + F)\cdots \\ &= (CF + DF + EF)(A + B + C + D)\cdots \\ &= (CF + DF + AEF + BEF)E\cdots \\ &= (CEF + DEF + AEF + BEF)(B + D + E)(A + C + F) \\ &= (CEF + DEF + AEF + BEF)(A + C + F) \\ &= CEF + DEF + AEF + BEF \end{aligned}$$

It follows that the chromatic number of this graph is $\kappa = 3$.

If we wish to obtain a proper coloring with only three colors, we will choose any of the products in the resulting expression for f (since they all are of size 3), say

$$CEF: \qquad C = \{a, d, g\} \qquad E = \{a, e, f\} \qquad F = \{b, c, g\}$$

These three subsets will cover V, but we have to back off to a partition:

$$C' = \{d, g\} \qquad E' = \{a, e, f\} \qquad F' = \{b, c\}$$

so as not to paint the same vertex twice. ■

This example has revealed an important simplification technique that is applicable to covering problems generally. We note that the cell E (and also F) is essential: It must appear in each of the irredundant coverings, since no other cell can cover the point e. In general, let (A, R, B) be a covering problem. The cell a is said to be **essential** (to the point b) if a is the only cell of A that covers b. In the corresponding incidence matrix, column b will have but a single 1 entry, in the row corresponding to a. The significance of these essential cells is revealed in the following rather obvious result:

Theorem 6.14 Let a be an essential cell in the covering problem (A, R, B). Then $C \cup \{a\}$ is an irredundant covering of (A, R, B) if and only if C is an irredundant covering of $(A \sim \{a\}, R, B \sim aR)$, where aR is the set of all points covered by a in the original problem.

Rather than provide a proof (it is not difficult, however), we comment on the application. Suppose an essential cell a is found. The theorem asserts that we then may consider a smaller problem, deleting row a and all columns in which that row has a 1 entry. If we find all irredundant coverings of the reduced problem, we have only to reinsert a in each of them to obtain all irredundant coverings of the original problem. The reader should try to apply this technique in reference to E and F in Example 6.66.

EXERCISES 6.6

1 In the diner's problem, suppose the costs of the individual items are $\#A = 3$, $\#B = 4$, $\#C = 2$, $\#D = 9$, $\#E = 1$, and $\#F = 5$. Determine the well-balanced meal of minimum cost.

2 Use the covering algebra to find the covering number and a list of all of the minimal covering sets of vertices in the following graphs.

(a) Exercise 1(a) of Exercises 6.1
(b) Exercise 1(b) of Exercises 6.1
(c) Exercise 1(c) of Exercises 6.1
(d) Exercise 1(d) of Exercises 6.1
(e) Exercise 3(a) of Exercises 6.1
(f) Exercise 3(b) of Exercises 6.1
(g) Exercise 3(c) of Exercises 6.1
(h) Figure 6.12(a)
(i) Figure 6.37
(j) Figure 6.38
(k) the cube with its (four) diagonals

3 Use the covering algebra to find the domination number and a list of all of the minimal dominating sets of vertices in the graphs of Exercise 2.

4 Find the independence number and a list of all of the maximal independent sets in the graphs of Exercise 2, using the results of Exercise 2 and the method of Example 6.64.

5 Do the same (as in Exercise 4), using the independence algorithm instead.

6 Find the clique number and a list of all the maximal cliques in the graphs of Exercise 2, using the modified independence algorithm.

7 Use the covering algebra to find the chromatic number and a minimal proper coloring in the graphs of Exercise 2.

8 Show how to use the covering algebra in order to find the edge covering number and a list of all of the minimal edge coverings in any graph. (*Note:* See Exercise 14 of Exercises 6.5.)

9 Show how to use the independence algorithm in order to find the edge independence number and a list of all the maximal independent sets of edges in any graph. (*Note:* See Exercise 14 of Exercises 6.5.)

10 Use the covering algebra to find the chromatic number and a minimal proper coloring in the graphs associated with the following maps.

(a) Exercise 8 of Exercises 6.4
(b) Exercise 9 of Exercises 6.4

11 Prove Theorem 6.14.

12 Find the edge covering number and a list of all the minimal edge coverings in the graphs of Exercise 2. (*Note:* See Exercise 8.)

13 Find the edge independence number and a list of all the maximal independent sets of edges in the graphs of Exercise 2. (*Note:* See Exercise 9.)

14 Give a counterexample to show that the converse of Theorem 6.12 is not true in general.

15 In a covering problem (A, R, B), b_j is said to *dominate* b_k if column j has ones in every row that column k has ones (and perhaps in others as well). Show that if we are looking for the irredundant coverings of (A, R, B), then we may delete any dominating column.

16 In a covering problem (A, R, B), a_i is said to *dominate* a_k if row i has ones in every column that row k has ones (and perhaps in others as well), and if furthermore, $\#a_i \leq \#a_k$. Show that if we are only looking for some minimal covering of (A, R, B), then we may delete any dominated row.

17 Use Theorem 6.14 and the results of Exercises 15 and 16 to obtain minimal coverings for the covering problems represented in Table 6.3.

TABLE 6.3

(a)

Cost		b_1	b_2	b_3	b_4	b_5	b_6
2	a_1	0	0	0	1	1	1
3	a_2	0	0	0	1	0	1
4	a_3	1	1	0	0	1	0
1	a_4	1	0	0	0	0	0
3	a_5	0	0	1	0	0	1
2	a_6	0	0	1	0	0	0

(b)

Cost		b_1	b_2	b_3	b_4	b_5	b_6	b_7	b_8
4	a_1	1	1	0	1	1	0	0	0
7	a_2	0	0	1	0	0	0	1	0
4	a_3	0	1	0	0	1	1	0	0
3	a_4	0	0	0	0	0	0	0	1
4	a_5	1	0	0	1	0	1	1	0
6	a_6	0	1	0	0	0	0	0	1
2	a_7	0	1	0	0	0	0	0	0

(c)

Cost		b_1	b_2	b_3	b_4	b_5	b_6	b_7
4	a_1	1	0	1	0	1	0	0
2	a_2	0	0	1	0	0	0	1
4	a_3	1	1	0	0	1	1	0
7	a_4	0	0	1	1	1	0	0
3	a_5	0	1	0	1	0	1	0
3	a_6	1	0	0	1	1	0	0
5	a_7	0	1	1	0	0	1	1

(d)

Cost		b_1	b_2	b_3	b_4	b_5	b_6	b_7	b_8	b_9	b_{10}	b_{11}
4	a_1	1	0	0	0	0	0	0	0	0	0	0
8	a_2	0	0	1	0	0	0	1	0	1	1	0
2	a_3	0	1	0	0	0	1	0	0	0	0	0
8	a_4	0	1	1	0	0	1	0	0	1	0	0
6	a_5	0	0	0	1	1	0	0	0	0	0	1
5	a_6	0	1	1	0	0	0	0	1	0	0	0
4	a_7	0	0	0	0	1	0	0	1	0	0	1
2	a_8	0	0	0	0	1	0	1	1	0	1	0
4	a_9	0	0	1	1	0	0	0	0	0	1	1

Sec. 6.7 DIRECTED GRAPHS

A directed graph is a graph in which we attach a direction to the edges.

In many of the important applications, it is more natural to deal with a graph in which a direction is associated with each edge. Actually there is a direction whenever we are thinking of a flow of information, the transport of some commodity, a transfer of control, etc. Though the definition of the directed graphs or *digraphs* is quite straightforward and perhaps more natural than that of the undirected graphs, the theory becomes somewhat cluttered; this is our reason for first concentrating on the latter.

Goals

After studying this section you should be able to:

- Define the notion of a *directed graph* and contrast this definition with that of the *undirected graph.*
- Discuss the use of *matrices* in treating the reachability problem for directed graphs.
- Describe the use of *matrix multiplication* in computing the iteration of the edge relation for directed graphs.
- Discuss and illustrate two algorithmic techniques for computing the *reachability matrix* of a directed graph.
- Use matrix techniques for determining the existence of paths of a given length in a directed graph.

Digraphs as Relations

The daily schedule offered by a particular airline will provide connections between various cities as illustrated in Figure 6.42. If we want to travel from A to B, a direct flight may not be available. But there may be a series of flights that will

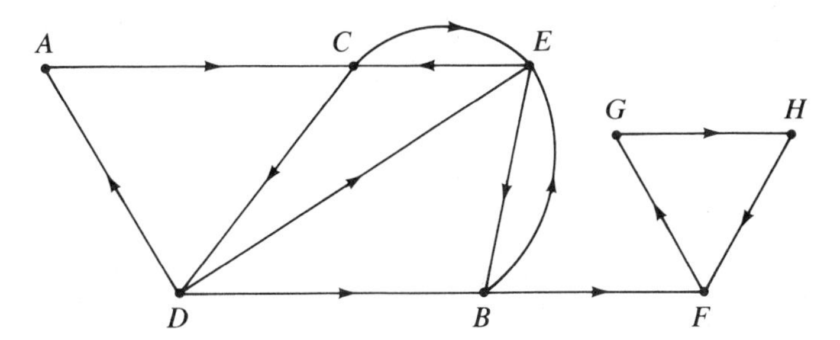

FIGURE 6.42

accomplish the same thing. It can be seen that a route (or path) from A (through C, then D) to B can be arranged, provided that the connections are reasonable—hopefully the connecting flight will not depart before we have landed! Moreover there is also the path from A through C, then E, to B to be considered. In choosing between these two routes, we would like to know the various flight times, distances, fares, etc. in order to help us to make a decision. If we wish to travel from G to C, however, then we had better consult another airline or wait until another day—as they say in Vermont, You can't get there from here.

In the simplified setting of this airline illustration, many of the important problems in the theory of directed graphs have already surfaced. The distinction between the digraphs and the graphs previously studied is immediately apparent. The edges here are oriented with arrows that indicate a direction. It then might appear that these directed graphs return us to the theory of relations all over again—relations on a set of vertices. Abstractly that is true enough; but here we emphasize the geometric point of view. Moreover we eventually superimpose enough additional structure (e.g., distances or other labels for the edges or vertices) as to render our new subject worthy of a separate treatment. After a suitable generalization of the concepts in the theory of undirected graphs (paths, connectedness, etc.), we will be led to a series of problems (minimum-distance paths, reachability, etc.) not ordinarily encountered in the abstract theory of relations, but which are the problems of greatest concern in the applications, as is easily understood.

In contrast with the definition of Section 6.1, a **directed graph** (or **digraph**, as it is abbreviated) is a system $G = (V, E)$ in which E is any relation on V. Either of the alternative notations $(v, w) \in E$ or $v \, E \, w$ will indicate a corresponding directed edge from v to w, as in the picture of a digraph. As before we make dual use of the symbol E to stand for the relation and/or the relation matrix. The context will always eliminate any possibility for confusion.

But for one simple change in notation—the representation of edges as (v, w) rather than $\{v, w\}$—all of our previous definitions of paths and circuits go through exactly as before. For instance, a **path** from v to w (of *length* k) in the digraph $G = (V, E)$ is a sequence $\langle v_0, v_1, \ldots, v_k \rangle$ of vertices v_i in V such that

$$\begin{matrix} v_0 = v \\ v_k = w \end{matrix} \quad \text{and} \quad (v_{i-1}, v_i) \in E$$

for all $i = 1, 2, \ldots, k$.

Example 6.67 As in the case of the undirected graphs, a relation matrix can serve to define a directed graph $G = (V, E)$. If

$$V = \{v_1, v_2, v_3, v_4, v_5\}$$

and E is the following relation:

$$\begin{array}{c} \\ v_1 \\ v_2 \\ v_3 \\ v_4 \\ v_5 \end{array} \begin{array}{c} \begin{array}{ccccc} v_1 & v_2 & v_3 & v_4 & v_5 \end{array} \\ \begin{bmatrix} 0 & 1 & 0 & 1 & 0 \\ 0 & 1 & 1 & 0 & 0 \\ 1 & 1 & 0 & 0 & 1 \\ 0 & 0 & 1 & 0 & 1 \\ 1 & 0 & 0 & 0 & 1 \end{bmatrix} \end{array}$$

then G is the digraph shown in Figure 6.43. Note that a reverse edge $w\ E\ v$ may or may not accompany an edge $v\ E\ w$. Note as well that we may or may not have edges $v\ E\ v$. Such edges (as $v_2\ E\ v_2$ and $v_5\ E\ v_5$ here) are called *loops*.

In this digraph, the sequence $\langle v_4, v_3, v_1, v_4, v_5, v_5, v_1 \rangle$ is a path of length 6 from v_4 to v_1. And again, $\langle v_2, v_3, v_4, v_1, v_2 \rangle$ is not a circuit, even though such a circuit does appear in the undirected graph obtained by the removal of arrows. It just happens that (v_3, v_4) and (v_4, v_1) are not edges of the digraph. ■

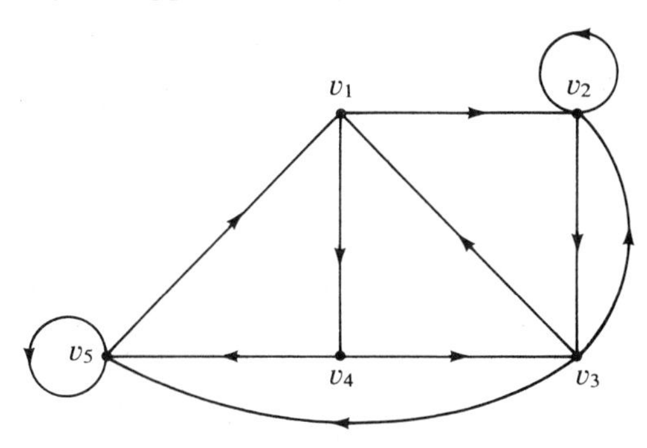

FIGURE 6.43

Example 6.68 In the airline digraph of Figure 6.42, $\langle A, C, D, B \rangle$ is a path of length 3 from A to B. There is no path from G to C, however. The sequence $\langle G, H, F, B, E, C \rangle$ is not a path from G to C because (F, B) is not an edge of the digraph; the arrow points the other way. We say that C is *not reachable* from G. ■

Example 6.69 In this same digraph the sequences $\langle A, C, D, A \rangle$ and $\langle G, H, F, G \rangle$ are circuits. Similarly, $\langle v_1, v_2, v_3, v_1 \rangle$ and $\langle v_1, v_4, v_3, v_5, v_1 \rangle$ are circuits in the digraph of Example 6.67 (Fig. 6.43). ■

Example 6.70 In the digraph shown in Figure 6.44, we have labeled the edges with numbers to represent perhaps the distances along various highways. In this way it is evident that the labeled digraphs provide an appropriate formalism for treating the various path problems, for example, the problem of finding a shortest path between two given points. We return to a study of such problems in the next section.

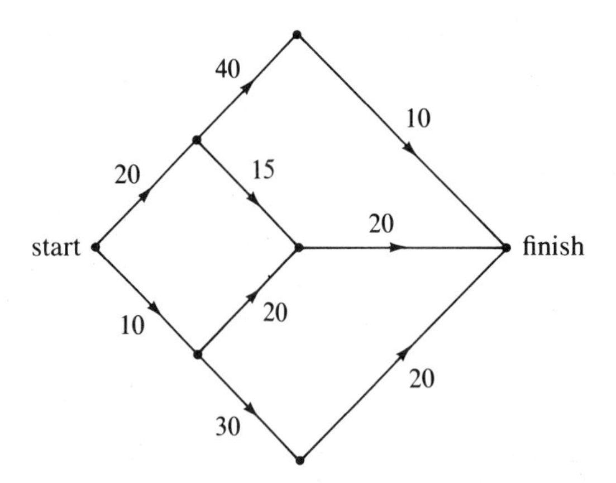

FIGURE 6.44

The same digraph might be used to depict a complex manufacturing process, from its start to its finish. Then the edges would correspond to activities or processes, with the understanding that an activity originating at a given node cannot start until all the activities terminating at that node have been completed. Thus if the labels designate the duration of a process, it would be the longest path from start to finish that would be important, that is, a *critical path* holding back the completion of the whole process. ■

Example 6.71 The skeleton of a flowchart is a digraph. The arrows on the edges indicate the flow of the algorithmic process from one statement to another. With given initial values for the input variables, the trace of the state of a computation (see Sec. 4.1) will correspond to a path in the digraph. ■

Matrix Computations

Before embarking on the analysis of these problems we would like to show how the study of digraphs can profit from the borrowing of ideas from the abstract theory of relations. These ideas mainly have to do with the concept or relational composition, and so a review of the composition of functions (originally presented in Sec. 1.3) would be helpful here. As a matter of fact we have purposely sought to avoid any mention of relational composition until now, realizing that this idea is most easily understood in the context of the intended applications—those that touch on questions of reachability and path problems in digraphs. The whole concept, however, is simple enough. If one can appreciate that "is an uncle of" is obtained by composing the two relations "is a brother of" and "is a parent of," then one is prepared for the forthcoming presentation.

Suppose that two relations are given:

R from the set U to the set V
S from the set V to the set W

Note that the range of R coincides with the domain of S. This is important in the eventual relational composition, just as it was in Section 1.3 for our discussion of the composition of functions. In these circumstances we can define the *composite relation* $R \circ S$ from U to W:

$$u(R \circ S)w \Leftrightarrow \text{there exists } v \text{ in } V \text{ with } u\ R\ v \text{ and } v\ S\ w$$

Observe particularly that if R and S happen to be functions, then our definition coincides with the idea of functional composition (again, see Section 1.3). The interesting question, however, is this: How do we construct $R \circ S$ when R and S are given? That is, can we obtain a graphical representation of the composite if we have the graphical representations of R and S? Can we obtain the relation matrix for $R \circ S$ if the matrices for R and S are at hand? We find that we can give affirmative answers for both questions.

If graphical representations for R and S are given as directed edges connecting certain vertices of U to V and, similarly, V to W, then the definition of composition states that a particular vertex u is joined by an arc to a particular vertex w if we can find an intermediate point $v \in V$ with edges (u, v) and (v, w). There may be more than one way in which this can come about, but that is all right. Of course, if there is none, if no such v exists, then we have $u(R \not\circ S)w$ and we do not join u to w in the composite relation.

Example 6.72 Two relations R and S are depicted graphically in Figure 6.45(a). The relation R is from U to V and S is from V to W, where

$$U = \{u_1, u_2, u_3, u_4\}$$
$$V = \{v_1, v_2, v_3, v_4, v_5\}$$
$$W = \{w_1, w_2, w_3\}$$

We recall that in the composite relation $R \circ S$ we are to draw an arc from u to w if there is a path $\langle u, v, w \rangle$ from u to w through an intermediate vertex $v \in V$. Thus

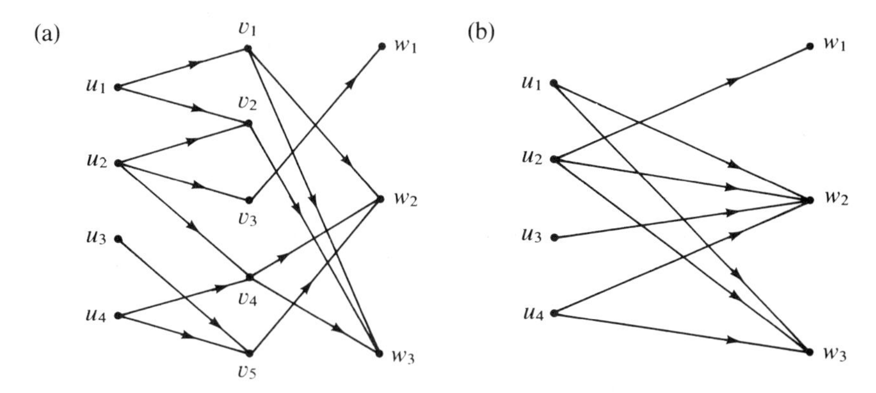

FIGURE 6.45

we draw the arc (u_1, w_3) in Figure 6.45(b) because of the path $\langle u_1, v_2, w_3 \rangle$ in Figure 6.45(a). There is also a path $\langle u_1, v_1, w_3 \rangle$, but that is all right. All we need is one. We do not draw an edge (u_1, w_1) in Figure 6.45(b) because there is no intermediate vertex $v \in V$ on a path $\langle u_1, v, w_1 \rangle$. In this way we obtain the graphical representation of the composite relation $R \circ S$ as given in Figure 6.45(b). ■

As for the equivalent matrix problem, we make use of the algebra B of Section 3.6. Thus suppose R and S denote the relation matrices $R = [r_{ij}]$ and $S = [s_{ij}]$. How would we be able to decide whether or not there is a path $\langle u_i, v, w_j \rangle$? We observe that the ith row of the matrix R gives the entry $r_{ik} = 1$ corresponding to the edge (u_i, v_k). Similarly, the jth column of S has an entry $s_{kj} = 1$ if (v_k, w_j) is an edge in the graphical representation of S. Thus what we need to find is a pair of entries:

'1' in position k of row $R_i = (r_{i1}, r_{i2}, \ldots, r_{in})$

'1' in position k of column $S_j = (s_{1j}, s_{2j}, \ldots, s_{nj})$

signifying the existence of an appropriate intermediate vertex v_k. Using the algebra B, we ask:

$$\begin{aligned} r_{i1} \cdot s_{1j} &= 1? \\ r_{i2} \cdot s_{2j} &= 1? \\ &\vdots \\ r_{in} \cdot s_{nj} &= 1? \end{aligned}$$

(Recall that only $1 \cdot 1 = 1$ in B.) If more than one of these calculations yields '1', that is all right; we still have $u_i(R \circ S)w_j$. Accordingly if we take the sum of these results, we obtain the correct answer (since $1 + 1 = 1$ in B). Altogether we have proved the following:

Theorem 6.15 The composite relation $R \circ S$ has the matrix

$$RS = T = [t_{ij}]$$

in which

$$t_{ij} = r_{i1}s_{1j} + r_{i2}s_{2j} + \cdots + r_{in}s_{nj}$$

(*Note:* Those of you who are familiar with ordinary matrix multiplication will note that we have a completely analogous formula here, except for the use of the algebra B.)

Example 6.73 In Example 6.72 the relations R and S have the matrices

$$R = \begin{bmatrix} 1 & 1 & 0 & 0 & 0 \\ 0 & 1 & 1 & 1 & 0 \\ 0 & 0 & 0 & 0 & 1 \\ 0 & 0 & 0 & 1 & 1 \end{bmatrix} \qquad S = \begin{bmatrix} 0 & 1 & 1 \\ 0 & 0 & 1 \\ 1 & 0 & 0 \\ 0 & 1 & 1 \\ 0 & 1 & 0 \end{bmatrix}$$

respectively. According to Theorem 6.15 the product matrix $RS = T$ is the relation matrix for the composite relation $R \circ S$. Thus we have the entry:

$$\begin{aligned} t_{13} &= r_{11}s_{13} + r_{12}s_{23} + r_{13}s_{33} + r_{14}s_{43} + r_{15}s_{53} \\ &= 1 \cdot 1 + 1 \cdot 1 + 0 \cdot 0 + 0 \cdot 1 + 0 \cdot 0 \\ &= 1 + 1 + 0 + 0 + 0 \\ &= 1 \end{aligned}$$

Note that the two 1's in the last expression correspond to the two paths from u_1 to w_3 in Figure 6.45(a), one through v_1 and the other through v_2.

Continuing in this way, we finally obtain the product matrix:

$$RS = T = \begin{bmatrix} 0 & 1 & 1 \\ 1 & 1 & 1 \\ 0 & 1 & 0 \\ 0 & 1 & 1 \end{bmatrix}$$

The reader can check that this is indeed the matrix for the composite relation of Figure 6.45(b). ■

Example 6.74 When we specialize to the case of relations on a single set V, the very definition of relational composition leads to the familiar properties:

(a) *associativity*: $R \circ (S \circ T) = (R \circ S) \circ T$

(b) *identity*: $R \circ 1 = 1 \circ R = R$

---

where 1 is the full relation, $v\ 1\ v$ for all v in V. What can we say about the corresponding relation matrices?

Suppose we demonstrate the associativity of the relational composition. If $u(R \circ (S \circ T))v$, then there exists an element x such that $u\ R\ x$ and $x(S \circ T)v$. The latter implies that there is an element y such that $x\ S\ y$ and $y\ T\ v$. A reconstitution of these relations gives $u(R \circ S)y$ and finally $u((R \circ S) \circ T)v$. And, obviously, the converse can be demonstrated in a similar fashion. ■

Now the question becomes, How do we use the relational composition and the computational result of Theorem 6.15 in the theory of directed graphs? Obviously we should specialize to the case where $U = W = V$, the set of vertices of a digraph $G = (V, E)$, and merely compose the edge relation E with itself. The composite $E \circ E$ will give us all the paths of length 2 in G. For longer paths, we just keep going. Thus we obtain the following result:

Theorem 6.16 There is a path (of length k) from v to w in the digraph $G = (V, E)$ if and only if

$$v(\underbrace{E \circ E \circ \cdots \circ E}_{k})w$$

Proof Note first of all that we can write the composites of E without parentheses, owing to the associativity of Example 6.74. Then we use induction on k. Clearly the theorem holds for $k = 1$ because an edge is a path of length 1. So suppose the result is true for some $k - 1 \geq 0$. Then if we have a path of length k from v to w, there is a sequence $\langle v_0, v_1, \ldots, v_k \rangle$ such that

$$\begin{matrix} v_0 = v \\ v_k = w \end{matrix} \quad \text{and} \quad (v_{i-1}, v_i) \in E$$

for $i = 1, 2, \ldots, k$. It follows that $\langle v_0, v_1, \ldots, v_{k-1} \rangle$ is a path of length $k - 1$ from v to v_{k-1}, and by the inductive hypothesis we have $v\ E^{k-1}\ v_{k-1}$. But also $v_{k-1}\ E\ w$, so that $v(E^{k-1} \circ E)w$, that is, $v\ E^k\ w$. Conversely, if $v\ E^k\ w$, then by the definition of composition there is a vertex u such that $v\ E^{k-1}\ u$ and $u\ E\ w$. By the inductive hypothesis again, there is a path of length $k - 1$ from v to u, and we use this to construct a path from v to w (of length k) in the obvious way. □

Corollary If $V = \{v_1, v_2, \ldots, v_n\}$ is the set of vertices for the digraph $G = (V, E)$, then there is a path from v_i to v_j (of length k) if and only if

$$e_{ij}^k = 1 \qquad \text{(the } i, j \text{ entry of the matrix } E^k \text{ is '1')}$$

Example 6.75 The digraph $G = (V, E)$ of Example 6.67 (Fig. 6.43) has the relation matrix

$$E = \begin{bmatrix} 0 & 1 & 0 & 1 & 0 \\ 0 & 1 & 1 & 0 & 0 \\ 1 & 1 & 0 & 0 & 1 \\ 0 & 0 & 1 & 0 & 1 \\ 1 & 0 & 0 & 0 & 1 \end{bmatrix}$$

We may use the result of Theorem 6.15 to compute

$$E^2 = \begin{bmatrix} 0 & 1 & 1 & 0 & 1 \\ 1 & 1 & 1 & 0 & 1 \\ 1 & 1 & 1 & 1 & 1 \\ 1 & 1 & 0 & 0 & 1 \\ 1 & 1 & 0 & 1 & 1 \end{bmatrix} \qquad E^3 = \begin{bmatrix} 1 & 1 & 1 & 0 & 1 \\ 1 & 1 & 1 & 1 & 1 \\ 1 & 1 & 1 & 1 & 1 \\ 1 & 1 & 1 & 1 & 1 \\ 1 & 1 & 1 & 1 & 1 \end{bmatrix}$$

as the relation matrices for the first two composites of E.

Then E^2 tells us whether there is a path of length 2 connecting the various vertices. Evidently there is a path of length 2 from v_1 to v_2, from v_1 to v_3, and from v_1 to v_5, but none from v_1 to v_1 or from v_1 to v_4, etc. Note that the path from v_1 to v_2 uses a loop. Similarly, E^3 shows that there are paths of length 3 connecting every pair of vertices except v_1 to v_4. If we are interested in paths of length 4, then we compute the matrix product E^4, and so on. We soon show how to bring an end to this seemingly endless computation if we are only interested in the existence or nonexistence of paths, whatever their length. ■

Reachability

In any transportation network, or in a directed graph indicating the flow of some commodity, it is often of interest to know whether one can travel or send goods from one point to another. Once the means for answering such existence questions are understood, we can move on to the more detailed optimality considerations, those of shortest or most economical routes.

We say that a vertex w is **reachable** from v in the directed graph $G = (V, E)$ if there is at least one path (of whatever length) from v to w. There may be several paths of various lengths, but that does not matter. The question is simply, Can I get there from here? A feasible, though not particularly efficient, procedure for answering these questions simultaneously (for all v, w in V) is immediately suggested by the corollary just presented. The matrices E^k give an explicit accounting of all of the vertices joined by paths of length k. Since we are not interested in the lengths at this point, we might try to form the **reachability matrix**

$$E^* = I + E + E^2 + E^3 + \cdots$$

in which corresponding elements of the various matrices are added together in the algebra B (again see Section 3.6). Here, I is the so-called **identity matrix**:

$$I = \begin{bmatrix} 1 & 0 & 0 & \cdots & 0 \\ 0 & 1 & 0 & \cdots & 0 \\ & \vdots & & & \\ 0 & 0 & \cdots & 0 & 1 \end{bmatrix}$$

having ones running down the diagonal, but zeros elsewhere. I will account for paths of length 0 from any vertex to itself. And while this is all well and good, it seems that we may be locked into an infinite procedure here, unless we observe the following fact:

Lemma

In any digraph $G = (V, E)$ with n vertices, we have

$$v\ E^*\ w \quad \text{iff} \quad v(I + E + E^2 + \cdots + E^{n-1})w$$

Proof In one direction, the proof of the lemma is obvious. So suppose $v\ E^*\ w$. This only means that there is some path from v to w in G. Now suppose that its length is k, so that the path can be described as a sequence $\langle v_0, v_1, \ldots, v_k \rangle$ with

$$\begin{aligned} v_0 &= v \\ v_k &= w \end{aligned} \quad \text{and} \quad (v_{i-1}, v_i) \in E$$

for $i = 1, 2, \ldots, k$. If $k > n - 1$, then according to the pigeonhole principle (Section 1.5), some vertex v_j must be encountered more than once in traversing this path. This shows that the path includes a circuit, that is,

$$\langle v_0, v_1, \ldots, v_k \rangle = \langle v_0, \ldots, v_j, \ldots, v_j, \ldots, v_k \rangle$$

(see Fig. 6.46). It follows that we can substitute a shorter path if the circuit is by-passed. Since we can continue to do this as long as the path length is greater than $n - 1$, we must eventually arrive at a path from v to w of length less than or equal to $n - 1$. And then, $v(I + E + E^2 + \cdots + E^{n-1})\ w$, as claimed. □

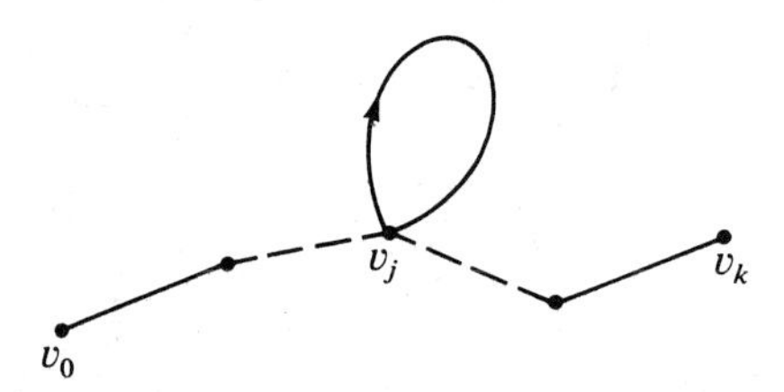

FIGURE 6.46

Theorem 6.17

The reachability matrix E^* is given by the finite computation

$$E^* = I + E + E^2 + \cdots + E^{n-1}$$

in any digraph with n vertices.

(*Note:* Again, those of you who are familiar with matrix algebra will observe that we are merely adding together the corresponding entries of the various matrices, as in ordinary matrix addition.)

Example 6.76 Let $G = (V, E)$ be the digraph shown in Figure 6.47. Then we have the relation matrix

$$E = \begin{bmatrix} 0 & 0 & 1 & 0 \\ 1 & 0 & 0 & 0 \\ 0 & 1 & 0 & 0 \\ 0 & 0 & 1 & 1 \end{bmatrix}$$

Using the ideas of Theorem 6.15, we compute

$$E^2 = \begin{bmatrix} 0 & 1 & 0 & 0 \\ 0 & 0 & 1 & 0 \\ 1 & 0 & 0 & 0 \\ 0 & 1 & 1 & 1 \end{bmatrix} \qquad E^3 = \begin{bmatrix} 1 & 0 & 0 & 0 \\ 0 & 1 & 0 & 0 \\ 0 & 0 & 1 & 0 \\ 1 & 1 & 1 & 1 \end{bmatrix}$$

And then, according to Theorem 6.17

$$\begin{aligned} E^* &= I + E + E^2 + E^3 \\ &= \begin{bmatrix} 1 & 0 & 0 & 0 \\ 0 & 1 & 0 & 0 \\ 0 & 0 & 1 & 0 \\ 0 & 0 & 0 & 1 \end{bmatrix} + \begin{bmatrix} 0 & 0 & 1 & 0 \\ 1 & 0 & 0 & 0 \\ 0 & 1 & 0 & 0 \\ 0 & 0 & 1 & 1 \end{bmatrix} + \begin{bmatrix} 0 & 1 & 0 & 0 \\ 0 & 0 & 1 & 0 \\ 1 & 0 & 0 & 0 \\ 0 & 1 & 1 & 1 \end{bmatrix} + \begin{bmatrix} 1 & 0 & 0 & 0 \\ 0 & 1 & 0 & 0 \\ 0 & 0 & 1 & 0 \\ 1 & 1 & 1 & 1 \end{bmatrix} \\ &= \begin{bmatrix} 1 & 1 & 1 & 0 \\ 1 & 1 & 1 & 0 \\ 1 & 1 & 1 & 0 \\ 1 & 1 & 1 & 1 \end{bmatrix} \end{aligned}$$

Our result indicates that v_4 is not reachable from v_1, v_2, or v_3, a fact that is consistent with the figure. ■

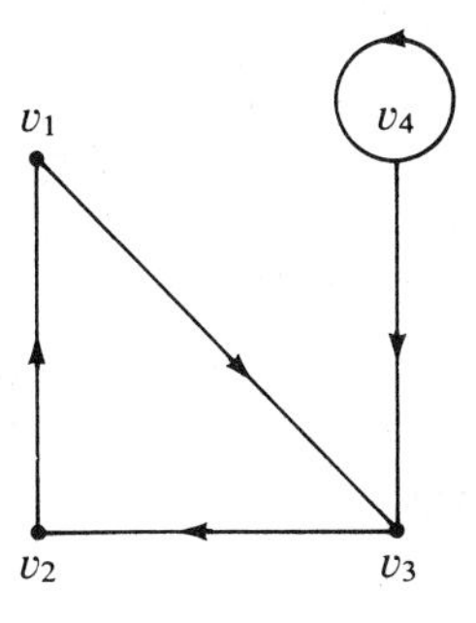

FIGURE 6.47

Example 6.77 For the digraph of Examples 6.67 and 6.75 (Fig. 6.43), we evidently obtain

$$E^* = \begin{bmatrix} 1 & 1 & 1 & 1 \\ 1 & 1 & 1 & 1 \\ 1 & 1 & 1 & 1 \\ 1 & 1 & 1 & 1 \end{bmatrix} = 1$$

Such a digraph is said to be **strongly connected** because each vertex is reachable from every other vertex. These digraphs are useful in the design of those communication networks where it is important that every pair of stations have a path of communication. ■

```
algorithm reachability (E: E*)
initialize E* as I + E
for k running from 1 to n
   for i running from 1 to n
      if i ≠ k and E*[i, k] = 1
         for j running from 1 to n
            add E*[k, j] to E*[i, j]
```

An even more efficient method for calculating the reachability matrix E^* is provided by the accompanying algorithm *reachability*. The main idea is fairly easy to grasp if we think in terms of the relations $E^{(k)}$:

$v\ E^{(k)}\ w \Leftrightarrow$ there is a path from v to w using only intermediate vertices from among $\{v_1, \ldots, v_k\}$

Thus, we begin with the matrix:

$$E^{(0)} = I + E \quad \text{(paths with no intermediate vertex)}$$

and because k runs up to n, we eventually obtain

$$E^{(n)} = E^* \quad \text{(paths having arbitrary intermediate vertices)}$$

Evidently the interior recomputation of E^* shows how $E^{(k)}$ is derived from $E^{(k-1)}$. The if statement signifies the existence of a path from v_i to v_k through intermediate vertices of the set $\{v_1, \ldots, v_{k-1}\}$ whenever $e_{ik}^{(k-1)} = 1$ (see Fig. 6.48). Then, obviously, there is a path from v_i to v_j involving intermediate vertices from among $\{v_1, \ldots, v_k\}$ in just the two cases:

---

1. There is a path from v_i to v_j using intermediate vertices from among $\{v_1, \ldots, v_{k-1}\}$;
2. There is a path from v_k to v_j using intermediate vertices from among $\{v_1, \ldots, v_{k-1}\}$.

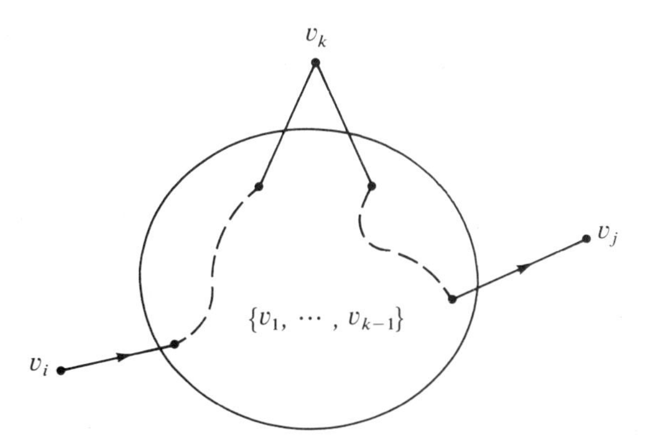

FIGURE 6.48

(Again, see Fig. 6.48.) The subsequent example will help to clarify any difficulties you may have in following the details.

Example 6.78 Suppose that we examine once more the digraph of Example 6.76 (Fig. 6.47). The algorithm begins with the assignment:

$$E^* \longleftarrow \begin{bmatrix} 1 & 0 & 1 & 0 \\ 1 & 1 & 0 & 0 \\ 0 & 1 & 1 & 0 \\ 0 & 0 & 1 & 1 \end{bmatrix} = \begin{bmatrix} 1 & 0 & 0 & 0 \\ 0 & 1 & 0 & 0 \\ 0 & 0 & 1 & 0 \\ 0 & 0 & 0 & 1 \end{bmatrix} + \begin{bmatrix} 0 & 0 & 1 & 0 \\ 1 & 0 & 0 & 0 \\ 0 & 1 & 0 & 0 \\ 0 & 0 & 1 & 1 \end{bmatrix}$$

From this point on, a trace of the state of the computation would appear as shown in Table 6.4.

TABLE 6.4

i	j	k	e^*_{11}	e^*_{12}	e^*_{13}	e^*_{14}	e^*_{21}	e^*_{22}	e^*_{23}	e^*_{24}	e^*_{31}	e^*_{32}	e^*_{33}	e^*_{34}	e^*_{41}	e^*_{42}	e^*_{43}	e^*_{44}
1		1	1	0	1	0	1	1	0	0	0	1	1	0	0	0	1	1
2	1						1											
	2							1										
	3								1									
	4									0								
3																		
4																		
1		2																
2																		
3	1										1							
	2											1						
	3												1					
	4													0				
									...									
			1	1	1	0	1	1	1	0	1	1	1	0	1	1	1	1

When $k = 1$ and $i = 2, j = 3$, the interior recomputation in E^* sets

$$e_{23}^* \longleftarrow 1 = 0 + 1 = e_{23}^* + e_{13}^*$$

(see Fig. 6.49(a)) because the statement is operating only under the assumption that $e_{ik}^* = e_{21}^* = 1$. Later on, with $k = 2$ and $i = 3, j = 1$, we have $e_{ik}^* = e_{32}^* = 1$, so we set

$$e_{31}^* \longleftarrow 1 = 0 + 1 = e_{31}^* + e_{21}^*$$

in accordance with Figure 6.49(b). ■

(a)

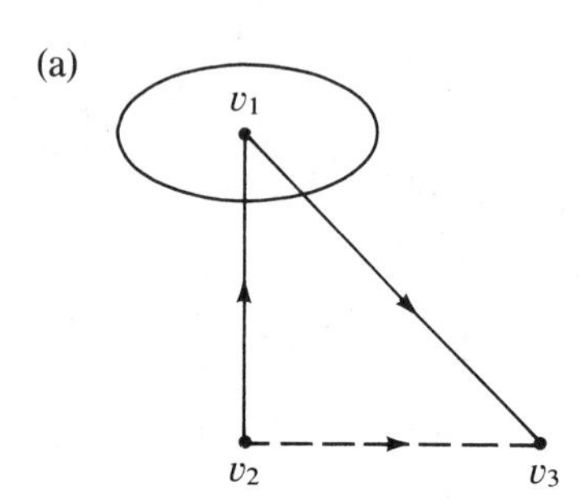

(b)

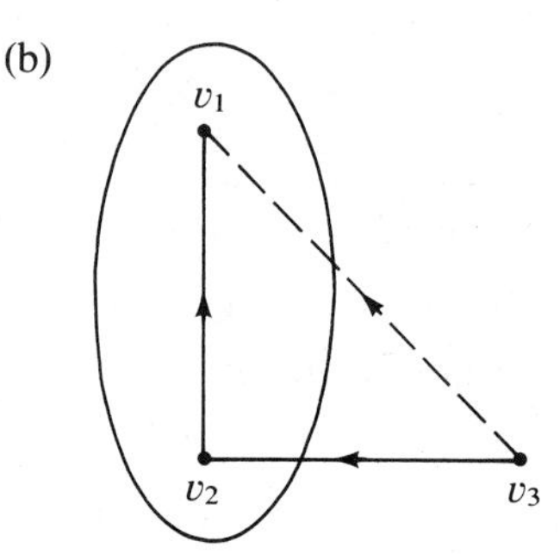

FIGURE 6.49

EXERCISES 6.7

1 Provide a pictorial representation of the digraphs represented by the following matrices:

(a)

	v_1	v_2	v_3	v_4	v_5
v_1	0	0	1	0	0
v_2	0	0	1	0	1
v_3	1	1	0	0	1
v_4	1	0	0	0	1
v_5	1	1	0	0	0

(b)

	v_1	v_2	v_3	v_4	v_5
v_1	0	1	1	0	1
v_2	0	0	0	1	1
v_3	0	0	0	0	0
v_4	1	0	1	0	1
v_5	0	0	1	0	0

(c)

	v_1	v_2	v_3	v_4	v_5	v_6
v_1	0	0	1	1	1	0
v_2	1	1	0	1	0	1
v_3	1	1	1	0	1	1
v_4	0	0	1	0	0	1
v_5	1	1	1	0	1	0
v_6	0	0	1	1	0	0

(d)

	v_1	v_2	v_3	v_4
v_1	1	1	1	1
v_2	0	1	1	1
v_3	1	1	0	1
v_4	1	1	1	0

2 Obtain matrices that represent the digraphs of Figure 6.50.

(a)

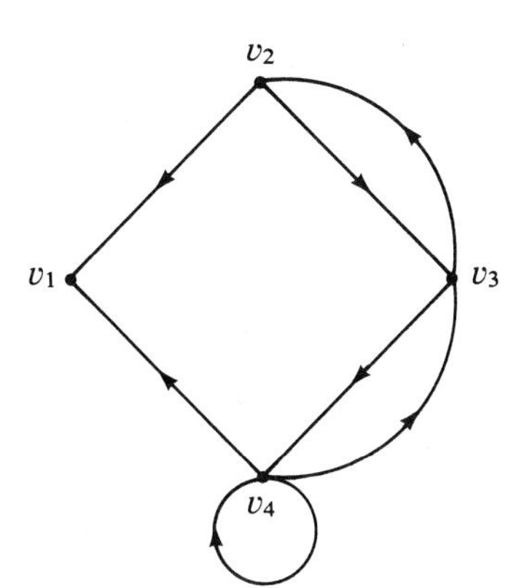

(b)

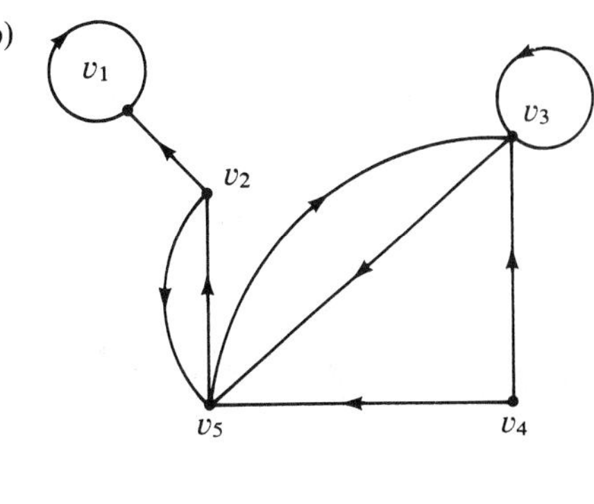

FIGURE 6.50

3 Obtain pictorial representations of the composite relations $R \circ S$ as depicted in Figure 6.51.

(a)

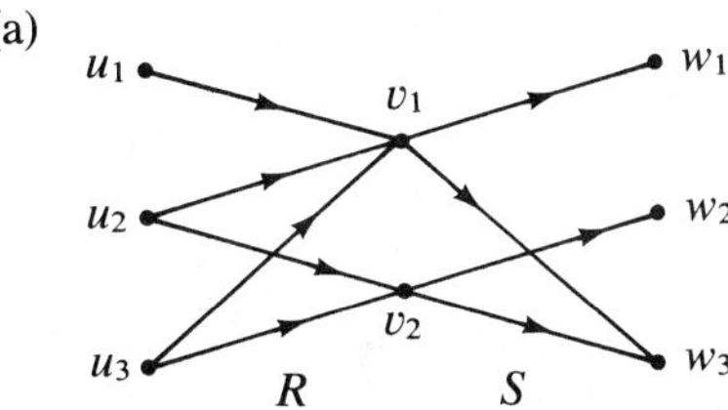

(b)

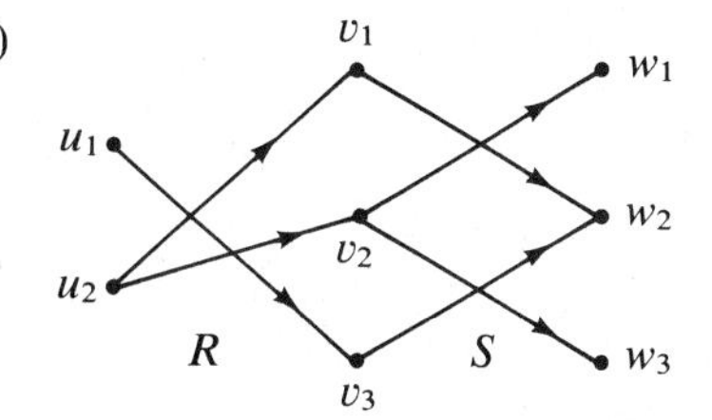

FIGURE 6.51

4 Compute the matrices corresponding to $R \circ S$ if R and S have the following matrix representations:

(a)
$$R = \begin{bmatrix} 1 & 0 & 1 & 1 \\ 0 & 1 & 1 & 0 \\ 0 & 1 & 0 & 1 \\ 0 & 0 & 1 & 0 \end{bmatrix} \qquad S = \begin{bmatrix} 1 & 1 & 0 \\ 0 & 1 & 1 \\ 1 & 0 & 0 \\ 0 & 1 & 1 \end{bmatrix}$$

(b)
$$R = \begin{bmatrix} 1 & 1 & 1 \\ 0 & 1 & 1 \\ 0 & 1 & 0 \\ 1 & 0 & 1 \end{bmatrix} \qquad S = \begin{bmatrix} 1 & 1 & 1 & 0 & 1 \\ 0 & 1 & 0 & 0 & 0 \\ 1 & 1 & 0 & 0 & 1 \end{bmatrix}$$

5 Use the corollary (of Theorem 6.16) to determine all paths of length 2 in the following digraphs.

(a) Exercise 1(a) **(b)** Exercise 1(b)

(c) Exercise 2(a) **(d)** Exercise 2(b)

6 Do the same (as in Exercise 5) for paths of length 3.

7 Compare the computational complexity of the reachability algorithm with that given by Theorem 6.17.

8 Use Theorem 6.17 to compute the reachability matrix E^* for the following digraphs.
(a) Exercise 1(a) (b) Exercise 1(b)
(c) Exercise 2(a) (d) Exercise 2(b)

9 Do the same (as in Exercise 8) using the reachability algorithm.

10 A digraph is said to be *strongly connected* if for each pair of vertices, v and w, there is a path from v to w (and from w to v). Give a matrix characterization of strongly connected digraphs.

11 Devise a formal algorithm for deciding whether or not a given digraph is strongly connected (see Exercise 10).

12 A digraph is said to be *unilaterally connected* if for each pair of vertices, v and w, there is a path joining them (in one direction or the other, or both). Give a matrix characterization of unilaterally connected digraphs.

13 Devise a formal algorithm for deciding whether or not a given digraph is unilaterally connected (see Exercise 12).

Sec. 6.8 PATH PROBLEMS

The study of path problems began with the Königsberg bridge problem, as introduced by Euler in 1736.

We already have alluded to the existence of various path problems in the theory of directed graphs. Actually, the undirected graphs offer path problems of their own, and because of their comparative simplicity we will use this setting to initiate our discussions.

In the Sunday supplement of the major newspapers you often will find a puzzle section to help you while away the weekend. It is surprising how many of these puzzles have an underlying graphical interpretation. Sometimes we are asked to find our way out of a maze. Another favorite puzzle asks us to trace just once around the edges of a certain line drawing without lifting our pencil from the page. This last example is a simple variation of a problem considered by Euler in 1736, the now famous Königsberg bridge problem. It is a fascinating example of the class of *path problems* that now directs our attention.

Goals

After studying this section you should be able to:

- Define the notion of an *Euler circuit* and describe a necessary and sufficient condition that such a circuit exists.

- Describe the idea of a *Hamiltonian path* and discuss the connection of this idea with the traveling salesperson problem.
- Discuss the algorithm for obtaining the shortest path between two vertices of a graph, and illustrate its use in the case of specific examples.
- Describe the variation of this algorithm for use with labeled digraphs.
- Use the shortest path algorithms for computing shortest paths in arbitrary given digraphs.

Eulerian and Hamiltonian Paths

In the city of Königsberg (now Kaliningrad) there were seven bridges connecting the banks of the Pregel River with its two islands, as shown in Figure 6.52(a). The famous Swiss mathematician, Leonard Euler, wondered whether it was possible to take a stroll in which one would start and end at the same point while crossing each bridge just once. (A similar problem arises in the design of parade routes.)

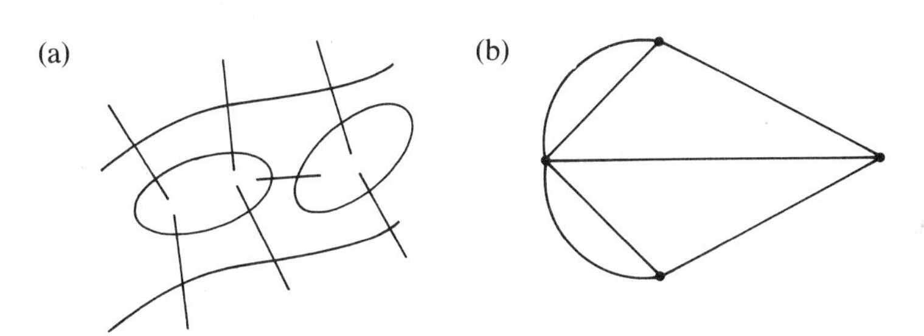

FIGURE 6.52

In the case of the Königsberg bridge problem we draw a corresponding *multigraph*, "multi" meaning that we allow more than one edge joining a given pair of vertices (see Fig. 6.52(b)). Then it is clear that we have returned to the pencil-pushing problem of the Sunday newspaper. In any case, we say that an **Euler path (circuit)** in a graph or multigraph is a path (circuit) that traverses each edge exactly once.

In the event that an Euler path can be traced, we will have transformed our graph or multigraph G into a digraph or multidigraph in simply observing the direction of traversal of the edges. Now suppose that in any digraph or multidigraph we generaliize the notion of the degree of a vertex as follows. We say that the vertex v has the **in-degree** and **out-degree**

$d^-(v)$ = number of directed edges entering v
$d^+(v)$ = number of directed edges leaving v

respectively. In some applications, a vertex with zero in-degree is called a *source*, and one with zero out-degree is called a *sink*. In the case of a digraph resulting from the tracing of an Euler circuit, the in-degree and the out-degree at each vertex must coincide. Every time we enter a vertex along one edge we must leave that vertex along a different edge. When phrased in terms of the degrees of the vertices in the original undirected graph or multigraph, this observation leads to the following conclusion.

Theorem 6.18 A connected (multi-) graph has an Euler circuit if and only if every vertex has an even degree.

Proof We already have shown the necessity of the condition. For the sufficiency, we use the fact (one that is easily verified) that every connected graph without vertices of odd degree has a circuit. In supposing that G has only vertices of even degree, we therefore may choose a circuit C_1. If $G = C_1$, then we are done. Otherwise we introduce an obvious temporary notation in claiming that $G–C_1$ again will have vertices only of even degree, since C_1 uses a pair of edges at each of its vertices. Continuing in this way, we can construct an inductive argument showing that we eventually will have an Euler circuit. Note, however, that this construction must be done with some care since there is the possibility that intermediate graphs $G–C_1–C_2–\cdots–C_k$ will not be connected. In addition we have to argue that the resulting sequence of circuits may yet be combined into a single Euler circuit. We leave you to work out the details of this solution. □

Corollary A connected (multi-) graph has an Euler path, not a circuit, if and only if it has exactly two vertices of odd degree.

Example 6.79 The Königsberg bridge problem does not have an Euler circuit because it has vertices of odd degree. In fact, all four of its vertices are of odd degree, so according to the corollary there does not even exist an Euler path. (Euler must have been disappointed.) ■

Example 6.80 The graph that is suggestive of a pair of spectacles in Figure 6.53 has an Euler path, not a circuit, since it has just two vertices of odd degree. ■

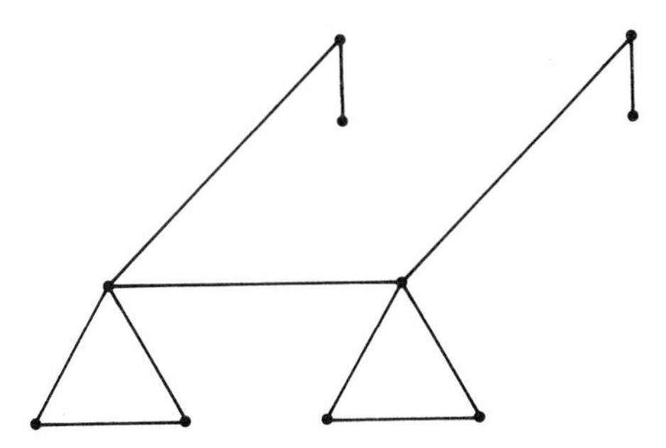

FIGURE 6.53

Example 6.81 In the digraph of Figure 6.44 the vertex labeled start is a source and the vertex labeled finish is a sink. The vertex in the center has in-degree two and out-degree one. ■

Example 6.82 The graph of Figure 6.29 has an Euler circuit because every vertex has an even degree. Can you trace around the figure, arriving finally at your point of origin, without lifting your pencil from the page? ■

If we revise Euler's problem by asking instead that every vertex be encountered once and only once, then we arrive at the **Hamiltonian paths (circuits)** of an undirected graph. Here we refer to W.R. Hamilton, a famous Irish mathematician of the nineteenth century. Being somewhat more ambitious than Euler, Hamilton sought to go around the world on a dodecahedron, one whose vertices were labeled by the names of 20 cities and, of course, the tour should encounter each city just once, excepting the origin, the point to which one should eventually return. If this particular polyhedron is projected onto the plane, according to the method described in Example 6.32, we will obtain the planar graph of Figure 6.54. In showing that Hamilton met with more success than Euler, we have indicated a Hamiltonian circuit for this graph.

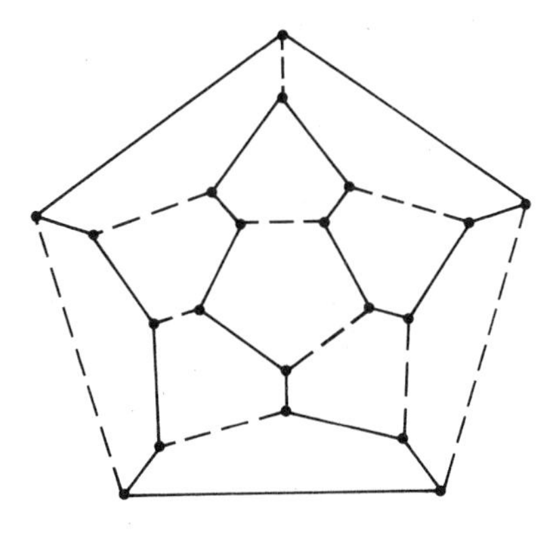

FIGURE 6.54

Although the problem of Hamiltonian paths bears a superficial resemblance to the Euler problem, no simple criterion is known for deciding whether such paths exist. This is unfortunate because the problem of arranging itineraries is quite real. The traveling salesperson, the circus, any traveling road company is faced with a problem not unlike that considered by Hamilton. What has come to be called the traveling salesperson problem, however, is actually more difficult still. The salesperson must visit a number of cities, and we suppose that the distances are given in advance. We then wish to arrange a minimum distance itinerary that will include a visit to each city just once, returning to the point of origin. It seems that such problems are quite difficult and, except in very special cases, no efficient solutions are known. In the case of a Hamiltonian path, most of the special solutions are to the effect that if the graph has enough edges, then a Hamiltonian path exists. That is true of the following result, one that you can verify inductively by extending the path.

Theorem 6.19 Every complete graph has a Hamiltonian path.

Shortest Paths in Graphs

We already have mentioned the importance of the various shortest path problems. We now describe a kind of propagation procedure for determining a shortest path connecting any two given vertices u, v in an undirected graph $G = (V, E)$. The algorithm should be studied quite closely, since it then can be taken as a model for understanding the subsequent shortest path algorithms in labeled digraphs. In the latter case, each directed edge $(x, y) \in E$ will be labeled with a *length* $\Delta(x, y)$. In the case of the undirected graphs, we simply assume that each edge is of length $\Delta = 1$.

In thinking of the vertex u as the origin, our algorithms actually define a (minimum) distance $d(w)$ from u, for each vertex w on the shortest path from u to v. The path itself then may be obtained by a separate *backstepping* algorithm, one whose broad outline may be described as follows.

We let the eventual shortest path be denoted by the sequence $\langle v_0, v_1, \ldots, v_k \rangle$ where, as usual,

$$\begin{matrix} v_0 = u \\ v_k = v \end{matrix} \quad \text{and each} \quad \{v_{i-1}, v_i\} \in E$$

$[(v_{i-1}, v_i) \in E$ in the case of the directed graphs$]$. Assuming that there is a path from u to v, the shortest path algorithm(s) will eventually determine $d(v) = d(v_k)$, so we choose $v_k = v$. In supposing that $v_k, v_{k-1}, \ldots, v_i$ already have been chosen, we then look for a vertex w with the properties:

$$w \; E \; v_i \quad \text{and} \quad d(w) = d(v_i) - \Delta(w, v_i)$$

and we choose $v_{i-1} = w$, continuing to backstep until we reach the origin u. We have not included the details of this backstepping procedure in our shortest path algorithm(s) because that would tend to obscure the main ideas involved. It is then left for you to formalize this technique and to tie the procedure in tandem with each of our shortest path algorithms, as indicated.

```
algorithm shortpath-graph (u, v: d)

set k to 0
initialize W as {u}, M as ∅
while v ∉ M and W ≠ ∅
   adjoin W to M
   for each w in W
      assign d[w] = k
      if v ∉ W
         increment k by one
         for each x ∈ W and y ∉ M
            if x E y
               adjoin y to W
            delete M from W
backstep to find shortest path
```

The idea behind our basic shortest-path algorithm is really quite simple. Formally the procedure may be phrased as in the accompanying algorithm called *shortpath-graph*. Like the ripples in a stream that emanate from the point at which we drop a pebble, we imagine a *wave* W, having its origin at u. The wave systematically traces all the paths of increasing distance from u, *marking* (with the set M) this distance as it is determined. In this way, circuits can be avoided. Obviously a shortest path would not contain a circuit. In the final block of statements (beginning with *increment*), we extend the wave front to a new set of unmarked vertices, whereas those that are already marked are then removed from the wave. The process continues until $v \in M$ (i.e., until our destination becomes *marked*), and its distance $d[v]$ is determined. One then applies the backstepping process, as discussed earlier.

Example 6.83 Consider the graph of Figure 6.12(a), redrawn in Figure 6.55(a). Suppose we try to use our algorithm to determine a shortest path from c to d. A trace of the execution of the algorithm then would appear as shown in Table 6.5.

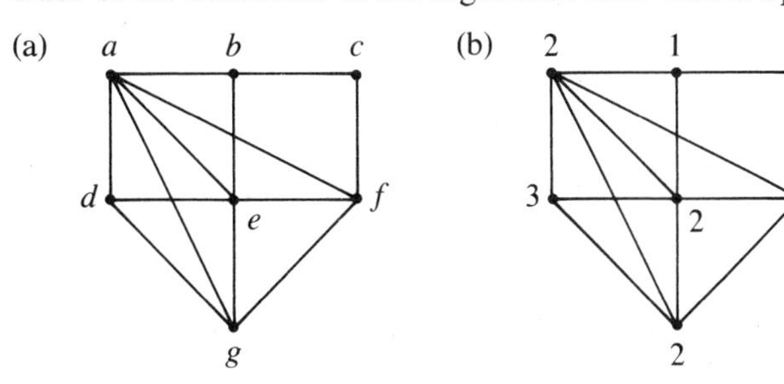

FIGURE 6.55

TABLE 6.5

u	v	k	M	W	d
c	d	0	$\varnothing$	$\{c\}$	
			$\{c\}$		$d[c] = 0$
		1		$\{b, c, f\}$	
				$\{b, f\}$	
			$\{b, c, f\}$		$d[b] = 1$
					$d[f] = 1$
		2		$\{a, b, e, f, g\}$	
				$\{a, e, g\}$	
			$\{a, b, c, e, f, g\}$		$d[a] = 2$
					$d[e] = 2$
					$d[g] = 2$
		3		$\{a, d, e, g\}$	
				$\{d\}$	
			$\{a, b, c, d, e, f, g\}$		$d[d] = 3$

In this particular example, all the vertices become marked and their minimum distances from c are determined as in Figure 6.55(b). A shortest path from c to d may then be found by using the backstepping procedure. ■

Example 6.84 In order to illustrate the backstepping procedure, as applied to the results of Example 6.83, we take $v_3 = d$ as the last vertex in our eventual path $\langle v_0, v_1, v_2, v_3 \rangle$ from c to d. Then we have three choices for v_2, namely $v_2 = a$, e, or g, because each of these satisfies the conditions:

$$v_2 \; E \; d \quad \text{and} \quad d[v_2] = d[d] - 1 = 2$$

Suppose we take $v_2 = g$. Then we have to look for a vertex v_1 such that

$$v_1 \; E \; g \quad \text{and} \quad d[v_1] = d[g] - 1 = 1$$

and here, only $v_1 = f$ is appropriate. Finally we obtain $v_0 = c$ such that

$$v_0 \; E \; f \quad \text{and} \quad d[v_0] = d[f] - 1 = 0$$

as required. Altogether this gives us the shortest path $\langle c, f, g, d \rangle$. ■

Shortest Paths in Digraphs

Now we are ready to consider shortest paths in the labeled digraphs. Because the technique is so similar to that just illustrated, we can afford to be a little more loose in our language, as in the description of the accompanying algorithm called *shortpath-digraph*.

algorithm shortpath-digraph (origin, destination: d)

set place to origin and d[origin] = 0
initialize wave and marked set
while place not destination and wave not empty
  mark the place
  for unmarked vertices directly accessible
    adjoin them to wave
    recompute tentative distances
  delete (old) place from wave
  choose (new) place from wave {of smallest tentative distance from origin}
backstep to find shortest path

Nevertheless there are a few distinctions worth noting. We imagine that all distances have first been initialized as ∞ (realizing that they will be recomputed later on). In the course of executing the algorithm, we visit various places u^* in the current wave. If a place has just become marked (as in Figure 6.56), we examine all vertices that are directly accessible along a directed edge from this place, and if they are unmarked, we adjoin them to the wave. At the same time we recompute their distance from the origin, using our knowledge of the minimum distance to this particular place u^*. In fact we set

$$d[w] \longleftarrow \min\{d[w], d[u^*] + \Delta(u^*, w)\}$$

replacing the old value of $d[w]$ if the indicated sum is smaller. We then choose a new place, one with the smallest distance estimate among all the vertices in the wave.

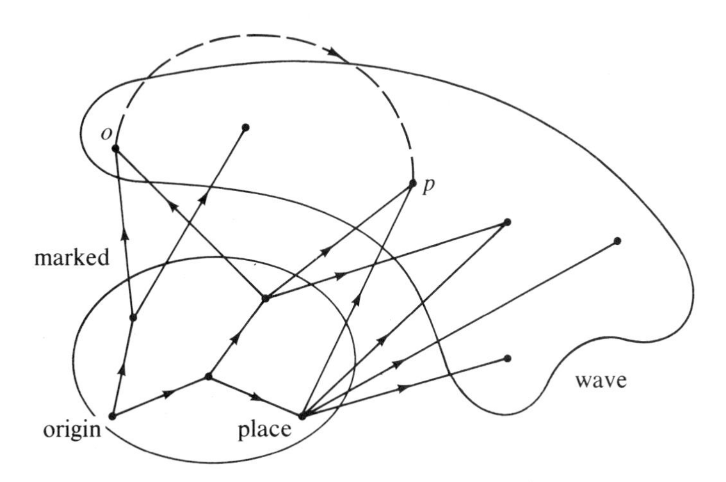

FIGURE 6.56

If this new place is p (in Fig. 6.56), then there cannot be a shorter path from the origin to p than the one that is implied in its selection. It deserves to be marked that its minimum distance has been determined. To see this we must observe that all the vertices that are directly accessible from marked vertices are in the wave (unless they are already marked). If the supposed shorter path to p leaves the unmarked vertices through o (in Fig. 6.56), and finally arrives along the dashed line at p, then o is in the wave and we could not have chosen p of smallest tentative distance in the wave! As indicated in the pseudocode algorithm, we continue moving from place to place until the place we select is our desired destination.

Example 6.85 In Figure 6.57 we have a digraph whose edges are labeled with various lengths. If we designate

$$u = \text{origin} = v_1$$
$$v = \text{destination} = v_8$$

then a trace of the execution of the algorithm is as shown in Table 6.6.

In reading this transcription, one must bear in mind that the distances $d[v_i]$ are not final, that is, not known to the minimal until the given vertex has become marked. Thus we have $d[v_6] = 11$, then $d[v_6] = 10$, and it is only a coincidence that this is already the minimum distance from the origin to v_6. Suppose there had been an edge of length 1 from v_3 to v_6!

In order to provide a few details, suppose we pick up the action at the point where we have arrived at the place $u^* = v_5$. In marking this place we adjoin $u^* = v_5$ to M. The unmarked vertices that are directly accessible from here are v_3 and v_7. We thus recompute:

$$d[v_3] \longleftarrow \min\{d[v_3], d[v_5] + \Delta(v_5, v_3)\} = \min\{7, 4 + 4\} = 7$$

$$d[v_7] \longleftarrow \min\{d[v_7], d[v_5] + \Delta(v_5, v_7)\} = \min\{7, 4 + 5\} = 7$$

and adjoin both vertices to W. (In this case they are already in W.) Then we delete our old place $u^* = v_5$ from W and select a new $u^* = w$ from W, of minimum

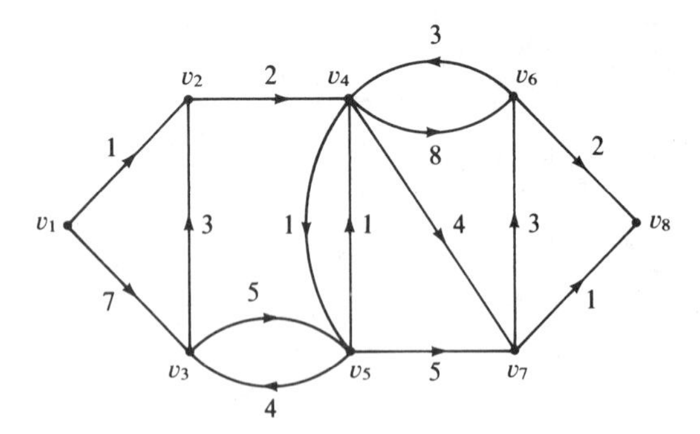

FIGURE 6.57

TABLE 6.6

u	v	u^*	M	W	d
v_1	v_8	v_1	$\varnothing$	$\{v_1\}$	$d[v_1] = 0$
			$\{v_1\}$	$\{v_1, v_2, v_3\}$	$d[v_2] = 1$
					$d[v_3] = 7$
				$\{v_2, v_3\}$	
		v_2	$\{v_1, v_2\}$	$\{v_2, v_3, v_4\}$	$d[v_4] = 3$
				$\{v_3, v_4\}$	
		v_4	$\{v_1, v_2, v_4\}$	$\{v_3, v_4, v_5, v_6, v_7\}$	$d[v_5] = 4$
					$d[v_6] = 11$
					$d[v_7] = 7$
				$\{v_3, v_5, v_6, v_7\}$	
		v_5	$\{v_1, v_2, v_4, v_5\}$	$\{v_3, v_5, v_6, v_7\}$	$d[v_3] = 7$
					$d[v_7] = 7$
				$\{v_3, v_6, v_7\}$	
		v_7	$\{v_1, v_2, v_4, v_5, v_7\}$	$\{v_3, v_6, v_7, v_8\}$	$d[v_6] = 10$
					$d[v_8] = 8$
				$\{v_3, v_6, v_8\}$	
		v_3	$\{v_1, v_2, v_3, v_4, v_5, v_7\}$		
				$\{v_6, v_8\}$	
		v_8			

(tentative) distance $d[w]$. In the present situation this selection involves a comparison from among

$$d[v_3] = 7 \qquad d[v_6] = 11 \qquad d[v_7] = 7$$

and we have chosen $u^* = v_7$. At this point we are ready to begin another cycle of the computation.

With this bit of detail you should be able to follow the execution of the algorithm to its completion. The termination comes as soon as $u^* = v = v_8$. You also should check that the backstepping procedure gives rise to the path $\langle v_1, v_2, v_4, v_7, v_8 \rangle$, a shortest path from v_1 to v_8. ■

EXERCISES 6.8

1 Determine all the in-degrees and out-degrees of vertices in the following digraphs.

(a) Exercise 1(a) of Exercises 6.7

(b) Exercise 1(b) of Exercises 6.7

(c) Exercise 1(c) of Exercises 6.7

(d) Exercise 1(d) of Exercises 6.7

(e) Exercise 2(a) of Exercises 6.7

(f) Exercise 2(b) of Exercises 6.7

2 Decide whether or not the following graphs have an Euler circuit, and explain your answers.

(a) Exercise 1(a) of Exercises 6.1

(b) Exercise 1(c) of Exercises 6.1

(c) $K_{2,4}$

(d) Figure 6.10

(e) K_6

(f) the cube with its (four) diagonals

3 Give an inductive proof of Theorem 6.19.

4 Use the shortpath algorithm for graphs to find a shortest path from u to v in each of the following situations.

(a) Exercise 1(c) of Exercises 6.1, $u = v_1$, $v = v_6$

(b) Exercise 3(c) of Exercises 6.1, $u = v_1$, $v = v_6$

(c) Exercise 1 of Exercises 6.4, $u =$ Malkuth, $v =$ Kether

(d) the star of David, u and v two opposite vertices

(e) Figure 6.9, $u = v_1$, $v = v_3$

(f) Figure 6.10, $u = v_1$, $v = v_9$

5 Devise a formal algorithm for implementing the backstepping procedure

(a) for undirected graphs, all edge lengths = 1

(b) for labeled digraphs

6 Use the shortpath algorithm for graphs to find shortest paths from start to finish in the mazes of Figure 6.58.

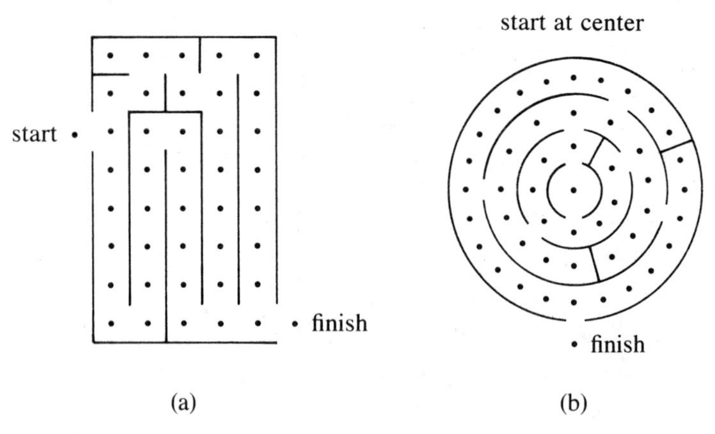

FIGURE 6.58

7 Use the shortpath algorithm for digraphs to find shortest paths from u to v in the following situations (assuming unit distance along each edge).

(a) Exercise 1(a) of Exercises 6.7, $u = v_1$, $v = v_5$
(b) Exercise 1(b) of Exercises 6.7, $u = v_2$, $v = v_3$
(c) Exercise 1(c) of Exercises 6.7, $u = v_6$, $v = v_5$
(d) Exercise 1(d) of Exercises 6.7, $u = v_2$, $v = v_1$
(e) Exercise 2(a) of Exercises 6.7, $u = v_2$, $v = v_1$
(f) Exercise 2(b) of Exercises 6.7, $u = v_4$, $v = v_1$

8 Use the shortpath algorithm for digraphs to find shortest paths from u to v in the labeled digraphs of Figure 6.59.

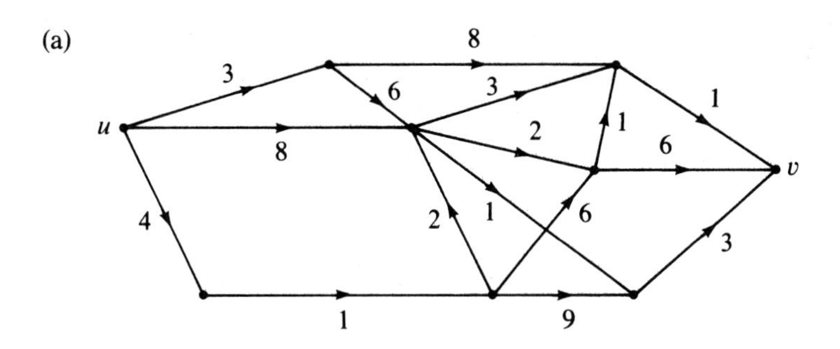

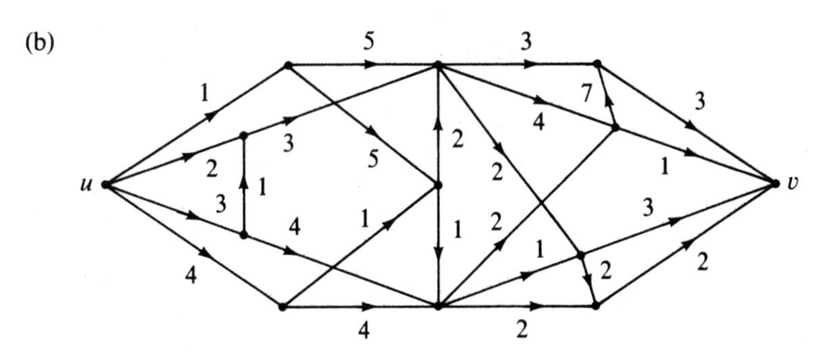

FIGURE 6.59

SOLUTIONS TO SELECTED EXERCISES

Section 1.1

1 **(a)** $A = \{\text{Truman, Eisenhower, Kennedy, Johnson, Nixon, Ford, Carter, Reagan}\}$
(c) $C = \{n: n \in Z, 0 \leq n \leq 1000\}$
(e) $E = \{\text{Connecticut, Maine, Massachusetts, New Hampshire, Rhode Island, Vermont}\}$

2 **(a)** All integers that are multiples of four
(e) All lowercase letters of our ordinary alphabet
(f) All logical operators of the Pascal language

7 Let p be the statement:

p: the barber shaves himself

and suppose that p is the case. Then, since he shaves all (and presumably only) those people who do not shave themselves, he doesn't shave himself (that is, "not p"). On the other hand, if he doesn't shave himself, then he does (that is, "not p" implies p, as well).

8 **(c)** $|E| = 6$ **(e)** $|B| = 4$ **(h)** $|G| = 1$

10 A and H are infinite.

Section 1.2

2 **(c)** In computing

$$\begin{aligned} A \cup H &= \{m: m = 4n, n \in Z\} \cup \{n: n \in N, n > 1000\} \\ &= \{\ldots, -8, -4, 0, 4, 8, \ldots, 996, 1000, 1001, 1002, \ldots\} \end{aligned}$$

we see that $A \cup H \nsubseteq N$ because, for example, $-4 \in A \cup H$ but $-4 \notin N$.

(d) A and H are not disjoint because, for example, $1004 \in A$ and $1004 \in H$.

3 **(a)** $A \cap B = \{a, b, d, f\} \cap \{a, c, f, g\} = \{a, f\}$

(d) $(A \cup B) \cap C = (\{a, b, d, f\} \cup \{a, c, f, g\}) \cap \{b, e, g\}$
$= \{a, b, c, d, f, g\} \cap \{b, e, g\} = \{b, g\}$

(h) $B \cup {\sim}C = \{a, c, f, g\} \cup {\sim}\{b, e, g\}$
$= \{a, c, f, g\} \cup \{a, c, d, f\} = \{a, c, d, f, g\}$

6 **(a)** To verify (S5b) we must establish the two inclusions:

$$A \cap (B \cup C) \subseteq (A \cap B) \cup (A \cap C)$$
$$A \cap (B \cup C) \supseteq (A \cap B) \cup (A \cap C)$$

For the first, suppose $x \in A \cap (B \cup C)$. Then $x \in A$ and $x \in B \cup C$ (i.e., $x \in B$ or $x \in C$), by the definition of intersection and union. In case $x \in A$ and $x \in B$ we have $x \in A \cap B$, whereas if $x \in A$ and $x \in C$ we have $x \in A \cap C$, again by the definition of intersection. It follows that $x \in (A \cap B) \cup (A \cap C)$ by the definition of union. Conversely, if $x \in (A \cap B) \cup (A \cap C)$, one uses similar techniques to show that $x \in A \cap (B \cup C)$, thus establishing the second inclusion.

(e) To verify (S9b), suppose $x \in \sim(A \cap B)$. Then either $x \notin A$ or $x \notin B$, i.e., $x \in \sim A$ or $x \in \sim B$, yielding $x \in \sim A \cup \sim B$ by the definition of union. This shows that

$$\sim(A \cap B) \subseteq \sim A \cup \sim B$$

Conversely, supposing that $x \in \sim A \cup \sim B$, we have $x \notin A$ or $x \notin B$, and in either case $x \notin A \cap B$ by the definition of intersection, showing $x \in \sim(A \cap B)$. This establishes the reverse inclusion:

$$\sim A \cup \sim B \subseteq \sim(A \cap B)$$

so that the desired set equality (S9b) follows.

9 $A \cap (C \cup (D \cup E)) \cap ((B \sim F) \cup (D \cap E))$

11 $(A \sim B) \cup (B \sim A)$

12 (a) $\{\{a\}, \{b\}, \{c\}\}$
$\{\{a, b\}, \{c\}\}$
$\{\{a, c\}, \{b\}\}$
$\{\{a\}, \{b, c\}\}$
$\{\{a, b, c\}\}$

Section 1.3

1 (a) $f(0) = 0^3 - 2 \cdot 0 + 1 = 1$

(b) $f(1) = 1^3 - 2 \cdot 1 + 1 = 1 - 2 + 1 = 0$

(c) $f(-1) = (-1)^3 - 2(-1) + 1 = -1 + 2 + 1 = 2$

(d) $f(2) = 2^3 - 2 \cdot 2 + 1 = 8 - 4 + 1 = 5$

3 (a) $x(7) = x_7 = (-1)^7 = -1$

(b) $x(7) = x_7 = 2 \cdot 7 = 14$

(c) $x(7) = x_7 =$ the 7th prime number $= 17$

4 (a) The function is 1–1 because $f(x) = f(y)$ implies $x + 1 = y + 1$ or $x = y$. It is also onto because every $z \in Z$ can be written

$$z = f(z - 1)$$

as the image of something $(z - 1)$ in the domain.

(d) Not 1–1 $[f(2) = f(-2), 2 \neq -2]$; not onto $[f(?) = -1]$

5 Denoting the function as g in each case, we have

(a) Not 1–1 $[g(2) = g(4), 2 \neq 4]$, not onto $[g(?) = d]$

(b) 1–1 and onto

(c) 1–1 but not onto $[g(?) = d]$

(d) Onto but not 1–1 $[g(1) = g(3), 1 \neq 3]$

6 (a) In Exercise 4(a), the inverse function g is given by $g(y) = y - 1$, for then we have

$$g \circ f(x) = g(x + 1) = (x + 1) - 1 = x$$
$$f \circ g(y) = f(y - 1) = (y - 1) + 1 = y$$

as required of an inverse function. Note the interchange in the roles of the function names, f and g, as compared with the discussion in the text.

(b) In Exercise 5(b) the inverse function f is given by

$$f(a) = 1$$
$$f(b) = 3$$
$$f(c) = 2$$

10 (a) $S \times T = \{(1, a), (1, b), (1, c), (2, a), (2, b), (2, c)\}$

(d) $S \times T = \{(a, \varnothing), (b, \varnothing), (c, \varnothing)\}$

12 (a) Denote the left-hand set as X and the right-hand set as Y. Then if $x \in X$, we have $x = (s, t)$ with $s \in A$ and $s \in B$, and $t \in C$ and $t \in D$. It follows that $x \in A \times C$ (since $s \in A$ and $t \in C$) and similarly, $x \in B \times D$ (since $s \in B$ and $t \in D$). Consequently, $x \in Y$, the intersection of these two sets. Conversely, if $y \in Y$, then $y \in A \times C$ and $y \in B \times D$, i.e., $y = (s, t)$ with $s \in A$ and $s \in B$, and also, $t \in C$ and $t \in D$. But according to the definition of intersection, this means that $s \in A \cap B$ and $t \in C \cap D$, i.e., $y \in X$. Having thus shown that $X \subseteq Y$ and $Y \subseteq X$, we conclude that $X = Y$, as we set out to prove.

16 It is again a one-to-one correspondence.

Section 1.4

1 Property (E2) fails, e.g., $3 \mid 12$ but $12 \nmid 3$.

4 If $a \sim b$, we show that $[a] = [b]$ by establishing that each set (equivalence class) is contained within the other. If $x \in [a]$, then $x \sim a$, using the definition of $[a]$. But $x \sim a$ and $a \sim b$ yields $x \sim b$ by (E3), and the latter means that $x \in [b]$. Similarly, if $y \in [b]$ so that $y \sim b$, we have $y \sim b$ and $b \sim a$ (using (E2)). From this it follows, again by (E3), that $y \sim a$, i.e., $y \in [a]$. Thus under the assumption that $a \sim b$ we have shown that $[a] \subseteq [b]$ and $[b] \subseteq [a]$, i.e., $[a] = [b]$.

6 (a) Yes

(c) No, (E3) fails

(e) No, (E3) fails as does (E1) in general

10 (b) Noting that more detailed arguments could be provided after the discussion in Section 2.1, we simply conclude the following:

E1. Every line is parallel to itself (by convention).

E2. If line l is parallel to line m, then line m is parallel to line l.

E3. If line l is parallel to line m and line m is parallel to line p, then l is parallel to p.

(e) Verifying the three properties in turn, we have the following:

E1. $(n, m) \sim (n, m)$ because $nm = mn$.

E2. If $(n, m) \sim (r, s)$, then $ns = mr$ or $rm = sn$, but the latter ensures that $(r, s) \sim (n, m)$.

E3. If $(n, m) \sim (r, s)$ and $(r, s) \sim (p, q)$, then we have

$$ns = mr \quad \text{and} \quad rq = sp$$

From this we conclude that

$$nq = (mr/s)(sp/r) = mp$$

and the latter establishes that $(n, m) \sim (p, q)$.

Section 1.5

1 Nine students subscribe to none of the magazines.

3 $26^2 = 676$

5 $(26)^2 (10)^4 + (26)^3 (10)^3 = 24{,}336{,}000$

7
$$\binom{n-1}{k-1} + \binom{n-1}{k} = \frac{(n-1)!}{(k-1)!(n-k)!} + \frac{(n-1)!}{k!(n-k-1)!}$$
$$= \frac{k(n-1)! + (n-k)(n-1)!}{k!(n-k)!}$$
$$= \frac{n!}{k!(n-k)!} = \binom{n}{k}$$

8 **(b)** 6,349,108,119,600
(c) 5005 (without respect to the order)
(d) 20!

11 **(b)** 36 **(e)** 181,440

12 Setting $a = b = 1$ in the binomial theorem, we obtain

$$2^n = (1+1)^n = \binom{n}{0}1^n 1^0 + \binom{n}{1}1^{n-1}1^1 + \cdots + \binom{n}{n}1^0 1^n$$
$$= \binom{n}{0} + \binom{n}{1} + \cdots + \binom{n}{n}$$

where the final terms on the right are the entries in the nth row of the Pascal triangle.

14 **(b)** $1 + 1 + 2 + 3 + 3 = 10$ (even)
(d) $1 + 1 + 3 + 1 + 5 = 11$ (odd)
(f) $1 + 0 + 3 + 1 + 4 = 9$ (odd)

Section 2.1

2 **(G1):** 1 and 2 are points and $1 \neq 2$

(G2): Only the line $\{1, 2\}$ passes through points 1 and 2
Only the line $\{1, 3\}$ passes through points 1 and 3
etc.

(G3): Point 3 is not on the line $\{1, 2\}$
Point 2 is not on the line $\{1, 3\}$
Point 3 is not on the line $\{1, 4\}$
Point 1 is not on the line $\{2, 3\}$
Point 1 is not on the line $\{2, 4\}$
Point 1 is not on the line $\{3, 4\}$

(G4): Point 3 not on line $\{1, 2\}$ has the parallel $\{3, 4\}$
Point 2 not on line $\{1, 3\}$ has the parallel $\{2, 4\}$
etc.

6 **(a)** Using (H4), the two lines l and m must intersect. If the lines intersect in more than one point, say with points of intersection $A, B, \ldots$, then by (H2), one and only one line passes through A and B, and this would contradict our assumption.

8 **(a)** P: Dreams come true
Q: I'll be with you

(f) P: A stone rolls
Q: It gathers no moss

(h) P: $A \subseteq C$ and $B \subseteq C$
Q: $A \cup B \subseteq C$

9 Using the law of contraposition, we have
(a) Dreams do not come true.
(b) This is not a sunny day.

Section 2.2

2 **(a)** I'll be with you (DET).
(d) The phone call wasn't long distance (CON).
(h) The flower is red (CAS).

3 **(a)** We had $x \in A \cup (A \cap B)$ and we wanted to show that $x \in A$. Assuming that $x \in A \cup (A \cap B)$, the cases are that $x \in A$ or $x \in A \cap B$, either of which will lead to the conclusion that $x \in A$.

6 If triangle ABC is to be isosceles, we use backward reasoning to note that we should try to establish that $a = b$, a and b being the sides opposite angles A and B, respectively. Now we use forward reasoning from the assumption that the triangle has area $c^2/4$. The area of a right triangle is $ab/2$, and thus we have $c^2 = 2ab$. However, using the Pythagorean Theorem ($c^2 = a^2 + b^2$) we are then able to write:

$$\begin{aligned} a^2 + b^2 &= 2ab \\ a^2 - 2ab + b^2 &= 0 \\ (a - b)^2 &= 0 \\ a - b &= 0 \\ a &= b \end{aligned}$$

thus establishing the desired implication.

10 **(c)** The following are a few possibilities:

ABCD is a square
A, B, C are right angles
ABCD is a parallelogram with equal diagonals

11 **(a)** *aaaaaaaa, ba, ab, aaaab, aabaab, abababab*
(b) *ba, baaaa, aaaab, bbbab, babbb*

Section 2.3

4 Any statement of the form $p \wedge \neg p$

5 When p is known to be absurd, we accept the validity of an implication $p \longrightarrow q$ on the basis of Brouwer's interpretation of the meaning of the implication symbol, i.e., if p is absurd, we can surely convert any arbitrary proof of p (there are none) into a proof of q.

7 **(a)**

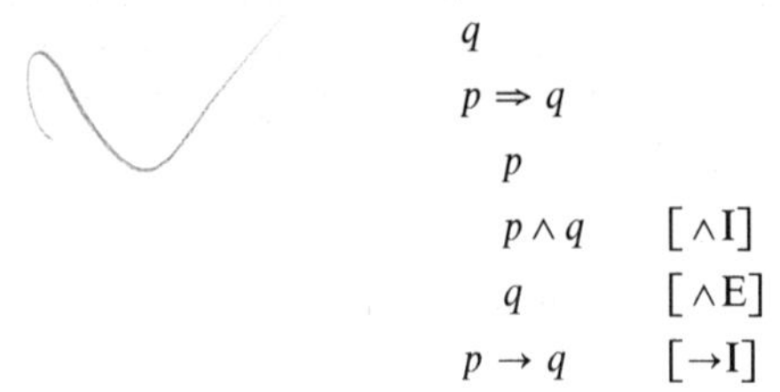

$$\begin{array}{ll} q & \\ p \Rightarrow q & \\ \quad p & \\ \quad p \wedge q & [\wedge \mathrm{I}] \\ \quad q & [\wedge \mathrm{E}] \\ p \rightarrow q & [\rightarrow \mathrm{I}] \end{array}$$

(c)

$p \wedge \neg p$
$p \wedge \neg p \rightarrow p \wedge \neg p$ [→A] (See Example 2.26)
$\neg(p \wedge \neg p)$ [¬I]
$p \wedge \neg p \rightarrow q$ [¬E]
q [→E]

8 (c)

$p \vee q \Rightarrow q \vee p$
$p \vee q$
$p \Rightarrow q \vee p$
p
$q \vee p$ [∨I]
$p \rightarrow q \vee p$ [→I]
$q \Rightarrow q \vee p$
q
$q \vee p$ [∨I]
$q \rightarrow q \vee p$ [→I]
$q \vee p$ [∨E]
$p \vee q \rightarrow q \vee p$ [→I]
$q \vee p \rightarrow p \vee q$ [Similarly]
$p \vee q \leftrightarrow q \vee p$ [↔I]

9

$p \wedge (p \vee q) \rightarrow p$ [Example 2.23]
$p \Rightarrow p \wedge (p \vee q)$
p
$p \vee q$ [∨I]
$p \wedge (p \vee q)$ [∧I]
$p \rightarrow p \wedge (p \vee q)$ [→I]
$p \wedge (p \vee q) \leftrightarrow p$ [↔I]

10 (b)

$s \rightarrow d$
s
d [→E]

(c)

$p \rightarrow n^2$
$n \rightarrow n^2$
$n \vee p$
n^2 [∨E]

(f)

$j \wedge s$
j [∧E]

(h)

$f \leftrightarrow l$
$f \rightarrow l$ [↔E]

Section 2.4

1 (a)

$p \Rightarrow p \vee \neg q$

p

$p \vee \neg q$ [∨I]

(d)

$p \wedge (q \wedge r) \Rightarrow q \vee r$

$p \wedge (q \wedge r)$

$q \wedge r$ [∧E]

q [∧E]

$q \vee r$ [∨I]

3 (a)

$p \rightarrow q \wedge r, p \Rightarrow r$

$p \rightarrow q \wedge r$

p

$q \wedge r$ [→E]

r [∧E]

(d)

$\neg p, (\neg p \rightarrow q) \vee [p \wedge (r \rightarrow q)] \Rightarrow r \rightarrow q$

$\neg p$

$(\neg p \rightarrow q) \vee (p \wedge (r \rightarrow q))$

$(\neg p \rightarrow q) \Rightarrow (r \rightarrow q)$

$\neg p \rightarrow q$

$r \Rightarrow q$

r

q [→E]

$r \rightarrow q$ [→I]

$(\neg p \rightarrow q) \rightarrow (r \rightarrow q)$ [→I]

$p \wedge (r \rightarrow q) \Rightarrow (r \rightarrow q)$

$p \wedge (r \rightarrow q)$

$r \rightarrow q$ [∧E]

$p \wedge (r \rightarrow q) \rightarrow (r \rightarrow q)$ [→I]

$r \rightarrow q$ [∨E]

4 (a)

$p \vee (q \wedge r), p \rightarrow s, (q \wedge r) \rightarrow s \Rightarrow s \vee p$

$p \vee (q \wedge r)$

$p \rightarrow s$

$(q \wedge r) \rightarrow s$

s [∨E]

$s \vee p$ [∨I]

6 (a) $(c \wedge k) \rightarrow (a \rightarrow m), \neg m \wedge a \Rightarrow (c \wedge \neg k)$

$(c \wedge k) \rightarrow (a \rightarrow m)$
$\neg m \vee a$
$c \Rightarrow \neg k$
  c
  $k \Rightarrow m \wedge \neg m$
    k
    $c \wedge k$ [∧I]
    $a \rightarrow m$ [→E]
    a [∧E]
    m [→E]
    $\neg m$ [∧E]
    $m \wedge \neg m$ [∧I]
  $k \rightarrow m \wedge \neg m$ [→I]
  $\neg k$ [¬I]
$c \rightarrow \neg k$ [→I]

7 (c) $p \wedge q \wedge (p \rightarrow r) \Rightarrow r \vee (q \rightarrow r)$

$p \wedge q \wedge (p \rightarrow r)$
  $q \rightarrow r$ [→I]
    q
    p [∧E]
    $p \rightarrow r$ [∧E]
    r [→E]
$r \vee (q \rightarrow r)$ [∨I]

(d) $\neg q \Rightarrow q \rightarrow \neg p$

$\neg q$
$q \rightarrow \neg p$ [¬E]

8 $p \rightarrow q, \neg q \Rightarrow \neg p$

$p \rightarrow q$
$\neg q$
$p \Rightarrow (q \wedge \neg q)$
  p
  q [→E]
  $q \wedge \neg q$ [∧I]
$p \rightarrow (q \wedge \neg q)$ [→I]
$\neg p$ [¬I]

9

$$p \to (r \to s) \Rightarrow r \to (p \to s)$$

$p \to (r \to s)$	
$r \to (p \to s)$	$[\to \mathrm{I}]$
$\quad r$	
$\quad p \to s$	$[\to \mathrm{I}]$
$\qquad p$	
$\qquad r \to s$	$[\to \mathrm{E}]$
$\qquad s$	$[\to \mathrm{E}]$

11 (b)

$p \to (q \to p \wedge q)$	$[\to \mathrm{I}]$
$\quad p$	
$\quad q \to p \wedge q$	$[\to \mathrm{I}]$
$\qquad q$	
$\qquad p \wedge q$	$[\wedge \mathrm{I}]$

Section 2.5

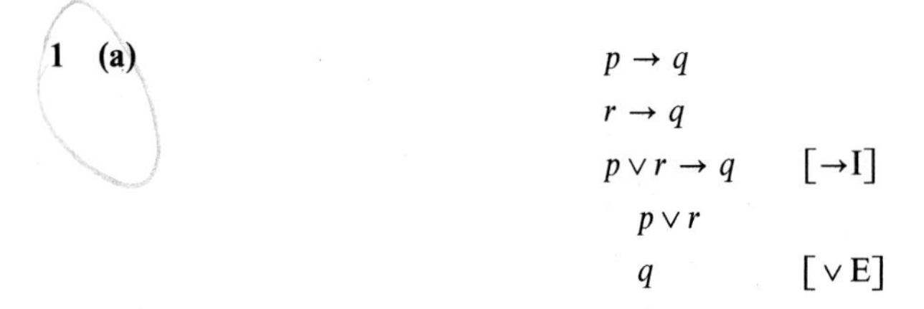

1 (a)

$p \to q$	
$r \to q$	
$p \vee r \to q$	$[\to \mathrm{I}]$
$\quad p \vee r$	
$\quad q$	$[\vee \mathrm{E}]$

2 We have shown that $p \to \neg p$ where p is the statement

p: n is a smallest integer between 0 and 1

Applying the rule $[\neg \mathrm{A}]$ we obtain

p: there is no (smallest) integer between 0 and 1

since n was arbitrary.

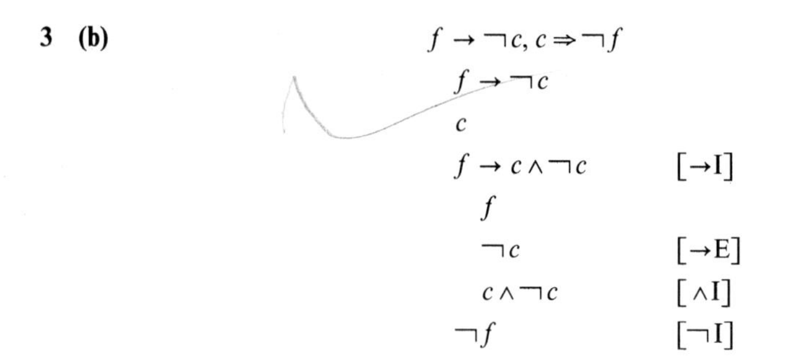

3 (b)

$$f \to \neg c, c \Rightarrow \neg f$$

$f \to \neg c$	
c	
$f \to c \wedge \neg c$	$[\to \mathrm{I}]$
$\quad f$	
$\quad \neg c$	$[\to \mathrm{E}]$
$\quad c \wedge \neg c$	$[\wedge \mathrm{I}]$
$\neg f$	$[\neg \mathrm{I}]$

(h)

$c \to e^2, o \to o^2, o \vee e \Rightarrow o^2 \vee e^2$

$c \to e^2$

$o \to o^2$

$o \vee e$

$o^2 \vee e^2$ [∨A]

4

$p \to q$

$q \to r$

$p \to r$ [→I]

  p

  q [→E]

  r [→E]

5 (d)

$\neg p \vee q$

$\neg p \to (p \to q)$ [→I]

  $\neg p$

  $p \to q$ [→I]

    p

    $p \wedge \neg p$ [∧I]

    q [∧C]

$q \to (p \to q)$ [→I]

  q

  $p \to q$ [→I]

    p

    $p \wedge q$ [∧I]

    q [∧E]

$p \to q$ [∨E]

6

$p \to q$

$\neg q$

$p \to q \wedge \neg q$ [→I]

  p

  q [→E]

  $q \wedge \neg q$ [∧I]

$\neg p$ [¬I]

8 (a)

$p \vee q \to \neg(p \vee q) \Rightarrow \neg p \wedge \neg q$

$p \vee q \to \neg(p \vee q)$

$\neg(p \vee q)$ [¬A]

$\neg p \wedge \neg q$ [Example 2.41]

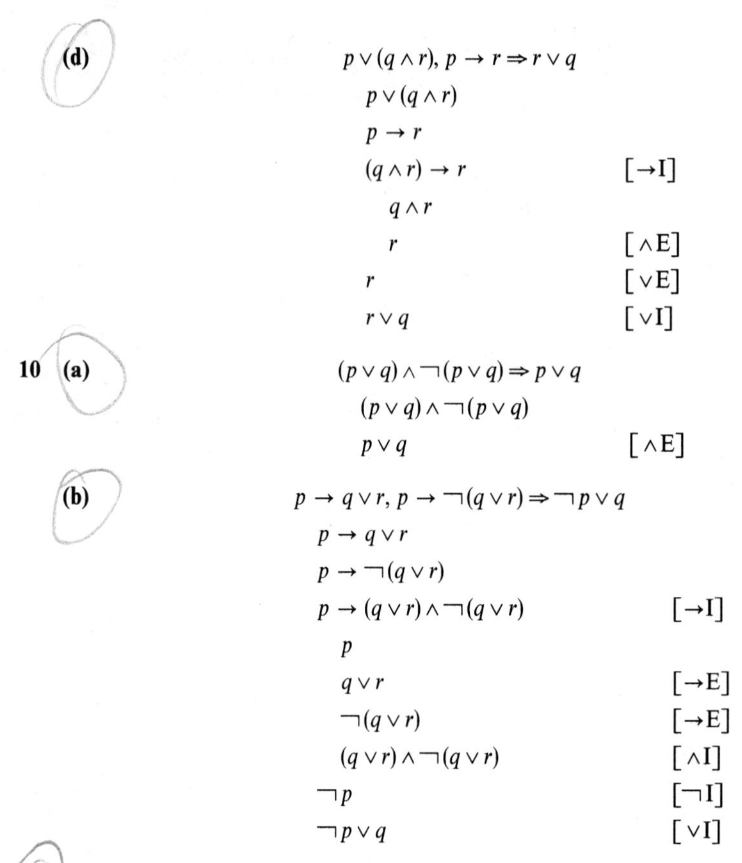

(d) $p \vee (q \wedge r), p \rightarrow r \Rightarrow r \vee q$

$\quad p \vee (q \wedge r)$

$\quad p \rightarrow r$

$\quad (q \wedge r) \rightarrow r$ [→I]

$\quad\quad q \wedge r$

$\quad\quad r$ [∧E]

$\quad r$ [∨E]

$\quad r \vee q$ [∨I]

10 (a) $(p \vee q) \wedge \neg(p \vee q) \Rightarrow p \vee q$

$\quad (p \vee q) \wedge \neg(p \vee q)$

$\quad p \vee q$ [∧E]

(b) $p \rightarrow q \vee r, p \rightarrow \neg(q \vee r) \Rightarrow \neg p \vee q$

$\quad p \rightarrow q \vee r$

$\quad p \rightarrow \neg(q \vee r)$

$\quad p \rightarrow (q \vee r) \wedge \neg(q \vee r)$ [→I]

$\quad\quad p$

$\quad\quad q \vee r$ [→E]

$\quad\quad \neg(q \vee r)$ [→E]

$\quad\quad (q \vee r) \wedge \neg(q \vee r)$ [∧I]

$\quad \neg p$ [¬I]

$\quad \neg p \vee q$ [∨I]

Section 2.6

1 Strictly speaking, $\neg\neg p$ should first be extracted from $\neg p \wedge \neg\neg p$ in order to apply [¬¬E], as follows:

$p \vee \neg p$

$\quad \neg(p \vee \neg p) \rightarrow p \wedge \neg p$ [→I]

$\quad\quad \neg(p \vee \neg p)$

$\quad\quad \neg p \wedge \neg\neg p$ [Example 2.41]

$\quad\quad \neg p$ [∧E]

$\quad\quad \neg\neg p$ [∧E]

$\quad\quad p$ [¬¬E]

$\quad\quad p \wedge \neg p$ [∧I]

$\quad \neg\neg(p \vee \neg p)$ [¬I]

$\quad p \vee \neg p$ [¬¬E]

2 In Exercise 1, we have shown that (EXM) holds if we can assume the validity of [¬¬E]. Conversely, we show that (EXM) permits [¬¬E] as follows:

$\neg\neg p$

$p \vee \neg p$ (EXM)

p [¬C]

4 (b)

$$
\begin{array}{ll}
p \vee 1 \to 1 & [\text{Definition of } 1] \\
1 \to p \vee 1 & [\to \text{I}] \\
\quad 1 & \\
\quad p \vee 1 & [\vee \text{I}] \\
p \vee 1 \leftrightarrow 1 & [\leftrightarrow \text{I}]
\end{array}
$$

(d)

$$
\begin{array}{ll}
p \wedge \neg p \to 0 & [\to \text{I}] \\
\quad p \wedge \neg p & \\
\quad 0 & [\wedge \text{C}] \\
0 \to p \wedge \neg p & [\text{Definition of } 0] \\
p \wedge \neg p \leftrightarrow 0 & [\leftrightarrow \text{I}]
\end{array}
$$

5 (a)

$$
\begin{array}{ll}
(r \wedge p) \vee (r \wedge q) \vee [(\neg p \vee \neg q) \wedge r] \vee \neg p & \\
= [r \wedge (p \vee q)] \vee [(\neg p \vee \neg q) \wedge r] \vee \neg p & \text{(L5b)} \\
= [r \wedge (p \vee q)] \vee [r \wedge (\neg p \vee \neg q)] \vee \neg p & \text{(L2b)} \\
= \{r \wedge [(p \vee q) \vee (\neg p \vee \neg q)]\} \vee \neg p & \text{(L5b)} \\
= \{r \wedge [(p \vee q) \vee (\neg q \vee \neg p)]\} \vee \neg p & \text{(L2a)} \\
= (r \wedge \{p \vee [q \vee (\neg q \vee \neg p)]\}) \vee \neg p & \text{(L3a)} \\
= \{r \wedge [p \vee ((q \vee \neg q) \vee \neg p)]\} \vee \neg p & \text{(L3a)} \\
= \{r \wedge [p \vee (1 \vee \neg p)]\} \vee \neg p & \text{(L8a)} \\
= [r \wedge (p \vee 1)] \vee \neg p & \text{(L7a)} \\
= (r \vee 1) \vee \neg p & \text{(L7a)} \\
= r \vee \neg p & \text{(L7a)}
\end{array}
$$

6 (a)

$$
\begin{array}{ll}
\neg(p \vee q) \to \neg p \wedge \neg q & [\to \text{I}] \\
\quad \neg(p \vee q) & \\
\quad \neg p \wedge \neg q & [\text{Example 2.41}] \\
\neg p \wedge \neg q \to \neg(p \vee q) & [\to \text{I}] \\
\quad \neg p \wedge \neg q & \\
\quad \neg p & [\wedge \text{E}] \\
\quad \neg q & [\wedge \text{E}] \\
\quad p \vee q \to q \wedge \neg q & [\to \text{I}] \\
\qquad p \vee q & \\
\qquad q & [\neg \text{C}] \\
\qquad q \wedge \neg q & [\wedge \text{I}] \\
\quad \neg(p \vee q) & [\neg \text{I}] \\
\neg(p \vee q) \leftrightarrow \neg p \wedge \neg q) & [\leftrightarrow \text{I}]
\end{array}
$$

7 The arguments are symbolized as follows:

(a) $\neg\neg t \Rightarrow t$

(i) $q \to [(a \wedge e) \to (r \wedge s)], \neg e \to \neg s, \neg a \to \neg(s \vee r) \Rightarrow \neg q \vee [(r \wedge e) \to s]$

8 (c) Using (L9a) we have

$$\neg(\neg p \vee q) \equiv \neg\neg p \wedge \neg q$$

so instead we prove:

$\neg\neg p \wedge \neg q \Rightarrow p$

$\neg\neg p \wedge \neg q$	
$\neg\neg p$	$[\wedge\text{E}]$
p	$[\neg\neg\text{E}]$

9 (a) Using (L9a) we have

$$\neg(\neg p \vee q) \equiv \neg\neg p \wedge \neg q$$

so instead we prove:

$\neg\neg p \wedge \neg q \Rightarrow p \wedge \neg q$

$\neg\neg p \wedge \neg q$	
$\neg\neg p$	$[\wedge\text{E}]$
p	$[\neg\neg\text{E}]$
$\neg q$	$[\wedge\text{E}]$
$p \wedge \neg q$	$[\wedge\text{I}]$

(d)

$p \Rightarrow q \vee \neg q$

p	
$q \vee \neg q$	(EXM)

Section 2.7

1 (a) Not well formed
(b) Well formed because

p, q, r are wff's	(2)
$r \wedge p$, $r \wedge q$ are wff's	(3(b))
$(r \wedge p)$, $(r \wedge q)$ are wff's	(3(f))
$\neg p$, $\neg q$ are wff's	(3(c))
$(\neg p \vee \neg q)$ is a wff	(3(a, f))
$[(\neg p \vee \neg q) \wedge r]$ is a wff	(3(b, f))
$[(\neg p \vee \neg q) \wedge r] \vee \neg p$ is a wff	(3(a))
$(r \wedge q) \vee [(\neg p \vee \neg q) \wedge r] \vee \neg p$ is a wff	(3(a))
$(r \wedge p) \vee (r \wedge q) \vee ((\neg p \vee \neg q) \wedge r) \vee \neg p$ is a wff	(3(a))

2 (a) $[r \wedge (\neg q)] \leftrightarrow \{(\neg p) \vee [0 \wedge (\neg 1)]\}$
(b) $[p \wedge (\neg q)] \vee [r \wedge (\neg q)]$

3 (a) Because

$$p \vee q = \neg(\neg p \wedge \neg q)$$

we have

$$0 \vee 0 = \neg(\neg 0 \wedge \neg 0) = \neg(1 \wedge 1) = \neg 1 = 0$$
$$0 \vee 1 = \neg(\neg 0 \wedge \neg 1) = \neg(1 \wedge 0) = \neg 0 = 1$$
$$1 \vee 0 = \neg(\neg 1 \wedge \neg 0) = \neg(0 \wedge 1) = \neg 0 = 1$$
$$1 \vee 1 = \neg(\neg 1 \wedge \neg 1) = \neg(1 \wedge 1) = \neg 0 = 1$$

4 (a) Not a tautology

p	q	r	$p \to q$	$p \to r$	$[(p \to q) \vee (p \to r)]$	$\neg q$	$(\neg q \vee r)$	*expression*
0	0	0	1	1	1	1	1	1
0	0	1	1	1	1	1	1	1
0	1	0	1	1	1	0	0	0
0	1	1	1	1	1	0	1	1
1	0	0	0	0	0	1	1	0
1	0	1	0	1	1	1	1	1
1	1	0	1	0	1	0	0	0
1	1	1	1	1	1	0	1	1

(d) Not a tautology

5 (d) Not a tautology

p	q	$\neg q$	$(p \wedge \neg q)$	$\neg p$	$\neg\neg p$	$(\neg\neg p \leftrightarrow q)$	*expression*
0	0	1	0	1	0	1	1
0	1	0	0	1	0	0	1
1	0	1	1	0	1	0	0
1	1	0	0	0	1	1	1

(f) A tautology

p	q	r	$q \to r$	$p \to q$	$p \to r$	$(p \to q) \to (p \to r)$	*expression*
0	0	0	1	1	1	1	1
0	0	1	1	1	1	1	1
0	1	0	0	1	1	1	1
0	1	1	1	1	1	1	1
1	0	0	1	0	0	1	1
1	0	1	1	0	1	1	1
1	1	0	0	1	0	0	1
1	1	1	1	1	1	1	1

6 We let propositional variables identify the arguments as follows:

p: Beethoven appreciators

q: Those who keep silence while the Moonlight Sonata is being played

r: Guinea pigs

s: Those who are hopelessly ignorant of music

Translating our argument into the statement calculus, we have

$$\begin{array}{l} \neg q \to \neg p \\ r \to s \\ \underline{q \to \neg s} \\ r \to \neg p \end{array}$$

where the latter is a semantic consequence of the three premises provided that

$$[(\neg q \to \neg p) \wedge (r \to s) \wedge (q \to \neg s)] \to (r \to \neg p)$$

is a tautology. As one shows by a truth table calculation, the answer is in the affirmative.

7

$$\begin{array}{ll}
[(\neg q \to \neg p) \wedge (r \to s) \wedge (q \to \neg s)] \to (r \to \neg p) & [\to \mathrm{I}] \\
\quad (\neg q \to \neg p) \wedge (r \to s) \wedge (q \to \neg s) & \\
\quad \neg q \to \neg p & [\wedge \mathrm{E}] \\
\quad r \to s & [\wedge \mathrm{E}] \\
\quad q \to \neg s & [\wedge \mathrm{E}] \\
\quad r \to \neg p & [\to \mathrm{I}] \\
\quad\quad r & \\
\quad\quad s & [\to \mathrm{E}] \\
\quad\quad \neg\neg s & [\neg\neg \mathrm{I}] \\
\quad\quad \neg q & [\neg \mathrm{B}] \\
\quad\quad \neg p & [\to \mathrm{E}]
\end{array}$$

10 **(a)** The two expressions are not equivalent:

p	q	$\neg q$	$(p \wedge \neg q)$	$(\neg\neg p \leftrightarrow q)$	$\neg q$	*exp.* 1	*exp.* 2
0	0	1	0	1	1	1	1
0	1	0	0	0	0	1	0
1	0	1	1	0	1	0	1
1	1	0	0	1	0	1	1

(b) Not equivalent

Section 3.1

1 As suggested, we let X_m denote the set

$$X_m = \{n: m \cdot n \text{ is defined}\}$$

and we first note that $1 \in X_m$ by virtue of (M1). If we now suppose that $n \in X_m$ so that

$m \cdot n$ is defined, then $m \cdot n + m$ is defined as well, because addition is defined for all pairs $m, n \in N$ according to the argument preceding Example 3.4 in the text. In considering (M2), we have

$$m \cdot n^+ = m \cdot n + m$$

thus showing that $n^+ \in X_m$. Having shown that (N1) and (N2) are satisfied for X_m, we use (N5) to conclude that $X_m = N$, i.e., that the operation $m \cdot n$ is defined for all $n \in N$ (and any $m \in N$).

2 **(a)** $3 + 5 = 3 + 4^+ = (3 + 4)^+$ **(c)** $3 \cdot 6 = 3 \cdot 5^+ = 3 \cdot 5 + 3$

5 (E1) $m^1 = m$; (E2) $m^{n^+} = m^n \cdot m$

8 **(b)**

$$\begin{aligned} (m \cdot n) \cdot 1 &= m \cdot n && \text{(M1)} \\ &= m \cdot (n \cdot 1) && \text{(M1)} \end{aligned}$$

$$\begin{aligned} (m \cdot n) \cdot p^+ &= (m \cdot n) \cdot p + m \cdot n && \text{(M2)} \\ &= m \cdot (n \cdot p) + m \cdot n && \text{(IH)} \\ &= m \cdot (n \cdot p + n) && \text{(I3)} \\ &= m \cdot (n \cdot p^+) && \text{(M2)} \end{aligned}$$

(a)

$$\begin{aligned} m \cdot 1 &= m && \text{(M1)} \\ &= 1 \cdot m && \text{(Lemma 1)} \end{aligned}$$

$$\begin{aligned} m \cdot n^+ &= mn + m && \text{(M2)} \\ &= nm + m && \text{(IH)} \\ &= n \cdot m + 1 \cdot m && \text{(Lemma 2)} \\ &= (n + 1) \cdot m && \text{(Exercise 7)} \\ &= n^+ \cdot m && \text{(A1)} \end{aligned}$$

Section 3.2

1 Because

$$\sum_{j=1}^{1} j^2 = \frac{1 \cdot (1 + 1)(2 \cdot 1 + 1)}{6} = 1$$

the statement is true for $n = 1$. Assuming it is true for some $n \geq 1$, we obtain

$$\begin{aligned} \sum_{j=1}^{n+1} j^2 &= (n + 1)^2 + \sum_{j=1}^{n} j^2 \\ &= (n + 1)^2 + \frac{n(n + 1)(2n + 1)}{6} \\ &= \frac{(6n^2 + 12n + 6) + (2n^3 + 3n^2 + n)}{6} \\ &= \frac{2n^3 + 9n^2 + 13n + 6}{6} \\ &= \frac{(n + 1)[(n + 1) + 1][2(n + 1) + 1]}{6} \end{aligned}$$

the same statement for $n + 1$.

3 If $|S| = 0$ then $S = \varnothing$ and

$$|\mathscr{P}(S)| = 1 = 2^0 = 2^{|S|}$$

as required ($\varnothing$ is the only subset of $\varnothing$). Now if the proposition is true for sets of size $n \geq 0$ and

$$S = \{x_1, x_2, \ldots, x_n, x_{n+1}\}$$

then we note that S has two types of subsets, those that include x_{n+1} and those that do not. Because the two types are equinumerous and the second type consists of subsets of $\{x_1, x_2, \ldots, x_n\}$, we have

$$|\mathscr{P}(S)| = 2 \cdot 2^n = 2^{n+1} = 2^{|S|}$$

once more, and this completes the proof.

4 When $n = 1$, we have

$$\sum_{j=1}^{1} 2^{j-1} = 2^{1-1} = 2^0 = 1 = 2 - 1 = 2^1 - 1$$

as required. If we then assume that the statement is true for some $n \geq 1$, we obtain

$$\sum_{j=1}^{n+1} 2^{j-1} = 2^{(n+1)-1} + \sum_{j=1}^{n} 2^{j-1}$$
$$= 2^n + (2^n - 1) = 2 \cdot 2^n - 1 = 2^{n+1} - 1$$

which is the same statement for $n + 1$.

12 When $n = 1$ we have

$$f(1 + 1)f(1 + 2) - f(1)f(1 + 3) = f(2)f(3) - f(1)f(4)$$
$$= 1 \cdot 2 - 1 \cdot 3 = 2 - 3 = -1 = (-1)^1$$

as required. If we then assume that the identity holds for some $n \geq 1$, we compute

$$f[(n + 1) + 1]f[(n + 1) + 2] - f(n + 1)f[(n + 1) + 3]$$
$$= f(n + 2)f(n + 3) - f(n + 1)f(n + 4)$$
$$= [f(n + 1) + f(n)]f(n + 3) - f(n + 1)[f(n + 3) + f(n + 2)]$$
$$= f(n + 1)f(n + 3) + f(n)f(n + 3) - f(n + 1)f(n + 3) - f(n + 1)f(n + 2)$$
$$= f(n)f(n + 3) - f(n + 1)f(n + 2)$$
$$= -[f(n + 1)f(n + 2) - f(n)f(n + 3)]$$
$$= -(-1)^n = (-1)^{n+1}$$

the same identity for $n + 1$.

14 If it is to be a constant function, the constant must be 3 (because $f(1) = 3$). If we now suppose that $f(n) = 3$ for some $n \geq 1$, we also have

$$f(n + 1) = 2f(n) - 3 = 2 \cdot 3 - 3 = 3$$

and indeed, f is a constant function.

16 First layer: ( )
Second layer: (( )), ( )( )
Third layer: ((( ))), (( )( )), ( )(( )), (( ))( ),
( )( )( ), (( ))(( )), ( )( )( )( ), (( ))( )( ), ( )( )(( ))

Section 3.3

7 Suppose

$$a = bq_1 + r_1 \quad \text{and} \quad a = bq_2 + r_2$$

with $0 \leq r_i < b$. Then

$$r_2 - r_1 = b(q_1 - q_2)$$

is numerically smaller than b, and yet, a multiple of b. It follows that $r_2 - r_1$ must be 0. Since $b \neq 0$, the above equation shows that $q_1 - q_2$ is 0 as well. Thus $r_1 = r_2$ and $q_1 = q_2$ as required.

9 A trace of the computation is as follows:

(b)

a	b	r	gcd
1350	297	162	
297	162	135	
162	135	27	
135	27	0	27

(f)

a	b	r	gcd
8024	412	196	
412	196	20	
196	20	16	
20	16	4	
16	4	0	4

11 If $c|r$ and $c|b$ then $r = nc$ and $b = mc$. Because

$$a = bq + r = mcq + nc = (mq + n)c$$

we see that a is also a multiple of c, i.e., $c|a$.

12 In tracing the computation:

(b)

a	b	r	gcd
48	51		
51	48	3	
48	3	0	3

we see that $\gcd(48, 51) = 3 \neq 1$, so 48 and 51 are not relatively prime.

14 **(b)** $\{1, 7, 61, 427\}$
(f) $\{1, 2, 3, 6, 7, 11, 14, 21, 22, 33, 42, 66, 77, 154, 231, 462\}$

Section 3.4

2 **(c)** 257 **(f)** 125
4 **(d)** 100111000 **(j)** 1010100000010
6 **(c)** 0.1875 **(f)** 0.10546875
7 **(d)** 0.10000001 **(j)** 0.0101010101...
9 If we divide each of the representations of n by 8 we obtain corresponding remainders:

$$e_0 = 4b_2 + 2b_1 + b_0$$

and in equating quotients, we may continue the process with

$$e_j 8^{j-1} + \cdots + e_1 8^0 = b_k 2^{k-3} + \cdots + b_6 2^3 + (b_5 2^2 + b_4 2^1 + b_3 2^0)$$

to obtain remainders:

$$e_1 = 4b_5 + 2b_4 + b_3$$

etc.
11 **(a)** 73 **(f)** 4115

Section 3.5

2 If

$$(a, b) \sim (c, d) \quad \text{and} \quad (c, d) \sim (e, f)$$

then $ad = bc$ and $cf = de$. It follows that

$$af = (bc/d)(de/c) = be$$

i.e., $(a, b) \sim (e, f)$, as required to verify (E3).
6 Because

$$(ad + bc)d = d(ad + bc) = dad + dbc < dbc + dbc = 2bdc$$

we have $(ad + bc)/2bd < c/d$ by the definition of order for the rational numbers.
8 The elements of $N \times N$ may be arranged in a tabulation:

$$\begin{array}{ccccc} (1,1) & (2,1) & (3,1) & (4,1) & \cdots \\ (1,2) & (2,2) & (3,2) & (4,2) & \cdots \\ (1,3) & (2,3) & (3,3) & (4,3) & \cdots \\ (1,4) & (2,4) & (3,4) & (4,4) & \cdots \\ \vdots & \vdots & \vdots & \vdots & \end{array}$$

and counted as in Figure 3.7, i.e.,

$$(1, 1), (2, 1), (1, 2), (1, 3), (2, 2), (3, 1), (4, 1), (3, 2), \ldots$$

10 **(b)** The mapping $n \to 2n - 1$ is the required one-to-one correspondence.
(e) The mapping $n \to 7n$ is the required one-to-one correspondence.

(f) The mapping

$$\begin{aligned}1 &\to \lambda\\ 2 &\to a\\ 3 &\to b\\ 4 &\to aa\\ 5 &\to ab\\ 6 &\to ba\\ 7 &\to bb\\ 8 &\to aaa\end{aligned}$$

etc., is the required one-to-one correspondence.

Section 3.6

2 **(a)** 0.3333×10^{4} **(d)** 0.8427×10^{-3}
5 **(a)** 0 1001101 100000100111000011001100
(d) 0 1001011 101001100011001101101010
6 **(a)** 0.6125×10^{2} **(d)** 0.8083×10^{7}
8 **(b)** 0.2488×10^{3} **(f)** 0.4093×10^{-1}

Section 4.1

4

```
find all divisors of n, e.g., with the factors algorithm
compute SUM of these divisors other than n itself
if SUM = n
   answer is 'yes'
else
   answer is 'no'
```

5

```
make a hot charcoal fire
set chicken on grill
mix a drink
as long as chicken not done
   turn chicken over
   apply barbeque sauce
   mix a drink
if chicken burnt
   throw it away
else
   serve
```

7

```
algorithm triangle (a, b, c: answer)
if (a < b + c) ∧ (b < a + c) ∧ (c < a + b)
   answer is 'true'
else
   answer is 'false'
```

8 **(c)** Input: n (positive integer)
Output: answer (Boolean)

(d) Input: chicken (uncooked)
Output: chicken (cooked)

9 In cancelling, we are dividing by zero.
14 **(a)** no **(b)** no **(c)** no **(d)** yes

Section 4.2

1

if $j \mid n$
  set nodivisors to 'false'
else
  increase j by 1

2 **(c)**

n	COUNT	j	answer
35	0	2	
		3	
		4	
	1	5	
		6	
	2	7	
		8	
		⋮	
		34	'false'

3 **(c)**

n	answer	j	nodivisors
35		2	'true'
		3	
		4	
		5	'false'
	'false'		

7

algorithm compound (deposit, years, rate, ppy: balance)
set periods to years × ppy
set interest to rate/ppy
initialize balance as deposit
for p running from 1 to periods
  add balance × interest to balance

8

deposit	*years*	*rate*	*ppy*	*periods*	*interest*	*p*	*balance*
1000.00	2	0.09	4	8	0.0225		1000.00
						1	1025.50
						2	1045.51
						3	1069.03
						4	1093.08
						5	1117.67
						6	1142.82
						7	1168.53
						8	1194.82

11

```
algorithm population (year)
assign pop as 200,000,000
assign rate as 0.05
initialize year as 1980
while pop ≤ 300,000,000
   increase year by 1
   add 0.05 × pop to pop
```

12 **(a)** S until $C = S \circ (S$ while $\neg C)$

17

```
algorithm big (fib)
set last at 1
set nexttolast at 1
while last ≤ 1000
   set next as sum of last and nexttolast
   replace nexttolast by last
   replace last by next
assign fib as last
```

27

```
algorithm perfect (n: answer)
set SUM to 0
set j at 1
while j < n
   if j | n
      increase SUM by j
   increase j by 1
if SUM = n
   answer is 'true'
else
   answer is 'false'
```

Section 4.3

2 **(c)**

```
algorithm sumrow (A, i: sum)
set sum to 0
for j running from 1 to 10
   add A_ij to sum
```

(e)

```
algorithm allsum (A: sum)
set sum to 0
for i running from 1 to 10
   for j running from 1 to 10
      add A_ij to sum
```

7

```
algorithm exchangesort (A)
for k running from 1 to n − 1
   for j running from 1 to n − k
      if A_j > A_{j+1}
         interchange A_j with A_{j+1}
```

8

```
algorithm fastexchangesort (A)
set unsorted to 'true'
set k to 1
while (k ≤ n − 1) ∧ unsorted
   set unsorted to 'false'
   for j running from 1 to n − k
      if A_j > A_{j+1}
         interchange A_j with A_{j+1}
         set unsorted to 'true'
increase k by 1
```

13

```
algorithm round (amount)
multiply amount by 100
replace amount by its integer part
divide amount by 100
```

15

```
function interest (principal, years, rate, ppy)
initialize balance as principal
for i running from 1 to years × ppy
   add balance × (rate/ppy) to balance
return interest as balance − principal
```

21 **(b)**

```
algorithm ratmpy (a, b: c)
set c_1 to a_1b_1
set c_2 to a_2b_2
compute d = gcd (c_1, c_2)
divide c_1 by d
divide c_2 by d
```

(d)

```
algorithm reduce (a)
compute d = gcd (a_1, a_2)
divide a_1 by d
divide a_2 by d
```

22

```
function power (x, n)
if n = 0
   return power as 1
else
   if n < 0
      replace x by 1/x
      replace n by -n
   initialize p as 1
   for i running from 1 to n
      multiply p by x
   return power as p
```

23

```
function factorial (n)
if n = 0
   return factorial as 1
else
   initialize p as 1
   for i running from 2 to n
      multiply p by i
   return factorial as p
```

Section 4.4

1 If I is an invariant assertion through repetitions of S (when C is false), and if $\{J\}\ S\ \{I\}$ holds, then each time the loop is traversed once more (because C is false) we have I holding again as we enter the condition C. It follows that if J is true at the beginning of execution of the compound statement S until C, we must have $I \wedge C$ at the conclusion of the execution, since we fail to complete another traversal only because C is true.

3 **(a)**

$$\{n > 0\} \text{ perfect } \left\{\text{answer} = \begin{array}{l}\text{“true” if } n \text{ is perfect} \\ \text{“false” otherwise}\end{array}\right\}$$

5 **(a)** $\{x > 0 \wedge y < 2\}$ interchange x and y $\{x < 2 \wedge y > 0\}$
(b) $\{x > 0 \wedge y < 0\}$ replace x by $y - 3$ $\{x < -3 \wedge y < 0\}$

7 Since S until C is identical to $S \circ (S \text{ while } \neg C)$, we may replace the until deduction rule with those of the sequence and the while construct (appropriately modified to account for the negation).

13 Algorithm *what* computes $f = n^n$. Note that in setting f to n and m to $n - 1$, we have the condition $\{fn^m = n^n\}$ as we enter the while construct. Arguing informally, if the condition $\{fn^m = n^n\}$ holds before the interior sequence is executed (with $m > 0$), then

$$fnn^{m-1} = n^n$$

holds afterwards, that is, $fn^m = n^n$ is an invariant assertion. It follows that when $m = 0$ (so that the loop is no longer traversed) we obtain $f = fn^0 = n^n$ as claimed.

Section 4.5

3 (c)

$$6! = 6 \cdot 5! = 6 \cdot 5 \cdot 4! = \cdots = 6 \cdot 5 \cdot 4 \cdot 3 \cdot 2 \cdot 1!$$
$$= 6 \cdot 5 \cdot 4 \cdot 3 \cdot (2 \cdot 1) = 6 \cdot 5 \cdot 4 \cdot (3 \cdot 2) = \cdots = 6 \cdot (5 \cdot 24) = 6 \cdot 120$$
$$= 720$$

5 Computing the midpoint repeatedly halves the difference last − first until finally last = first or last = first + 1 (i.e., last − first ≤ 1).

6 For any integer n, expressed as a sequence of digits, tahw(n) is the integer with this sequence spelled backwards.

7 (a)

```
function F(n)
if n = 1 ∨ n = 2
   return F as 1
else
   set last and nexttolast to 1
   for i running from 3 to n
      set next to sum of last and nexttolast
      replace nexttolast by last
      replace last by next
   return F as last
```

(b)

```
function F(n)
if n = 1 ∨ n = 2
   return F as 1
else
   return F as F(n − 1) + F(n − 2)
```

9

```
function gcd (a, b)
if a < b
   interchange a and b
compute the remainder r on dividing a by b
if r = 0
   return gcd as b
else
   return gcd as gcd (b, r)
```

Section 4.6

1 (a)

$$T(n) = \log_2 n = 100 \Longrightarrow 2^{\log_2 n} = 2^{100} \Longrightarrow n = 2^{100}$$
$$T(n) = 2n^2 = 100 \Longrightarrow n^2 = 50 \Longrightarrow n \approx 7$$
$$T(n) = n^3/4 = 100 \Longrightarrow n^3 = 400 \Longrightarrow n \approx 7$$
$$T(n) = 2^n = 100 \Longrightarrow n \approx 6$$

3 (b) $T(n) = 18 = 18 \cdot 1 \sim 1$

(c) $T(n) = \log_{10} n = \log_{10} 2 \cdot \log_2 n \sim \log n$

(d) $T(n) = n(\frac{1}{2}n^2 + 4) = \frac{1}{2}n^3 + 4n \sim n^3$

4 (c) Since $T(R) \le c_1 n$ and $T(S) \le c_2 \log n$ we have

$$T(R \circ S) = T(R) + T(S) \le c_1 n + c_2 \log n$$
$$\le (c_1 + c_2)n$$

Therefore we have $T(R \circ S) \sim n$.

6 (b) Since $T_C(S) \sim \log n$ with at most $\frac{1}{2}n$ executions, we have

$$T(S \text{ while } C) = |C|T(C) + \sum^{|C|} T_C(S)$$
$$= |C| + \tfrac{1}{2}n \log n \sim n \log n$$

7 (c) Working from the inside out, we compute

$$T(\text{if-else}) = 1 + \max(1, 1) = 1 + 1 = 2$$

$$T(\text{while}) = (n-2)\cdot 2 + (n-2)T(\text{if-else})$$
$$= 4(n-2)$$

$$T(\text{else}) = 3 + T(\text{while}) = 4n - 5$$

and finally

$$T(\text{prime}) = 1 + \max(1, T(\text{else}))$$
$$= 4n - 4$$
$$\sim n$$

Section 5.1

3 (c) $6 - 2x + x^2$ **(d)** $10 - 8x + 5x^2$

7 $A = -5$ and $B = 5$

9 $\deg(a(x)\cdot b(x)) = \deg a(x) + \deg b(x)$

10 If

$$a(x)b(x) = a(x)c(x)$$

then we have:

$$a(x)b(x) - a(x)c(x) = 0$$
$$a(x)(b(x) - c(x)) = 0$$

If $a(x) \neq 0$, we conclude (using the cancellation law) that

$$b(x) - c(x) = 0$$

that is, $b(x) = c(x)$.

13 (b)

$$a(x)b(x) = c(x)$$
$$= \frac{1}{4} - \frac{47}{8}x + \frac{239}{16}x^2 + \frac{87}{8}x^3 - \frac{27}{2}x^4 + 27x^5$$

For example,

$$\begin{aligned} c_3 &= a_0b_3 + a_1b_2 + a_2b_1 + a_3b_0 \\ &= 2(0) + 1(9) - \frac{1}{2}(-3) + 3\left(\frac{1}{8}\right) \\ &= 0 + 9 + \frac{3}{2} + \frac{3}{8} \\ &= \frac{87}{8} \end{aligned}$$

(c)

$$\begin{aligned} a(x)b(x) &= c(x) \\ &= -6x + 7x^2 + 15x^3 - 20x^4 + 31x^5 - 42x^6 \end{aligned}$$

For example,

$$\begin{aligned} c_3 &= a_0b_3 + a_1b_2 + a_2b_1 + a_3b_0 \\ &= -3(-6) + 5(1) - 4(2) + 7(0) \\ &= 18 + 5 - 8 + 0 \\ &= 15 \end{aligned}$$

16 **(c)**

$$\begin{aligned} a(x) &= -3 + 5x - 4x^2 + 7x^3 \\ &= -3 + x(5 + x(-4 + x(7))) \end{aligned}$$

$$a(-2) = -3 - 2(5 - 2(-4 - 2(7))) = -85$$

or alternatively, using a tabulation as in Example 5.12 we have

x	k	$aofx$
-2		7
	2	-14
		-18
	1	36
		41
	0	-82
		-85

(d) $a(-2) = -58$

19 $c \neq 0$ and $d \neq 0 \Rightarrow cd \neq 0$

Section 5.2

3 **(b)** "+ −" and $y = -4$ at $x = 0$ **(c)** "− −" and $y = 6$ at $x = 0$

6 **(a)**

$$\begin{array}{r|rrrrr} 5 & 1 & -4 & 1 & 0 & 7 \\ & & 5 & 5 & 30 & 150 \\ \hline & 1 & 1 & 6 & 30 & \boxed{157} \end{array}$$

$(x - 5)(x^3 + x^2 + 6x + 30) + 157$

(b)

-4	-2	4	-3	0	2	7
		8	-48	204	-816	3256
	-2	12	-51	204	-814	3263

$$(x + 4)(-2x^4 + 12x^3 - 51x^2 + 204x - 814) + 3263$$

8 **(a)** Using Theorem 5.1, the tabulations

1:	2	1	-3	-1	-1:	2	-3	-1	3
2:	2	3	2	6	-2:	2	-5	6	-10

show that all roots lie between -2 and 2. Furthermore, the Change of Sign Principle shows that there are roots between the pairs of integers: $(-2, -1)$, $(0, 1)$, $(1, 2)$.

(d) Using Theorem 5.1, the tabulations

1:	3	13	11	7	-1:	3	7	-9	5
					-2:	3	4	-10	16
					-3:	3	1	-5	11
					-4:	3	-2	6	-28

show that all roots lie between -4 and 1. Furthermore, the Change of Sign Principle shows that there are roots between the pairs of integers: $(-4, -3)$, $(-1, 0)$, $(0, 1)$.

11 **(b)** At most 2 positive roots; at most 2 negative roots.

(d) At most 2 positive roots; at most 1 negative root.

13 Division by $x - b$ yields

$$a(x) = (x - b)q(x) + r$$

where r is a constant. Assuming $a(b) = 0$ we obtain

$$0 = a(b) = (b - b)q(x) + r = 0 + r = r$$

so that $r = 0$ and $a(x) = (x - b)q(x)$, i.e., $x - b$ is a factor of $a(x)$.

15 **(a)** Among the possibilities:

$$\frac{p}{q} = 1, -1, 2, -2, 3, -3, 6, -6$$

we find that 1, 2, -1, and -3 are roots.

(e) Among the possibilities:

$$\frac{p}{q} = 1, -1, \frac{1}{2}, -\frac{1}{2}, \frac{1}{4}, -\frac{1}{4}, 2, -2, 3, -3, \frac{3}{2}, -\frac{3}{2}, \frac{3}{4}, -\frac{3}{4}$$

we find that $-1, \frac{1}{2}, -2$, and $-\frac{3}{2}$ are roots.

18 **(d)** -1.58 **(e)** -0.62

26 The argument is one of induction on the degree of the given polynomial. Thus, if $\deg a(x) = 1$, $a(x) = a_1x + a_0$. If both coefficients are of the same sign, then there can be no positive roots, whereas there is just one positive root if the coefficients are of opposite sign. In any case the statement "The number of positive roots is bounded by the number of variations in sign," holds when $\deg a(x) = 1$. So we suppose that the statement holds for any polynomial $q(x)$ of degree n. We then consider any polynomial $a(x)$ of degree $n + 1$. If $a(x)$ has no positive roots, then of course the desired statement holds trivially. If $a(x)$ has a positive root at $x = b$, however, then by the factor lemma we may write:

$$a(x) = (x - b)q(x)$$

where $\deg q(x) = n$. The number of positive roots of $a(x)$ is one more than the number of positive roots of $q(x)$. This argument shows that

$$\begin{aligned}\#(\text{positive roots of } a) &= 1 + \#(\text{positive roots of } q) \\ &\leq 1 + \#(\text{variations in sign of } q) \\ &\leq \#(\text{variations in sign of } a)\end{aligned}$$

where the first inequality follows from the inductive hypothesis and the second inequality is a result of the proof of Theorem 5.2 as presented in the text. And the overall inequality then completes the inductive argument.

Section 5.3

1 **(a)** $x^3 - 4x^2 - 17x + 60$ **(c)** $x^4 + x^3 - 19x^2 - 49x - 30$

3 **(a)** $-\frac{1}{3}x^2 + \frac{2}{3}x + 4$ **(b)** $-\frac{119}{936}x^2 + \frac{105}{936}x + \frac{3968}{936}$

7 **(a)** $\frac{1}{108}(x^3 - 15x^2 + 36x + 140)$

(c) $-\frac{1}{16}(x^2 + 3x - 70)$

(f) $-\frac{1}{4080}(x^4 - 3x^3 - 112x^2 + 12x + 432)$

13

$$\begin{aligned} y_0\frac{x - x_1}{x_0 - x_1} + y_1\frac{x - x_0}{x_1 - x_0} &= \frac{-y_0(x - x_1) + y_1(x - x_0)}{x_1 - x_0} \\ &= \frac{-y_0x + y_0x_1 + y_1x - y_1x_0 + y_0x_0 - y_0x_0}{x_1 - x_0} \\ &= \frac{y_0(x_1 - x_0) + (x - x_0)(y_1 - y_0)}{x_1 - x_0} \\ &= y_0 + \frac{(x - y_0)}{(x_1 - y_0)}(y_1 - y_0)\end{aligned}$$

14

$$\begin{aligned} y_0\frac{(x - x_1)(x - x_2)(x - x_3)}{(x_0 - x_1)(x_0 - x_2)(x_0 - x_3)} &+ y_1\frac{(x - x_0)(x - x_2)(x - x_3)}{(x_1 - x_0)(x_1 - x_2)(x_1 - x_3)} \\ + y_2\frac{(x - x_0)(x - x_1)(x - x_3)}{(x_2 - x_0)(x_2 - x_1)(x_2 - x_3)} &+ y_3\frac{(x - x_0)(x - x_1)(x - x_2)}{(x_3 - x_0)(x_3 - x_1)(x_3 - x_2)}\end{aligned}$$

Section 5.4

2 **(a)** $q(x) = x^2 - 2x - 8, \quad r(x) = -10x + 55$

(b) $q(x) = x^3 - 4x^2 + 27x - 137, \quad r(x) = 782x - 1231$

3 **(a)** $-\frac{347}{2310}x^3 + \frac{534}{385}x^2 - \frac{1699}{2310}x - \frac{248}{77}$

5 **(b)** $\{\pm 1, \pm 2, \pm 3, \pm 4, \pm 5, \pm 6, \pm 10, \pm 12, \pm 15, \pm 20, \pm 30, \pm 60\}$

6 **(a)** $s = -4, \quad t = 23$ **(c)** $s = 1, \quad t = -19$

7 **(a)** $s = \frac{2}{3}x + \frac{5}{3}$, $t = -\frac{2}{3}x^2 - \frac{5}{3}x + 4$ **(d)** $s = 1$, $t = -x^2$

13 **(a)** $(x-4)(x^4 + 2x^3 + 4x^2 + 2x + 3)$

(c) $(x+1)(x^2 - x + 3)$

(f) $(x-2)(x+1)(x+2)(x^2 - x + 3)$

14 **(a)** Irreducible, by Eisenstein's criterion.

(c) Reducible, by Kronecker's algorithm (alternatively, since -1 is a root, $x + 1$ is a factor).

(e) Irreducible, by Eisenstein's criterion.

20 Using the division algorithm, we may write:

```
function factors (y)
S = ∅
for k running from 1 to √y
   division (y, k: q, r)
   if r = 0
      adjoin k and −k to S
assign factors as S
```

Section 5.5

1 **(a)** "↑↓" and $y = 0$ at $x = 0$ **(b)** "− +"

10 Through a change in variable, one sees that the behavior of $(x - x_0)^s$ near the root x_0 is identical to that of x^s near $x = 0$. When s is odd, x^s is of opposite sign to either side of $x = 0$, whereas x^s is of the same sign to either side of $x = 0$ when s is even.

12 **(a)** $\dfrac{2}{(x-3)^2} - \dfrac{1}{(x-3)^3}$ **(c)** $\dfrac{2}{x^2-2} + \dfrac{3}{(x^2-2)^2}$

14 **(a)** $\dfrac{-\frac{1}{3}}{x+2} + \dfrac{\frac{1}{3}}{x-1}$ **(d)** $\dfrac{-x}{x^2+x+1} + \dfrac{x^3 - x^2 + 1}{x^4 + 1}$

15 **(a)** $\dfrac{2}{x-1} - \dfrac{2}{x} - \dfrac{1}{x^2}$ **(b)** $\dfrac{\frac{1}{16}}{x} + \dfrac{\frac{1}{8}}{x^2} - \dfrac{\frac{1}{16}}{x-2} + \dfrac{\frac{1}{4}}{(x-2)^3}$

17 When $f = a/p^m$ we define the sequences of polynomials a_k, r_k, and f_k $(k \geq 0)$ according to the fraction algorithm as follows:

$$a_0 = a \qquad f_0 = f$$

$$a_k = pa_{k+1} + r_k \qquad f_{k+1} = f_k - \frac{r_k}{p^{m-k}}$$

And we claim that for each integer k from 0 to m, the identity

$$f_k = \frac{a_k}{p^{m-k}}$$

holds. When $k = 0$, we certainly have

$$f_0 = f = \frac{a}{p^m} = \frac{a_0}{p^{m-0}}$$

Moreover, if we assume that the identity holds for some $k \geq 0$, then we also have

$$\begin{aligned} f_{k+1} &= f_k - \frac{r_k}{p^{m-k}} \\ &= \frac{a_k}{p^{m-k}} - \frac{r_k}{p^{m-k}} \\ &= \frac{pa_{k+1} + r_k - r_k}{p^{m-k}} \\ &= \frac{a_{k+1}}{p^{m-(k+1)}} \end{aligned}$$

that is, we obtain the same identity for $k + 1$. It follows that

$$\begin{aligned} f = f_0 = f_1 &+ \frac{r_0}{p^m} \\ &= f_2 + \frac{r_1}{p^{m-1}} + \frac{r_0}{p^m} \\ &= \cdots = f_m + \frac{r_{m-1}}{p} + \cdots + \frac{r_1}{p^{m-1}} + \frac{r_0}{p^m} \\ &= a_m + \frac{r_{m-1}}{p} + \cdots + \frac{r_1}{p^{m-1}} + \frac{r_0}{p^m} \end{aligned}$$

as generated in the fraction algorithm.

Section 5.6

1 **(a)** $3x^2 + 9x + 2$ **(b)** $8x - 1$

4 $\Delta 2^x = 2^{x+1} - 2^x = 2^x(2 - 1) = 2^x$

7 **(a)**

$$\begin{aligned} \Delta(f + g)(x) &= (f + g)(x + 1) - (f + g)(x) \\ &= f(x + 1) + g(x + 1) - f(x) - g(x) \\ &= f(x + 1) - f(x) + g(x + 1) - g(x) \\ &= \Delta f(x) + \Delta g(x) \\ &= (\Delta f + \Delta g)(x) \end{aligned}$$

(d)

$$\begin{aligned} \Delta\left(\frac{f}{g}\right)(x) &= \frac{f}{g}(x + 1) - \frac{f}{g}(x) \\ &= \frac{f(x + 1)}{g(x + 1)} - \frac{f(x)}{g(x)} \\ &= \frac{f(x + 1)g(x) - f(x)g(x + 1)}{g(x)g(x + 1)} \\ &= \frac{f(x + 1)g(x) - f(x)g(x) + f(x)g(x) - f(x)g(x + 1)}{g(x)g(x + 1)} \\ &= \frac{g(x)\,\Delta f(x) - f(x)\,\Delta g(x)}{g(x)g(x + 1)} \\ &= \frac{g\,\Delta f - f\,\Delta g}{gg^+}(x) \end{aligned}$$

10 **(b)** $15x^{(4)} - 15x^{(2)} + 4x^{(1)} - 9$
(d) $18x^{(5)} - 20x^{(4)} + 36x^{(3)} - 15x^{(2)} + 2$
13 **(a)** $\frac{3}{2}x^2 + \frac{3}{2}x + 5$
14 **(a)** $x^{(4)} + 9x^{(3)} + 8x^{(2)} - 2x^{(1)} - 3$
(b) $6x^{(5)} + 57x^{(4)} + 123x^{(3)} + 44x^{(2)} - 7x^{(1)} + 1$

23

$$\Delta \sum_{i=0}^{n} f(x) = \sum_{i=0}^{n+1} f(x) - \sum_{i=0}^{n} f(x) = f(n+1) = f^+(n)$$

24

$$\begin{aligned}\sum_{x=0}^{n} (3x^2 + 3x - 4) &= \sum_{x=0}^{n} \Delta(x^3 - 5x + 2) \\ &= (x^3 - 5x + 2)^+ - 2 \\ &= (n+1)^3 - 5(n+1) \\ &= n^3 + 3n^2 - 2n - 4 \\ &= (n+1)[(n+3)(n-1) - 1]\end{aligned}$$

25 **(b)**

$$\begin{aligned}\sum_{x=0}^{n} (x-1)^3 &= \frac{1}{4}\sum_{x=0}^{n} \Delta(x^4 - 6x^3 + 13x^2 - 12x) \\ &= \frac{1}{4}(x^4 - 6x^3 + 13x^2 - 12x)^+ - 0 \\ &= \frac{1}{4}(n^4 - 2n^3 + n^2 - 4) \\ &= \frac{1}{4}(n-2)(n^3 + n + 2)\end{aligned}$$

29 After first computing the difference:

$$\Delta a^x = a^{x+1} - a^x = a^x(a-1)$$

we obtain

$$\begin{aligned}\sum_{x=0}^{n} a^x &= \frac{1}{a-1}\sum_{x=0}^{n} \Delta a^x = \frac{1}{a-1}\left((a^n)^+ - a^0\right) \\ &= \frac{a^{n+1} - 1}{a-1}\end{aligned}$$

Section 5.7

1 **(a)** $(x-3)^2 + 1$ **(e)** $(x + \frac{1}{2})^2 + \frac{11}{4}$

5 If p_i is a linear combination of $p_1, \ldots, p_{i-1}, p_{i+1}, \ldots, p_k$ so that

$$p_i = c_1p_1 + \cdots + c_{i-1}p_{i-1} + c_{i+1}p_{i+1} + \cdots + c_kp_k$$

then we may transpose to obtain

$$c_1p_1 + \cdots + c_{i-1}p_{i-1} + c_ip_i + c_{i+1}p_{i+1} + \cdots + c_kp_k = 0$$

where the c's are not all zero, since $c_i = -1$.

6 **(b)**

$$\begin{aligned}\mathbf{a} + \mathbf{b} &= (a_1, a_2, \ldots, a_n) + (b_1, b_2, \ldots, b_n) \\ &= (a_1 + b_1, a_2 + b_2, \ldots, a_n + b_n) \\ &= (b_1 + a_1, b_2 + a_2, \ldots, b_n + a_n) \\ &= (b_1, b_2, \ldots, b_n) + (a_1, a_2, \ldots, a_n) = \mathbf{b} + \mathbf{a}\end{aligned}$$

(g) $$\begin{aligned}(k+l)\mathbf{a} &= (k+l)(a_1, a_2, \ldots, a_n) \\ &= ((k+l)a_1, (k+l)a_2, \ldots, (k+l)a_n) \\ &= (ka_1 + la_1, ka_2 + la_2, \ldots, ka_n + la_n) \\ &= (ka_1, ka_2, \ldots, ka_n) + (la_1, la_2, \ldots, la_n) \\ &= k(a_1, a_2, \ldots, a_n) + l(a_1, a_2, \ldots, a_n) = k\mathbf{a} + l\mathbf{a}\end{aligned}$$

7 Only (a), (f) are in echelon form.

11 (a) Yes. We need to see if there exist scalars c_1, c_2 with

$$\begin{aligned}(-1, 11, -1) &= c_1(2, -3, 4) + c_2(3, 5, 7) \\ &= (2c_1, -3c_1, 4c_1) + (3c_2, 5c_2, 7c_2) \\ &= (2c_1 + 3c_2, -3c_1 + 5c_2, 4c_1 + 7c_2)\end{aligned}$$

Using the definition for vector equality, we have

$$\begin{aligned}2c_1 + 3c_2 &= -1 \\ -3c_1 + 5c_2 &= 11 \\ 4c_1 + 7c_2 &= -1\end{aligned}$$

Since this system has the solution $c_1 = -2$, $c_2 = 1$, we have

$$(-1, 11, -1) = -2(2, -3, 4) + (3, 5, 7)$$

(c) No **(f)** Yes

14 (a) Dependent **(c)** Independent **(f)** Independent

15 (a)

$$\begin{bmatrix} 2 & 3 & -4 & 2 \\ 4 & 0 & 3 & 1 \\ 1 & 1 & 1 & 1 \\ -3 & 6 & -7 & 3 \end{bmatrix} \to \begin{bmatrix} 1 & 1 & 1 & 1 \\ 4 & 0 & 3 & 1 \\ 2 & 3 & -4 & 2 \\ -3 & 6 & -7 & 3 \end{bmatrix} \to \begin{bmatrix} 1 & 1 & 1 & 1 \\ 0 & -4 & -1 & -3 \\ 0 & 1 & -6 & 0 \\ 0 & 9 & -4 & 6 \end{bmatrix}$$

$$\to \begin{bmatrix} 1 & 1 & 1 & 1 \\ 0 & 1 & -6 & 0 \\ 0 & -4 & -1 & -3 \\ 0 & 9 & -4 & 6 \end{bmatrix} \to \begin{bmatrix} 1 & 1 & 1 & 1 \\ 0 & 1 & -6 & 0 \\ 0 & 0 & -25 & -3 \\ 0 & 0 & 50 & 6 \end{bmatrix}$$

$$\to \begin{bmatrix} 1 & 1 & 1 & 1 \\ 0 & 1 & -6 & 0 \\ 0 & 0 & 1 & 3/25 \\ 0 & 0 & 0 & 0 \end{bmatrix} \qquad \text{rank} = 3$$

(d) rank = 2

Section 6.1

3 (b)

v_2	1			
v_3	1	1		
v_4	1	0	0	
v_5	0	1	1	1
	v_1	v_2	v_3	v_4

4 (b) $d(v_1) = 3$, $d(v_2) = 3$, $d(v_3) = 3$, $d(v_4) = 2$, $d(v_5) = 3$ The number (4) of vertices of odd degree is even.

9 (b) simple: $v_1v_3v_5v_2$ nonsimple: $v_5v_2v_3v_1v_2v_5v_4$

11 (a) $v_1v_2v_3v_4v_2v_6$

13 (b) $V = \{v_1, v_2, v_3, v_4, v_5\}$ $E = \{\{v_1, v_5\}, \{v_2, v_4\}, \{v_3, v_4\}\}$

x	A	B	p
			0
v_1	$\{v_1\}$	$\varnothing$	
	$\{v_1, v_5\}$	$\{v_1, v_5\}$	1
v_2	$\{v_2\}$		
	$\{v_2, v_4\}$		
	$\{v_2, v_3, v_4\}$	$\{v_1, v_2, v_3, v_4, v_5\}$	2

14

```
algorithm edges (G:m)
initialize m as 0
for j running from 1 to n − 1
   for i running from j + 1 to n
      if e_ij = 1
         increase m by 1
```

Section 6.2

1 The revised equations become:

$$L_1: 7(i_1 + i_3) + 10 + 6(i_1 + i_4) - 20 = 0$$
$$L_2: 3(i_2 + i_3 - i_4) + 4i_2 + 5i_2 - 10 = 0$$
$$L_3: -5 + 1(i_3 - i_4) + 15 + 8(i_3 - i_4) + 7(i_1 + i_3) + 3(i_2 + i_3 - i_4) = 0$$
$$L_4: 2i_4 + 10 + 6(i_1 + i_4) - 20 - 8(i_3 - i_4) - 15 = 0$$

2 We conclude that the collections are dependent by showing that the corresponding echelon matrices have a zero row:

(a)
$$\begin{bmatrix} 1 & 1 & 0 & 1 & 0 & 0 & 1 \\ 0 & 0 & 0 & 0 & 1 & 1 & 1 \\ 1 & 1 & 0 & 1 & 1 & 1 & 0 \end{bmatrix} \to \begin{bmatrix} 1 & 1 & 0 & 1 & 0 & 0 & 1 \\ 0 & 0 & 0 & 0 & 1 & 1 & 1 \\ 0 & 0 & 0 & 0 & 1 & 1 & 1 \end{bmatrix}$$
$$\to \begin{bmatrix} 1 & 1 & 0 & 1 & 0 & 0 & 1 \\ 0 & 0 & 0 & 0 & 1 & 1 & 1 \\ 0 & 0 & 0 & 0 & 0 & 0 & 0 \end{bmatrix}$$

(c)
$$\begin{bmatrix} 1 & 1 & 0 & 1 & 1 & 1 & 0 \\ 0 & 0 & 1 & 1 & 0 & 1 & 1 \\ 1 & 1 & 1 & 0 & 1 & 0 & 1 \\ 1 & 1 & 1 & 0 & 0 & 1 & 0 \end{bmatrix} \to \begin{bmatrix} 1 & 1 & 0 & 1 & 1 & 1 & 0 \\ 0 & 0 & 1 & 1 & 0 & 1 & 1 \\ 0 & 0 & 1 & 1 & 0 & 1 & 1 \\ 0 & 0 & 1 & 1 & 1 & 0 & 0 \end{bmatrix}$$
$$\to \begin{bmatrix} 1 & 1 & 0 & 1 & 1 & 1 & 0 \\ 0 & 0 & 1 & 1 & 0 & 1 & 1 \\ 0 & 0 & 0 & 0 & 0 & 0 & 0 \\ 0 & 0 & 0 & 0 & 1 & 1 & 1 \end{bmatrix} \to \begin{bmatrix} 1 & 1 & 0 & 1 & 1 & 1 & 0 \\ 0 & 0 & 1 & 1 & 0 & 1 & 1 \\ 0 & 0 & 0 & 0 & 1 & 1 & 1 \\ 0 & 0 & 0 & 0 & 0 & 0 & 0 \end{bmatrix}$$

3 Consider the collection:

$$c_6 = (1, 1, 1, 0, 1, 0, 1)$$
$$c_2 = (0, 0, 1, 1, 1, 0, 0)$$
$$c_3 = (0, 0, 0, 0, 1, 1, 1)$$

Since the matrix

$$\begin{bmatrix} 1 & 1 & 1 & 0 & 1 & 0 & 1 \\ 0 & 0 & 1 & 1 & 1 & 0 & 0 \\ 0 & 0 & 0 & 0 & 1 & 1 & 1 \end{bmatrix}$$

is already in echelon form (and has no zero rows) the collection is independent.

6 **(d)** $\rho(G) = m - n + p = 7 - 5 + 1 = 3$

(h) $\rho(G) = m - n + p = 16 - 8 + 1 = 9$

11 Take $c = c_1 + c_2$, using the coordinatewise addition in B^m.

Section 6.3

1 **(c)** Remove $\{v_1, v_2\}$, $\{v_3, v_6\}$, $\{v_2, v_4\}$, and $\{v_2, v_6\}$.

(d) Remove $\{v_1, v_3\}$, $\{v_4, v_5\}$, and $\{v_2, v_5\}$.

4 Select the edges *cf, af, be, ef, df* in order.

6 Since

$$\rho(G) = m - n + p = m - m + 1 = 1$$

such a graph must be a single circuit with trees (perhaps) growing at the corners.

7 **(d)**

i	x	T
1	λ	$T[\lambda]$ = warning
2	0	$T[0]$ = the
3	00	$T[00]$ = surgeon
4	000	$T[000]$ = general
5	0001	$T[0001]$ = has
6	0000	$T[0000]$ = determined
7	001	$T[001]$ = that
8	00000	$T[00000]$ = cigarette
9	00011	$T[00011]$ = smoking
10	000110	$T[000110]$ = is
11	000001	$T[000001]$ = dangerous
12	01	$T[01]$ = to
13	1	$T[1]$ = your
14	0001100	$T[0001100]$ = health

11 Since k comparisons are necessary, the search complexity is $k \sim \log n$.

13 **(c)** $A/((B + C) \cdot (D + E)) \to ABC + DE + \cdot /$

(d) $(A/(B + C)) \cdot (D + E) \to ABC + /DE + \cdot$

17 **(c)** $((A \cdot B)/C) \cdot (D + E)$ **(d)** $A \cdot ((B + C) \cdot (D/E))$

19

```
algorithm inorder (T)
if root is not null
   inorder (left (T))
   print root
   inorder (right (T))
```

Section 6.4

1 The graph is planar.

2 **(a)** If it were planar, then

$$r = m - n + 2 = 13 - 6 + 2 = 9$$

and Theorem 6.6 would give

$$3r \leq 2m \quad (\text{or } 27 \leq 26)$$

a contradiction. So the graph is nonplanar.

(b) If it were planar, then

$$r = m - n + 2 = 16 - 8 + 2 = 10$$

and Theorem 6.6 would give (since there are no triangles)

$$4r \leq 2m \quad (\text{or } 40 \leq 32)$$

a contradiction. So the graph is nonplanar.

(c) The graph is planar.

4 We have

$$n = 10 \qquad m = 22 \qquad r = 14$$
$$n - m + r = 10 - 22 + 14 = 2$$

5 **(a)** $4 - 6 + 4 = 2$ **(b)** $20 - 30 + 12 = 2$

Section 6.5

1 There are

$$1 + 1 + 2 + 3 + 2 + 1 + 1$$

or 11 in all.

3 The following invariants:

	n	m
(a)	5	6
(b)	5	7
(c)	5	8
(d)	6	

show that no two of these graphs are isomorphic.

4 The invariant m distinguishes all but the pair (a), (d). These can be distinguished with the degree spectrum.

6 **(a)** $\langle 0, 0, 0, 4, 1 \rangle$ **(i)** $\langle 0, 0, 0, 0, 8, 0, 0, 0 \rangle$

7 **(a)** $\beta = 2\ \{v_1, v_2\}$ **(d)** $\beta = 3\ \{v_1, v_2, v_3\}$ **(f)** $\beta = 3\ \{v_1, v_3, v_5\}$

9 **(a)** $\alpha = 3\ \{v_3, v_4, v_5\}$ **(c)** $\alpha = 5\ \{v_1, v_2, v_3, v_4, v_5\}$ **(f)** $\alpha = 3\ \{v_2, v_4, v_6\}$

10 **(a)** $\kappa = 3$ **(b)** $\kappa = 3$ **(c)** $\kappa = 4$

18 Let $\{e_j\}$ be a maximum-sized independent set of edges and let $\{v_i\}$ be a minimum-sized covering set of vertices. Corresponding to each e_j is a vertex v_i covering it. This correspondence is one-to-one since the set $\{e_j\}$ is independent. It follows that $\beta_e \leq \alpha$ as stated.

Section 6.6

2 (e)

	e_1	e_2	e_3	e_4	e_5	e_6	e_7	e_8	e_9
v_1	1	0	0	0	0	0	0	1	0
v_2	1	1	0	0	0	1	0	0	1
v_3	0	1	1	0	0	0	1	0	0
v_4	0	0	1	1	0	0	0	1	1
v_5	0	0	0	1	1	0	0	0	0
v_6	0	0	0	0	1	1	1	0	0

$$\begin{aligned} f &= (v_1 + v_2)(v_2 + v_3)(v_3 + v_4) \\ &\quad (v_4 + v_5)(v_5 + v_6)(v_2 + v_6)(v_3 + v_6)(v_1 + v_4)(v_2 + v_4) \\ &= v_2v_4v_6 + v_1v_2v_3v_5 + v_1v_3v_4v_6 + v_2v_3v_4v_5 \end{aligned}$$

$\alpha = 3$ and $\{v_2, v_4, v_6\}$ is a minimum-sized covering set

4 (e) Since $\{v_2, v_4, v_6\}$ is a minimal covering set, $\{v_1, v_3, v_5\}$ is a maximal independent set and $\beta = 3$.

7 (e) Denote the maximal independent sets

$$A = \{v_1, v_3, v_5\}$$
$$B = \{v_1, v_6\}$$
$$C = \{v_4, v_6\}$$
$$D = \{v_2, v_5\}$$

	v_1	v_2	v_3	v_4	v_5	v_6
A	1	0	1	0	1	0
B	1	0	0	0	0	1
C	0	0	0	1	0	1
D	0	1	0	0	1	0

$$\begin{aligned} f &= (A + B)DAC(A + D)(B + C) \\ &= ACD \end{aligned}$$

$\kappa = 3$ and $A' = \{v_1, v_3\}$, $C = \{v_4, v_6\}$, $D = \{v_2, v_5\}$ is a minimal proper coloring

11 In replacing the covering problem (A, R, B) by $(A \sim \{a\}, R, B \sim aR)$ we are assured that the points of aR are covered by a. Thus if C is an irredundant covering of the reduced problem, $C \cup \{a\}$ will be a covering of the original problem. Certainly a cannot be removed (because it is essential). And if some other cell can be removed from $C \cup \{a\}$, then it can be removed from C, contradicting the fact that C is irredundant. So $C \cup \{a\}$ is an irredundant covering as well. The converse argument is similar.

15 In covering the point b_κ we automatically cover b_j as well.

Section 6.7

2 (a)

	v_1	v_2	v_3	v_4
v_1	0	0	0	0
v_2	1	0	1	0
v_3	0	1	0	0
v_4	1	0	1	1

4 (a) $\begin{bmatrix} 1 & 1 & 1 \\ 1 & 1 & 1 \\ 0 & 1 & 1 \\ 1 & 0 & 0 \end{bmatrix}$

5 (a)
$$E^2 = \begin{bmatrix} 1 & 1 & 0 & 0 & 1 \\ 1 & 1 & 0 & 0 & 1 \\ 1 & 1 & 1 & 0 & 1 \\ 1 & 1 & 1 & 0 & 0 \\ 0 & 0 & 1 & 0 & 1 \end{bmatrix}$$

6 (a)
$$E^3 = \begin{bmatrix} 1 & 1 & 1 & 0 & 1 \\ 1 & 1 & 1 & 0 & 1 \\ 1 & 1 & 1 & 0 & 1 \\ 1 & 1 & 1 & 0 & 1 \\ 1 & 1 & 0 & 0 & 1 \end{bmatrix}$$

showing that there are paths of length 3 except to v_4 and from v_5 to v_3.

Section 6.8

1 (e) $d^-(v_1) = 2 \quad d^+(v_1) = 0$
$d^-(v_2) = 1 \quad d^+(v_2) = 2$
$d^-(v_3) = 2 \quad d^+(v_3) = 2$
$d^-(v_4) = 2 \quad d^+(v_4) = 3$

3 Begin by choosing an arbitrary edge. Continuing inductively, let $\langle v_1, v_2, \ldots, v_k \rangle$ be a path that does not duplicate any vertices. Then examine the two possibilities:
(a) Every vertex of the graph occurs on the path;
(b) There exists a vertex v not in the path.
In case (b), use the fact that the graph is complete to extend the path.

7 (c)

u	v	u^*	M	W	d
v_2	v_5	v_2	$\varnothing$	$\{v_2\}$	$d[v_2] = 0$
			$\{v_2\}$	$\{v_1, v_2, v_4, v_6\}$	$d[v_1] = d[v_4]$
				$\{v_1, v_4, v_6\}$	$= d[v_6] = 1$
		v_1	$\{v_1, v_2\}$	$\{v_1, v_3, v_4, v_6\}$	$d[v_3] = 2$
				$\{v_3, v_4, v_6\}$	$d[v_4] = 1$
		v_4	$\{v_1, v_2, v_4\}$	$\{v_3, v_4, v_6\}$	$d[v_3] = 2$
				$\{v_3, v_6\}$	$d[v_6] = 1$
		v_6	$\{v_1, v_2, v_4, v_6\}$	$\{v_3, v_6\}$	$d[v_3] = 2$
				$\{v_3\}$	
		v_3	$\{v_1, v_2, v_3, v_4, v_6\}$	$\{v_3, v_5\}$	$d[v_5] = 3$
				$\{v_5\}$	
		v_5			

Using the backstepping procedure, we find that $\langle v_2, v_1, v_3, v_5 \rangle$ is a shortest path (length 3) from v_2 to v_5.

INDEX

Abbreviation, proof, 102–104, 105
Absolute error, 198
Absolute value function, 26
Absorption Law, 15, 95, 445
Abstraction
  definition by, 1
  and equivalence, 426
  relations and, 456, 458–459
  vector space, 363
Absurdity, logical
  contradiction and, 124
  negation and, 117
  reduction to, 112
Acyclic graph, 398, 404. *See also* Circuit(s)
Addition operation on N
  defined recursively, 159
  and induction, 147, 148
  of polynomials, 280–282
Algebra
  covering problem, 443–446
  history of, 277
  linear, 357–358, 361
Algebraic numbers, 304–306, 363
Algorithm(s), 158, 203, 204
  approxroot, 303
  area, 239
  auxiliary, 315, 320–324
  bisection, 236, 240
  computational complexity of, 263, 265–268
  connectivity, 383, 400
  correctness, 246
    deductive rules for, 249, 250–251, 252, 254
    testing, 247
    verification, 247–248
  definition of an, 210–211
  derived constructs for, 222–225, 234
  division, 168, 169, 320, 332
    polynomial, 319–320
  Euclidean, 171–172, 211, 212, 213, 216, 240, 247, 255–256, 320, 322, 337
  factor, 325–326
  factors, 172
  fraction, 336, 339
  gauss, 366
  hypotenuse, 208
  independence, 448
  interpolate, 312, 320, 327
  Kronecker, 325–326
  largest, 220
  minmax, 262–263
  as modules, 238–242
  partial, 338
  polyeval, 287, 320
  polyproduct, 285
  prime, 225, 227, 242
  pseudolanguage, 215, 216–220, 225–227
  ratadd, 241–242
  reachability, 466
  recursive, 231, 259–262, 273
  selectionsort, 237–238
  set computation, 232
  shortpath-digraph, 477
  shortpath-graph, 475
  sorting, 233, 235–236, 237
  state, computation, 213–215, 248, 249
  using subscripted variables, 233–236
  treesort, 406
  triangles, 223, 238
Algorithmic modules, 238
Alphabet set(s), 4
  and equivalence relations, 42
  sequences, 27–28
  See also n-set
Alternative statements, 134
Annotations, 101, 105

Antecedent, 67
  in forward and backward reasoning, 80–82
  and implication, 132
Approximate roots, 302–304
Approximation, 198–199
Approxroot algorithm, 303
Archimedean principle, 152
Archimedes, 152
Area
  algorithm, 239
  formula for, 238
Argument
  of algorithmic modules, 239–240
  classical vs. constructive, 116, 119–120
  formal proof, 101, 104
  of the function, 24
  logical, 81, 84, 104
  negation of, 93, 106, 118–119
  polynomial, 285–286
Aristotle, 62, 65
Arithmetic operations, 146–148, 170
  algorithms for, 158, 168, 171
  computerized, 192, 196–197
    errors of, 197–199
  differencing and, 344
  fractional, 168
  function call in algorithm and, 241
  laws of, 148–152, 187, 199
  of polynomials, 280–285, 286–287
  of rational numbers, 186–187, 241
Arrays, 231, 234, 406, 408. *See also* Subscripted variables
Aspects of Constructivism (Bishop), excerpt, 63
Associative law, 15, 37, 95
  and induction, 149, 150
  irredundant coverings and, 444
  and relational composition, 461
Assumption(s), mathematical, 7, 8, 11, 73, 152, 398
  disproving, 45
  in formal proofs, 92–93, 100, 103, 104
  inductive hypothesis, 149–150, 156, 157
  and logic, 69, 72, 74, 81, 117
  multiple, 103
  and rules for disjunction, 92
  and rules of implication, 91, 109
  See also Axioms; Contradiction; Implication(s)
Auxiliary polynomials, 315, 320–324
Auxiliary variables, 210, 212
Axiomatic method, 64, 123, 124
  and Euclidean geometry, 71
Axioms, 65, 72
  and Boolean algebra, 123
  geometry, 73
  for the natural numbers, 145, 155
  of ordered sets, 66

Base conversion, 175, 177–179
  reverse, 178
  table of, 178
Base notation, 176–177
Base, exponential, 193
Basis for circuit space, 392
Bell, E. T., 143
Binary system, 28, 175, 176, 196, 199–200, 375
  See also Computer programming languages
Binary tree, 404
Binomial coefficients, 56–57
Binomial theorem, 56
Bipartite graph, 378
Bisection, 236, 238, 262–263
  for approximate roots, 302
  See also Divide and conquer
Bisection algorithm, 236, 240, 272, 302
Bishop, E., 63, 93
  Aspects of Constructivism (excerpt), 63
  interpretation of negation, 117
Bits, data, 176, 179
  fields of, 194
Boole, George, 123, 128, 199
Boolean algebra, 123, 127
Boolean number system, 199–200, 205
  coordinate vector spaces and, 363, 364
  pseudolanguage algorithms and, 227, 235, 236, 242
Bound on roots theorem, 296, 298
Boundary conditions, 57, 416
Bracing notation, 2, 8. *See also* Set-builder notation
Brouwer, Luitzen E. J., 62, 93, 124, 142
  Collected Works (excerpt), 63
  and interpretation of negation, 117

Calculus, defined, 10
  difference, 345–347, 353
  differential, 343, 345
  functions and, 28
  logical, 95–97
    classical, 120–123, 124–125, 343
    constructive, 63
  of sets, 12, 16, 120, 122, 123
  statement, 127, 128
  summation, 352–355
Cancellation, 281, 282
  Law of, 283
Cardinality, 3, 5
  and characterization of equicardinality, 58, 188
  of an infinite set, 188
  of multiple sum sets, 50
  and one-to-one correspondence, 29, 188–189
  of permutations, 51, 52
  in power sets, 58
  in product operations, 34
  of rational numbers, 188–189
Carroll, Lewis, 135, 136
Cases (CAS)
  and disjunctive elimination, 92
  proof by, 86–87
Cause and effect, 67
Change of sign principle, 291, 297
Characteristic table, 130–133
  tautology and, 138
  See also Truth table
Characteristics, exponential, 195
Characterization, defined, 79
  of acyclic connected graph trees, 398
Chords, 400
Chromatic number, 435, 451–452
Church, Alonzo, 202
Circuits, 381–382, 384, 398
  Euler path, 471, 472
  independent, 389–392, 401
  of planar graphs, 416, 418, 419
  rank, 392
    formula for, 393–395
Circuit space, 389–392
Claim of consistency, 137
Class. *See* Set(s)
Classical laws
  vs. constructive, 116, 118, 120, 123, 124–125
  extended, 120–123
  tautologies and, 138
Clique number, 431–432
  independence and, 447–451
Coefficients, 278–279, 328
  in Kronecker algorithm, 325
  positive polynomial, 293–294, 296, 299, 335
Collected Works (excerpt)
  Brouwer, 63
  Kronecker, 143
Collecting formula, 285, 328
Collection(s), 14, 19
  and covering problems, 440–441
  and equivalence relations, 40, 45
  ordered, 65
  planarity and, 416
  rank of, 364–368, 392, 401
  sorting a, 233
Collinear points, 72
Columns, matrix, 367
Combinations, 51, 52–53, 56
  graph, 380
  linear, 358–359, 367
Combinatorial mathematics, 47. *See also* Graphs
  pigeon-hole principle, 60
Common denominator, 170
  in algorithmic modules, 241
Common notions, 64, 65, 72. *See also* Axioms
Commutative Law, 15, 95, 119
  and induction, 149, 151
  and irredundant coverings, 444
  rational number system and, 187
Compiler, computer language, 28
Complementary graph, 378
Complement
  of the graph, 423–424
  set, 16–17
Complete graphs, 377–381
  and Hamiltonian paths, 474
Completed infinity, 124
Completing the square, 359
Components, 362, 390
  and connectivity, 382, 400
  matrix, 364
Composite function, 30–32
Composition, 30
  functional, 30–33
  relational, 459–460

Compound processes, 216–220
algorithmic modularity and, 242
deductive rules and, 249
Computable numbers, 306, 363
Computational complexity, 263, 264–265, 269–273
Computer programming language(s)
absolute values in, 26
and algorithm modularity, 239
algorithm running time and growth rate in, 265, 266–268
alternative statements in, 134
and arithmetic operations, 192, 196–197, 284–285
arrays in, 232, 234
and errors of computation, 197–199
global and scope references in, 100
and indentation format, 101–102
and integer division, 168
logical equivalence in, 95
logical expressions, precedence of, 130
memory storage and, 193, 194–195, 196, 232
pseudolanguage and, 215, 225, 227, 232
sequence and, 28, 96–97
Conclusion, 69, 91
Condition, algorithm state, 213
Conditional process, 218
Congruence (modulo n), 40, 41, 184, 426
Conjunction, 86
negating a, 122
operation, 131, 132
priority and, 130
rules of, 91–92, 104
derived, 110–111
Connectives, use of, 63, 85–86, 91, 128
and logical equivalence, 94
See also Logical reasoning
Connectivity, graph, 382–384, 397, 398
algorithm, 383
and graphical invariants, 429
Consequent, 67
in forward and backward reasoning, 80–82
and implication, 132
Constant, 42–43, 344
polynomial, 280, 282
in recursive functions, 158
Constant of proportionality, 266
Constructibility, 6–7, 306
empty set and, 12
and product operations, 33–34
Constructive proofs, 116, 118
vs. nonconstructive proofs, 119–120
See also Elimination rules; Introduction rules
Constructs, pseudolanguage, 219, 222–225, 248, 253
and subscripted variables, 234
Containment set, 11, 15
in union and intersection characterizations, 14, 16
Contours, 416, 418
Contradiction, 74, 75, 106, 120, 124, 398
classical laws and, 119–120
and Law of excluded middle, 119
proof by, 328, 421
well-ordering principle, 152
Contraposition (CON), law of, 69, 70, 109
and derived disjunction rules, 111
and derived negation rules, 112
Control variable, 224, 233
Converse implication, 78–79
Conversion
base, 175, 177–179
of fractional parts, 179–181
problem, 347–348, 349, 351–352
Coordinate vector spaces, 362–364, 391
theorem, 363
Coordinates, 34, 291
Corollary
classical logic, 123
complementary graphs, 423
Euler circuit, 472
finite set products, 49, 52
Kuratowski graphs, 421
and Law of detachment, 85
path length, 462
permutations of n-set, 53, 55
tree characterization, 399
Correspondence. *See* One-to-one correspondence
Countability, 188, 190
Counting numbers, 3, 4
Counting techniques, 47–48, 49, 53, 54, 55, 56, 57, 58
base conversion, 177, 178
for rational numbers, 189–190
successor function and, 146–147

Covering algebra, 443–446
Covering problems, 439–443
Covering set, 431–432, 447
Creative definition, principle of, 1
Cubic polynomial, 315–316
Current, electrical, 387
Cyclical order, 66

Decimal system
  conversion between binary and, 177–179, 180
  exponential representation of, 193–194
    negative powers in, 180
  in integer division, 168–169, 170
Decision problem, 205
Dedekind, Julius W. R., 1
Deductive logic, 69, 70, 98
  and claim of consistency, 137
  classical, 118, 122
  and Euclidean geometry, 72
  induction and, 155–156
  Law of detachment, 84–85
  notion of adequacy and, 137
  and problem-solving, 203
  and proof development, 104–105, 256
  propositional rules of, 89, 90–93, 94–95, 99
  rules for algorithm correctness, 249, 250–251, 252, 254
  and truth table evaluation, 127
Defined inductively. *See* Recursion
Defining property, 8
  classical logic of, 120, 121
  in extended number system, 152
Degree
  polynomial, 279, 281, 319, 350
  spectrum, 430
  vertex, 380–381
Delimiter, 91
Delta, use of, 343, 346, 474
DeMorgan's Laws, 17, 122, 123
deMorgan, Augustus, 17, 122
Denominator(s), 170, 185
  powers as, 335–337
Dependency
  and loop equations, 389
  polynomial, 359, 360
Derivative, 343
Derived constructs, 222–225
Derived rules, 108
  conjunction, 110–111
  disjunction, 111–112
  equivalence, 114
  implication, 109–110
  negation, 112–114
  See also Propositional rules
Descartes, René, 206, 277
Descartes' Rule of Signs, 298
Design specification, 247, 250, 251
  verification of, 256
Detachment, Law of, 84–85
Deterministic operations, 211
Difference, set, 16
Difference operator, 342–345, 346–347
  in Newton expansion, 348
Digraphs. *See* Directed graphs
Diophantus, 277
Directed graphs, 374, 455–458
  reachability of, 463–468
  relational composition in, 458–460, 462, 465
  shortest paths in, 476–479
  strongly connected, 466
Discrete mathematics, 190, 200, 342–343
Discrete number systems, 144, 154, 168, 176, 184, 193, 278, 305–306, 363
Disjoint sets, 14
  of infinite subsets, 40
  partitions and, 20, 434
Disjunction, 85, 119
  negating a, 122
  operation, 132
  rules of, 91–92
    derived, 111–112
Disjunctive elimination, 92
Distinct variables, 69, 70
Distributive law, 15, 95
  and induction, 149, 151
  irredundant coverings and, 444
Divide and conquer, 236, 238, 262–263
Dividend, 294, 296, 319
Divisibility, 38, 71, 82
  algorithm factors of, 172–173
  of integers, 166–169
  of related pairs, 38–39, 41
Division
  long, 167, 169, 295, 319
  synthetic, 295–296, 361

Division algorithm, 168, 169
and base conversion, 179
polynomial, 319, 332
Divisor, 294, 319, 320–321
Domain, 11
of a function, 24
and infinite sequence, 26
Dominating set, 431–432, 442
covering and, 446–447
Double negation, 117. *See also* Negation

Echelon form, 365, 392
Economy tree, 402
theorem, 403
Edge(s), linear graph, 373, 375–376, 389
deletion of, 399–400, 418
in directed graphs, 456, 457, 462
formula for, 380
and graphical invariants, 429, 453
maximality of, 398
multigraph, 471
and network analysis, 386–387
and planarity, 416
and vertex degrees summation, 380–381
See also Circuits; Paths
Eisenstein's Irreducibility Criterion, 328–329
Eisenstein, Ferdinand Gotthold, 328
Electrical network, 386–389, 418, 424
Elementary circuit, 381–382, 390
Elementary factorial polynomials, 345–347, 351
Elementary processes, 211–212, 213
analysis of, 248–249
and pseudolanguage, 216, 224, 232, 248–249
in sorting algorithm, 237
time complexity and, 269
See also Compound processes
Elementary row operations, 365
Elements, 2, 3
general set concept and, 1
as image, 29
n-tuples, 34, 59
in set-builder notation, 7
Elements (Euclid), 64
Elimination rule(s), 91–92, 102, 105–106, 123
of double negation, 117
summary of, 99
See also Derived rules; Introduction rule(s)
Ellipses, 3, 4
Empty set, 12, 13
and complementation operations, 17
negation and the, 93
partitions and, 20
in power set constructions, 58–59
and treesort algorithm, 408
Equality
compared to equivalence, 43
congruence and, 41
factorial polynomial, 351–352
of functions, 37
and inductive hypothesis, 157
of lines, 72
and mathematical abstraction, 1
polynomial, 280
of sets, 15, 16, 18
Equations
defining functions with, 24
linear, 36, 387
loop, 388–389
recurrence, 57
of the second degree, 277
straight line, 311–312, 315
See also Algorithm(s); Polynomial(s); Set(s)
Equicardinality, 58, 59, 188. *See also* Cardinality
Equivalence class, 43–45
rational number as, 186
Equivalence relations, 38, 41–43
characterization of, 45–46, 185
compared to equality, 43
and congruence, 41
and connectivity, 382
among functions, 43
logical, 79–80, 94, 110, 114, 123
operation, 132
properties of, 41, 43
rational numbers and, 185–186
reciprocal rules of, 94–95
derived, 114
and spanning trees, 400
Errors
absolute, 198

Errors *(Continued)*
in computerized arithmetic operations, 197–199
relative, 198
truncation, 198
Essential cells, 452
Euclid, 64, 71, 72, 113, 171
Euclidean algorithm, 171, 211, 212, 213, 320, 322
compound process in, 216, 217
as gcd function, 240–241
revised, 337
testing the, 247–248
verification of, 255–256
Euclidean geometry, 71–76
Euler, Leonhard, 373, 374, 470, 471
Euler's formula, 418–420, 423. *See also* Planar graphs
Euler path, 471
Execution, of algorithm, 212–213
and generation of binary trees, 405, 407, 409
trace of, 384
Exponential systems, normalized, 193–194
Exponentiation operation, 153
Exponents, 176–177, 193
characteristics of, 195
degree, 279
negative, 180–181, 194
Expression, set, 18
Extended number system, 151–152
Extension by definition, 70–71, 72

Factorial notation, 53
Factorial polynomials, 345–347. *See also* Elementary factorial polynomials
Factorization, 168, 172, 320
and Kronecker algorithm, 326
of polynomials, 318, 324–329
Falsity, 130
Boolean, 242, 248, 251–252
contradiction and, 124
in elementary processes, 213
truth table evaluation and, 127
Fibonacci, Leonardo, 166
Fibonacci numbers, 166, 229
Finite sequence, 27, 47
and binary trees, 404
coordinate vector spaces of, 362–364
polynomial, 361
See also Infinite sequence; Sequences
Finite set(s), 3, 4
cardinality of, 3, 29
composite functions in, 31
constructibility of, 6
counting techniques, 47–48, 49, 51, 53, 55
ellipses in, 3
relations, 34–35
subsets and, 12
Flexibility
process, 218
pseudolanguage, 231
Floating point arithmetic, 196
Flowcharts, 418
and directed graphs, 458
Forest, spanning, 400
Formal proof, 92–93, 99–101, 104
FORTRAN programming language, 28, 168
Forward and backward proofs, 81–83, 104
See also Logical reasoning
Four-color map theorem, 373, 436
Fraction algorithm, 336, 339
Fractional parts, conversion of, 179–181
Fraction(s)
in integer division, 168–169
partial, expansion of, 336, 337–340
reducible, 170, 185
in straight line equations, 312
See also Rational numbers
Frege, Gottlob, 1
Function algorithm
module as, 240–242
recursive, 259–262
Function eval algorithm, 320
Function factors algorithm, 320
Function notation, 24
Function(s), defined, 23
composite, 30–32
computing values of a, 25
elementary factorial polynomial, 345–347
use of equation as rule in, 24, 25
eval, 320–321
factorial, 159, 260

factors, 320–322
gcd, 241
graphing, 289, 290–292, 332–334
inverse, concept of, 28
invertible, 32
log, 266, 343
one-to-one, 29, 32, 59
one-to-one correspondence in, 29, 30, 32, 59
onto, 29, 32, 59
polynomial, 279, 285–286, 290–292
power, 261–262
power set, 59
rational, 331, 332–334, 335, 337–340, 345
by recursion, 158–160
and relations, distinguishing between, 35
sequences, 26–28
successor, 146, 147
well-defined, 28–29
Fundamental theorem, of the difference and summation calculus, 353

Gauss algorithm, 366
Gauss, Karl F., 366
Gcd function, 241
Gödel, Kurt, 202
Graphing
polynomial functions, 289–290, 291, 292–294, 300, 304, 311, 332
rational functions, 332–334
Graphs and Their Uses (Ore), excerpt, 373
Graphs, 373, 374–377
bipartite, 378
complementary, 378
complete, 377–381
connectivity of, 382–384
directed, 455–458, 462, 463, 465, 466
shortest paths in, 476–479
and independent circuits, 389–392
isomorphism and invariants, 426, 427, 429
pictorial representation of, 375
planar, 415, 416–418, 420–424
and network analysis, 386–389
trees, 382, 397, 398, 402, 404, 408
undirected, 456, 473, 474
shortest paths in, 474–476

Greatest common divisor (gcd), 170–172
Euclidean algorithm and, 171–173, 212–213, 247, 322
Growth rate, algorithm, 266–268, 292

Hamiltonian paths, 473–474
Hamilton, W. R., 473
Hamlet, 118
Heuristics, 206
Horizontal intercept. *See* Roots, polynomial
Horner's method, 286–287
and change of sign principle, 292
and function eval algorithm, 320, 321
and synthetic polynomial division, 296
How To Solve It (Polya), 206
Hypotenuse algorithm, 208

Idempotent law, 15, 95
and irredundant coverings, 444
Identifiers
absolute value, 26
function, 25
inverse, 33
sequence, 27
set, 4
Identity. *See* Properties
Identity Law, 461
Identity matrix, 463
Image, function, 24, 29
in invertible functions, 32
and pigeon-hole principle, 60
planar graph projection of, 420
Implication(s), logical, 67–68
circuit rank formula and, 398
classical logic and, 122
and equivalence, 78–80
introduction and elimination of the, 91–92, 102
and Law of detachment, 84–85
operation, 132
truth of, 132–133, 136
proving an, 80–81, 96, 100
rules of, 90–91
derived, 109
semantic, 135–136
See also Propositional rules
Incidence relation, 440, 452
Inclusion and exclusion formula, 50

Indentation format, proof, 100–101
  in algorithm pseudolanguage, 226
  conventions on, 101–102
Independence
  algorithm, 448
  circuit, 392, 394, 401
  numbers, 431–432
    and clique numbers, 447–451
  polynomial, 359–362, 368
Independent set, 431, 434
  and chromatic numbers, 451–452
Indeterminate x, 278–279
Index
  in polynomial multiplication, 284
  sorting algorithm, 237
  summation, 352
Indirect proofs. *See* Contradiction, proof by
Induction, 144, 147. *See also* Recursion
  and algorithm proofs, 246, 248, 322
  and Descartes' Rule of Signs, 298
  directed graphs and, 462
  and labeled binary trees, 409, 410
  principles of, 155–158
  proof by, 149–151, 154, 158
  and recursion, 155, 159, 164
Inductive hypothesis, 149, 150, 156, 157
  and planar graph contours, 418
  recursive functions and, 159
  and relational composition, 462
  and summation, 161
Inequality. *See* Relation(s)
Infinite sequence, 26
  and summation, 160
  See also Finite sequence; Sequences
Infinite set(s), 3
  and cardinality, 3
  constructibility of, 6–7
  constructive vs. nonconstructive proofs of, 119
  ellipses and, 3
  recursively defined, 164–165
Infinite subset, 38, 40
Infix form, 411
Initialization, 224
Input, 205, 211
Instance, 204, 205
Integer quotient, 169
Integers
  base conversion of, 175, 177–179, 180
  divisibility of, 167–169
  and graphical invariants, 429, 430
  ordering of the, 152–153
  vector spaces and, 363
  See also Numbers
Intercepts, 292–294. *See also* Limiting values
Interchange, 248–249
Internal Revenue Service tax information on depreciation, summation in, 161
Interpolating polynomial, 309, 311, 312–316, 326–327
Intersection(s), 13
  characterization of union and set, 14–15
  complete graph, 378
  and disjoint sets, 14
  generalized, 19–20
  of lines, 72
  operations, properties of set, 15–16, 17–18
  and planarity, 416, 420
  in sum sets, 49, 50
Introduction rule(s), 91–92, 102, 105, 123
  of double negation, 117
  summary of, 99
  See also Derived rule(s); Elimination rules
Introduction to Metamathematics (Kleene), excerpt, 203
Intuition, mathematical, 62, 65, 124, 152
Invariant assertion, 253, 254, 255–256
Invariants, graphical, 426, 429. *See also* Isomorphism
Inverse function. *See* Invertible function
Inversion, permutation, 54–55
Invertibility, 32, 79
Invertible functions, 28
  characterization of, 32–33
  See also Function(s)
Irrational roots, 302, 304–305
Irreducibility, 327, 328–329
  partial fraction expansion and, 337
Irredundant covering, 440, 441, 444
Isomorphism, 426–428. *See also* Invariants, graphical
Iteration, 222, 224

Justifications
  for axiomatic methods, 70
  and formal proof format, 101, 102
  for k-combinations in n-set, 55
  for k-permutations in n-set, 43
  for the product rule, 49
  for the sum rule, 49

k-combinations, 52, 53, 55. *See also* Combinations
  Pascal triangle and, 56–57
k-permutations, 51, 53, 55. *See also* Permutations
k-subsets. *See* k-combinations
Kepler, Johannes, 228
Kirchhoff's Law, 387, 388
Kleene, Stephen, 202
  Introduction to Metamathematics (excerpt), 203
Königsberg bridge problem, 470, 471–472
Kronecker, Leopold, 62, 124, 142, 144, 146, 324
  Collected Works (excerpt), 143
Kronecker algorithm, 325–326
Kronecker, The Doubter (E.T. Bell), 143
Kuratowski graphs, 421, 423
Kuratowski theorem, 422

Labeled binary tree, 405–409
LaGrange Interpolation formula, 314–315, 316, 321
  and linear polynomial combinations, 358–359
  and Newton expansion, 350
LaGrange, Joseph-Louis, 276
  Lectures on Elementary Mathematics (excerpt), 277
Law of cancellation, 283
Law of Contraposition (CON), 69, 70, 84, 109, 111, 112
Law of Detachment (DET), 84–85
  and rules of implication, 91
Law of excluded middle (EXM), 62, 63
  and double negation elimination rule, 118–119
  and logical calculus, 121, 124
Leading coefficient, 280
Leading term, 280, 283, 292
Least common multiples (lcm), 170–171, 325
Leaves, tree, 404, 451
Lectures on Elementary Mathematics (LaGrange), excerpt, 277
Leibnitz, Gottfried Wilhelm von, 206
Lemma(s)
  path length, 463
  planar graph, 418
  polynomial growth rate, 267, 292
  proof by induction, 149–151, 299
Less than inequalities
  in linear order, 65
  logical implications of, 70–79
  relations on set, 38, 152
  See also Relations
Less than or equal to inequalities
  logical implications of, 78–79
  relations on a set, 38
  See also Relations
Limiting values, 292–294. *See also* Intercepts
Linear algebra, 357–358, 361
  and circuit rank, 389
Linear combinations, 358–359
  in matrix collections, 367
Linear graphs, 373
Linear order, 65
Linearity, 357
Lines
  equality of, 72
  intersection of, 72
  parallel, 71, 72
  and straight line equation, 311–312
Log function, property of, 343
  and treesort algorithm, 408–409
Logarithmic growth rate, 266–267
Logical calculus, 95–97
  classical, 120–123, 124–125
  constructive, 63
  well-formed formulas of, 127, 128–130
    hierarchical convention in, 129
Logical equivalence, 94, 110, 123
  derived rules of, 114
Logical implication. *See* Implication(s), logical
Logical reasoning
  classical methods of, 116, 119, 120–123
  compared to set operations, 120, 122, 123
  constructive, 63
  contraposition, 69, 109

Logical reasoning *(Continued)*
descriptions of, 63
forward and backward, 81–83
operations of, 95–97, 108, 109, 111, 112, 114, 130
propositional rules of, 89, 90–93
substitution and, 67, 69
and theorem proofs, 66–67, 74–75, 78, 85
Long division, 167, 169, 295
in polynomial division algorithm, 319
Loop current analysis, 387
Loop equations, 388–389
Loop invariants, 253, 254
Loops, 457. *See also* Edges
Lowest terms, 170
Lukasiewicz, Jan, 411

Mantissa(s), 193
negative, 194
truncated, 196
Mappings. *See* Function(s)
Marking path distance, 475
Mathematical abstraction
equality relations and, 1
Mathematical induction. *See* Induction
Mathematical language, 63, 67, 69, 71, 72, 79
Matrix, 364
computations, 458–463
covering, 451
graph circuit structure, 392
identity, 463
incidence, 440, 452
independence algorithm, 449
and isomorphism, 426
linear graph, 376, 384
reachability, 463, 464, 465
relation, 456–457, 461, 462–463, 465
Maximality, edge, 398
Mean, 243
Member. *See* Element
Membership, defined, 3, 5
and binary tree sorting, 408–409
in constructible sets, 6–7
defining property of, 8
and equality of sets, 15
and equivalence relation, 38, 44
k-permutation, 51
in one-to-one correspondence function, 30
in ordered collections, 65
product set, 49
and Russell's paradox, 5–6
union and intersection, 13, 14–15
Minimal connector problem, 402
Minimal covering, 441
Minmax algorithm, 262–263, 272
Modular problem-solving, 220, 238
Modules, algorithmic, 238–242
function, 240–241, 259–262
Multigraph, 471
Multiples, 170
of roots and poles, 334–335
Multiplication operation on N, 148
defined recursively, 159
of polynomials, 282–285, 286
Mutually disjoint sets, 14
and partitions, 20

n-set, 51, 53
equicardinality of subsets in, 59
and Pascal triangle, 56–57
n-tuples, 34, 59
ordered, 362
Natural numbers, 4, 65, 144, 155, 160, 168
and arithmetic operations, 146–148
axioms for, 144–146, 155
division algorithm for, 169
infinite sequence and, 26, 160
and one-to-one correspondence with rational numbers, 188–189
and mathematical induction, 155, 160–161
Negation, 69, 85, 86
classical logic and, 122
double, rules of, 117–118
operation, 131
priority and, 130
in proof development, 106
rules of, 93–94
derived, 112–114
Negative polynomial coefficients, 293–294, 299
Nested form, 286, 287
Network analysis, 386–389
Newton expansion formula, 348–350, 351
Nonconstructive proofs, 119–120. *See also* Classical laws

Nonplanar graphs, 417, 422. *See also* Planar graphs
Normalization condition, 193
Normalized exponential systems, 193–195
Notion of adequacy, 137
Null word sequence (of length 0), 27, 44
  and labeled binary trees, 404
  recursion and, 164
Numbers
  algebraic, 304–306, 363
  binary, 175–181
  Boolean, 199–200, 363
  chromatic, 435, 451–452
  computable, 306, 363
  connectivity, 382
  counting, 3, 4
  decimal, 175–181
  Fibonacci, 166, 329
  graphical invariant, 431, 446–447, 447–451
  hexadecimal, 183
  integers, 152, 167, 175, 177, 180, 363
  natural, 4, 65, 144, 155, 160, 168
  negative, 152
  octal, 182
  perfect, 119
  prime, 27, 173
  square-free, 229
  rational, 168, 183, 184–187, 188–191, 363
  relatively prime, 172
  triangular, 329
Numerator, 170, 185
Numerical invariants, 431–432, 434

Ohm's Law, 387, 388
One-to-one correspondence, 29
  equicardinality of, 58, 59, 188
  and invertible functions, 32, 33
  and isomorphism, 427, 428
  between rational and algebraic numbers, 305
  between rational and natural numbers, 188–189
One-to-one function, 29, 32, 33
  and pigeon-hole principle, 60
Onto function, 29, 32, 33
Operations. *See* Arithmetic operations; Logical reasoning, operations of; Set operations
Optimal tree, 402–404
Optimization problems, 208
Order
  of algorithm growth rate, 266–268
  axioms of, 65–67, 152–153
  cyclical, 66
  increasing, of base conversion, 179
  linear, 65
  notational, of polynomials, 279
  in permutations and combinations, 51, 52
  of the rationals, 190–191
  sorting, 233
  tree traversal, 408
Ordered collections, 65
Ordered n-tuples, 362
Ordered pairs, 33
  coordinate vector spaces of, 362–364
  in graph of the function, 291
  relations and, 34
  in rule of the product, 48–49
Ordered triples, 362
Ore, Oystein, 372
  Graphs and Their Uses (excerpt), 373
Outline, 208
Output, 205, 211

Parabola, 312
Paradox, 124. *See also* Contradictions; Russell's paradox
Parallel lines, 71, 72
Parameters, 193
  graphical, 384, 438–439
Parenthesis-free expressions, 411
Parity (even or odd), 54
Partial algorithm, 338
Partial fraction expansion, 336, 337–340
Partitions, 19–20, 38, 40
  bipartite, 378
  case, 86–87
  chromatic number and, 435
  and divisibility, 41
  of equivalence classes, 44, 45
  infinite subsets of, 40
  planarity and, 416
  See also Equivalence relations

Pascal programming language
  alternative statements in, 134
  empty set in, 12
  format, 102
  global and scope references in, 100
  identifiers for, 4
  integer division and, 168
  into machine language, 28
  pseudolanguage and, 215, 227
  and set operations, 10
Pascal triangle, 56–57
Path(s), 382, 394–395
  in directed graphs, 456
  problems, 470–474
  See also Circuits
Percentage, 198
Percent number, 119
  odd, 119–120, 124
Permutations, 51–52, 53–55
  and inversions, 54–55
  parity of, 54, 55
Philosophy of Mathematics and Natural Science (Weyl), excerpt, 1
Pi, 306
Pigeonhole principle, 60, 464
Pivoting, 367
Planar graphs, 415, 416–418
  testing, 420–424
  See also Euler formula
Pole, rational function, 334
  multiplicity of, 334–335
Polish form, 411. *See also* Lukasiewicz, Jan
Polya, George, 206
Polyeval algorithm, 287
Polyhedral formula. *See* Euler's formula
Polynomial growth rate, 267–268, 292
Polynomial(s), 278
  arithmetic operations of, 280–285
  conversion, 347, 348, 349, 351–352
  covering problems and, 444, 445
  evaluation, 285–287, 296, 302–303
  factorization, 324–327
  graphing, 289, 290–294, 300, 304, 311
  independent, 360–362
  interpolation, 309, 312–316
  irreducible, 328–329
  linear combination of, 358–359
  and rational functions, 331, 332–334, 335, 337–340
Polyproduct algorithm, 285
Positional notation, 176–177
Positive integers. *See* Natural numbers
Positive polynomial coefficient, 293–294, 296, 299, 335
Post, Emil L., 202
Postfix form. *See* Polish form
Post-order traversal, 412
Postulates, 64, 65, 72. *See also* Axiom(s)
Power set constructions, 58–59
Powers, 176–177
  as denominators, 335–337, 339
  negative, 180–181
  See also Exponents
Precision
  and approximate roots, 302, 306
  exponential, 193
Premise, 69, 91, 106
Prescribed root, 310–311, 313
Prime algorithm, 225, 227, 242
Prime factor, 113
Prime number(s), 27, 68, 71, 82, 173, 205
  and Euclid's assertion, 113
  relatively, 172
Primitive terms, 65
Principle of inclusion and exclusion, 50
Principle of mathematical induction, 155
  of recursively defined summation, 160–161
Principle(s)
  Archimedean, 152–153
  change of sign, 291, 297
  characterization of set unions and intersections, 14
  of inclusion and exclusion, 50
  of mathematical induction, 155
  pigeonhole, 60, 464
  for problem-solving, 207
  of proof development, 105–106
  well-ordering, 152
Printed circuit, 418
Priority, 129
Problem, defined, 205
  optimization, 208
  sub-, 238
Problem-solving, 203, 204–206, 236
  algorithms, 211, 212–213
    recursive, 259, 262–263
  modular, 220, 238
  principles of, 206–210

pseudolanguage and, 216, 220
recursive solutions to, 162–164
Processes. *See* Compound processes; Elementary processes
Product, 33–34
and relations, 34–35
rule of the, 48
Proof(s)
deductive rules for algorithmic, 249, 250–251, 252, 254
development of, 104–106
formal, 92–93, 99–101, 104
format, 100–104
by induction, 149–151, 154, 155–158
logical, 74–75, 93, 100–101
techniques, 78, 79, 80, 81, 84, 85, 86–87, 89, 99–101, 108, 110, 111, 112, 114, 246
testing vs. verification, 246, 255–256
See also Derived rules; Propositional rules
Properties
edge maximality (minimality), 398
empty set, 12
equivalence relations, 41
of functions, 28–30
of integer arithmetic by induction, 149–151
log function, 343
ordered collection, 65, 66
of polynomials, 278–279
in set-builder notation, 7
of set operations, 15–16, 17–18
of set unions and intersections, 14–15
of Weyl's set theory, 1
Propositional rules, 89, 90–93, 94–95
derived rules, 108, 109, 110, 111, 112, 114
of double negation, 117–118
and proof development, 104–105, 156
summary of, 99
Propositional variables, 128
and recursion, 164
in truth table evaluation, 133, 134
Propositions. *See* Theorems
Pseudolanguage, 215
of algorithmic modules, 239–242
of algorithm recursion, 259–263
compound processes in, 216–220
and computational complexity, 269
elementary processes in, 216, 224, 232, 248–249, 269
extendable and flexible, 231
guidelines, 225–227
subscripted variables in, 232, 237
and top-down methodology, 218, 220–221, 225
Pythagorean theorem, 207–208

Quadratic polynomial, 359
Quotient, 167, 168, 169
of polynomial division algorithm, 319–320
rational function, 332
and relative error, 198
set, 186
and synthetic division, 294

Range
exponential, 193
function, 24
Rank,
circuit, 392, 393–395
of a collection, 364–368
Ratadd algorithm, 241–242. *See also* Rational numbers
Rational arithmetic, 185, 241
Rational functions, 331, 332–334, 335, 337–340, 345
Rational numbers, 168, 183
algorithmic modules of, 241–242
arithmetic laws for, 187
cardinality of, 188–189
construction of, 184–186
and coordinate vector spaces, 363
Rational roots, 300–302
Reachability, 463–468
algorithm, 466
Reasoning. *See* Logical reasoning
Reciprocity, logical
in derived conjunction rules, 110
of double negation rules, 117
and reciprocal rules of equivalence, 94–95
Recurrence equation, 56–57
Recursion, 155, 158–160
algorithm, 241, 259–262
and collection rank, 367
using inductive hypothesis, 159
post-order traversal, 413

Recursion *(Continued)*
in problem-solving solutions, 162–164
of sets, 164–165
of summation symbol, 160
and treesort algorithm, 407–408
Reduction
to an absurdity, 112
algorithm ratadd, 241
to lowest terms, 170
matrix, to echelon form, 365–366, 392
of rational function, 333, 336
Redundancy, 360, 364
circuit, 402
polynomial covering problems and, 444
Reflexive property
and equivalence relations, 41, 43, 70
implication and, 96, 109, 113
Regions, planar, 416, 418
Reinsertion, edge or chord, 400–401
Relations, 34–35
Archimedean principle of, 152
binary, 375
circular, 46
compared to functions, 35
congruence, 40, 41
directed graphs as, 455–457, 458–460
of divisibility, 38
equivalence, 38, 41–43, 45–46, 79–80, 132, 185
implication, 109
inductive proof of, 157–158
matrix, 378, 380
ordered collections and, 66
planar graph, 422
recursive, 163
on a set, 38–40
together, 45, 47
vertex-edge incidence, 442
Relative correctness, 252
Relative error, 198
Relatively prime
numbers, 172
polynomials, 324, 337
Remainders, 169
in Euclidean algorithm, 171, 211, 255
of polynomial division algorithm, 319–320
rational arithmetic and, 185
in synthetic division, 294
Repetition, pseudolanguage, 216
constructs, 219, 222, 224, 248, 253
flexibility of, 218
sorting algorithm, 237
Resistance, electrical, 387
Result, formal proof, 100
Roots
approximate, 302–304
of graph trees, 404, 405
multiplicity of, and poles, 334–335
of polynomial function, 294, 296–302
prescribed, 310–311, 313
rational, 300–302
square, 207–208, 305, 321
Rows, matrix, 365
Rule(s)
absolute-value function, 26
composition, 30
contraposition, 69
Descartes', of signs, 298
enumeration, 48, 49, 53, 54, 55, 56–57
use of, in functions, 24, 25
power set, 59
propositional, 89, 90–93, 94–95, 101–105
derived, 108, 109, 111, 112, 114
double negation, 117–118
summary of, 99
of substitution, 67
Running time algorithm, 265–267. *See also* Growth rate
Russell's paradox, 5–6, 124
and derived negation rules, 112
Russell, Lord Bertrand, 1, 5

Scalar quantities, 361, 362, 363
multiplication by, 362
Scientific notation, 193
Scope, 100
Searching, data, 235, 236, 272, 406
Selection, compound process, 217
derived constructs, 222, 251, 270
sort algorithm, 237
Selectionsort algorithm, 237–238, 239–240
time complexity of, 271, 409
Semantic consequence, 135
Separation, polynomial, 337, 338–339
Sequences, 26–28
in base conversion, 179

in compound processes, 217, 222
elementary polynomial, 345
graph path, 382, 393–394, 401, 430, 474
interpolate, 309, 312, 321
iterated, 222
of permutations, 51, 53–54
of recursive algorithms, 261–262
summation of, 160–161, 345, 352
terms of the, 26
and time complexity, 270
Set formation, 1
Set operations, 10, 12–14
generalizations of, 19–20
product, 33–34
properties of, 15–16, 17–18
Set(s)
cardinality of, 3, 34, 50, 51, 58
collection, 14, 19, 40, 45
complement, 16–17
constructible, 6–7, 12
containment, 11, 14
counting techniques, 47, 48, 49, 53, 55, 59
creative description of, 1
defined, 2, 6
and defining property, 8
disjoint, 14, 40
equality of, 15, 16
and function factors algorithm, 321
of graph edges, 389
hierarchies of, 5, 6, 124
intersection of, 13, 14–15
notation, 2–3
ordered, 66
partitions, 19–20, 38, 40, 45
power, 58–59
recursively defined, 164–165
relations, 34–35, 38–39, 40, 45
sequence of a, 27
standard, 4
successor, 145
union, 13, 14, 45
Set-builder notation, 7–8, 13
constructibility and, 7
and product operations, 33
Shortpath-digraph algorithm, 475
Shortpath-graph algorithm, 477
Significant digits, 193, 197
Simple path, 382. *See also* Elementary circuit
Simplifying expressions, 121
Sink, vertex, 472
Sorting, data, 233, 235–236, 237, 265, 397, 406–410
Source, vertex, 472
Spaces, coordinate vector, 362–364
Spanning subgraph, 377. *See also* Graphs; Subsets
Spanning tree, 399–402
Sphere, projected, 420
Square root, 207–208, 305, 321. *See also* Roots
Standard deviation, 243
Statement of a theorem, 99–100
States,
computational, 212, 248, 249, 467
condition of, 213
transformation of, 213
Statistical enumeration, 50–51
Stereographic projection, 420
Straight line equation, 311–312, 315
Subgraphs, 377, 393, 401, 422. *See also* Graphs; Subsets
Subproofs, 100, 101
indentation of, 101–102
See also Proofs
Subscripted variables, 231, 232–236
Subsets, defined, 11
in binomial coefficient recurrence, 56
and equicardinality, 58, 59
linear graph, 375, 389, 404
partitions and, 19–20, 38, 44
permutation or combination, 52
and product operations, 33
and relations, 34, 38, 40, 44, 45
and well-ordering principle, 152
Substitution, 67, 69, 70
argument, 286, 287
operation, 168
rule of, 67
Subtractive cancellation, 197
Successor, 145, 147
in proofs by induction, 149
well-ordered, 152
Sum operations
in covering problems, 445–446
and inclusion and exclusion principle, 50

Sum operations *(Continued)*
rational function, 336, 337–338, 340
and recurrence equation, 56–57
rule of the, 49
of two polynomials, 281
Summation, 160–161
calculus, 348, 352–355
operator, 342, 352
polynomial, 286
vertex degrees, 380–381
Supreme Intellect, 125
Symbolic computation, 325
Symbols, mathematical
absolute value, 26
angle brackets, 27
braces, 2, 12
cardinality, 3
case, 87
comparison of, 86
complement, 16
conjunction, 86
contraposition, 69
delta, 343, 474
detachment, 84
disjunction, 85
ellipses, 3, 4
empty set, 12
function, 24
generalization, 19
generalized product set, 34
Greek, history of, 277
identifiers, 4
implication, 68, 90
intersection, 13
introduction and elimination, 91
large negative or positive, 332
logical equivalence, 79, 94, 95
membership, 3
negation, 69
null word, 27
parallel, 71
pi, 306
power set, 58
priority of logical, 129
relation on a set, 41
set collection, 14
set operation, summary of, 5
subset, 11
successor, 145
summation, 160
superset, 11
triple, validity of, 253
union, 13
word, 27
Symmetric property, 41, 43, 45
Synthetic division, 294–296, 361
and interpolating polynomials, 316
in polynomial conversion, 351–352
of rational roots, 301, 302

Tautological statement, 118, 128, 137
and truth-table, 134
Taylor expansion, 347
Terminating decimal, 181
Termination condition, 163
Termination mechanism, 262
Terms
cancellation of, 281, 282
of the polynomial, 270
sequence, 26
Testing, algorithm, 247–248
Theorem(s), 128
bound on roots, 296, 298
characterization of trees, 398
circuit rank formula, 393
composite relation, 460
coordinate vector spaces, 363
countability, of the rationals, 188, 190
Descartes' Rule of Signs, 298
economy tree, 403
Eisenstein's Irreducibility Criterion, 328–329
essential cell, 452
Euler circuit, 472
formal proof of a, 100–101
four-color, 373, 436
fundamental, 353
graph relations, 434
Hamiltonian path, 474
independence, 434
integer ordering, 152, 190
irrational root, 305
irredundant covering sums, 446
Kuratowski, 422
linear combination of vectors, 367
path length, 462
pi, nonalgebraic, 306

proving, with logic, 66–67, 74–75
Pythagorean, 207–208
rational root, 300
reachability matrix, 464
relative error, 198
tautologies and, 137
vertex degrees, 380
See also Proofs
Tie-breaking rule, 129–130. *See also* Priority
Time complexity, 265
computation of, 269–273
Top-down methodology, 218, 220–221, 225
Top-level design pseudolanguage, 237
Total order. *See* Linear order
Towers-of-Hanoi puzzle, 162–164
Transfinite theories, 124
Transformation, state, 213
Transitive closure, 47
Transitive property
equivalence relations, 41, 42, 43, 45, 66
implications and, 109
Transposition, 398
Tree, linear graph, 382, 397, 415
characterization of, 398–399
labeled binary, 404–409
optimal, 402–404
spanning, 399–402
traversal of, 408
post-order, 412
Treesort algorithm, 406
Triangle algorithm, 223, 238
Trichotomy, 101, 261
Triples, algorithm, 249–250, 408
True values, 198
Truncation, 168, 196, 197, 367
Truncation error, 198
Truth, 118, 130, 133, 137
Boolean, 242, 248, 252
induction and, 156
See also Falsity; Validity
Truth table evaluation, 127, 133, 137
Truth value, 130, 131
characteristic tables and, 132
Turing, Alan Mathison, 202
Twenty Questions, 235–236
Type, instance, 205
inferred, 227
set, 232
Underlining procedure, 412
Undirected graphs, 456, 473, 474. *See also* Directed graphs
Union(s)
characterization of, and intersections, 14–15
of equivalence classes, 44, 45
generalized, 19–20
operations, properties of, 15–16, 17–18
of sum sets, 49, 50
Universal contradictory statement, 120
Universal valid statement, 120
Universe of discourse, 8
and empty set, 12
product operations and, 33
and set complementation, 16–18
subsets of, 11–12

Validity, logical, 80, 81
of algorithmic triple, 250, 251, 256
of arithmetic laws by induction, 149, 150
conjunction and disjunction, 92
derived rules and, 114
and law of detachment, 84, 91
negation and, 117, 118
principle of induction and, 155, 156
Value of a function, 24
denoting terms in, 26
infinite sequence and, 26
limiting, 292–294
well-defined, 28–29
Variable(s)
auxiliary, 210, 212, 224
control, 224, 233
input and output, 205, 211
propositional, 128
subscripted, 231, 232–236
Vectors, 34, 361
addition of, 362
linear combination, 367
multiplication of, by a scalar, 362
subtraction, 363
zero, 363, 391
Verification, proof, 246–247
of Euclidean algorithm, 255–256
See also Testing, algorithm
Vertex, 374–375
binary tree, 404

Vertex *(Continued)*
degrees, 380–381, 471–472
-edge incidence, 442
and isomorphism, 428
planar graph, 420
and reachability, 463, 464
set, 393–394
Vertical intercept, 293. *See also* y-intercept
Vieta, Franciscus, 203, 277
Voltage, 387

Well-defined function, 28–29, 186
Well-formed formulas, 127, 128–130
and recursion, 164
semantic consequence of, 135
semantics for, 130–133
syntax for, 128–130
Well-formed parentheses strings, 167
Well-ordering principle, 152
Weyl, Hermann, 1, 7
Philosophy of Mathematics and Natural Sciences (excerpt), 1
Word sequence (of length n), 27
and computer memory storage, 193
and equivalence relations, 42
and recursion, 164–165

y-intercept, 293, 310, 333

Zero
and Euclidean algorithm, 211
in extended number system, 152
and Kirschhoff's Law, 387
polynomial, 280, 282
vector, 363, 391